AF291882

Cellular Senescence in Health, Disease and Aging: Blessing or Curse?

Cellular Senescence in Health, Disease and Aging: Blessing or Curse?

Editor

Markus Riessland

MDPI • Basel • Beijing • Wuhan • Barcelona • Belgrade • Manchester • Tokyo • Cluj • Tianjin

Editor
Markus Riessland
Stony Brook University,
USA

Editorial Office
MDPI
St. Alban-Anlage 66
4052 Basel, Switzerland

This is a reprint of articles from the Special Issue published online in the open access journal *Life* (ISSN 2075-1729) (available at: http://www.mdpi.com).

For citation purposes, cite each article independently as indicated on the article page online and as indicated below:

LastName, A.A.; LastName, B.B.; LastName, C.C. Article Title. *Journal Name* **Year**, *Volume Number*, Page Range.

ISBN 978-3-0365-2175-6 (Hbk)
ISBN 978-3-0365-2176-3 (PDF)

Cover image courtesy of Markus Riessland

Contents

About the Editor

Riessland, Markus is a trained molecular biologist with a background in neuroscience, human genetics and neurodegenerative diseases. Dr. Riessland received his PhD from the Clinic of the University of Cologne, Institute for Human Genetics, Germany. Early in his career, Dr. Riessland was involved in several internationally funded projects, where he performed and published studies on epigenetic modifiers as a potential therapy for the neurodegenerative disease spinal muscular atrophy (SMA). His research is particularly focused on the identification and characterization of neuron-specific disease-modifying factors that may facilitate the development of novel therapeutic strategies for degenerative disorders of the central nervous system. Dr. Riessland focuses on the understanding of cellular senescence. Cellular senescence is a common biological process in which mitotic cells may shut down the cell cycle when they recognize they have suffered DNA damage during division. This process causes a generation of "undead cells" (also known as "Zombie Cells"). This helps to prevent damaged cells from growing uncontrollably and causing problems such as cancer. Undead cells are, in fact, common and they are found all over the body. However, senescence is not typically seen in the nerve cells of the brain. Unlike most other cells in the body, neurons stop dividing once they are fully formed. In the lab of Nobel Laureate Paul Greengard at Rockefeller University, Dr. Riessland discovered that, surprisingly, post-mitotic dopaminergic neurons—which regulate motivation, memory, and movement by producing the chemical messenger dopamine—can nevertheless become senescent. This finding could have widespread implications for the understanding of many age-related neurodegenerative disorders (e.g., Parkinson's disease) and the aging process itself. Currently, his lab in the Center for Nervous System Disorders at the Department of Neurobiology and Behavior at Stony Brook University uses stem cell-based approaches as well as mouse models and next generation sequencing techniques (TRAP-seq, RNA-seq, ATAC-seq, scRNA-seq, etc.) to tackle the questions where and how cellular senescence in the brain could occur and spread, which cell types are involved and what the molecular triggers are. Additionally, research in the lab focuses on the identification and molecular characterization of genetic modifiers that influence the vulnerability of neuronal subtypes. The knowledge of molecular modifiers helps us to understand the underlying reasons of vulnerability that could be leveraged to protect cells from neurodegeneration. Moreover, the lab's research aims to interfere with the aging process by ameliorating the unwanted negative effects of cellular senescence.

Editorial

Cellular Senescence in Health, Disease and Aging: Blessing or Curse?

Markus Riessland [1,2]

[1]   Department of Neurobiology and Behavior, Stony Brook University, Stony Brook, NY 11794, USA;
      markus.riessland@stonybrook.edu
[2]   Center for Nervous System Disorders, Stony Brook University, Stony Brook, NY 11794, USA

check for
updates

Citation: Riessland, M. Cellular Senescence in Health, Disease and Aging: Blessing or Curse?. *Life* **2021**, *11*, 541. https://doi.org/10.3390/life11060541

Received: 28 May 2021
Accepted: 7 June 2021
Published: 9 June 2021

Publisher's Note: MDPI stays neutral with regard to jurisdictional claims in published maps and institutional affiliations.

Sixty years ago (1961), Hayflick and Moorhead reported that primary cells terminate their growth and stop dividing after ~50 passages or one year in culture. This seminal study described the phenomenon that we now refer to as "cellular senescence" [1]. More specifically, the description by Hayflick and Moorhead unraveled "replicative senescence", which is caused by cell-division-dependent telomere attrition. Since then, increasing numbers of additional senescence-inducing factors have been identified. In parallel, a plethora of cell types have been recognized to possess the ability to enter a state of cellular senescence. These studies revealed diverse senescence-related cellular phenotypes and identified various metabolic changes, gene-activity alterations and other molecular markers [2–4]. Although some gene expression changes are characteristic hallmarks of cellular senescence, a single molecular marker has not been identified. Accordingly, the univocal identification of a senescent cell remains challenging. To address this problem, the International Cell Senescence Association (ICSA) assembled a list of key features observed in senescent cells [2].

A particularly interesting feature of senescent cells is the so-called senescence-associated secretory phenotype (SASP), which remodels the gene epression profile of a senescent cell causing the secretion of proinflammatory molecules to signal to the immune system "come here and remove me". During development, and in organisms with fully functional immune systems, senescent cells are usually detected and cleared from the tissue [5]. In case where immune cells do not remove the senescent cells, they remain in the tissue and continue to express the SASP. In turn, this would cause a damaging local inflammation and could also induce remodeling of the surrounding tissue as well as the spreading of senescence. Aged organisms possess a significantly reduced regenerative potential and immune function resulting in the accumulation of senescent cells [5]. Interestingly, this accumulation has also been observed in age-related disorders, neurodegenerative diseases, cardiovascular diseases, and others [6,7]. Because of its detrimental effect on the surrounding tissue, the accumulation of senescent cells is not just a consequence, but can instead be understood as a major driver of aging. Accordingly, recent studies described that the removal of senescent cells showed beneficial effects on healthspan and lifespan [8]. This exciting research led to the discovery of "senolytics", drugs which can kill senescent cells. Moreover, because of the heterogeneity of cell types that show senescence-like phenotypes, including cardiovascular cells and post-mitotic neuronal cells [6,9,10], further research is required to unravel the molecular background that renders a cell type vulnerable to senescence and to determine the pathways that induce senescence in a cell type-specific manner.

Given that there are many open questions in the field, this Special Issue of *Life* was created to shed light on the molecular pathways of cellular senescence, inflammaging, and the possible strategies to interfere with these processes. The work published in this Special Issue of *Life*, entitled "Cellular Senescence in Health, Disease and Aging: Blessing or Curse?", mirrors the broad interest in the field of cellular senescence since the presented studies highlight quite diverse aspects of senescence and related pathways from various areas of research.

The manuscript by Panchanathan et al. reports observations that identify the interferon inducible POP3 PYHIN protein as a potential negative regulator of the AIM2 inflammasome and SASP in senescent human prostate epithelial cells. This study provides insight into the age-related development of prostatic inflammatory diseases [11].

Senescence DNA damage foci (SDF) and telomere-dysfunction-induced foci (TIF) can be identified by the histone marker γH2AX for cellular senescence and DNA damage, respectively, which makes γH2AX a useful tool for the identification of these traits in diverse tissues [12]. In this Special Issue, Siddiqui and colleagues determine the feasibility of using γH2AX as a molecular biomarker of DNA damage in Alzheimer's disease (AD). The authors report a protocol that employs laser scanning cytometry (LSC) to measure endogenous γH2AX in buccal cell nuclei from mild cognitive impairment (MCI) patients, AD patients, and healthy controls [13].

Secreted protein acidic and rich in cysteine (SPARC), a molecule that has been described to be overexpressed in senescent cells [14], was the topic of an Opinion manuscript by Ghanemi et al. [15]. The authors emphasize that SPARC not only acts as a regeneration factor but also counteracts the aging-related decrease in regeneration ability, and thus can be seen as a potential factor for preventing age-related conditions.

p16$^{\text{INK4A}}$, which is often highly upregulated in many types of cellular senescence, acts as a tumor suppressor and is frequently reduced in human cancers. In this Special Issue, Leon et al. review the potential role of p16 in the regulation of immunological surveillance. In brief, the authors discuss the hypothesis that a p16-positive tumor would foster immunosurveillance by inviting immune cells into the tumor microenvironment, whereas a p16-null tumor would reduce immunosurveillance and promote tumor growth [16].

Finally, two reviews from the Orr lab highlight the importance of cellular senescence in the human brain. Gillispie et al. summarize the role of mitotic cells in brain senescence and discuss implications in neurodegenerative diseases and cancer [17]. The second manuscript reviews the recent discovery of post-mitotic senescence in the brain. In short, Sah et al. provide a comprehensive overview of the current knowledge of the cellular senescence of brain cells, including neurons [18]. Additionally, this manuscript gives an elegant introduction into the field of cellular senescence.

Generally, I hope that this Special Issue of *Life* will capture the attention of both specialists and non-specialists who are interested in understanding the molecular processes involved in cellular senescence and inflammaging. As seen in the diverse articles in this Special Issue, cellular senescence and the molecules that are crucial in its underlying pathways are of high interest in many areas of research. The rising interest in a more thorough understanding of cellular senescence is reflected by the fact that the National Institutes of Health (NIH) have recently established the Common Fund's Cellular Senescence Network (SenNet) Program to identify and characterize the differences in senescent cells within the body, across various states of human health, and throughout lifespan. It is an exciting time for researchers working on senescence and aging, and overall, there is great hope that the outcome of this research can translate into strategies that provide beneficial effects on healthspan and lifespan in humans.

Funding: This research received no external funding.

Acknowledgments: I would like to thank all the contributors of the Special Issue of *Life* (ISSN 2075-1729): "Cellular Senescence in Health, Disease and Aging: Blessing or Curse?"), belonging to the section "Cell Biology and Tissue Engineering".

Conflicts of Interest: The author declares no conflict of interest.

References

1. Hayflick, L.; Moorhead, P.S. The serial cultivation of human diploid cell strains. *Exp. Cell Res.* **1961**, *25*, 585–621. [CrossRef]
2. Gorgoulis, V.; Adams, P.D.; Alimonti, A.; Bennett, D.C.; Bischof, O.; Bishop, C.; Campisi, J.; Collado, M.; Evangelou, K.; Ferbeyre, G.; et al. Cellular Senescence: Defining a Path Forward. *Cell* **2019**, *179*, 813–827. [CrossRef] [PubMed]
3. Dodig, S.; Cepelak, I.; Pavic, I. Hallmarks of senescence and aging. *Biochem. Med. (Zagreb)* **2019**, *29*, 030501. [CrossRef] [PubMed]
4. Hernandez-Segura, A.; Nehme, J.; Demaria, M. Hallmarks of Cellular Senescence. *Trends Cell Biol.* **2018**, *28*, 436–453. [CrossRef] [PubMed]
5. van Deursen, J.M. The role of senescent cells in ageing. *Nature* **2014**, *509*, 439–446. [CrossRef] [PubMed]
6. Childs, B.G.; Li, H.; van Deursen, J.M. Senescent cells: A therapeutic target for cardiovascular disease. *J. Clin. Investig.* **2018**, *128*, 1217–1228. [CrossRef] [PubMed]
7. Martinez-Cue, C.; Rueda, N. Cellular Senescence in Neurodegenerative Diseases. *Front. Cell Neurosci.* **2020**, *14*, 16. [CrossRef] [PubMed]
8. Baker, D.J.; Childs, B.G.; Durik, M.; Wijers, M.E.; Sieben, C.J.; Zhong, J.; Saltness, R.A.; Jeganathan, K.B.; Verzosa, G.C.; Pezeshki, A.; et al. Naturally occurring p16(Ink4a)-positive cells shorten healthy lifespan. *Nature* **2016**, *530*, 184–189. [CrossRef] [PubMed]
9. Riessland, M.; Kolisnyk, B.; Kim, T.W.; Cheng, J.; Ni, J.; Pearson, J.A.; Park, E.J.; Dam, K.; Acehan, D.; Ramos-Espiritu, L.S.; et al. Loss of SATB1 Induces p21-Dependent Cellular Senescence in Post-mitotic Dopaminergic Neurons. *Cell Stem Cell* **2019**. [CrossRef] [PubMed]
10. Musi, N.; Valentine, J.M.; Sickora, K.R.; Baeuerle, E.; Thompson, C.S.; Shen, Q.; Orr, M.E. Tau protein aggregation is associated with cellular senescence in the brain. *Aging Cell* **2018**, *17*, e12840. [CrossRef]
11. Panchanathan, R.; Ramalingam, V.; Liu, H.; Choubey, D. Human Prostate Epithelial Cells Activate the AIM2 Inflammasome upon Cellular Senescence: Role of POP3 Protein in Aging-Related Prostatic Inflammation. *Life* **2021**, *11*, 366. [CrossRef]
12. Martínez-Zamudio, R.I.; Robinson, L.; Roux, P.-F.; Bischof, O. SnapShot: Cellular Senescence Pathways. *Cell* **2017**, *170*, 816–816.e1. [CrossRef]
13. Siddiqui, M.S.; Francois, M.; Rainey-Smith, S.; Martins, R.; Masters, C.L.; Ames, D.; Rowe, C.C.; Macaulay, L.S.; Fenech, M.F.; Leifert, W.R. Evaluation of GammaH2AX in Buccal Cells as a Molecular Biomarker of DNA Damage in Alzheimer's Disease in the AIBL Study of Ageing. *Life* **2020**, *10*, 141. [CrossRef]
14. Wang, S.; Moerman, E.J.; Jones, R.A.; Thweatt, R.; Goldstein, S. Characterization of IGFBP-3, PAI-1 and SPARC mRNA expression in senescent fibroblasts. *Mech. Ageing Dev.* **1996**, *92*, 121–132. [CrossRef]
15. Ghanemi, A.; Yoshioka, M.; St-Amand, J. Secreted Protein Acidic and Rich in Cysteine as A Regeneration Factor: Beyond the Tissue Repair. *Life* **2021**, *11*, 38. [CrossRef] [PubMed]
16. Leon, K.E.; Tangudu, N.K.; Aird, K.M.; Buj, R. Loss of p16: A Bouncer of the Immunological Surveillance? *Life* **2021**, *11*, 309. [CrossRef] [PubMed]
17. Gillispie, G.J.; Sah, E.; Krishnamurthy, S.; Ahmidouch, M.Y.; Zhang, B.; Orr, M.E. Evidence of the Cellular Senescence Stress Response in Mitotically Active Brain Cells—Implications for Cancer and Neurodegeneration. *Life* **2021**, *11*, 153. [CrossRef]
18. Sah, E.; Krishnamurthy, S.; Ahmidouch, M.Y.; Gillispie, G.J.; Milligan, C.; Orr, M.E. The Cellular Senescence Stress Response in Post-Mitotic Brain Cells: Cell Survival at the Expense of Tissue Degeneration. *Life* **2021**, *11*, 229. [CrossRef]

Article

Human Prostate Epithelial Cells Activate the AIM2 Inflammasome upon Cellular Senescence: Role of POP3 Protein in Aging-Related Prostatic Inflammation

Ravichandran Panchanathan [1], Vaikundamoorthy Ramalingam [1,2], Hongzhu Liu [1] and Divaker Choubey [1,*]

[1] Department of Environmental Health, University of Cincinnati, Cincinnati, OH 45221, USA; pancharn@ucmail.uc.edu (R.P.); ramalingam@iict.res.in (V.R.); liuhongzhuk@hotmail.com (H.L.)

[2] Centre for Natural Products and Traditional Knowledge, CSIR—Indian Institute of Chemical Technology, Hyderabad 500 007, India

[*] Correspondence: divaker.choubey@uc.edu; Tel.: +1-(513)-558-1014

Abstract: Increased levels of type I (T1) interferon (IFN)-inducible POP3 protein in myeloid cells inhibit activation of the AIM2 inflammasome and production of IL-1β and IL-18 proinflammatory cytokines. The *AIM2* mRNA levels were significantly higher in benign prostate hyperplasia (BPH) than the normal prostate. Further, human normal prostate epithelial cells (PrECs), upon becoming senescent, activated an inflammasome. Because in aging related BPH senescent PrECs accumulate, we investigated the role of POP3 and AIM2 proteins in pre-senescent and senescent PrECs. Here we report that the basal levels of the POP3 mRNA and protein were lower in senescent (*versus* young or old) PrECs that exhibited activation of the T1 IFN response. Further, treatment of PrECs and a BPH cell line (BPH-1) that expresses the androgen receptor (AR) with the male sex hormone dihydrotestosterone (DHT) increased the basal levels of POP3 mRNA and protein, but not AIM2 and inhibited activation of the AIM2 inflammasome. Of interest, a stable knockdown of POP3 protein expression in the BPH-1 cell line increased cytosolic DNA-induced activation of AIM2 inflammasome. These observations suggest a potential role of POP3 protein in aging-related prostatic inflammation.

Keywords: prostate; senescence; inflammation; AIM2 inflammasome; POP3

Citation: Panchanathan, R.; Ramalingam, V.; Liu, H.; Choubey, D. Human Prostate Epithelial Cells Activate the AIM2 Inflammasome upon Cellular Senescence: Role of POP3 Protein in Aging-Related Prostatic Inflammation. *Life* **2021**, *11*, 366. https://doi.org/10.3390/life11040366

Academic Editor: Markus Riessland

Received: 26 January 2021
Accepted: 3 March 2021
Published: 20 April 2021

Publisher's Note: MDPI stays neutral with regard to jurisdictional claims in published maps and institutional affiliations.

1. Introduction

Molecular mechanisms that contribute to the development of aging-related prostatic inflammatory diseases, including benign prostate hyperplasia (BPH), remain largely unknown [1–4]. Notably, studied using biopsies from patients with BPH and informative animal models have indicated a role for prostatic inflammation (PI) in the development of BPH [5–9].

Prostatic infections induce production of T1 interferons (IFN-α/β) through activation of the cGAS-STING pathway as a part of innate immune response [4,10,11]. The T1 IFNs, upon binding to a cell surface receptor, activate the JAK/STAT signaling in cells, resulting in stimulation of the expression of T1 IFN-inducible proteins [12]. The T1 IFN-inducible PYHIN protein family includes human IFI16 proteins, pyrin-only protein 3 (POP3), and AIM2 protein [13–15]. The proteins in the family share the *N*-terminal PYRIN domain (PYD) and the C-terminal HIN domain [13]. The PYD allows homotypic protein-protein interactions and the HIN domain allows sequence-independent binding to DNA [13,14]. The POP3 protein lacks the HIN domain [15].

We have reported earlier that treatment of human normal prostate epithelial cells (PrECs) and normal prostate stromal cells (PrSCs) with T1 IFN increased the levels of the androgen receptor (AR) and stimulated the transcription of AR-regulated gene [16]. Activation of AR in PrECs and PrSCs by the male sex hormone, dihydrotestosterone (DHT), regulates cell proliferation and survival [17,18]. Further, we noted earlier that activation of the AR in human prostate cancer cell line PC-3 stimulated the expression of IFI16 PYHIN proteins [19]. Increased expression of the IFI16 proteins in human normal PrECs, PC-3 prostate cancer cell line, and human normal diploid fibroblasts (HDFs) associated with the onset of cellular senescence [20–23]. Of interest, AR also drives human PrECs to cellular senescence [24]. Although the senescent cells exit the cell cycle permanently and do not divide, these cells secrete proteases and proinflammatory cytokines (e.g., IL-6, IL-1β, and IL-18) [25,26]. This phenotype of senescent cells has been termed senescence-associated secretory phenotype (SASP) and the phenotype is thought to contribute to aging-related chronic inflammation [25,26]. Notably, senescent PrECs accumulate in BPH [27] and their SASP promotes BPH [28]. However, the molecular mechanisms that contribute to the development of SASP in the senescent PrECs remain unclear.

The AIM2 protein senses cytoplasmic dsDNA in a variety of cell types and recruits an adaptor protein ASC through its PYD to form the AIM2 inflammasome [13,14]. The activated AIM2 inflammasome through activation of caspase-1 protease proteolytically cleaves the gasdermin D protein, pro-IL-1β, and pro-IL-18 [29]. Activated gasdermin D induces cell death by pyroptosis [29]. Proteolytic cleavage of pro-IL-1β, and pro-IL-18 promotes the secretion of the mature IL-1β and IL-18 proinflammatory cytokines [13,14,29]. Notably, increased levels of the POP3 protein in macrophages bound with PYD of AIM2 protein and the binding diminished the ability of the AIM2 protein to bind with ASC adaptor protein and to form AIM2 inflammasome [15].

Given that human prostatic infections are associated with chronic inflammation [4,5], and the development of BPH is associated with an accumulation of senescent PrECs with SASP [27,28], we investigated the role of POP3 and AIM2 proteins in senescent PrECs. Here we report that levels of the POP3 protein decreased in senescent PrECs as compared with pre-senescent proliferating or old cells. Further, DHT-mediated activation of the AR in human PrECs and in a benign prostate hyperplasia (BPH) cell line (BPH-1) up-regulated the expression of POP3 protein and inhibited cytosolic DNA-induced activation of the AIM2 inflammasome. Further, a knockdown of POP3 protein expression in BPH-1 cells activated the activity of the AIM2 inflammasome. Our observations have important implications for the development of aging-related prostatic inflammatory diseases.

2. Materials and Methods

2.1. Reagents

Synthetic double-stranded DNA (Poly (dA:dT)) in complex with transfection reagent (LyoVec) and LyoVec were from InvivoGen (San Diego, CA, USA) and EDTA-free protease inhibitor cocktail was from Roche Applied Science (Indianapolis, IN, USA). Dihydrotestosterone (DHT) was purchased from Sigma-Aldrich (St. Louis, MO, USA) and a stock (100 mM) was prepared in 100% ethanol and stored at −20 °C.

2.2. Primary PrECs, Prostate Cell Line, and Treatments

Human primary prostate epithelial cells (PrECs; at passage 2) from different donors (age 19 to 37) in culture (or frozen vials) were purchased from Lonza (Houston, TX, USA). Cells were maintained in culture as suggested by the supplier in the presence of medium supplements that were provided by the supplier as a part of the PrEGM™ Bulletkit™. Immortalized BPH-1 cell line was originally provided by Dr. Simon Hayward (Vanderbilt University Medical Center, Nashville, TN, USA) [30].

When indicated, cells in culture were treated with the indicated concentrations of dihydrotestosterone (DHT; stock in 100% ethanol) in phenol-free culture medium that was supplemented with charcoal/Dextran-treated fetal bovine serum (to decrease the endogenous levels of the steroid hormones) from the US source (from HyClone).

The asynchronous onset of cellular senescence in the primary cultures of human PrECs in late passages (passage 7 and higher) was assessed using well-described criteria for cellular senescence, including cell morphological changes and positivity to senescence-associated acidic β-galactosidase (SA-β-gal) as we have described [20]. In senescent cultures of PrECs, >90% cell population tested positive for the SA-β-gal, exhibited a large and flat cell morphology, and stopped cell proliferation [20].

2.3. Antibodies

Following antibodies were used to specifically detect proteins in immunoblotting: AR (sc-816), IFI16 (sc-8023), ASC (sc-22514), IL-1β (sc-7884), and IL-18 (sc-7954) from Santa Cruz Biotech (Santa Cruz, CA, USA); Caspase-1 (AHZ0082) from Invitrogen (Grand Island, NY, USA); Anti-STAT1 (cat # 9172), p-STAT1 (cat # 9171), and β-actin (cat # 4967) from Cell Signaling Technology (Danvers, MA, USA). Rabbit polyclonal antibodies that we raised against the C-terminal AIM2 peptide that specifically detected two human hAIM2 isoforms have been described [31]. Specific custom anti-peptide rabbit polyclonal antibodies were raised against a peptide (REEQETGICGSPSSARSV) in the POP3 protein, which detected an IFN-inducible POP3 protein of an expected size (~18 kDa) in total cell extracts from IFN-treated THP-1 cells as described [15]. Horseradish peroxidase (HRP) conjugated secondary anti-mouse (NXA-931) and anti-rabbit (NA-934) antibodies were from GE Healthcare Biosciences (Piscataway, NJ, USA).

2.4. Immunoblotting

Total cell lysates were prepared in radio-immunoprecipitation assay (RIPA) buffer (50 mM Tris-Cl (pH 7.4), 150 mM NaCl, 1% Nonidet P-40, 0.5% sodium deoxycholate, 0.1% sodium dodecyl sulfate) that was supplemented with complete mini EDTA-free protease inhibitor cocktail and phosphatase inhibitors (Cell Signaling, Danvers, MA, USA) as described [31]. The lysates containing approximately equal amounts of total proteins (~25−50 μg) were subjected to immunoblotting [31]. When indicated, actin protein was used as an internal control (because levels of actin did not change after DHT-treatment of cell types that we used). Enhanced chemiluminescence (ECL) signals of proteins were measured by the Molecular Imager Gel Doc XR⁺ System (Bio-Rad, Hercules, CA, USA). Immunoblots that were used for quantification of protein levels are shown in Supplementary Figures and the quantification values in an accompanied Table.

2.5. Inflammasome Assay

Activation of inflammasome activity in PrECs or BPH-1 cells was assessed using the criteria described earlier [32]. In brief, we subjected the total cell lysates or proteins from cell culture medium to immunoblotting and assessed (i) a decrease in the cellular levels of pro-caspase-1 (p45); (ii) an increase in the cellular levels of activated caspase 1 (p20) and/or (p10); (iii) a decrease in the cellular levels of pro-IL-1β (p31); and (iv) an increase in the cellular levels of the mature IL-1β (p17). Notably, in contrast to macrophages [32], in PrECs and BPH-1 cell line, activation of the inflammasome activity was accompanied by moderate to appreciable changes in the cellular levels of pro-caspase-1 (p45) and pro-IL-1β (p31) under our experimental conditions as described [31]. Further, when indicated, we detected the secreted levels of the mature IL-1β and IL-18 in the culture medium after precipitation of the proteins from the medium.

2.6. RNA Isolation and PCR

Cells were collected by centrifugation and the pellets were suspended into the Trizol reagent (Invitrogen) to isolate total RNA as described [31]. cDNA synthesis and semi-quantitative RT-PCR were performed as described [31]. The following primers were used for RT-PCR: the human AR (forward: 5′-CATCTGTGAAATAGAGCCTATCATATCCAC-3′; backward: 5′-TAACGCCTGCCTAGTGGCTTTGGAG-3′), IFI16 (forward: 5′-CCAAGACT GAAGAC TGAA-3′; backward: 5′-ATGGTCAATGACATCCAG-3′), POP3 (forward: 5′-ATGGAGA GTAAATATAAGGAG-3′; backward: 5′-TCAACATGCATTCCCA GAAAT-3′), AIM2 (forward: 5′-ATGTGAAGCCGTCCAGA-3′; backward: 5′-CATCATT TCTGATGG CTGCA-3′), and actin (forward: 5′-GCTCGTCGT CGACAACGGCTC-3′; backward: 5′-CATG ATCTG GGTCATCTTCTC-3′). Levels of actin mRNA were used as an internal control. To determine the fold change (FC) in the levels of an mRNA following a treatment, the intensity of the actin DNA band (an internal control) on the agarose gel and the DNA band of a gene of interest were measured by the Molecular Imager Gel Doc XR$^+$ System (Bio-Rad, Hercules, CA, USA) with Image Lab Software. Next, the ratio was calculated using the DNA band intensity value for the gene of interest and actin DNA band. This ratio in control cells was indicated as 1 and the FC for DHT-treated samples was calculated by calculating the ratio between the value from treated samples (calculated as in the case of control sample) and the control value 1.

For quantitative real-time TaqMan PCR assays, Applied Biosystems's (Foster City, CA, USA) technique was used [31]. The PCR cycling program consisted of denaturing at 95 °C for 10 min and 40 cycles at 95 °C for 15 s, and annealing and elongation at 60 °C for 1 min. The TaqMan assays for *IFI16* (assay Id #Hs00194216_mL), human interferon-β (*IFNB*; assay Id # Hs01077958 _s1), and for the endogenous control β-actin (assay Id# Hs99999903_mL) were purchased from Applied Biosystems (Foster City, CA, USA) and used as suggested by the supplier. The POP3 TaqMan assay was custom designed: (forward: 5′-AGCACGAGTAGCCAACTT GATT-3′; backward: 5′-GGTCTTCCTCACTGCAGACA-3′).

2.7. Transfection

Sub-confluent cultures of PrECs or BPH-1 cells were either treated with vehicle (ethanol) or with the indicated concentrations of DHT (in ethanol) as noted. Following the treatment, cells were "primed" with TNF-α for 3 has described [31]. Control or "primed" cells were either transfected with LyoVec (control) or poly (dA:dT)/LyoVec (5 μg/mL) for the indicated time. At the end of incubations, cells were harvested to prepare total cell lysates.

2.8. Stable Knockdown of POP3 Expression

To knockdown POP3 protein expression in BPH-1 cell line, cells were either transfected with an empty vector (pcDNA3.1) or pcDNA3.1-POP3(AS) plasmid (a PCR fragment was cloned in the multiple cloning site in the vector in the reverse orientation), thus, allowing the expression of an antisense mRNA. The transfected cells were selected using G418 (500 μg/mL) for two weeks and the G418-resistant colonies (>300 colonies) were pooled. To maintain cells in culture, a reduced concentration (250 μg/mL) of the G418 was used. The transfected cells were cultured without G418 in the medium for two days prior to the experiments.

2.9. Statistical Methods

Experiments involving immunoblotting and semi-quantitative RT-PCR techniques were repeated at least 3-times. A representative result is shown. For quantitative PCR, the assays were performed in triplicates. Fold-changes in the levels of certain proteins and mRNAs are indicated based on the quantitation of signal in independent experiments. The statistical measurement values, when indicated, were presented as means ± SEM. The statistical significance of differences in the measured mean frequencies between the two experimental groups was calculated using the Student two-tailed *t*-test.

3. Results

3.1. Activation of Type I Interferon Signaling in Senescent PrECs Differentially Regulated the Expression of POP3 and AIM2 Proteins

Senescent human diploid fibroblasts (HDFs), as compared with young proliferating or old HDFs, expressed higher basal levels of the IFN-β and activated the type I IFN-signaling [22]. Further, activation of the type I IFN-signaling in senescent HDFs increased the levels of AIM2 protein but decreased IFI16 protein levels [22]. Therefore, we examined the expression of IFN-β and the IFN-β-inducible PYHIN-family proteins in proliferating (passage 2), old (passage 5), and senescent PrECs (passage 8). As shown in Figure 1A, the levels of IFN-β mRNA were significantly higher in senescent vs. proliferating or old PrECs.

Consistent with our previous observations [22], senescent PrECs exhibited activation of T1 IFN response as compared with proliferating or old PrECs as determined by increases in the levels of type I IFN-inducible STAT1 protein and its activating phosphorylation on Tyr-701 residue (Figure 1B). Interestingly, in contrast to senescent HDFs, the levels of type I IFN-inducible IFI16 proteins were higher in senescent PrECs than the young or old PrECs. Expectedly [16], the levels of AR were also higher in senescent PrECs than the young or old PrECs. Further, the levels of AIM2 protein were higher in senescent vs. young or old PrECs. However, the levels of POP3 protein were lower in senescent vs. young or old PrECs. Because POP3 protein inhibited activation of the AIM2 inflammasome [15], we also examined the levels of the mature IL-1β (p17) and IL-18 (p18) in the culture medium. We found that the levels of IL-1β and IL-18 were higher in the culture media of the senescent PrECs than young proliferating cells. These observations thus suggested activation of an inflammasome in senescent PrECs.

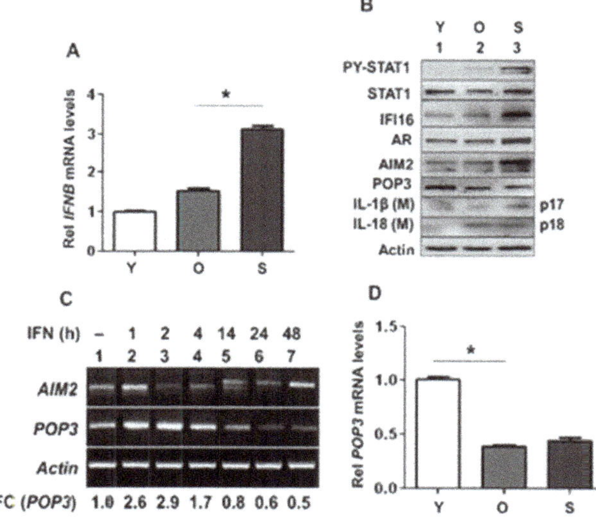

Figure 1. Activation of type I interferon signaling in human senescent PrECs differentially regulated the expression of POP3 and AIM2. (**A**) Total RNAs isolated from young proliferating (Y; passage-2), old (O; passage-5), or senescent (S; passage-8) human PrECs were subjected to quantitative real-time PCR using the TaqMan assay specific for the human *IFNB* mRNA. The RNA levels were normalized using *ACTIN* mRNA. The relative levels of the *IFNB* mRNA in young PrECs are indicated as 1. The values indicated as SEM (* $p < 0.05$). (**B**) Total cell extracts prepared from young (Y; passage-2), old (O; passage-5), or senescent (S; passage-8) human PrECs were analyzed by immunoblotting using the antibodies specific to the indicated proteins. The IL-1β(M) and IL-18(M) indicate the cleaved forms of the pro-IL-1β and pro-IL-18 that were detected in the culture medium. The experiments were repeated at least two times from cells derived from two different donors of different ages. Immunoblots that were used for quantification of protein levels are shown in Supplementary Figure S1 and the quantification of protein levels in an accompanied Table. (**C**) Young proliferating (Y; passage-2) PrECs were either left untreated or treated with 1000 u/mL of IFN-β for indicated times (h). Total RNA was isolated and subjected to RT-PCR for the indicated mRNAs as described in Material and Methods. Fold changes (FC) in the levels of POP3 mRNA were calculated as noted in Materials and Methods. (**D**) Total RNAs isolated from young, old, or senescent PrECs, as described in the panel (**A**), were subjected to the quantitative real-time PCR using the TaqMan assays (in triplicates) specific for the human *POP3* mRNA. The RNA levels were normalized using *ACTIN* mRNA. The relative levels of the *POP3* mRNA in young PrECs are indicated as 1. The values indicated as SEM (* $p < 0.05$).

IFN-β treatment of human macrophages for increasing length of time (0 to 48 h) differentially regulated the expression levels of POP3 and AIM2 mRNA [15]. Therefore, to investigate the potential role of AIM2 inflammasome activation in SASP of PrECs that activated the T1 IFN-signaling (Figure 1A), we treated proliferating PrECs with increasing length of time (0–48 h) and compared the levels of POP3 and AIM2 mRNA levels. As shown in Figure 1C, the treatment of PrECs with IFN-β increased the levels of AIM2 and POP3 mRNAs within an hour. However, the levels of AIM2 mRNA decreased after an hour of treatment but increased again after 48 h of treatment. In contrast, the levels of POP3 mRNA stayed higher after an hour of the treatment of PrECs but stayed lower after 14 h of treatment. Accordingly, a quantitative PCR revealed that old and senescent PrECs expressed lower basal levels of POP3 mRNA than proliferating cells (Figure 1D). These observations are consistent with a chronic activation of the T1 IFN-signaling in senescent PrECs, contributing to an increased AIM2/POP3 protein ratio through a transcriptional mechanism.

3.2. Androgen Receptor Activation in Proliferating PrECs Increased the Expression of POP3

Human primary PrECs express detectable levels of the androgen receptor (AR) [16]. Further, treatment of primary PrECs with type I IFN increased the levels of AR and stimulated the transcriptional activity of AR [16]. Because senescent PrECs exhibited activation of T1 IFN response and expressed higher basal levels of AR (Figure 1B), we tested whether activation of the AR in proliferating PrECs could regulate the expression of POP3 and AIM2. Consistent with our previous observations [19], treatment of proliferating PrECs with the male sex hormone DHT (10 nM) for 14 h increased the levels of *IFI16* mRNA (Figure 2A). Further, the treatment increased the levels of POP3 mRNA ~ 4-fold. However, the levels of AIM2 mRNA remain unchanged. Therefore, we performed quantitative PCR to assess the extent of increase in the levels of POP3 mRNA by DHT in PrECs. As shown in Figure 2B, treatment of cells with DHT significantly increased the levels of POP3 and IFI16 mRNAs. Accordingly, we also noted measurable increases in the levels of IFI16 and POP3 proteins in extracts from DHT-treated proliferating PrECs (Figure 2C). Consistent with these observations, treatment of LNCaP human prostate cancer cells, which express abundant levels of AR (as compared with normal proliferating PrECs) [33], with 10 nM concentration of DHT also increased the levels of POP3 mRNA (Figure 2D). Similarly, treatment of human benign prostate hyperplasia cell line BPH-1 with 10 nM DHT also increased the levels of IFI16 and POP3 proteins, but not AIM2 protein (data not shown). However, treatment of androgen independent human prostate cancer cell line PC-3 with 10 nM DHT did not increase the levels of POP3 protein (data not shown). Together, these observations are consistent with activation of T1 IFN-signaling in human senescent PrECs potentiating stimulation of the AR-mediated increases in the levels of IFI16 and POP3 proteins, but not the AIM2 protein.

3.3. Androgen Treatment of PrECs Inhibited Cytosolic DNA-Induced Activation of the AIM2 Inflammasome

To determine whether activation of androgen receptor in human proliferating PrECs, which increased the levels of POP3 protein (Figure 2), could inhibit AIM2 inflammasome activity, we compared the inflammasome activation in proliferating PrECs after vehicle (alcohol) or DHT treatment. As shown in Figure 3, DHT treatment of human primary PrECs (passage 2), as compared with control cells (vehicle treated), appreciably increased the levels of the POP3 protein (compare lane 3 with 1). Further, the basal levels of procaspase-1 (p45) were lower in control cells that were stimulated with the synthetic DNA poly [dA:dT] (compare lane 2 with 4). Accordingly, levels of the activated caspase-1 (p20) were higher in the control cells than DHT-treated cells that were stimulated with synthetic DNA. Similarly, the secreted IL-1β (p17) and IL-18 protein levels were higher in the culture medium of control cells than DHT-treated cells. Together, these observations indicated that activation of AR by DHT in proliferating normal PrECs up-regulated the levels of POP3 and the up-regulation associated with a decrease in cytosolic DNA-induced activation of the AIM2 inflammasome activity.

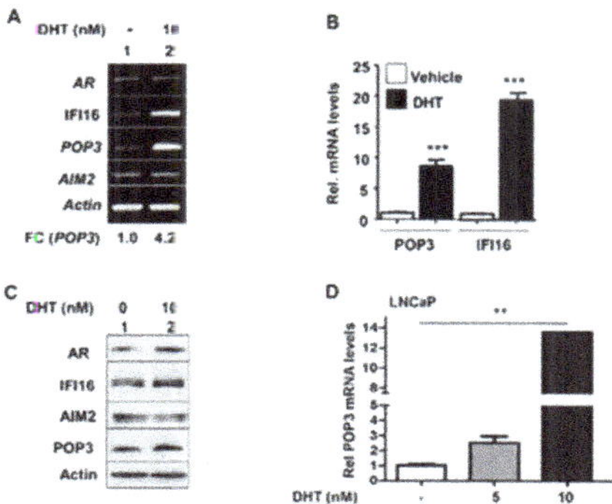

Figure 2. Androgen receptor activation in human proliferating PrECs increased the expression of POP3. (**A**) Sub-confluent cultures of proliferating young (passage 2) human PrECs were either treated with vehicle (ethanol) or 10 nM DHT for 18 h. After the treatment, total RNA was isolated and subjected to RT-PCR for the levels of mRNAs for the indicated genes. The fold change (FC) in the levels of POP3 mRNA in response to DHT-treatment of cells as compared with vehicle treated cells was estimated as described in methods. (**B**) Total mRNA isolated in panel (**A**) was subjected to quantitative real-time PCR (in triplicates) using the TaqMan assay specific for the indicated mRNA. The mRNA levels were normalized using *ACTIN* mRNA. The relative levels of the mRNA in vehicle treated PrECs are indicated as 1. The values indicated as SEM (* $p < 0.05$, ** $p < 0.01$, *** $p < 0.001$). (**C**) Cultures of young proliferating (passage 2) PrECs were either treated with vehicle (lane 1) or 10 nM DHT for 18 h as described in methods. After the treatment, total cell lysates containing equal amounts of proteins were subjected to immunoblotting using the antibodies specific to the indicated proteins. Immunoblots that were used for quantification of protein levels are shown in Supplementary Figure S2 and the quantification of protein levels in an accompanied Table. The experiment was repeated two times using cells derived from a single donor. (**D**) Cultures of the LNCaP cells were either treated with vehicle or the indicated concentration of DHT for 18 h. Total mRNA was isolated and subjected to quantitative real-time PCR using the TaqMan assay specific for the *POP3* mRNA. The *POP3* mRNA levels in all samples were normalized using *ACTIN* mRNA. The relative levels of the mRNA in vehicle treated LNCaP cells are indicated as 1. The values indicated as SEM (* $p < 0.05$, ** $p < 0.01$, *** $p < 0.001$).

3.4. A Stable Knockdown of POP3 Protein Expression in BPH-1 Cell Line Increased Cytosolic DNA-Induced AIM2 Inflammasome Activation

We also investigated whether a knockdown of POP3 protein expression in BPH-1 cells could increase activation of the AIM2 inflammasome without or after DHT treatment. As shown in Figure 4A, stable transfection of BPH-1 cells with an expression vector that allowed the expression of the antisense POP3 mRNA appreciably reduced the basal levels of POP3 mRNA in cells as compared with cells that were transfected with an empty vector. Further, androgen-treatment of vector transfected control cells and their stimulation with cytosolic synthetic DNA did not result in appreciable activation of the inflammasome activity as determined by the lack of detection of proteolytically cleaved caspase-1 (p20) in cell lysates and secreted mature IL-1β (p17) and IL-18 (p13) in the culture medium (Figure 4B). However, a knockdown of the POP3 protein expression in cells and their treatment with DHT did not result in a measurable increase in POP3 protein levels. Importantly, treatment of cells with the synthetic DNA robustly activated the inflammasome activity in cells, as measured by increases in the proteolytically cleaved and activated caspase-1 (p20) levels in cell lysates and the secreted levels of the mature IL-1β (p17) and IL-18 (p18) in the culture medium. Together, these observations indicated that a stable knockdown of

POP3 protein expression in BPH-1 cell line activated cytosolic DNA-induced activity of the AIM2 inflammasome.

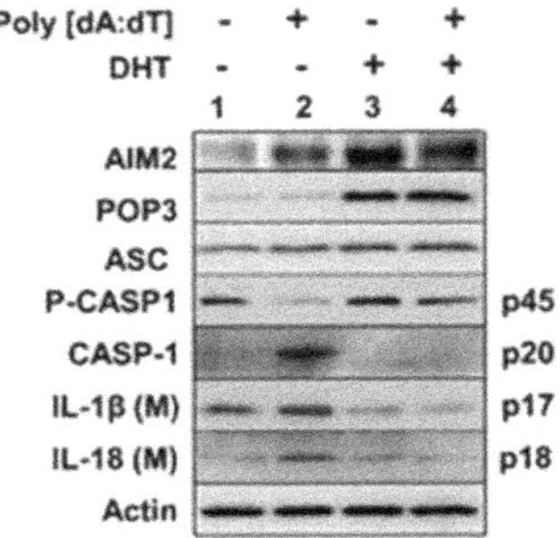

Figure 3. Androgen treatment of PrECs inhibited cytosolic DNA-induced activation of the AIM2 inflammasome. Cultures of human proliferating PrECs were either treated with vehicle (lanes 1 and 2) or with 10 nM DHT (lanes 3 and 4) for 18 h. Cells treated with either vehicle or DHT were further treated with 10 ng/mL TNF-α for 3 h to "prime" cells. The primed cells were either treated with LyoVec (lanes 1 and 3) or poly(dA:dT)/LyoVec (5 µg/mL; lanes 2 and 4) for 4 h. After the treatment, total cell lysates and culture medium (after precipitation of proteins) were subjected to immunoblotting using the antibodies specific to the indicated proteins. The IL-1β (M) and IL-18 (M) indicate the cleaved forms of the pro-IL-1β and pro-IL-18 that were detected in the culture medium. The experiment was repeated using proliferating PrECs from a single donor. Immunoblots that were used for quantification of protein levels are shown in Supplementary Figure S3 and the quantification in an accompanied Table.

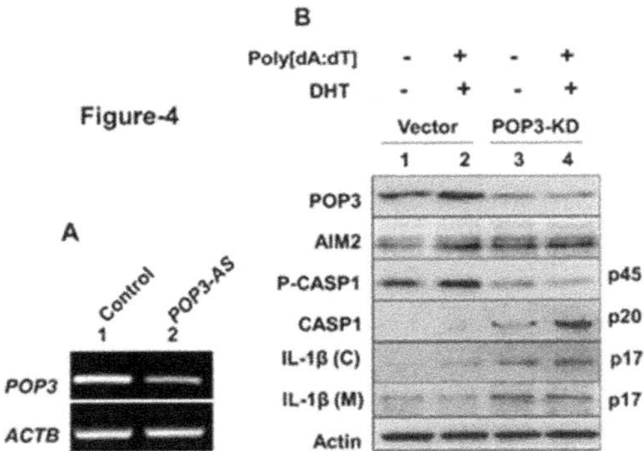

Figure 4. A stable knockdown of POP3 protein expression in BPH-1 cell line increased AIM2 inflammasome activation. (**A**) Total RNA isolated from vector transfected control BPH-1 cells (lane 1) or cells transfected with the pcDNA3.1-POP3(AS) vector allowing the expression of the antisense POP3 mRNA were analyzed by RT-PCR for the indicated genes. The experiment was repeated two times. (**B**) Control BPH-1 cells in panel (**A**) or cells transfected with pcDNA3.1-POP3(AS) vector were either treated with vehicle (lanes 1 and 3) or 10 nM DHT (lanes 2 and 4) for 18 h. Cells were further treated with 10 ng/mL TNF-α for 3 h to "prime" cells. The primed cells were incubated with either LyoVec (lanes 1 and 3) or poly (dA:dT)/LyoVec (5 µg/mL; lanes 2 and 4) for 4 h. After the treatment, total cell lysates and cell culture medium (after precipitation of proteins) were subjected to immunoblotting using the antibodies specific to the indicated proteins. The IL-1β (C), the cleaved IL-1β (p17) within the cell; IL-1β (M), cleaved IL-1β detected in the culture medium. The experiment was repeated two times. Immunoblots that were used for quantification of protein levels are shown in Supplementary Figure S4 and the quantification in an accompanied Table.

4. Discussion

Senescent PrECs accumulate in BPH [27]. Further, SASP is associated with activation of an inflammasome activity and an increase in the production of proinflammatory cytokines [34,35]. Therefore, our observations that (i) human senescent PrECs expressed higher basal levels of the AIM2 protein and lower basal levels of POP3 protein (Figure 1); and (ii) reduced basal levels of the POP3 protein in senescent PrECs (Figure 1) associated with activation of cytosolic DNA-responsive AIM2 inflammasome are consistent with a role of POP3 protein in the suppression of SASP in senescent PrECs.

Serum androgen levels decrease in men with aging [36,37]. Further, androgen receptor levels increase in PrECs in certain parts of the prostate [37]. Therefore, our observations that senescent PrECs that activated type I IFN signaling expressed higher basal levels of the AR (Figure 1B) and activation of androgen receptor in human PrECs increased the levels of POP3 protein (Figure 2) and increased levels of POP3 in PrECs inhibited cytosolic DNA-induced activation of the AIM2 inflammasome activity (Figure 3) support the idea that aging-related reduced serum levels of androgens in men contribute to a decrease in the levels of POP3 protein in PrECs, thus leading to an increase in activation of the AIM2 inflammasome. Because our observations implicate a role for the POP3 protein in aging-related prostatic inflammation, further work will be needed to examine the role of androgen-AR/POP3/AIM2 axis in the development of aging-related prostatic diseases.

POP3 protein also bound with the IFI16 proteins and inhibited activation of the IFI16 inflammasome [15]. Notably, androgens mediated activation of AR in human normal proliferating PrECs also increased the expression of *IFI16* gene [19]. As increased levels of IFI16 proteins in human proliferating PrECs potentiated the p53-mediated cell cycle arrest that is associated with cellular senescence [23], it is conceivable that androgens-mediated up-regulation of the POP3 protein in human PrECs also affects the cell cycle inhibitory functions of the IFI16 proteins. Because AR also drives human PrECs to cellular senescence [24], further work is needed to determine whether androgens-mediated increased levels of POP3 protein in human PrECs modulate the p53-mediated functions.

Activation of certain inflammasomes contributes to the development of prostatic diseases in animal models and humans [38–41]. These diseases include chronic prostatitis and chronic pelvic pain syndrome [38], BPH associated prostatic inflammation [39], and prostate cancer [40,41]. However, it remains unclear whether androgens-mediated activation of the AR in PrECs regulates the activity of the inflammasomes. Therefore, our observations that activation of AR in human PrECs suppressed activation of the AIM2 inflammasome are likely to serve the basis for further studies.

Androgen deprivation therapy (ADT) in prostate cancer patients is often associated with increased production of pro-inflammatory cytokines (e.g., IL1β) [28]. However, it remains unknown whether ADT in prostate cancer patients promote prostatic inflammation through activation of inflammasomes. Therefore, our observations that activation of AR in human PrECs up-regulated the expression of POP3 protein, an inhibitor of the production of inflammatory cytokines (IL-1β and IL-18) through activation of inflammasomes, are of significance.

The 5′-regulatory region of the *POP3* gene remains uncharacterized. Although the expression of *POP3* gene is induced by type I IFN [15], the IFN-responsive *cis*-element(s) remain unknown. Therefore, our observations that treatment of human PrECs (and BPH-1) cell line with androgen DHT increased levels of the POP3 mRNA and protein will require further work to identify the molecular mechanisms through which AR activation in PrECs increases the levels of POP3 mRNA and protein.

In summary, our observations identify the IFN-inducible POP3 PYHIN protein as a potential negative regulator of the AIM2 inflammasome and SASP in human senescent PrECs. These observations also suggest that aging-related reduced levels of androgens in men through reduced basal activation of the AR in PrECs increase the activation of the AIM2 inflammasome. Thus, our observations have important implications for aging-related development of prostatic inflammatory diseases.

Supplementary Materials: The following are available online at https://www.mdpi.com/article/10.3390/life11040366/s1, Figure S1: Protein levels in immunoblots in Figure 1B were quantified and fold changes in protein levels were calculated as noted in the Materials and Methods. The table shows the values for the protein band intensity fractions using the GelQuant.NET software. The values in the parenthesis in the table indicate fold changes in protein levels, Figure S2: Protein levels in immunoblots in Figure 1C were quantified and fold changes in protein levels were calculated as noted above. The values in the parenthesis in the table indicate fold changes in protein levels, Figure S3: Protein levels in immunoblots in Figure 3 were quantified and fold changes in protein levels were calculated as noted above. The values in the parenthesis in the table indicate fold changes in protein levels, Figure S4: Protein levels in immunoblots in Figure 4B were quantified and fold changes in protein levels were calculated as noted above. The values in the parenthesis in the table indicate fold changes in protein levels.

Author Contributions: Conceptualization, D.C.; methodology, R.P., V.R. and H.L.; writing-original draft, D.C. and R.P.; writing—review and editing, D.C. All authors have read and agreed to the published version of the manuscript.

Funding: This work was in part supported by a Merit Review Award (# I01 BX001133) from the United States Department of Veterans Affairs Biomedical Laboratory Research & Development Service to D.C.

Institutional Review Board Statement: Not applicable.

Informed Consent Statement: Not applicable.

Data Availability Statement: Not applicable.

Acknowledgments: We thank Christian Stehlik, Northwestern University, Chicago, for generously providing an expression vector for POP3 protein and antiserum to detect POP3 protein. We also thank Larissa Ponomareva for the technical assistance with human primary prostate epithelial cells.

Conflicts of Interest: The authors declare no conflict of interest.

References

1. De Marzo, A.M.; Nakai, Y.; Nelson, W.G. Inflammation, atrophy, and prostate carcinogenesis. *Urol. Oncol.* **2007**, *25*, 398–400. [CrossRef] [PubMed]
2. De Marzo, A.M.; Platz, E.A.; Sutcliffe, S.; Xu, J.; Grönberg, H.; Drake, C.G.; Nakai, Y.; Isaacs, W.B.; Nelson, W.G. Inflammation in prostate carcinogenesis. *Nat. Rev. Cancer* **2007**, *7*, 256–269. [CrossRef] [PubMed]
3. Sutcliffe, S.; Platz, E.A. Inflammation and prostate cancer: A focus on infections. *Curr. Urol. Rep.* **2008**, *9*, 243–249. [CrossRef] [PubMed]
4. Klein, E.A.; Silverman, R. Inflammation, infection, and prostate cancer. *Curr. Opin. Urol.* **2008**, *18*, 315–319. [CrossRef]
5. Sfanos, K.S.; Isaacs, W.B.; De Marzo, A.M. Infections and inflammation in prostate cancer. *Am. J. Clin. Exp. Urol.* **2013**, *1*, 3–11.
6. Nakai, Y.; Nonomura, N. Inflammation and prostate carcinogenesis. *Int. J. Urol.* **2013**, *20*, 150–160. [CrossRef]
7. Wallach, D.; Kang, T.B.; Kovalenko, A. Concepts of tissue injury and cell death in inflammation: A historical perspective. *Nat. Rev. Immunol.* **2014**, *14*, 51–59. [CrossRef]
8. Kwon, O.J.; Zhang, L.; Ittmann, M.M.; Xin, L. Prostatic inflammation enhances basal-to-luminal differentiation and accelerates initiation of prostate cancer with a basal cell origin. *Proc. Natl. Acad. Sci. USA* **2014**, *111*, E592–E600. [CrossRef]
9. Vignozzi, L.; Maggi, M. Prostate cancer: Intriguing data on inflammation and prostate cancer. *Nat. Rev. Urol.* **2014**, *11*, 369–370. [CrossRef]
10. Ho, C.H.; Fan, C.K.; Yu, H.J.; Wu, C.C.; Chen, K.C.; Liu, S.P.; Cheng, P.C. Testosterone suppresses uropathogenic Escherichia coli invasion and colonization within prostate cells and inhibits inflammatory responses through JAK/STAT-1 signaling pathway. *PLoS ONE* **2017**, *12*, e0180244.
11. Fischer, K.; Tschismarov, R.; Pilz, A.; Straubinger, S.; Carotta, S.; McDowell, A.; Decker, T. Cutibacterium acnes infection induces type I interferon synthesis through the cGAS-STING pathway. *Front. Immunol.* **2020**, *11*, 571334. [CrossRef]
12. Stark, G.R.; Kerr, I.M.; Williams, B.R.; Silverman, R.H.; Schreiber, R.D. How cells respond to interferons? *Annu. Rev. Biochem.* **1998**, *67*, 227–264. [CrossRef]
13. Choubey, D.; Duan, X.; Dickerson, E.; Ponomareva, L.; Panchanathan, R.; Shen, H.; Srivastava, R. Interferon-inducible p200-family proteins as novel sensors of cytoplasmic DNA: Role in inflammation and autoimmunity. *J. Interferon Cytokine Res.* **2010**, *30*, 371–380. [CrossRef]
14. Connolly, D.J.; Bowie, A.G. The emerging role of human PYHIN proteins in innate immunity: Implications for health and disease. *Biochem. Pharmacol.* **2014**, *92*, 405–414. [CrossRef]

15. Khare, S.; Ratsimandresy, R.A.; de Almeida, L.; Cuda, C.M.; Rellick, S.L.; Misharin, A.V.; Wallin, M.C.; Gangopadbyay, A.; Forte, E.; Gottwein, E.; et al. The PYRIN domain-only protein POP3 inhibits ALR inflammasomes and regulates responses to infection with DNA viruses. *Nat. Immunol.* **2014**, *15*, 343–353. [CrossRef]

16. Basrawala, Z.; Alimirah, F.; Xin, H.; Mohideen, N.; Campbell, S.C.; Flanigan, R.C.; Choubey, D. Androgen receptor levels are increased by interferons in human prostate stromal and epithelial cells. *Oncogene* **2006**, *25*, 2812–2817. [CrossRef]

17. Coffey, D.S.; Isaacs, J.T. Control of prostate growth. *Urology* **1981**, *17*, 17–24.

18. Zhou, Y.; Bolton, E.C.; Jones, J.O. Androgens and androgen receptor signaling in prostate tumorigenesis. *J. Mol. Endocrinol.* **2015**, *54*, R15–R29. [CrossRef]

19. Alimirah, F.; Chen, J.; Xin, H.; Choubey, D. Androgen receptor auto-regulates its expression by a negative feedback loop through upregulation of IFI16 protein. *FEBS Lett.* **2006**, *580*, 1659–1664. [CrossRef]

20. Xin, H.; Curry, J.; Johnstone, R.W.; Nickoloff, B.J.; Choubey, D. Role of IFI16, a member of the IFN-inducible p200-protein family, in prostate epithelial cellular senescence. *Oncogene* **2003**, *22*, 4831–4840. [CrossRef]

21. Xin, H.; Pereira-Smith, O.M.; Choubey, D. Role of IFI 16 in cellular senescence of human fibroblasts. *Oncogene* **2004**, *23*, 6209–6217. [CrossRef]

22. Duan, X.; Ponomareva, L.; Veeranki, S.; Panchanathan, R.; Dickerson, E.; Choubey, D. Differential roles for the interferon-inducible IFI16 and AIM2 innate immune sensors for cytosolic DNA in cellular senescence of human fibroblasts. *Mol. Cancer Res.* **2011**, *9*, 589–602. [CrossRef]

23. Choubey, D.; Panchanathan, R. IFI16, an amplifier of DNA-damage response: Role in cellular senescence and aging-associated inflammatory diseases. *Ageing Res. Rev.* **2016**, *28*, 27–36. [CrossRef]

24. Mirochnik, Y.; Veliceasa, D.; Williams, L.; Maxwell, K.; Yemelyanov, A.; Budunova, I.; Volpert, O.V. Androgen receptor drives cellular senescence. *PLoS ONE* **2012**, *7*, e31052. [CrossRef]

25. Campisi, J.; Andersen, J.K.; Kapahi, P.; Melov, S. Cellular senescence: A link between cancer and age-related degenerative disease? *Semin. Cancer Biol.* **2011**, *21*, 354–359. [CrossRef]

26. Coppé, J.P.; Patil, C.K.; Rodier, F.; Sun, Y.; Muñoz, D.P.; Goldstein, J.; Nelson, P.S.; Desprez, P.Y.; Campisi, J. Senescence-associated secretory phenotypes reveal cell-nonautonomous functions of oncogenic RAS and the p53 tumor suppressor. *PLoS Biol.* **2008**, *6*, 2853–2868. [CrossRef]

27. Castro, P.; Giri, D.; Lamb, D.; Ittmann, M. Cellular senescence in the pathogenesis of benign prostatic hyperplasia. *Prostate* **2003**, *55*, 30–38. [CrossRef]

28. Vital, P.; Castro, P.; Tsang, S.; Ittmann, M. The senescence-associated secretory phenotype promotes benign prostatic hyperplasia. *Am. J. Path.* **2014**, *184*, 721–731. [CrossRef]

29. Kumari, P.; Russo, A.J.; Shivcharan, S.; Rathinam, V.A. AIM2 in health and disease: Inflammasome and beyond. *Immunol. Rev.* **2020**, *297*, 83–95. [CrossRef]

30. Hayward, S.W.; Dahiya, R.; Cunha, G.R.; Bartek, J.; Deshpande, N.; Narayan, P. Establishment and characterization of an immortalized but non-transformed human prostate epithelial cell line: BPH-1. *In Vitro Cell. Dev. Biol. Anim.* **1995**, *31*, 14–24. [CrossRef]

31. Ponomareva, L.; Liu, H.; Duan, X.; Dickerson, E.; Shen, H.; Panchanathan, R.; Choubey, D. AIM2, an IFN-inducible cytosolic DNA sensor, in the development of benign prostate hyperplasia and prostate cancer. *Mol. Cancer Res.* **2013**, *11*, 1193–1202. [CrossRef] [PubMed]

32. Martinon, F.; Burns, K.; Tschopp, J. The inflammasome: A molecular platform triggering activation of inflammatory caspases and processing of pro-IL-beta. *Mol. Cell* **2002**, *10*, 417–426. [CrossRef]

33. Veldscholte, J.; Berrevoets, C.A.; Ris-Stalpers, C.; Kuiper, G.G.; Jenster, G.; Trapman, J.; Brinkmann, A.O.; Mulder, E. The androgen receptor in LNCaP cells contains a mutation in the ligand binding domain which affects steroid binding characteristics and response to antiandrogens. *J. Steroid Biochem. Mol. Biol.* **1992**, *41*, 665–669. [CrossRef]

34. Acosta, J.C.; O'Loghlen, A.; Banito, A.; Guijarro, M.V.; Augert, A.; Raguz, S.; Fumagalli, M.; Da Costa, M.; Brown, C.; Popov, N.; et al. Chemokine signaling via the CXCR2 receptor reinforces senescence. *Cell* **2008**, *133*, 1006–1018. [CrossRef]

35. Hoare, M.; Narita, M. Transmitting senescence to the cell neighborhood. *Nat. Cell Biol.* **2013**, *15*, 887–889. [CrossRef]

36. Maggio, M.; Basaria, S.; Ceda, G.P.; Ble, A.; Ling, S.M.; Bandinelli, S.; Valenti, G.; Ferrucci, L. The relationship between testosterone and molecular markers of inflammation in older men. *J. Endocrinol. Investig.* **2005**, *28*, 116–119.

37. Vickman, R.E.; Franco, O.E.; Moline, D.C.; Vander Griend, D.J.; Thumbikat, P.; Hayward, S.W. The role of the androgen receptor in prostate development and benign prostatic hyperplasia: A review. *Asian J. Urol.* **2020**, *7*, 191–202. [CrossRef]

38. Chen, C.S.; Chang, P.J.; Lin, W.Y.; Huang, Y.C.; Ho, D.R. Evidences of the inflammasome pathway in chronic prostatitis and chronic pelvic pain syndrome in an animal model. *Prostate* **2013**, *73*, 391–397. [CrossRef]

39. Kashyap, M.; Pore, S.; Wang, Z.; Gingrich, J.; Yoshimura, N.; Tyagi, P. Inflammasomes are important mediators of prostatic inflammation associated with BPH. *J. Inflamm.* **2015**, *12*, 37. [CrossRef]

40. Karan, D.; Tawfik, O.; Dubey, S. Expression analysis of inflammasome sensors and implication of NLRP12 inflammasome in prostate cancer. *Sci. Rep.* **2017**, *7*, 4378. [CrossRef]

41. Karan, D.; Dubey, S. From inflammation to prostate cancer: The role of inflammasomes. *Adv. Urol.* **2016**, *2016*, 3140372. [CrossRef]

Article

Evaluation of GammaH2AX in Buccal Cells as a Molecular Biomarker of DNA Damage in Alzheimer's Disease in the AIBL Study of Ageing

Mohammad Sabbir Siddiqui [1,2], Maxime Francois [1,3], Stephanie Rainey-Smith [4], Ralph Martins [4,5], Colin L. Masters [6], David Ames [7], Christopher C. Rowe [8], Lance S. Macaulay [1], Michael F. Fenech [1] and Wayne R. Leifert [1,3,*]

[1] CSIRO Health and Biosecurity, Molecular Diagnostic Solutions, Adelaide SA5005, Australia; mohammad.siddiqui@mq.edu.au (M.S.S.); maxime.francois@csiro.au (M.F.); lance.macaulay@csiro.au (L.S.M.); Michael.Fenech@unisa.edu.au (M.F.F.)

[2] School of Agriculture, Food & Wine, the University of Adelaide, Urrbrae 5064, Australia

[3] School of Biological Sciences, the University of Adelaide, Adelaide SA 5005, Australia

[4] Centre of Excellence for Alzheimer's Disease Research & Care, School of Medical Sciences, Edith Cowan University, Joondalup 6027, Australia; s.rainey-smith@ecu.edu.au (S.R.-S.); r.martins@ecu.edu.au (R.M.)

[5] Sir James McCusker Alzheimer's Disease Research Unit (Hollywood Private Hospital), Nedlands 6009, Australia

[6] The Florey Institute of Neuroscience and Mental Health, The University of Melbourne, Parkville 3052, Australia; c.masters@unimelb.edu.au

[7] National Ageing Research Institute, Parkville 3052, Australia; dames@unimelb.edu.au

[8] Department of Nuclear Medicine & Centre for PET, Austin Health, Heidelberg 3084, Australia; crowe@unimelb.edu.au

* Correspondence: wayne.leifert@csiro.au; Tel.: +61-8-8303-8821

Received: 17 July 2020; Accepted: 4 August 2020; Published: 6 August 2020

Abstract: In response to double-stranded breaks (DSBs) in chromosomal DNA, H2AX (a member of histone H2A family) becomes phosphorylated to form γH2AX. Although increased levels of γH2AX have been reported in the neuronal nuclei of Alzheimer's disease (AD) patients, the understanding of γH2AX responses in buccal nuclei of individuals with mild cognitive impairment (MCI) and AD remain unexplored. In the current study, endogenous γH2AX was measured in buccal cell nuclei from MCI ($n = 18$) or AD ($n = 16$) patients and in healthy controls ($n = 17$) using laser scanning cytometry (LSC). The γH2AX level was significantly elevated in nuclei of the AD group compared to the MCI and control group, and there was a concomitant increase in P-trend for γH2AX from the control group through MCI to the AD group. Receiver-operating characteristic curves were carried out for different γH2AX parameters; γH2AX in nuclei resulted in the greatest area under the curve value of 0.7794 ($p = 0.0062$) with 75% sensitivity and 70% specificity for the identification of AD patients from control. In addition, nuclear circularity (a measure of irregular nuclear shape) was significantly higher in the buccal cell nuclei from the AD group compared with the MCI and control groups. Additionally, there was a positive correlation between the nuclear circularity and γH2AX signals. The results indicated that increased DNA damage is associated with AD.

Keywords: γH2AX; Alzheimer's disease; DNA damage; mild cognitive impairment; senescence

---

1. Introduction

Alzheimer's disease (AD) is a neurodegenerative disease that is characterised clinically by severe memory loss, cognitive deterioration, and behavioural changes [1,2]. AD is the most common cause of dementia in old age, representing approximately 60–80% of all dementia cases [3–5]. According to the World Health Organization, 46.8 million people were affected by dementia in the year 2015 [6]. It has been estimated that by the year 2030, 74.7 million people will be affected by AD unless effective interventions are implemented [6]. This increase in the prevalence of AD not only reduces the quality of life, health, and wellbeing of those affected, but also causes a significant financial burden at both the social and economic levels [7].

The classic neuropathological lesions in AD consist of (i) aggregated amyloid plaques containing extracellular hydrophobic deposition of amyloid β peptides (Aβ) in the neuronal body, and (ii) neurofibrillary tangles composed of aggregates of hyperphosphorylated and misfolded tau protein (a microtubule-associated protein) that appear within the neurons [8]. AD patients are usually identified by neuropsychological assessment when the disease has progressed to an advanced stage of cognitive impairment when it is already too late to cure [9,10]. Currently, the ability to detect the early stage of AD and track the different stages of AD progression to guide the choice of therapy is limited. The Mini-Mental State Examination (MMSE) is a validated research-based set of 30 questions assessing memory loss, cognitive decline, and visuospatial and language impairment that is currently used as a standard tool for the clinical diagnosis of AD [11,12]. However, the test lacks accuracy for the diagnosis of AD in living subjects, and diagnostic confirmation can only be achieved post-mortem by the examination of the senile plaques and neurofibrillary tangles in the cerebral tissue [13,14]. The most validated AD disease-related established diagnostic biomarkers are from cerebrospinal fluid (CSF) (aβ1-42, total tau, and phosphorylated tau), structural magnetic resonance imaging (MRI) (e.g., hippocampal volumetry), and amyloid positron emission tomography and fluorodeoxyglucose positron emission tomography imaging [15,16]. Mild cognitive impairment (MCI) is an intermediate state between the cognitive changes of normal ageing and the earliest clinical signs of dementia and is represented as a declining cognition that does not meet the diagnostic criteria of dementia [17]. Individuals affected by MCI have a higher risk of developing AD with an annual conversion rate of approximately 10–15% per year [18–20]. Recent evidence indicates that AD is a systemic disorder that can be mirrored by subclinical pathologies in various peripheral tissues other than the brain, thereby rationalising the grounds for investigating cellular biomarkers in peripheral tissues for the diagnosis of MCI/AD risk [21–25]. A recent study has shown that salivary Aβ42 levels can be used to diagnose AD as well as to predict the risk of its future onset [26]. There is a need for non-invasive biomarkers and inexpensive diagnostic approaches with high specificity and sensitivity to identify individuals at increased risk of developing MCI and AD so that early diagnosis and the initiation of preventative therapy is commenced to halt progression to irreversible neurological impairment.

Human buccal mucosa has considerable potential as an easily accessible source of cells that can be collected in a minimally invasive manner. Defects in buccal mucosa cells may reflect systemic changes in pathology in other tissues of ectodermal origin, such as the nervous system [27–29]. It has been suggested that the ubiquitous presence and different expression of β-amyloid precursor protein (APP) in the buccal mucosa could be a useful means to estimate the regenerative status of tissue [30]. Accumulation of tau protein in the brain is the major component of neurofibrillary tangles, and is the hallmark of AD pathogenesis [31,32]. The amount of buccal cell tau protein was observed at higher levels in AD subjects and correlated with the levels of tau protein in the CSF [33]. AD is associated with genomic DNA damage, and lack of DNA repair capacity could potentially lead to genomic instability [34–39].

The buccal micronucleus cytome assay has been developed to score the cytological markers of DNA damage, cell death, and regenerative capacity of buccal mucosa cells [34,40]. Individuals who had just been diagnosed with AD, but had not yet taken medication for their condition, had significantly reduced basal buccal cell frequency compared to unaffected age-matched controls suggesting reduced regenerative capacity. Aneuploidy (abnormal chromosomal number) has been investigated in buccal cells of AD patients in comparison with respective controls, with the results showing a higher aneuploidy level in chromosomes 17 and 21, which are known to encode tau and APP, respectively [34,41,42]. A recent study showed abnormal DNA content (e.g., hyperploidy in nuclei; a marker of aneuploidy) in buccal mucosa cells of AD patients [28]. The same study also demonstrated decreased amount of neutral lipids as measured by Oil Red-O staining in buccal cells from MCI patients [28]. Buccal samples of AD patients were tested for telomere shortening and displayed a significantly shorter telomere length when compared to healthy older controls [43]. A previous study suggested that DNA strand breaks may be increased in lymphocytes of MCI and AD patients [44].

In response to double-stranded breaks (DSBs) in chromosomal DNA, H2AX (a member of histone H2A family and part of the chromatin structure) becomes phosphorylated to form γH2AX [45]. γH2AX has also been found to be increased in neuronal cells of AD and with ageing in lymphocytes [46–48]. While H2AX is distributed uniformly throughout chromatin, only H2AX molecules located in close vicinity to DSBs become phosphorylated [45,49,50]. The association of astrocyte degeneration and DNA damage with AD has been elucidated by investigating γH2AX signals in astrocytes from the hippocampus, which is known to be the most vulnerable region affected by AD [46]. The results showed a significantly increased number of γH2AX-immunopositive nuclei in the astrocytes of AD patients in comparison to healthy controls, suggesting that astrocytes may be associated with impaired neuronal function and contribute to the pathogenesis of AD [46]. Additionally, a recent study reported elevated γH2AX levels in the hippocampal tissue of individuals with both AD pathology and clinical dementia than those seen in a normal ageing group [47]. γH2AX has been used as a DSB marker in irradiated human buccal cells and was found to be dose responsive in different buccal cell types [51,52]. However, buccal cell DNA damage involving γH2AX, an important marker of DNA damage and DNA damage response, has not been reported in neurodegenerative disorders such as AD.

Taken together, the evidence outlined above forms the basis of the hypothesis we tested that buccal cells from individuals with MCI and AD exhibit elevated levels of γH2AX compared to buccal cells from healthy controls. To test this hypothesis, the endogenous levels of γH2AX in buccal cells from participants in the Australian Imaging, Biomarkers and Lifestyle Flagship Study of Ageing (AIBL) who were either healthy controls, MCI cases, or AD cases were measured. An automated laser scanning cytometry (LSC) γH2AX protocol was used to measure multiple parameters (area, integral, MaxPixel) of γH2AX signals, as well as the ploidy and nuclear shapes and senescent cells in thousands of buccal cells per subject.

2. Materials and Methods

2.1. Human Ethics and Clinical Assessment of the Participants

Approval for the Australian Imaging, Biomarkers and Lifestyle Flagship Study of Ageing (AIBL) was from the institutional ethics committees of Austin Health (Parkville, Vic, Australia), St Vincent's Health (Fitzroy, Vic, Australia), Hollywood Private Hospital (Nedlands, WA, Australia), Edith Cowan University (Perth, WA, Australia), and CSIRO Australia. All volunteers were informed of the purpose of the study and gave written consent before participating in the study. The demographic and health characteristics of participants included in this study have been well characterised and reported previously [53]. Diagnosis of MCI and AD was performed and confirmed by experienced AIBL clinicians using a battery of neuropsychological tests that were selected on the basis that together covered the main domains of cognition that are affected by AD and other dementias [53]. Data reported in this study are from a total of 51 randomly sub-sampled participants, including: (1) the cognitively healthy control group (n = 17); (2) the MCI group (n = 18) clinically diagnosed with MCI; and the (3) AD group (n = 16) clinically diagnosed with AD. Full blood pathology testing was conducted as described previously [54,55]. There were no blood pathology data available for 10 participants.

2.2. Buccal Cell Collection and Microscope Slide Preparation

Prior to buccal cell collection, each participant was first required to rinse their mouth twice with water. Small flat-headed toothbrushes were rotated 20 times against the inner part of the cheeks in a circular motion. Both cheeks were sampled using separate toothbrushes. Heads of the brushes were transferred into a 25 mL tube containing 20 mL of Saccomano's fixative solution and agitated vigorously to dislodge cells into the solution. Cells were then centrifuged at $1000\times g$ for 10 min before discarding and replacing supernatant with fresh 5 mL of buccal cell buffer (10 mM Tris (hydroxymethyl) aminomethane, 0.1 M ethylenediaminetetraacetic acid, 20 mM NaCl, pH 7.0. The cell suspension was drawn up and down five times into a 10 mL syringe using a 21G needle in order to maximise the likelihood of dispersing cell aggregates into a single cell suspension. The cell suspension was then passed through a 100 μm filter in a Swinex filter holder to remove clumps of cells. Cell concentration was assessed using a haemocytometer and cells were then cytocentrifuged for 5 min at 600 rpm onto microscope slides to a final number of 3000 cells per cytospot using a Shandon CytospinVR 4 (ThermoFisher Scientific, Waltham, MA, USA). Slides were washed once with distilled water and air-dried for 1 h and subsequently transferred to ethanol:acetic acid (3:1) fixative for 10 min. The slides were air-dried for 1 h and stored in sealed microscope boxes with desiccant at $-80\,^\circ$C until the staining procedure was performed.

2.3. Preparation of Buccal Cells for Immunofluorescence

A circle was drawn around each cytospot using a hydrophobic PAP pen (Dako, Australia) and air-dried at $22\,^\circ$C for 10 min. Slides were rinsed in Dulbecco's phosphate-buffered saline (DBPS) for 15 min at $22\,^\circ$C, incubated in 70% ethanol ($4\,^\circ$C) for 20 min and washed in DPBS for 15 min at $22\,^\circ$C. Buccal cell cytospots were then treated with 150 μL of prewarmed ($37\,^\circ$C) pepsin solution (containing 750 U/ml of porcine gastric mucosa pepsin) in 0.01 M HCl and then covered with parafilm for 30 min at $37\,^\circ$C in a humidified box. The slides were then washed twice with DPBS for 5 min. Buccal cells were then permeabilised with 1% Triton X-100 for 15 min at room temperature. Slides were then rinsed three times in DPBS, and a blocking step was performed by incubating cells in 10% goat serum for 1 h at room temperature before being washed once with DPBS. The anti-γH2AX antibody was added to each cytospot at a dilution of 2 μg/mL in DPBS containing 10% goat serum and covered with parafilm overnight at $4\,^\circ$C in a humidified box. Slides were washed three times in DPBS for 5 min and a secondary antibody Alexa Fluor 488 Goat antimouse IgG was added to each cytospot at 2 μg/mL in DPBS containing 10% FBS and covered with parafilm for 1 h at room temperature. Slides were washed three times in DPBS for 5 min and nuclei were counterstained with 4,6-diamidino-2-phenylindole (DAPI) at a concentration of 1 μg/mL for 10 min at room temperature. The excess DAPI was removed by rinsing the slides with a solution containing 300 mM NaCl and 34 mM sodium citrate. Slides were then mounted with coverslips and DPBS: glycerol (1:1) medium. The edges of coverslips were sealed with nail polish to prevent drying prior to performing laser scanning cytometry.

2.4. Laser Scanning Cytometry Measurements of γH2AX

Laser scanning cytometry (LSC) measurements were carried out with an iCyte® Automated Imaging Cytometer (Thorlabs, Sterling, VA, USA) with full autofocus function as well as 405 nm and 488 nm lasers for excitation of DAPI and Alexa Fluor 488, respectively. Fluorescence from DAPI (blue) and Alexa Fluor 488 (green) was collected with a photomultiplier tube. Samples were scanned in separate passes (consecutively) to prevent spectral overlap. The nuclei and γH2AX events were contoured using empirically determined thresholds to exclude the scoring of false positives (e.g., small fluorescent debris). The frequency (%) of nuclei containing γH2AX signal was recorded as well as multiple parameters within each nucleus, including the total γH2AX integral (a function of γH2AX intensity and size) and the MaxPixel value (the value of the most intense γH2AX signal/pixel within nuclei). These parameters were generated using the iCyte® 3.4 software and subsequently transferred into excel, then GraphPad Prism for further statistical analyses. Nuclei were also classified into round, long, or oval shapes by utilising the iCyte software parameters which included area, circularity, perimeter, and diameter as described in the legend of Figure 1.

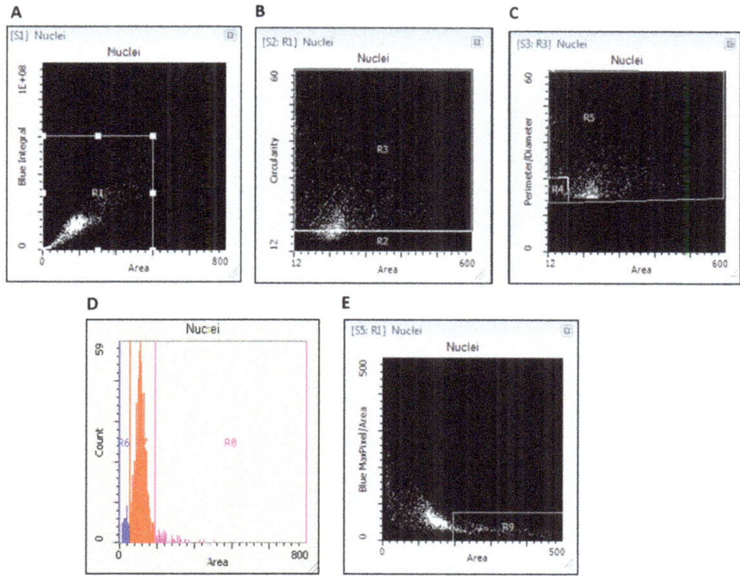

Figure 1. Scattergram and histogram for separation of buccal cell nuclei types by laser scanning cytometry (LSC). A representative example of DNA content scattergram and histogram for a participant from the control group. (**A**) A scattergram was generated to separate cells based on differences in nuclear staining and area by plotting their blue integral versus the area. Nuclei having area values that ranged from 0 to 600 μm^{-2} and blue integral values that ranged from 0 to 4×10^7 (arbitrary units) were separated in Region 1 (R1). (**B**) Nuclei in R1 were analysed by plotting their circularity (*y*-axis) versus nuclear area (*x*-axis) where "round" nuclei were identified in Region 2 (R2). (**C**) Nuclei from Region 3 (R3) were further analysed by plotting their perimeter/diameter ratio (*y*-axis) versus nuclear area (*x*-axis). Two new groups were identified from R3; long nuclei were identified in R4 and oval nuclei in R5. (**D**) A histogram plot of the same data in R1 showing the <2N, 2N, and >2N peaks as represented in R6, R7, and R8, respectively, and the respective frequency of DNA content events scored, showing majority of buccal cells being scored as 2N. (**E**) Nuclei in R1 were plotted against nuclear area versus the ratio of the maximal pixel intensity/area of DAPI fluorescence per nucleus. The cells in R9 had morphometric characteristics of cellular senescence (i.e., increased nuclear size (area) combined with decreased intensity of MaxPixel of DNA-associated fluorescence per nucleus, after DNA staining with DAPI).

2.5. Statistical Analysis

GraphPad Prism 6.01 (GraphPad Prism, San Diego, CA, USA) was used to statistically analyse the data. LSC γH2AX data were checked for normality using the D'Agostino and Pearson omnibus normality test. Differences in relative γH2AX signals in the lymphocytes from control, MCI and AD groups were compared using the Kruskal–Wallis test for non-Gaussian distributed data followed by Dunn's multiple comparisons test. Correlation coefficients were obtained using Pearson's correlation coefficients for Gaussian distributed data and Spearman's rho for non-Gaussian distributed data. Analysed data are reported as mean ± standard error of the mean (SEM) with $p < 0.05$ considered statistically significant. Receiver-operating characteristic curves (ROC) were prepared for selected γH2AX parameters between the control and MCI or AD groups to obtain the area under the curve (AUC), sensitivity, specificity, confidence interval, and *p*-value.

3. Results

3.1. Clinical Characteristics of Participants

The mean age, gender distribution (male/female), body mass index (BMI), and MMSE score of AIBL participants in the control, MCI, and AD groups is shown in Table 1. There were no significant differences for gender ratio and BMI between the groups, while there was a significant difference in age ($p = 0.0039$) between control and AD group; however, there was no correlation of age with γH2AX ($r = 0.08$). As expected, there was a significant decrease in the MMSE scores of both the MCI ($p = 0.0126$) and AD ($p = 0.0001$) groups compared with the control group.

Table 1. Clinical characteristics of participants.

	Control n = 18	MCI n = 17	AD n = 16
Sex (M:F)	12:6	11:6	9:7
Age (years)	72.2 ± 1.5	78.7 ± 1.9	81.0 ± 1.8 **
BMI	27.0 ± 1.3	23.4 ± 1.3	24.8 ± 1.1
MMSE score	29.1 ± 0.2	26.0 ± 0.8 *	12.8 ± 1.8 ***

Means and standard error of the mean (SEM) are reported for each group. Significance was accepted at $p < 0.05$. Abbreviations: AD, Alzheimer's disease; F, female; M, male; MCI, mild cognitive impairment; MMSE, Mini-Mental State Examination score. * $p < 0.05$, ** $p < 0.01$, *** $p < 0.0001$.

3.2. Scoring of γH2AX Signals in Buccal Cells by LSC

All nuclei were separated and analysed according to their ploidy status (DNA content) as follows; <2N, 2N, and >2N as well as different nuclear shapes (round nuclei, long nuclei, oval nuclei) and cellular senescence status (see Figure 1). For 2N nuclei, the peak of the nuclei count coincided with the mean DAPI integral.

Fluorescence images of buccal cell nuclei containing discrete or diffuse γH2AX foci within nuclei were categorised into round, long, and oval nuclei as shown in Figure 2 [51]. Figure 3 summarises the data for the different γH2AX parameters measured (integral, MaxPixel, area, and foci/nucleus) for all nuclei from the control, MCI, and AD groups. Cells were also scored by their ploidy status (i.e., the data for <2N nuclei, 2N nuclei, >2N nuclei, round nuclei, long nuclei, and oval nuclei are shown in Supplementary Table S1). There was a significant increase in the γH2AX integral ($p = 0.0332$) in AD cells compared to control cells in all nuclei (Supplementary Table S1). Consistent with the increase in the γH2AX integral, a significant increase in the γH2AX MaxPixel value ($p = 0.0199$) and the number of γH2AX foci/nucleus ($p = 0.0234$) were also observed in AD cells compared to control cells (Figure 3A,B) and MCI vs. AD ($p = 0.0458$) as shown. Additionally, there was also a significant increase in the linear trend for the γH2AX MaxPixel value ($p = 0.0124$) across the groups (i.e., AD > MCI > control) in all nuclei (Figure 3A). However, there was no significant difference in the area of γH2AX foci (Supplementary Table S1).

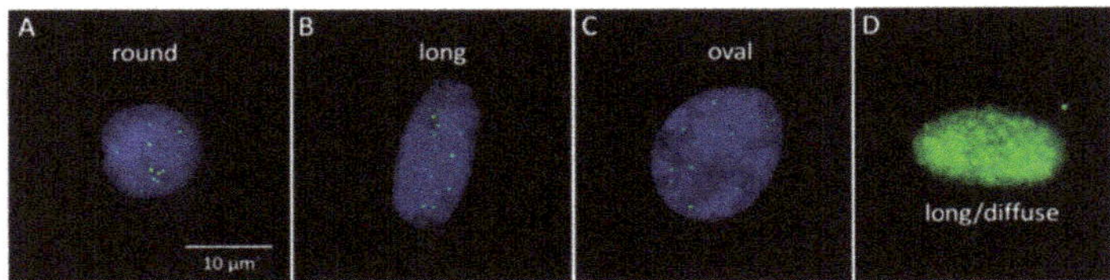

Figure 2. Representative shapes of buccal nuclei and γH2AX foci. Example of buccal cell nuclei visualised (stained with DAPI) with a fluorescence microscope. Nuclei were classified into three categories, i.e., round nuclei (**A**), long nuclei (**B**), and oval nuclei (**C**). Discrete γH2AX foci (green signal) were observed in (**A**–**C**) in these representative images. (**D**) A diffuse γH2AX signal within a nucleus.

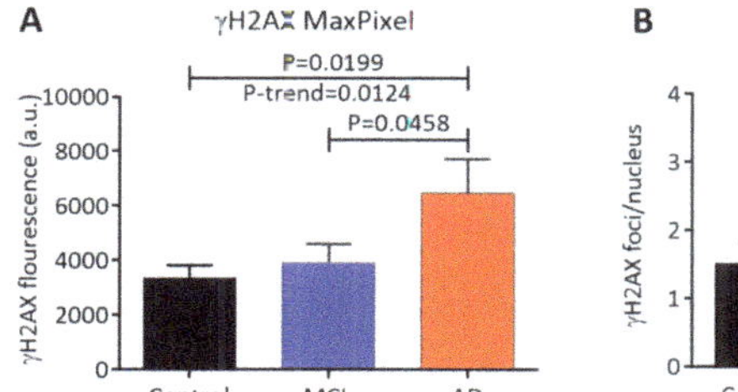
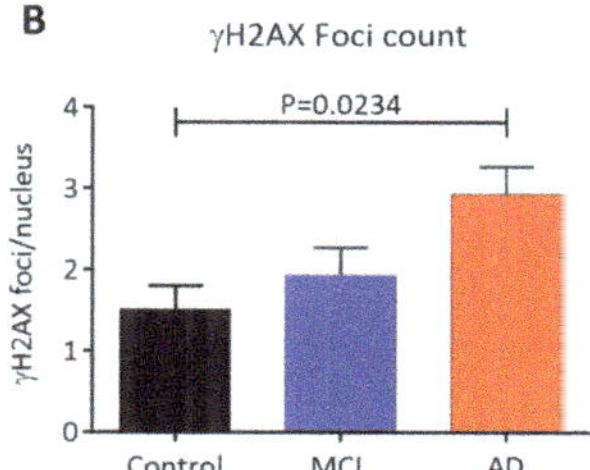

Figure 3. γH2AX MaxPixel and number of foci/nucleus in all cells. (**A**): γH2AX MaxPixel; (**B**): γH2AX foci/nucleus. These parameters were measured by LSC for control (n = 17), MCI (n = 18), and AD (n = 16). Abbreviations: a.u., arbitrary units; AD, Alzheimer's disease; MCI, mild cognitive impairment. Data are means ± SEM. *p*-values are shown.

The frequency (%) of round, long, and oval shaped nuclei was not significantly different between control, MCI, and AD groups (not shown). Supplementary Table S2 shows a significant increase was observed for the γH2AX integral (p = 0.0123), γH2AX MaxPixel (p = 0.0014), γH2AX area (p = 0.0062), and γH2AX foci/nucleus (p = 0.0015) in putative senescent cells when comparing AD versus control cells. The significant increase was also observed for the γH2AX integral (p = 0.0349), γH2AX MaxPixel (p = 0.0134), and γH2AX area (p = 0.0345) in AD senescent cells compared to MCI senescent cells (Supplementary Table S2). There were no differences in the percentage of senescent cells across the groups (Supplementary Table S3).

3.3. Nuclear (Morphology Characteristics) Circularity, Integral, and Area in Buccal Cells

The circularity of buccal cell nuclei in the control, MCI, and AD groups was also measured using the circularity feature available with the iCyte. A high circularity value indicates more irregular shaped nuclei; in contrast, the lowest circularity value indicates a perfect circle. There was a significant increase in nuclear circularity (p = 0.0075) in all nuclei of AD cells compared to control cells. In addition, a significant increase in nuclear circularity (p = 0.0257) was also observed in AD cells compared to MCI cells. A significant increase in the linear p-trend for the nuclear circularity value (p = 0.0027) was observed across the groups (i.e., AD > MCI > control) in all nuclei (Figure 4). For the nuclear integral and area, no significant difference was found between the control, MCI, and AD groups.

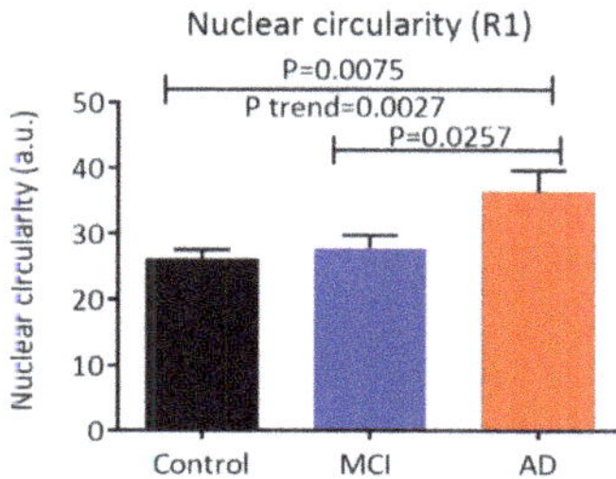

Figure 4. Circularity of buccal cell nuclei. Circularity of all buccal cell nuclei was measured in the control (n = 17), MCI (n = 18), and AD (n = 16) groups. Abbreviations: a.u., arbitrary units; AD, Alzheimer's disease; MCI, mild cognitive impairment. Data are means ± SEM.

3.4. Receiver-Operating Characteristic Curve

Since the γH2AX parameters (e.g., integral, γH2AX MaxPixel, γH2AX foci/nucleus) were significantly higher in AD compared to the control group for each category of nuclei with evaluation of the diagnostic values of these parameters for discriminating AD patients from controls, receiver operating characteristic (ROC) curves were generated. The area under the curve (AUC) values for γH2AX integral, MaxPixel, and foci/nucleus were 0.7353 (p = 0.2118), 0.7794 (p = 0.0062), and 0.7684 (p = 0.0086), respectively (Figure 5A–C). Of all parameters analysed using ROC curves, the γH2AX MaxPixel value showed the greatest value for the identification of AD, with 75% sensitivity and 70% specificity.

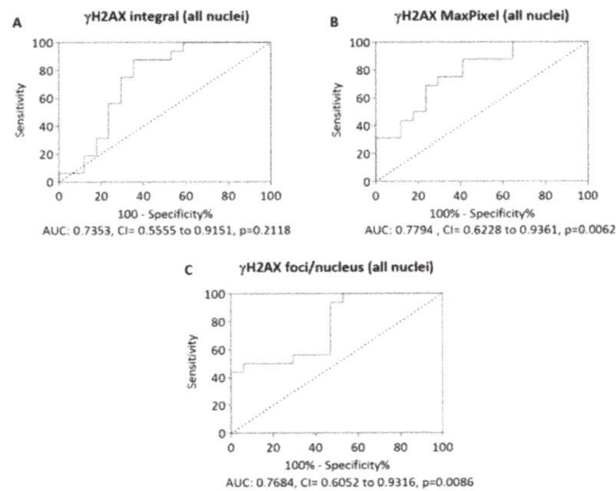

Figure 5. Receiver operating characteristic (ROC) curves for selected LSC-measured γH2AX parameters for control and AD nuclei. ROC curves were generated for the γH2AX integral (**A**), γH2AX MaxPixel (**B**), and γH2AX foci/nucleus (**C**) using measurements in buccal cells from control and AD cells. Abbreviations: AUC (area under the curve), CI (confidence interval).

3.5. Correlation of γH2AX Signals (Integral, MaxPixel) in Different Types of Buccal Cell Nuclei with the MMSE Score

To investigate whether the γH2AX signals in different types of buccal cell nuclei were related to the advancement of cognitive decline in the subjects, the correlations between the γH2AX integral, γH2AX MaxPixel, and MMSE scores were tested. Table 2 summarises the correlation coefficient (r) and p-values for each of the γH2AX parameters analysed in different types of buccal cell nuclei. The parameters highlighted in bold indicate that the γH2AX integral or MaxPixel negatively correlated with the MMSE score, to varying degrees, dependent on cell type analysed. For example, senescent cells had a strong negative correlation of γH2AX integral with the MMSE score (r = −0.5229, p = 0.0002).

Table 2. Summary of correlations between LSC scored γH2AX signals vs. MMSE score.

	Parameters	Correlation (r)	CI	p-Value
All nuclei	γH2AX integral	−0.1899	−0.4014–0.0408	0.0959
	γH2AX MaxPixel	−0.2266	−0.4331–0.0024	**0.0460**
Round	γH2AX integral	−0.3535	−0.5816 to −0.0737	**0.0148**
	γH2AX MaxPixel	−0.4550	−0.6565 to −0.1930	**0.0013**
Long	γH2AX integral	−0.3039	−0.5437 to −0.0183	**0.0378**
	γH2AX MaxPixel	−0.4141	−0.6268 to −0.1440	**0.0038**
Oval	γH2AX integral	−0.3534	−0.5816 to −0.0736	**0.0148**
	γH2AX MaxPixel	−0.4678	−0.6656 to −0.2086	**0.0009**
Senescent	γH2AX integral	−0.5229	−0.7044 to −0.2773	**0.0002**
	γH2AX MaxPixel	−0.5156	−0.6993 to −0.2680	**0.0002**

Parameters highlighted in bold text were considered statistically significant. All are Spearman's rho correlation. CI: 95% confidence interval.

3.6. Correlation of γH2AX Integral with Blood Parameters

Correlation tests were carried out between each of the blood parameters shown in Table 3 and the γH2AX integral values in all nuclei. Of all blood parameters analysed for correlation with γH2AX integral, only total protein was significantly correlated ($r = 0.332$, $p = 0.0389$). In addition, correlation tests were also performed between each of these blood parameters and the γH2AX MaxPixel values. There was no correlation of γH2AX MaxPixel with any blood parameters when data from all nuclei were analysed.

Table 3. Summary of the correlations tested between the γH2AX integral in buccal cells and blood measurements from the Australian Imaging, Biomarkers and Lifestyle Flagship Study of Ageing (AIBL) cohort.

Parameters	Correlation (r)	95% Confidence Interval	p-Value
Homocysteine	0.0092	−0.1537–0.4472	0.9541
Serum folate	0.1617	−0.377–0.198	0.3125
Vitamin B12	−0.1295	−0.4205–0.1856	0.4195
Red cell folate	0.0005	−0.3151–0.3161	0.9975
Calcium	0.0422	−0.2770–0.3531	0.7985
Cholesterol	−0.0270	−0.1924–0.4290	0.4261
Triglycerides	−0.118	−0.3397–0.2911	0.8704
HDL	−0.1846	−0.4726–0.1391	0.2606
LDL	0.2371	−0.08484–0.5142	0.1461
Albumin	0.0305	−0.2879–0.3428	0.8539
Bilirubin	−0.2013	−0.4860–0.1220	0.2191
Urea	−0.0181	−0.3318–0.2992	0.9131
Creatinine	0.0134	−0.3035–0.3276	0.9354
eGFR	0.0427	−0.2766–0.3535	0.7964
Glucose	−0.2302	−0.5088–0.09207	0.1586
Total protein	**0.332**	**0.01837–0.5862**	**0.0389**
ALT	0.0088	−0.3077–0.3234	0.9579
AP	0.0101	−0.3065–0.3247	0.9514
GGT	0.0708	−0.2504–0.3779	0.6684
Ceruloplasmin	−0.2476	−0.5224–0.07374	0.1286
Fe	−0.2834	−0.5498–0.03533	0.0804
Transferrin	0.170	−0.1539–0.4608	0.3009
Trsat	−0.2688	−0.5387–0.05111	0.0980
Ferritin	−0.0201	−0.3336–0.2973	0.9031
Insulin	−0.1066	−0.4084–0.2163	0.5185
Testosterone	0.1546	−0.1692–0.4483	0.3472
LH	0.0245	−0.2933–0.3375	0.8822
FT4	0.1808	−0.1429–0.4696	0.2707
TSH	0.1425	−0.1812–0.4384	0.3868
FT3	0.1999	−0.1234–0.4849	0.2223
Cl	0.04746	−0.2722–0.3577	0.7742
AST	−0.1123	−0.4132–0.2108	0.4961
PCV	−0.0888	−0.3933–0.2334	0.5911
Mg	0.1919	−0.1317–0.4785	0.2418
RCC	−0.0009	−0.3165–0.3147	0.9952
MCV	−0.226	−0.5055–0.09647	0.1665
MCH	−0.2427	−0.5185–0.07897	0.1366
MCHC	−0.1327	−0.4303–0.1909	0.4206
RDW	−0.208	−0.4913–0.1152	0.2039
ESR	−0.1164	−0.4167–0.2068	0.4803
Platelets	−0.05805	−0.3669–0.2623	0.7255
MPV	−0.1251	−0.4239–0.1983	0.4481
WCC	−0.2584	−0.5307–0.06222	0.1122
Neutrophils	−0.2226	−0.5028–0.1001	0.1733
Lymphocytes	−0.1001	−0.4030–0.2225	0.5442
Monocytes	−0.2631	−0.5343–0.05722	0.1056
Eosinophils	−0.1277	−0.4261–0.1958	0.4386
Basophils	−0.2012	−0.4859–0.1222	0.2194

Parameters highlighted in bold text were considered statistically significant. Abbreviations: ALT, alanine aminotransferase; AP, alkaline phosphatase; AST, aspartate aminotransferase; Cl, chloride; eGFR, estimated glomerular filtration rate; ESR, erythrocyte sediment rate; Fe, iron; FT3, free thyroxine; FT4, free triiodothyronine; GGT, gamma-glutamyl transferase; HDL, high-density lipoprotein; LDL, low-density lipoprotein; LH, luteinising hormone; MCH, mean cell haematocrit; MCHC, mean corpuscular haemoglobin concentration; MCV, mean corpuscular volume; Mg, magnesium; MPV, mean platelet volume; PCV, packed cell volume; RCC, red blood cell count; RDW, red cell volume distribution; Trsat, transferrin saturation; TSH, thyroid stimulation hormone; WCC, white cell count.

3.7. Correlation of γH2AX in Control, MCI and AD Nuclei with Blood Parameters

Correlation tests were carried out between each of these blood parameters and the γH2AX integral or the γH2AX MaxPixel values in control, MCI, and AD nuclei. Table 4 summarises the *r* and *p*-values obtained for γH2AX integral with each of the blood parameters. *p*-values highlighted in bold text indicates significant correlations. Albumin, AP, testosterone, and MCV positively correlated with γH2AX integral (Table 4) in MCI nuclei. Total protein, transferrin, LH, FT4, MCH, and MCHC correlated with γH2AX integral in control nuclei. There was no correlation of any of the blood parameters with γH2AX in the AD group.

Table 4. Summary of the correlations tested between γH2AX integral scores in buccal cells and blood parameters in the control, MCI, and AD groups from the AIBL cohort.

	Control	MCI	AD
Homocysteine	$r = -0.070, p = 0.804$	$r = 0.514, p = 0.106$	$r = -0.175, p = 0.518$
Serum folate	$r = 0.193, p = 0.491$	$r = 0.256, p = 0.448$	$r = 0.134, p = 0.635$
Vitamin B12	$r = -0.041, p = 0.883$	$r = -0.293, p = 0.382$	$r = -0.243, p = 0.383$
Red cell folate	$r = 0.288, p = 0.299$	$r = 0.003, p = 0.993$	$r = -0.149, p = 0.595$
Calcium	$r = -0.041, p = 0.884$	$r = -0.433, p = 0.244$	$r = 0.065, p = 0.817$
Cholesterol	$r = 0.467, p = 0.079$	$r = -0.279, p = 0.467$	$r = -0.072, p = 0.799$
Triglycerides	$r = 0.114, p = 0.685$	$r = -0.516, p = 0.155$	$r = -0.033, p = 0.906$
HDL	$r = 0.194, p = 0.489$	$r = -0.266, p = 0.488$	$r = -0.292, p = 0.292$
LDL	$r = 0.465, p = 0.080$	$r = -0.016, p = 0.968$	$r = 0.292, p = 0.802$
Albumin	$r = 0.209, p = 0.454$	$r= 0.724, p = 0.027$	$r = -0.018, p = 0.951$
Bilirubin	$r = -0.286, p = 0.300$	$r = -0.173, p = 0.656$	$r = -0.187, p = 0.504$
Urea	$r = 0.500, p = 0.058$	$r = -0.181, p = 0.640$	$r = -0.326, p = 0.236$
Creatinine	$r = -0.276, p = 0.320$	$r = 0.407, p = 0.277$	$r = -0.038, p = 0.893$
eGFR	$r = 0.186, p = 0.508$	$r = -0.259, p = 0.502$	$r = 0.092, p = 0.745$
Glucose	$r = -0.457, p = 0.087$	$r = 0.112, p = 0.775$	$r = -0.175, p = 0.534$
Total protein	$r= 0.557, p = 0.031$	$r = 0.127, p = 0.745$	$r = 0.133, p = 0.636$
ALT	$r = -0.224, p = 0.421$	$r = 0.109, p = 0.779$	$r = -0.035, p = 0.901$
AP	$r = -0.189, p = 0.498$	$r= 0.681, p = 0.043$	$r = -0.046, p = 0.870$
GGT	$r = -0.108, p = 0.700$	$r = -0.087, p = 0.824$	$r = 0.025, p = 0.931$
Ceruloplasmin	$r = -0.133, p = 0.638$	$r = -0.149, p = 0.703$	$r = -0.294, p = 0.287$
Fe	$r = -0.298, p = 0.280$	$r = -0.385, p = 0.306$	$r = -0.309, p = 0.261$
Transferrin	$r= 0.628, p = 0.012$	$r = -0.225, p = 0.560$	$r = -0.034, p = 0.904$
Trsat	$r = -0.344, p = 0.209$	$r = -0.294, p = 0.442$	$r = -0.282, p = 0.308$
Ferritin	$r = -0.252, p = 0.366$	$r = 0.025, p = 0.949$	$r = -0.100, p = 0.721$
Insulin	$r = -0.162, p = 0.565$	$r = 0.013, p = 0.975$	$r = 0.280, p = 0.310$
Testosterone	$r = -0.162, p = 0.565$	$r= 0.684, p = 0.042$	$r = 0.175, p = 0.532$
LH	$r= 0.522, p = 0.046$	$r = -0.235, p = 0.542$	$r = -0.177, p = 0.527$
FT4	$r= 0.648, p = 0.009$	$r = -0.078, p = 0.842$	$r = 0.155, p = 0.582$
TSH	$r = 0.228, p = 0.411$	$r = 0.056, p = 0.887$	$r = 0.146, p = 0.603$
FT3	$r = 0.431, p = 0.109$	$r = -0.014, p = 0.972$	$r = 0.115, p = 0.684$
Cl	$r = -0.173, p = 0.650$	$r = -0.269, p = 0.485$	$r = 0.173, p = 0.538$
AST	$r = -0.173, p = 0.536$	$r = 0.032, p = 0.935$	$r = -0.185, p = 0.508$
PCV	$r = -0.267, p = 0.335$	$r = 0.074, p = 0.850$	$r = -0.061, p = 0.829$
Mg	$r = -0.016, p = 0.954$	$r = 0.263, p = 0.495$	$r = 0.255, p = 0.359$
RCC	$r = -0.081, p = 0.773$	$r = 0.279, p = 0.467$	$r = -0.071, p = 0.799$
MCV	$r = -0.425, p = 0.115$	$r= -0.678, p = 0.045$	$r = -0.045, p = 0.871$
MCH	$r= -0.658, p = 0.008$	$r = -0.657, p = 0.055$	$r = 0.054, p = 0.848$
MCHC	$r= -0.689, p = 0.005$	$r = -0.479, p = 0.193$	$r = 0.307, p = 0.265$
RDW	$r = -0.197, p = 0.481$	$r = 0.213, p = 0.582$	$r = -0.378, p = 0.165$
ESR	$r = -0.157, p = 0.577$	$r = -0.209, p = 0.589$	$r = -0.186, p = 0.507$
Platelets	$r = 0.049, p = 0.861$	$r = 0.265, p = 0.490$	$r = -0.158, p = 0.576$
MPV	$r = 0.057, p = 0.844$	$r = -0.143, p = 0.713$	$r = -0.438, p = 0.103$
WCC	$r = -0.163, p = 0.563$	$r = 0.369, p = 0.327$	$r = -0.473, p = 0.075$
Neutrophils	$r = -0.292, p = 0.291$	$r = 0.588, p = 0.096$	$r = -0.496, p = 0.059$
Lymphocytes	$r = 0.412, p = 0.127$	$r = -0.356, p = 0.347$	$r = -0.206, p = 0.460$
Monocytes	$r = -0.420, p = 0.119$	$r = 0.091, p = 0.815$	$r = -0.335, p = 0.223$
Eosinophils	$r = 0.015, p = 0.958$	$r = -0.517, p = 0.154$	$r = -0.218, p = 0.435$
Basophils	$r = -0.171, p = 0.542$	$r = 0.408, p = 0.275$	$r = -0.331, p = 0.226$

Parameters highlighted in bold text were considered statistically significant. Abbreviations: ALT, alanine aminotransferase; AP, alkaline phosphatase; AST, aspartate aminotransferase; Cl, chloride; eGFR, estimated glomerular filtration rate; ESR, erythrocyte sediment rate; Fe, iron; FT3, free thyroxine; FT4, free triiodothyronine; GGT, gamma-glutamyl transferase; HDL, high-density lipoprotein; LDL, low-density lipoprotein; LH, luteinising hormone; MCH, mean cell haematocrit; MCHC, mean corpuscular haemoglobin concentration; MCV, mean corpuscular volume; Mg, magnesium; MPV, mean platelet volume; PCV, packed cell volume; RCC, red blood cell count; RDW, red cell volume distribution; Trsat, transferrin saturation; TSH, thyroid stimulation hormone; WCC, white cell count.

4. Discussion

The objective of this study was to investigate whether buccal cells from MCI and AD patients have higher levels of endogenous γH2AX (a biomarker of double-strand DNA breaks) compared with healthy controls, with the ultimate aim of testing whether the buccal cell γH2AX assay might be useful as a diagnostic test for those with cognitive impairment and or AD. The γH2AX assay offers an excellent opportunity to robustly measure the levels of DNA double-strand breaks and cellular response in individuals or populations and test its suitability for clinical purposes [56–58]. The LSC method was used to quantify endogenous γH2AX in buccal cells from individuals who met the clinical criteria for MCI or AD and in healthy controls. The results of this study showed increased levels of γH2AX (thus DNA damage) in the buccal cells of patients with AD compared to those in cells from MCI patients or healthy controls, and there was a concomitant increase with a linear trend from the control group through MCI to the AD group. This result was further supported by the significantly increased negative correlation between γH2AX signals and MMSE scores when the analysis included all subjects. The LSC protocol developed here simultaneously quantifies different γH2AX parameters (integral, MaxPixel, area, foci/nucleus) in cells with different nuclear DNA content (ploidy status) as well as cells with different morphological features such as nuclear shapes, based on their area, perimeter, diameter, and circularity. Nuclear circularity (irregular nuclear shape) was increased significantly in AD cells compared to control cells and there was a concomitant increase with a linear trend from controls through MCI to AD. A significant positive correlation was also observed between nuclear circularity and γH2AX signals in the different types of nuclei analysed. The results of this study demonstrate that buccal cells exhibit increased levels of endogenous γH2AX in AD cells relative to those from MCI patients and healthy controls, and suggest the possibility of using γH2AX as a potential marker for determining those individuals with MCI that may be progressing to AD.

At present, the analysis of Aβ (1–42), total tau, and phospho-tau-181 in CSF allows reliable, sensitive, and specific diagnosis of AD, but the collection of CSF is an invasive procedure with potential random variation in AD-specific biomarker measurements [59–61]. Thus, there is a clear need to search for inexpensive and minimally invasive surrogate markers to diagnose and monitor AD progression. The use of surrogate cells, and particularly exfoliated buccal cells, is of particular interest since buccal cell collection is reliable, fast, relatively simple, cheap, minimally invasive, and painless. Since both the human nervous system and buccal cells are of ectodermal origin, the regenerative potential of the brain, which has been found to be altered in AD, may be mirrored in the buccal mucosa. Studying the buccal mucosa cells from healthy individuals revealed decreased nuclear diameter and cell diameter with increasing age [62]. Another study showed a decrease in the thickness of the epidermis and underlying cell layer with increasing age [63]. It is possible that the lack of regenerative potential of buccal cells from MCI and AD patients may be a consequence of accelerated ageing. A previous study has investigated the formation of micronuclei (a cytogenetic marker of either chromosome segregation or breakage) in buccal mucosa cells. An increased micronuclei frequency was observed in patients with AD compared to age- and gender-matched controls [64]. The same study also reported an abnormal cytome profile characterised by a lower frequency of basal cells, condensed chromatin, and karyorrhectic cells in AD patients, suggesting reduced regenerative capacity in buccal cells from AD patients. Another study showed a significant 1.5-fold increase in trisomy 21 and a significant 1.2-fold increase in trisomy 17 in buccal cells of AD patients compared to matched controls [34], providing further evidence of abnormalities in buccal cells in AD patients. A number of studies have been conducted to assess the association between astrocyte degeneration and DNA damage in AD by investigating the γH2AX signals in astrocytes from the hippocampal region [46,47]. The results from these studies demonstrated increased γH2AX signal in the nuclei of cells from AD patients compared to those from healthy controls. To the best of the researcher's understanding, there are no earlier reports investigating the levels of γH2AX in buccal cells and their ability to distinguish those individuals with MCI and AD from those of control patients. Since the level of DNA DSBs in buccal cells, as marked by γH2AX immunostaining, has not been previously used to investigate the pathogenesis of AD, the findings from this study support the previous observation of increased γH2AX signals in nuclei of astrocytes from AD patients relative to those of healthy controls [46,47]. In the present study, there was an increasing linear trend in the γH2AX MaxPixel values observed in control through to MCI and AD cells, suggesting that buccal cells from MCI patients may be more susceptible to DNA damage than those from healthy controls. There are no reports investigating γH2AX in buccal cells from MCI patients compared to those from healthy controls; however, the insights from our previous studies carried out in lymphocytes are in line with the observations of the current study, and demonstrate a significant increase in oxidative DNA damage (oxidised DNA bases) in lymphocytes from an MCI group compared with a control group [44]. It is of interest to explore whether MaxPixel γH2AX in AD nuclei represent some unique type of DNA damage (e.g., a site of clustered DSBs).

ROC curve analysis was carried out to assess the diagnostic accuracy of γH2AX assay in identifying individuals with AD from controls. ROC curve for LSC scored γH2AX MaxPixel yielded the area under the ROC curve value of 0.7794 with 75% sensitivity and 70% specificity for the AD ($p = 0.0062$) group, suggesting that measurement of γH2AX MaxPixel in the buccal cell might be useful in discriminating AD and control. Although the good sensitivity and specificity achieved in this study are promising for the value of γH2AX assay in identifying AD from control, given the relatively low number of participants tested within each group, and the lack of defined γH2AX thresholds for determining of test positivity, we cannot currently recommend its routine use in clinical practice. Therefore, it is important to clearly demonstrate its accuracy involving larger numbers of participants tested within each group and standardise the γH2AX assay by validating the stringent cutoff point of test positivity prior to it being widely used routinely for differentiating AD from non-AD and from control.

In this study, irregular nuclear shapes (circularity) were measured using the circularity parameter of LSC in different types of nuclei (e.g., all nuclei, <2N nuclei, 2N nuclei, >2N nuclei). A higher circular value indicates a more irregular nuclear shape, and correspondingly, normal ageing affects nuclear shape that may involve defects in lamins [65]. The results showed a significantly higher circularity in all nuclei of AD cells compared to control cells, as well as in AD cells compared to MCI cells. The higher circularity in AD cells compared to control and MCI cells might be due partly to the accumulation of DNA damage leading to morphometric and cytometric alterations in the buccal mucosa cells of AD patients. Previously, the morphological and cytometric parameters of buccal cells have been assessed using microscopy and ImageJ analysis, respectively, following Papanicolaou staining [66]. The results from that study showed a significant decrease in the number of intermediate buccal cells in the AD group compared to the control group [66]. In addition, evidence of increased levels of DNA damage, indicated by the formation of micronuclei (a biomarker of chromosome mis-segregation) has been previously detected in buccal cells from AD patients and Down syndrome cases who have a high risk of developing AD [64,67]. In our study, the γH2AX integral and MaxPixel values were positively correlated with nuclear circularity in the different types of buccal cell nuclei analysed (data not shown), which may reflect the fact that DNA damage in these cells is associated with an irregular nuclear shape. It is possible that the increased DNA damage in those irregularly shaped nuclei is associated with altered nuclear lamina structure. The nuclear lamina is a filamentous structure under the inner nuclear membrane composed of A-type and B-type lamins [68,69]. Recent studies show that the deficient A-type lamin is associated with altered structural nuclear proteins with a variety of human diseases, including severe premature ageing syndromes. Indeed A-type-lamin-deficient cells have been associated with impaired DNA repair capacity and maintaining telomere localisation, structure, length, and function [70,71]. Moreover, loss of A-type-lamin-leads to localisation of telomeres away from the nuclear membrane towards the center of the nucleus [71]. Colocalisation of γH2AX with a telomere DNA probe allowed visualisation of dysfunctional telomeres [72–74]. A previous report in human buccal cells of AD patients showed significantly shortened telomeres in an older AD group in comparison with older controls [43]. Therefore, it is plausible that the positive correlation between nuclear circularity and γH2AX in buccal cells of AD patients observed in this study may be linked with deficient nuclear lamin contributing to telomere dysfunction. Future studies should explore whether the γH2AX signals in buccal cells of AD patients are mostly in the nuclear periphery or aggregated centrally and associated with dysfunctional telomeres which may be due to deficient A-type lamin coupled with increased nuclear circularity. It is possible that irregular nuclear shape caused by a defect in lamins lead to telomere dysfunction and/or shortening. Taken together, altered nuclear morphology, cellular structure, and increased levels of DNA damage associated with dysfunctional telomeres in buccal cells may contribute to the irregular nuclear shape observed in buccal cells of AD patients. A further study of changes in nuclear circularity coupled with multiple DNA damage markers (e.g., γH2AX, 8HOdG) associated with telomere dysfunction and AD-specific markers (e.g., putative tau, Aβ) in buccal cells from a large patient cohort will better assess the likelihood of discriminating AD and MCI patients from healthy controls using these tests.

Cellular senescence is elicited in damaged cells and characterised by the presence of γH2AX, and senescence-associated β-galactosidase (SA-β-gal) activity, and is detectable by immunocytochemistry [75,76]. Previous studies have shown increased number of senescent nuclei during ageing and in age-related diseases [75,76]. It is accepted that older animals exhibit more cellular senescence than younger animals as demonstrated by increased p16 (INK4a), senescence associated β-galactosidase activity, and γH2AX positive signals [73,77,78]. The morphological features of senescent nuclei in cultured fibroblasts after methotrexate (Mtx) treatment have been assessed using the features available in the iCyteR software for LSC [79,80]. In a recent study [79], senescent nuclei were isolated based on the criteria of decreased levels of DAPI staining (MaxPixel staining) paralleled by increases in nuclear size (area) and the simultaneous expression of senescence markers (e.g., the p21WAF1, p16INK4a, or p27KIP1 cyclin kinase inhibitors), and demonstrating that senescent nuclei are flattened and larger in size. To date, the morphological features of senescent nuclei in buccal cells have not been assessed using the features available in LSC. In this study, putative senescent nuclei were identified by plotting the ratio of MaxPixel intensity of DAPI fluorescence per nucleus to nuclear area versus the nuclear size (area). A significant increase in the γH2AX signal was observed in senescent nuclei of AD cells compared to control and MCI cells for all individual γH2AX parameters measured by LSC, suggesting that accumulation of DNA DSBs may contribute to cellular senescence and impaired repairing capacity of senescent nuclei may ultimately contribute to the risk of developing AD. Although previous studies in cultured fibroblasts have characterised the morphological features of senescent nuclei using immunocytochemical analysis of the expression of additional senescent markers, such as the p21WAF1, p16INK4a, or p27KIP1 cyclin kinase inhibitors, our study did not confirm this, but rather attempted for the first time to identify senescent nuclei of control, MCI, and AD cells by their morphometric features alone. It is important to note that senescent cells showed the strongest negative correlation for γH2AX integral and γH2AX MaxPixel in relation to MMSE scores. While investigating the morphological features of senescent buccal cells is important, it is also important for future research to simultaneously measure the expression of senescence markers in conjunction with DNA damage markers (e.g., γH2AX) and AD-specific markers (e.g., aβ1-42, total tau, and phosphorylated-tau) in buccal cells in order to discriminate AD and MCI patients from healthy controls.

In the present study, from all of the blood parameters examined, only total protein showed a positive correlation with buccal cell γH2AX signals when all samples were analysed together. Correlations between blood parameters and buccal cell γH2AX signals in the control, MCI and AD groups were further assessed in three separate tests. Although a significant correlation between buccal cell γH2AX signals and several blood parameters (e.g., albumin, total protein, transferrin, FT4, FT3, MCH, MCV) in control and MCI group was observed, in the AD group, no blood parameters showed a significant correlation with buccal cell γH2AX signals. The negative correlation with MCV and MCH are important because these are biomarkers of anemia, which was previously shown to be a risk factor for MCI and AD in the AIBL study [81]. In this study, the positive correlation between transferrin and γH2AX signals suggests that the plasma transferrin levels may have a role in increasing γH2AX signals in AD. However, a previous study showed lower serum transferrin levels in AD patients compared with controls [82]. These results strongly suggest that the development of pathological features of AD is not restricted to the brain but is associated with multiple metabolic changes occurring in peripheral cells [27].

5. Conclusions

To date, no studies have assessed the presence of γH2AX in the buccal cells of AD patients relative to control and MCI patients, and the available literature on the use γH2AX as a DNA DSB marker in ageing populations is not yet sufficient to understand the association between DNA DSBs and AD. Identification of reliable biomarkers in non-invasive samples will be useful for early diagnosis and treatment of AD, which may prevent the onset of irreversible AD and reduce the overall economic and human cost of the disease. Buccal cells offer a sample source that is easily obtained in a relatively non-invasive manner. The LSC-based γH2AX protocol may be converted to an ELISA-type format or other simpler analytical technique for cellular γH2AX and therefore may provide a practical tool for assessing DNA DSBs in buccal cells of control, MCI, and AD patients. The levels of γH2AX in buccal cells quantified by LSC may have prognostic implications to understand the pathogenesis of AD better and offer the opportunity to monitor disease progression and the bioefficacy of potential preventative measures (i.e., diet, lifestyle, and therapeutics). Moreover, LSC provides identification and quantification of buccal cell subtypes based on cellular features that were previously not measurable (e.g., nuclear shape, DNA content, nucleus size, nucleus MaxPixel value). Scoring of buccal cell nuclear parameters in conjunction with multiple DNA damage parameters and AD-specific markers will be useful to establish a potential biomarker panel with high specificity for AD patients. Thus, the combination of cytome and proteome approaches to a single sampling of buccal cells may significantly increase the sensitivity and/or specificity for AD diagnosis, which will have relevance not only for future clinical practice but also for the reliable prediction of those individuals who are likely to develop MCI and AD and also to monitor the bioefficacy of a preventative strategy. The buccal cell γH2AX assay may provide a useful method for AD and MCI diagnosis, particularly when sample collection must occur remotely and/or in disadvantaged communities unable to attend more expensive prognostic or diagnostic testing facilities. In this study, a small sample was analysed; therefore, comprehensive studies using large prospective cohorts are warranted in order to validate the suitability of the LSC-based buccal cell γH2AX assay, particularly to identify those in the early stages of AD.

Supplementary Materials: The following are available online at http://www.mdpi.com/2075-1729/10/8/141/s1, Table S1. Summary of one-way ANOVA tests for different γH2AX parameters measured using LSC in different types of buccal cell nuclei; Table S2. Summary of the one-way ANOVA tests for different γH2AX parameters in putative senescent nuclei; Table S3. Summary of the one-way ANOVA tests for % of senescent nuclei across Control, MCI, and AD.

Author Contributions: Conceptualisation, M.S.S., M.F.F., and W.R.L.; methodology, M.S.S., M.F., and W.R.L.; formal analysis, M.S.S., M.F., M.F.F., and W.R.L.; resources, S.R.-S., R.M., C.L.M., D.A., C.C.R., and L.S.M.; data curation, M.S.S., M.F.F., and W.R.L.; writing—original draft preparation, M.S.S., M.F., M.F.F., and W.R.L.; writing—review and editing, all authors; supervision, W.R.L. and M.F.; project administration, M.F. and L.S.M.; funding acquisition, M.F. and W.R.L. All authors have read and agreed to the published version of the manuscript.

Funding: The AIBL study received funding support from the Science and Industry Endowment Fund, National Health and Medical Research Council (NHMRC) and Dementia Collaborative Research Centres (DCRC), Alzheimer's Australia and the McCusker Alzheimer's Research Foundation, as well as industry, including Pfizer, Merck, Janssen, and GE Healthcare. Financial support from the CSIRO's Preventative Health Flagship is gratefully acknowledged. W.R.L. received a grant from The JO & JR Wicking Trust, which is managed by ANZ Trustees (Australia).

Acknowledgments: We thank Tori Nguyen and Maryam Hor for assisting with the preparation of microscope slides. We thank all the participants who took part in this study and the clinicians who referred participants. The Australian Imaging, Biomarker & Lifestyle Flagship Study of Ageing (AIBL) study is a collaboration between CSIRO Australia, Edith Cowan University (Perth, WA, Australia), The Florey Institute of Neuroscience and Mental Health (Melbourne, Victoria, Australia), National Ageing Research Institute, and Austin Health (Parkville, Vic, Australia). It also involves support from CogState Ltd. (Melbourne, Vic, Australia), Hollywood Private Hospital, and Sir Charles Gairdner Hospital (Nedlands, WA, Australia).

Conflicts of Interest: The authors declare no conflict of interest.

References

1. Burns, A.; Byrne, E.J.; Maurer, K. Alzheimer's disease. *Lancet* **2002**, *13*, 163–165. [CrossRef]
2. Alzheimer's Association; Thies, W.; Bleiler, L. Alzheimer's disease facts and figures. *Alzheimers Dement* **2011**, *77*, 208–244.
3. Lobo, A.; Launer, L.J.; Fratiglioni, L.; Andersen, K.; Di Carlo, A.; Breteler, M.M.; Copeland, J.R.M.; Dartigues, J.F.; Jagger, C.; Martinez-Lage, J.; et al. Prevalence of dementia and major subtypes in Europe: A collaborative study of population-based cohorts. Neurologic diseases in the elderly research group. *Neurology* **2000**, *54* (Suppl. S5), S4–S9.
4. Tijms, B.M.; Moller, C.; Vrenken, H.; Wink, A.M.; de Haan, W.; van der Flier, W.M.; Starr, C.J.; Scheltens, P.; Barkhof, F. Single-Subject grey matter graphs in Alzheimer's disease. *PLoS ONE* **2013**, *88*, e58921. [CrossRef] [PubMed]
5. Forlenza, O.V.; Diniz, B.S.; Stella, F.; Teixeira, A.L.; Gattaz, W.F. Mild cognitive impairment. Part 1: Clinical characteristics and predictors of dementia. *Rev. Bras. Psiquiatr.* **2013**, *35*, 178–185. [CrossRef] [PubMed]
6. Prince, M.; Wimo, A.; Guerchet, M.; Claire Ali, G.; Wu, Y.; Prina, M. The global impact of dementia. In *World Alzheimer's Disease Report 2015*; Alzheimer's Disease International: London, UK, 2015.
7. Sloane, P.D.; Zimmerman, S.; Suchindran, C.; Reed, P.; Wang, L.; Boustani, M.; Sudha, S. The public health impact of Alzheimer's disease, 2000–2050: Potential implication of treatment advances. *Annu. Rev. Public Health* **2002**, *23*, 213–231. [CrossRef] [PubMed]
8. Ittner, L.M.; Gotz, J. Amyloid-Beta and tau—A toxic pas de deux in Alzheimer's disease. *Nat. Rev. Neurosci.* **2011**, *12*, 65–72. [CrossRef]
9. Weintraub, S.; Wicklund, A.H.; Salmon, D.P. The neuropsychological profile of Alzheimer disease. *Cold Spring Harb. Perspect. Med.* **2012**, *22*, a006171. [CrossRef]
10. Storandt, M. Neuropsychological assessment in Alzheimer's disease. *Exp. Ageing Res.* **1991**, *17*, 100–101.
11. Tombaugh, T.N.; McIntyre, N.J. The mini-mental state examination: A comprehensive review. *J. Am. Geriatr. Soc.* **1992**, *40*, 922–935. [CrossRef]
12. Mitchell, A.J. A meta-analysis of the accuracy of the mini-mental state examination in the detection of dementia and mild cognitive impairment. *J. Psychiatr. Res.* **2009**, *43*, 411–431. [CrossRef] [PubMed]
13. Armstrong, R.A. Plaques and tangles and the pathogenesis of Alzheimer's disease. *Folia Neuropathol.* **2006**, *44*, 1–11. [PubMed]
14. Nelson, P.T.; Alafuzoff, I.; Bigio, E.H.; Bouras, C.; Braak, H.; Cairns, N.J.; Castellani, R.J.; Crain, B.J.; Davies, P.; Tredici, K.D.; et al. Correlation of Alzheimer disease neuropathologic changes with cognitive status A review of the literature. *J. Neuropathol. Exp. Neurol.* **2012**, *71*, 362–381. [CrossRef] [PubMed]
15. Humpel, C. Identifying and validating biomarkers for Alzheimer's disease. *Trends Biotechnol.* **2011**, *29*, 26–32. [CrossRef] [PubMed]
16. Henry, M.S.; Passmore, A.P.; Todd, S.; McGuinness, B.; Craig, D.; Johnston, J.A. The development of effective biomarkers for Alzheimer's disease: A review. *Int. J. Geriatr. Psychiatry* **2013**, *28*, 331–340. [CrossRef]
17. Gauthier, S.; Reisberg, B.; Zaudig, M.; Petersen, R.C.; Ritchie, K.; Broich, K.; Belleville, S.; Brodaty, H.; Bennett, D.; Chertkow, H.; et al. Mild cognitive impairment. *Lancet* **2006**, *367*, 1262–1270. [CrossRef]
18. Fischer, P.; Jungwirth, S.; Zehetmayer, S.; Weissgram, S.; Hoenigschnabl, S.; Gelpi, E.; Broich, K.; Belleville, S.; Brodaty, H.; Bennett, D.; et al. Conversion from subtypes of mild cognitive impairment to Alzheimer dementia. *Neurology* **2007**, *68*, 288–291. [CrossRef]
19. Farias, S.T.; Mungas, D.; Reed, B.R.; Harvey, D.; DeCarli, C. Progression of mild cognitive impairment to dementia in clinic- vs. community-based cohorts. *Arch. Neurol.* **2009**, *66*, 1151–1157. [CrossRef]
20. Petersen, R.C.; Roberts, R.O.; Knopman, D.S.; Boeve, B.F.; Geda, Y.E.; Ivnik, R.J.; Smith, G.E.; Jack, C.R. Mild cognitive impairment: Ten years later. *Arch. Neurol.* **2009**, *66*, 1447–1455. [CrossRef]
21. Gasparini, L.; Racchi, M.; Binetti, G.; Trabucchi, M.; Solerte, S.B.; Alkon, D.; Etcheberrigaray, R.; Gibson, G.; Blass, J.; Paoletti, R.; et al. Peripheral markers in testing pathophysiological hypotheses and diagnosing Alzheimer's disease. *FASEB J.* **1998**, *12*, 17–34. [CrossRef]
22. Joachim, C.L.; Mori, H.; Selkoe, D.J. Amyloid beta-protein deposition in tissues other than brain in Alzheimer's disease. *Nature* **1989**, *341*, 226–230. [CrossRef] [PubMed]
23. Soininen, H.; Syrjanen, S.; Heinonen, O.; Neittaanmaki, H.; Miettinen, R.; Paljarvi, L. Amyloid beta-protein deposition in skin of patients with dementia. *Lancet* **1992**, *339*, 245. [CrossRef]

24. Khan, T.K.; Alkon, D.L. Peripheral biomarkers of Alzheimer's disease. *J. Alzheimer's Dis.* **2015**, *44*, 729–744. [CrossRef] [PubMed]

25. Goldstein, L.E.; Muffat, J.A.; Cherny, R.A.; Moir, R.D.; Ericsson, M.H.; Huang, X.; Mavros, C.; Coccia, J.A.; Faget, K.Y.; Fitch, K.A.; et al. Cytosolic beta-amyloid deposition and supranuclear cataracts in lenses from people with Alzheimer's disease. *Lancet* **2003**, *361*, 1258–1265. [CrossRef]

26. Lee, M.; Guo, J.P.; Kennedy, K.; McGeer, E.G.; McGeer, P.L. A method for diagnosing Alzheimer's disease based on salivary amyloid-beta protein 42 levels. *J. Alzheimer's Dis.* **2017**, *55*, 1175–1182. [CrossRef]

27. Francois, M.; Leifert, W.; Martins, R.; Thomas, P.; Fenech, M. Biomarkers of Alzheimer's disease risk in peripheral tissues; focus on buccal cells. *Curr. Alzheimer Res.* **2014**, *11*, 519–531. [CrossRef]

28. Francois, M.; Leifert, W.; Hecker, J.; Faunt, J.; Martins, R.; Thomas, P.; Fenech, M. Altered cytological parameters in buccal cells from individuals with mild cognitive impairment and Alzheimer's disease. *Cytom. Part A* **2014**, *85*, 698–708. [CrossRef]

29. Leifert, W.R.; Francois, M.; Thomas, P.; Luther, E.; Holden, E.; Fenech, M. Automation of the buccal micronucleus cytome assay using laser scanning cytometry. *Methods Cell Biol.* **2011**, *102*, 321–339.

30. Kummer, C.; Wehner, S.; Quast, T.; Werner, S.; Herzog, V. Expression and potential function of beta-amyloid precursor proteins during cutaneous wound repair. *Exp. Cell Res.* **2002**, *280*, 222–232. [CrossRef]

31. Braak, H.; Braak, E. Demonstration of amyloid deposits and neurofibrillary changes in whole brain sections. *Brain Pathol.* **1991**, *11*, 213–216. [CrossRef]

32. Khan, S.S.; Bloom, G.S. Tau: The center of a signaling nexus in Alzheimer's disease. *Front. Neurosci.* **2016**, *10*, 31. [CrossRef] [PubMed]

33. Hattori, H.; Matsumoto, M.; Iwai, K.; Tsuchiya, H.; Miyauchi, E.; Takasaki, M.; Kamino, K.; Munehira, J.; Kimura, Y.; Kawanishi, K.; et al. The tau protein of oral epithelium increases in Alzheimer's disease. *J. Gerontol. Ser. A Biol. Sci. Med. Sci.* **2002**, *57*, M64–M70. [CrossRef] [PubMed]

34. Thomas, P.; Fenech, M. Chromosome 17 and 21 aneuploidy in buccal cells is increased with ageing and in Alzheimer's disease. *Mutagenesis* **2008**, *23*, 57–65. [CrossRef] [PubMed]

35. Fraga, C.G.; Shigenaga, M.K.; Park, J.W.; Degan, P.; Ames, B.N. Oxidative damage to DNA during aging: 8-hydroxy-2′-deoxyguanosine in rat organ DNA and urine. *Proc. Natl. Acad. Sci. USA* **1990**, *87*, 4533–4537. [CrossRef]

36. Goukassian, D.; Gad, F.; Yaar, M.; Eller, M.S.; Nehal, U.S.; Gilchrest, B.A. Mechanisms and implications of the age-associated decrease in DNA repair capacity. *FASEB J.* **2000**, *14*, 1325–1334.

37. Wilson, D.M., III; Bohr, V.A.; McKinnon, P.J. DNA damage, DNA repair, ageing and age-related disease. *Mech. Aging Dev.* **2008**, *129*, 349–352. [CrossRef]

38. Bucholtz, N.; Demuth, I. DNA-Repair in mild cognitive impairment and Alzheimer's disease. *DNA Repair (Amst.)* **2013**, *12*, 811–816. [CrossRef]

39. Lovell, M.A.; Xie, C.; Markesbery, W.R. Decreased base excision repair and increased helicase activity in Alzheimer's disease brain. *Brain Res.* **2000**, *855*, 116–123. [CrossRef]

40. Thomas, P.; Holland, N.; Bolognesi, C.; Kirsch-Volders, M.; Bonassi, S.; Zeiger, E.; Knasmueller, S.; Fenech, M. Buccal micronucleus cytome assay. *Nat. Protoc.* **2009**, *44*, 825–837. [CrossRef]

41. Iqbal, K.; Grundke-Iqbal, I.; Smith, A.J.; George, L.; Tung, Y.C.; Zaidi, T. Identification and localization of a tau peptide to paired helical filaments of Alzheimer disease. *Proc. Natl. Acad. Sci. USA* **1989**, *86*, 5646–5650. [CrossRef]

42. Koo, E.H. The beta-amyloid precursor protein (APP) and Alzheimer's disease: Does the tail wag the dog? *Traffic* **2002**, *33*, 763–770. [CrossRef] [PubMed]

43. Thomas, P.; O'Callaghan, N.J.; Fenech, M. Telomere length in white blood cells, buccal cells and brain tissue and its variation with ageing and Alzheimer's disease. *Mech. Ageing Dev.* **2008**, *129*, 183–190. [CrossRef] [PubMed]

44. Migliore, L.; Fontana, I.; Trippi, F.; Colognato, R.; Coppede, F.; Tognoni, G.; Nucciarone, B.; Siciliano, G. Oxidative DNA damage in peripheral leukocytes of mild cognitive impairment and AD patients. *Neurobiol. Aging* **2005**, *26*, 567–573. [CrossRef] [PubMed]

45. Rogakou, E.P.; Pilch, D.R.; Orr, A.H.; Ivanova, V.S.; Bonner, W.M. DNA double-stranded breaks induce histone H2AX phosphorylation on serine 139. *J. Biol. Chem.* **1998**, *273*, 5858–5868. [CrossRef]

46. Myung, N.H.; Zhu, X.; Kruman, I.I.; Castellani, R.J.; Petersen, R.B.; Siedlak, S.L.; Perry, G.; Smith, M.A.; Lee, H.G. Evidence of DNA damage in Alzheimer disease: Phosphorylation of histone H2AX in astrocytes. *Age* **2008**, *30*, 209–215. [CrossRef]

47. Silva, A.R.; Santos, A.C.F.; Farfel, J.M.; Grinberg, L.T.; Ferretti, R.E.; Campos, A.H.J.F.M.; Cunha, I.W.; Begnami, M.D.; Rocha, R.M.; Carraro, D.M.; et al. Repair of oxidative DNA damage, cell-cycle regulation and neuronal death may influence the clinical manifestation of Alzheimer's disease. *PLoS ONE* **2014**, *9*, e99897. [CrossRef]

48. Schurman, S.H.; Dunn, C.A.; Greaves, R.; Yu, B.; Ferrucci, L.; Croteau, D.L.; Seidman, M.M.; Bohr, V.A. Age-Related disease association of endogenous gamma-H2AX foci in mononuclear cells derived from leukapheresis. *PLoS ONE* **2012**, *7*, e45728. [CrossRef]

49. Rogakou, E.P.; Boon, C.; Redon, C.; Bonner, W.M. Megabase chromatin domains involved in DNA double-strand breaks in vivo. *J. Cell Biol.* **1999**, *146*, 905–916. [CrossRef]

50. Savic, V.; Yin, B.; Maas, N.L.; Bredemeyer, A.L.; Carpenter, A.C.; Helmink, B.A.; Yang-Iott, K.S.; Sleckman, B.P.; Bassing, C.H. Formation of dynamic gamma-H2AX domains along broken DNA strands is distinctly regulated by ATM and MDC1 and dependent upon H2AX densities in chromatin. *Mol. Cell* **2009**, *34*, 298–310. [CrossRef]

51. Siddiqui, M.S.; Francois, M.; Fenech, M.F.; Leifert, W.R. GammaH2AX responses in human buccal cells exposed to ionizing radiation. *Cytom. Part A* **2015**, *87*, 296–308. [CrossRef]

52. Gonzalez, J.E.; Roch-Lefevre, S.H.; Mandina, T.; Garcia, O.; Roy, L. Induction of gamma-H2AX foci in human exfoliated buccal cells after in vitro exposure to ionising radiation. *Int. J. Radiat. Biol.* **2010**, *86*, 752–759. [CrossRef] [PubMed]

53. Ellis, K.A.; Bush, A.I.; Darby, D.; De Fazio, D.; Foster, J.; Hudson, P.; Lautenschlager, N.T.; Lenzo, N.; Martins, R.N.; Maruff, P.; et al. The Australian imaging, biomarkers and lifestyle (AIBL) study of aging: Methodology and baseline characteristics of 1112 individuals recruited for a longitudinal study of Alzheimer's disease. *Int. Psychogeriatr.* **2009**, *21*, 672–687. [CrossRef] [PubMed]

54. Doecke, J.D.; Laws, S.M.; Faux, N.G.; Wilson, W.; Burnham, S.C.; Lam, C.P.; Mondal, A.; Bedo, J.; Bush, A.I.; Brown, B.; et al. Blood-Based protein biomarkers for diagnosis of Alzheimer disease. *Arch. Neurol.* **2012**, *69*, 1318–1325. [CrossRef] [PubMed]

55. Faux, N.G.; Ellis, K.A.; Porter, L.; Fowler, C.J.; Laws, S.M.; Martins, R.N.; Pertile, K.K.; Rembach, A.; Rowe, C.C.; Rumble, R.L.; et al. Homocysteine, vitamin B12, and folic acid levels in Alzheimer's disease, mild cognitive impairment, and healthy elderly: Baseline characteristics in subjects of the Australian imaging biomarker lifestyle study. *J. Alzheimer's Dis.* **2011**, *27*, 909–922. [CrossRef] [PubMed]

56. Shukkur, E.A.; Shimohata, A.; Akagi, T.; Yu, W.; Yamaguchi, M.; Murayama, M.; Chui, D.; Takeuchi, T.; Amano, K.; Subramhanya, K.H.; et al. Mitochondrial dysfunction and tau hyperphosphorylation in Ts1Cje, a mouse model for Down syndrome. *Hum. Mol. Genet.* **2006**, *15*, 2752–2762. [CrossRef] [PubMed]

57. Bader Lange, M.L.; St Clair, D.; Markesbery, W.R.; Studzinski, C.M.; Murphy, M.P.; Butterfield, D.A. Age-Related loss of phospholipid asymmetry in APP(NLh)/APP(NLh) x PS-1(P264L)/PS-1(P264L) human double mutant knock-in mice: Relevance to Alzheimer disease. *Neurobiol. Dis.* **2010**, *38*, 104–115. [CrossRef]

58. DelleDonne, A.; Kouri, N.; Reinstatler, L.; Sahara, T.; Li, L.; Zhao, J.; Dickson, D.W.; Ertekin-Taner, N.; Leissring, M.A. Development of monoclonal antibodies and quantitative ELISAs targeting insulin-degrading enzyme. *Mol. Neurodegener.* **2009**, *4*, 1–6. [CrossRef]

59. Fagan, A.M.; Roe, C.M.; Xiong, C.; Mintun, M.A.; Morris, J.C.; Holtzman, D.M. Cerebrospinal fluid tau/beta-amyloid (42) ratio as a prediction of cognitive decline in nondemented older adults. *Arch. Neurol.* **2007**, *64*, 343–349. [CrossRef]

60. Ingber, A.P.; Hassenstab, J.; Fagan, A.M.; Benzinger, T.L.; Grant, E.A.; Holtzman, D.M.; Morris, J.C.; Roe, C.M. Cerebrospinal fluid biomarkers and reserve variables as predictors of future "non-cognitive" outcomes of Alzheimer's disease. *J. Alzheimer's Dis.* **2016**, *52*, 1055–1164. [CrossRef]

61. Zetterberg, H. Cerebrospinal fluid biomarkers for Alzheimer's disease: Current limitations and recent developments. *Curr. Opin. Psychiatry* **2015**, *28*, 402–409. [CrossRef]

62. Donald, P.M.; George, R.; Sriram, G.; Kavitha, B.; Sivapathasundharam, B. Hormonal changes in exfoliated normal buccal mucosal cells. *J. Cytol.* **2013**, *30*, 252–256. [CrossRef] [PubMed]

63. Williams, D.M.; Cruchley, A.T. Structural aspects of ageing in the oral mucosa. In *The Effect of Ageing in the Oral Mucosa and Skin*; Squier, C.A., Hill, M.W., Eds.; CRC Press: London, UK; Bocca Raton, FL, USA, 1994; pp. 65–74.

64. Thomas, P.; Hecker, J.; Faunt, J.; Fenech, M. Buccal micronucleus cytome biomarkers may be associated with Alzheimer's disease. *Mutagenesis* **2007**, *22*, 371–379. [CrossRef] [PubMed]

65. Chan, Y.K.; Tsai, M.H.; Huang, D.C.; Zheng, Z.H.; Hung, K.D. Leukocyte nucleus segmentation and nucleus lobe counting. *BMC Bioinform.* **2010**, *11*, 558. [CrossRef] [PubMed]

66. De Oliveira, R.M.; Lia, E.N.; Guimaraes, R.M.; Bocca, A.L.; Cavalcante Neto, F.F.; da Silva, T.A. Cytologic and cytometric analysis of oral mucosa in Alzheimer's disease. *Anal. Quant. Cytol. Histol.* **2008**, *30*, 113–118. [PubMed]

67. Thomas, P.; Harvey, S.; Gruner, T.; Fenech, M. The buccal cytome and micronucleus frequency is substantially altered in Down's syndrome and normal ageing compared to young healthy controls. *Mutat. Res.* **2008**, *638*, 37–47. [CrossRef] [PubMed]

68. Hozak, P.; Sasseville, A.M.; Raymond, Y.; Cook, P.R. Lamin proteins form an internal nucleoskeleton as well as a peripheral lamina in human cells. *J. Cell Sci.* **1995**, *108*, 635–644.

69. Dechat, T.; Pfleghaar, K.; Sengupta, K.; Shimi, T.; Shumaker, D.K.; Solimando, L.; Goldman, R.D. Nuclear lamins: Major factors in the structural organization and function of the nucleus and chromatin. *Genes Dev.* **2008**, *22*, 832–853. [CrossRef]

70. Gonzalo, S. DNA damage and lamins. *Adv. Exp. Med. Biol.* **2014**, *773*, 377–399.

71. Gonzalez-Suarez, I.; Redwood, A.B.; Perkins, S.M.; Vermolen, B.; Lichtensztejin, D.; Grotsky, D.A.; Morgado-Palacin, L.; Gapud, E.J.; Sleckman, B.P.; Sullivan, T.; et al. Novel roles for A-type lamins in telomere biology and the DNA damage response pathway. *EMBO J.* **2009**, *28*, 2414–2427. [CrossRef]

72. Smogorzewska, A.; Karlseder, J.; Holtgreve-Grez, H.; Jauch, A.; de Lange, T. DNA ligase IV-dependent NHEJ of deprotected mammalian telomeres in G1 and G2. *Curr. Biol.* **2002**, *12*, 1635–1644. [CrossRef]

73. Fumagalli, M.; Rossiello, F.; Clerici, M.; Barozzi, S.; Cittaro, D.; Kaplunov, J.M.; Bucci, G.; Dobreva, M.; Matti, V.; Beausejour, C.M.; et al. Telomeric DNA damage is irreparable and causes persistent DNA-damage-response activation. *Nat. Cell Biol.* **2012**, *14*, 355–365. [CrossRef] [PubMed]

74. Hewitt, G.; Jurk, D.; Marques, F.D.; Correia-Melo, C.; Hardy, T.; Gackowska, A.; Anderson, R.; Taschuk, M.; Mann, J.; Passos, J.F. Telomeres are favoured targets of a persistent DNA damage response in ageing and stress-induced senescence. *Nat. Commun.* **2012**, *3*, 708. [CrossRef] [PubMed]

75. Dimri, G.P.; Lee, X.; Basile, G.; Acosta, M.; Scott, G.; Roskelley, C.; Medrano, E.E.; Linskens, M.; Rubelj, I.; Pereira-Smith, O. A biomarker that identifies senescent human cells in culture and in aging skin in vivo. *Proc. Natl. Acad. Sci. USA* **1995**, *92*, 9363–9367. [CrossRef] [PubMed]

76. Sikora, E.; Arendt, T.; Bennett, M.; Narita, M. Impact of cellular senescence signature on ageing research. *Ageing Res. Rev.* **2011**, *10*, 146–152. [CrossRef] [PubMed]

77. Berkenkamp, B.; Susnik, N.; Baisantry, A.; Kuznetsova, I.; Jacobi, C.; Sörensen-Zender, I.; Broecker, V.; Haller, H.; Melk, A.; Schmitt, R. In vivo and in vitro analysis of age-associated changes and somatic cellular senescence in renal epithelial cells. *PLoS ONE* **2014**, *9*, e88071. [CrossRef]

78. Wang, C.; Jurk, D.; Maddick, M.; Nelson, G.; Martin-Ruiz, C.; von Zglinicki, T. DNA damage response and cellular senescence in tissues of aging mice. *Aging Cell* **2009**, *8*, 311–323. [CrossRef]

79. Zhao, H.; Darzynkiewicz, Z. Biomarkers of cell senescence assessed by imaging cytometry. *Methods Mol. Biol.* **2013**, *965*, 83–92.

80. Zhao, H.; Halicka, H.D.; Traganos, F.; Jorgensen, E.; Darzynkiewicz, Z. New biomarkers probing depth of cell senescence assessed by laser scanning cytometry. *Cytom. Part A* **2010**, *77*, 999–1007. [CrossRef]

81. Faux, N.G.; Rembach, A.; Wiley, J.; Ellis, K.A.; Ames, D.; Fowler, C.J.; Martins, R.N.; Pertile, K.K.; Rumble, R.L.; Trounson, B.; et al. An anemia of Alzheimer's disease. *Mol. Psychiatry* **2014**, *19*, 1227–1234. [CrossRef]

82. Fischer, P.; Gotz, M.E.; Danielczyk, W.; Gsell, W.; Riederer, P. Blood transferrin and ferritin in Alzheimer's disease. *Life Sci.* **1997**, *60*, 2273–2278. [CrossRef]

Review

Loss of p16: A Bouncer of the Immunological Surveillance?

Kelly E. Leon [1,2,†], Naveen Kumar Tangudu [1,†], Katherine M. Aird [1] and Raquel Buj [1,*]

[1] UPMC Hillman Cancer Center, Department of Pharmacology & Chemical Biology, University of Pittsburgh School of Medicine, Pittsburgh, PA 15213, USA; kel170@pitt.edu (K.E.L.); tangudunk@upmc.edu (N.K.T.); katherine.aird@pitt.edu (K.M.A.)

[2] Biomedical Sciences Graduate Program, Penn State College of Medicine, Hershey, PA 15213, USA

* Correspondence: bujgomezr@upmc.edu; Tel.: +1-412-623-3241; Fax: +1-412-623-7709

† These authors contributed equally to this work.

Abstract: p16[INK4A] (hereafter called p16) is an important tumor suppressor protein frequently suppressed in human cancer and highly upregulated in many types of senescence. Although its role as a cell cycle regulator is very well delineated, little is known about its other non-cell cycle-related roles. Importantly, recent correlative studies suggest that p16 may be a regulator of tissue immunological surveillance through the transcriptional regulation of different chemokines, interleukins and other factors secreted as part of the senescence-associated secretory phenotype (SASP). Here, we summarize the current evidence supporting the hypothesis that p16 is a regulator of tumor immunity.

Keywords: senescence-associated secretory phenotype (SASP); senescence; cell-cycle; melanoma; pancreatic adenocarcinoma; tumor infiltration; chemotherapy resistance

Citation: Leon, K.E.; Tangudu, N.K.; Aird, K.M.; Buj, R. Loss of p16: A Bouncer of the Immunological Surveillance? *Life* **2021**, *11*, 309. https://doi.org/10.3390/life11040309

Academic Editor: Markus Riessland

Received: 10 March 2021
Accepted: 30 March 2021
Published: 2 April 2021

Publisher's Note: MDPI stays neutral with regard to jurisdictional claims in published maps and institutional affiliations.

1. Introduction

Since the 19th century, immunologists have been speculating about the idea that the immune system may be a strong, efficient, and specific weapon against cancer (reviewed in [1]). However, it was not until the beginning of the 21st century that scientists began to understand the mechanisms behind tumor immunity and to develop immunotherapy regimens [2]. In parallel, the senescence field discovered that senescent cells, although in a stable state of cell cycle arrest [3], are highly active and acquire a pro-inflammatory microenvironment termed the senescence-associated secretory phenotype (SASP) [4,5]. Through the SASP, senescent cells can modify their microenvironment and regulate other cells, including cells of the immune system [5–10].

Among the multiple pathways that are commonly deregulated in cancer and senescence [11], the p16[INK4A] (hereafter called p16) pathway is particularly intriguing. On one hand, loss of p16 is a common feature of cancer that causes an increase in the proliferative capacity of the cell [12]; on the other hand, upregulation of p16 is a hallmark of senescence that contributes to the characteristic state of cell cycle arrest [13]. Interestingly, recent publications demonstrate that suppression of p16 correlates with decreased activity of immune cells [14–17], and our recent publication shows that p16 suppression decreases expression of the SASP [18]. Altogether these data suggest that p16 may have a, yet unknown, role in the regulation of tumor immunity that might have important implications in the treatment of cancers with low or null p16 expression. In this review, we describe the roles of p16 and the SASP in both senescence and cancer and dissect the latest publications that support the hypothesis that p16 may be a regulator of tumor immunity.

2. p16 in Cancer and Senescence

Cyclin Dependent Kinase Inhibitor 2A (*CDKN2A*) located in chromosome 9p21 is a tumor suppressor gene that encodes p14[ARF] (p19[ARF] in mice, hereafter p14 and p19) and p16 proteins using two different open reading frames. While p14[ARF] is involved in the regulation of the p53 pathway (reviewed in [19]), p16's canonical role is to inhibit the assembly and activation of the cyclin-dependent kinases CDK4/6, impairing the hyperphosphorylation of the retinoblastoma (RB) protein and the E2F-mediated expression of proliferation-promoting genes [20]. Due to its role as cell cycle brake, it is not surprising that ~50% of human cancer shows decreased expression of *CDKN2A* [12]. Interestingly, although the most common alterations of p16 are deletions and promoter hypermethylation affecting both p16 and p14 [21–23], cancer-associated mutations are more commonly found in p16 than p14 [24], suggesting a critical regulatory role of p16 in the cell. Importantly, loss of p16 alone is not enough to produce cancer, mainly because normal cells have other mechanisms to abrogate the cell cycle progression (e.g., p53, CHK1/2, APC) (reviewed in [25]). However, it has been shown that the suppression of p16 facilitates malignant transformation of cells upon different hyperproliferative signals and stressors such as oncogenes, oxidative stress, ionizing radiation, and others. All together these data suggest that tight regulation of p16 expression is critical to maintain healthy cellular proliferation.

On the other hand, p16 is known to be highly expressed during cellular senescence [26–28]. Senescence is a stable state of cell cycle arrest acquired upon different stressors such as aberrant proliferative signals or DNA damage among others [29]. Since cells are continuously affected by a wide variety of stressors, it is not surprising that a wide variety of senescence inducers exist, including: oncogenes, ionizing radiation, genotoxic chemicals, reactive oxygen species, chemotherapeutic agents or shortage in dNTPs among others (reviewed in [30]). Interestingly, in the vast majority of cases, the distinctive cell cycle arrest is achieved by upregulation of the p16 protein as a direct consequence of the pathways (e.g., p38 and ERK) [31–34], epigenetic factors (e.g., polycomb) [35–37] and transcription factors (e.g., ETS2 and AP-1) [38–41] altered by the senescence inducers. In this regard, several groups including our laboratory have shown that suppression of p16 can bypass senescence [42–46], indicating that p16 is critical to maintain the senescence phenotype.

In addition to its canonical role regulating the cell cycle, an increasing amount of evidence indicates that p16 has non-canonical, RB-independent roles. For instance, our laboratory found that suppression of p16 bypasses oncogene-induced senescence in part by promoting an increase in nucleotide and deoxyribonucleotide levels in a mechanism mediated by the mTORC1 complex and in an RB-independent manner (i.e., operating outside the cell cycle) [42]. Additionally, p16 has been found to regulate tumor suppressive miRNAs, mitochondria biogenesis, oxidative stress, transcription factors such as AP-1 and NF-κB or protein translation though EEF1A2 (reviewed in [47]). Altogether this shows that p16 is not a simple cell cycle brake, but also a regulator of other processes. Indeed, the observation that the large majority of cancer-associated mutations targeting *CDKN2A* are mainly affecting the p16 open reading frame, reinforces the importance of p16 as a regulator of cellular physiology and stresses the necessity to further investigate p16-mediated regulatory processes.

3. Role of SASP in Senescence and Cancer

Together with the signature cell cycle arrest, the SASP is one of the most prominent phenotypes of senescent cells. The SASP is composed of various soluble and non-soluble factors including cytokines, chemokines, and proteases that are highly expressed and secreted by senescent cells, creating a pro-inflammatory microenvironment that affects themselves and other non-senescent cells in an autocrine and paracrine fashion, respectively [48,49]. In part because of its ability to modify the environment and impact the behavior of other cells, the SASP is tightly regulated at multiple levels. Transcriptionally, various factors (C/EBPβ and NF-κβ) [9,10], upstream regulators (p38, MAPK, GAT4A, p53, or ATM) [5,50–53] and non-coding RNAs (miRNAs, lncRNAs and circRNAs) [54] have been described to regulate SASP expression. Additionally, mTORC1-mediated translational regulation of MAPKAPK2 and IL1A has been shown to impact several SASP factors [55,56]. Finally, during senescence, there is a rearrangement of the genomic architecture leading to a new TAD (topologically associated domain) landscape [57,58]. This rearrangement is induced by changes in DNA methylation, nucleosome organization and histone modification, giving rise to the so-called senescence-associated heterochromatin foci (SAHF) [59–64]. There is a close relationship between the SAHF and the SASP and multiple epigenetic modifiers such as HMGB2, BRD4, MLL1, macroH2A1 or SIRT1, among others, have been shown to impact expression of the SASP [65–69]. In this regard, recent work from our laboratory shows that increased expression of the histone methyltransferase DOT1L upon oncogene-induced senescence, drives the expression of the major SASP-inducer IL1A though increased deposition of the active histone marks H3K79me2/3 at the *IL1A* gene loci [70].

Due to its inherent arrest of the cell-cycle described above, cellular senescence has been considered a bona fide tumor suppressor mechanism [71,72]. However, in the late 90s and early 2000s new experiments demonstrated that senescent cells can promote cellular proliferation and tumor growth due to acquisition of the SASP [7,73–75]. The SASP is highly dynamic and variable depending on several factors such as genetic background, cell type, the inducer of senescence, and the time of which senescence has occurred [76–79]. Due to its complex and variable nature, physiological roles of the SASP have not been well delineated. On one side, the SASP has been shown to contribute to the immunological surveillance, i.e., the process whereby the cells of the innate and adaptative immune system detect and destroy damaged cells [80–82]. For example, SASP secretion by senescence hepatocytes promotes T-cell mediated immunological surveillance within the liver, promoting the clearance of pre-malignant senescent cells and hence avoiding tumor progression [83]. On the other hand, the SASP has also been shown to contribute to numerous detrimental effects such as tumor promotion and progression and therapy resistance (reviewed in [84]). For instance, it has been described that some SASP secreted by senescent stromal cells promote an immunosuppressive microenvironment increasing the number of myeloid-derived suppressors cells (MDSCs), thus impairing immunological surveillance and promoting tumor growth [85,86]. In this regard, the dynamic nature of the SASP and its time-dependent regulation seems to be key to understanding the interaction with the immune system and hence the positive or negative outcome in the tumoral area [76,87]. Additionally, the SASP is not exclusive to senescent cells, and inflammatory phenotypes similar to the SASP, known as "SASP-like", have been described in tumor cells [67,88]. Indeed our laboratory has found that similar to senescent cells, different tumor types display different SASP (SASP-like) profiles [18], further demonstrating the wide variability of the SASP. Whether this indicates that those tumors bypassed senescence at some point during their malignant transformation is still unknown and more research is needed. However, it is becoming clearer that both the senescent cells in the tumor environment and the tumor cells themselves contribute to the maintenance of a SASP-driven inflammatory microenvironment. Therefore, it is imperative to delineate the mechanisms of SASP expression and secretion as well as map the SASP composition upon different conditions, tissue types and timeframes to design efficient and personalized immunotherapies.

4. p16 Regulation of Tumor Immunity

We previously described that canonical and non-canonical roles of p16 in part regulate cellular homeostasis. However, is it possible that p16 regulates other processes that impact the cellular microenvironment? Recent evidence demonstrates that loss of *CDKN2A* expression in tumor cells correlates with different immunological processes within the tumor that may impair immunological surveillance suggesting that p16 not only regulates cellular homeostasis but also tissue homeostasis. Below we will dissect the current evidence that implicates p16 as a regulator of intratumor immunity.

The first piece of evidence was published by Balli et al., [14]. These authors used a previously published expression signature based on two key cytolytic effectors (*GZMA* and *PRF1*) upregulated upon CD8+ T cell activation [89] to assess the intratumoral cytolytic T-cell activity in pancreatic adenocarcinoma samples from The Cancer Genome Atlas (TCGA).

Interestingly, non-silent mutations and deletions of *CDKN2A* correlated with decreased cytolytic activity. One year later, Wartenberg et al. [15] investigated the immune cell composition within the microenvironment in a series of pancreatic ductal adenocarcinoma and cross compared it with a high-throughput analysis of somatic mutations. They found that high mutation rates of *CDKN2A* correlate with a so-called immune-escape microenvironment, which is a microenvironment poor in T and B cells and enriched in FOXP3+ Tregs. Consistent with this result, Morrison et al. [16] found that loss of *CDKN2A* significantly correlates with immune deserts, defined by a profile of 394 immune transcripts. These pieces of evidence suggest that low *CDKN2A* expression both impacts the number and the activity of the intratumoral immune cells. Moreover, suppression of *CDKN2A* in mesenchymal stromal cells has been shown to decrease CD11b+ Gr-1^{hi} neutrophils, CD11b+ Gr1^{low} monocytes and CD45-CD31-Integrinα7+ satellite cells in a model of chronic inflammatory myopathy [17]. Although all those investigations are based on correlative analysis, these data may indicate that *CDKN2A* is necessary for regulation of the physiological immune response upon different inflammatory events. However, it should be noted that there are some publications that disagree with this thesis [90–92]. Thus, further mechanistic studies are necessary to determine in which circumstances suppression of *CDKN2A* in non-immune cells decreases immunological surveillance.

How does the suppression of p16 abrogate immunological surveillance? A recent publication from our laboratory demonstrates that the specific knockdown of p16 in oncogene-induced senescent cells leads to decreased expression of several SASP genes, including the most well characterized cytokines *IL6*, *CXCL8* and *CSF3*, the proteases *MMP3*, *PLAU* and *PLAT*, the growth factors *AREG*, *EREG* and *VEGFA* and the glycoprotein *ICAM1* [18]. Early suppression of p16 bypasses senescence in vitro and in vivo [18,42,43,93,94]; however, we demonstrated that knockdown of p16 at late time points upon oncogene-induced senescence does not bypass senescence but still decreases the expression of *IL6* and *CXCL8* [18], suggesting that this is uncoupled from the senescence-associated cell cycle arrest. Additionally, we found that low *CDKN2A* expression in tumors of 6 different types, including melanoma and pancreatic adenocarcinoma, correlated with a decreased SASP signature [18], further demonstrating the role of p16 in regulation of the SASP. Consistent with our observation, other articles have found decreased expression of SASP factors upon suppression of p16 in a murine model of intervertebral disc regeneration [95] and liver fibrosis [96]. Interestingly, induction of senescence through p16 overexpression does not increase SASP gene expression [50], suggesting that p16 is necessary but not sufficient to induce the SASP. As we have previously discussed, the SASP has pleiotropic and context-dependent effects that can both promote tumor progression and enhance anti-tumor immunity (Reviewed in [97]); thus, it is plausible that the observed decreased of immunological surveillance in tumors with suppression of *CDKN2A* is mediated by decreased SASP. More studies are needed to understand the exact mechanism whereby suppression of p16 leads to decreased SASP gene expression and whether this leads to decreased immunological surveillance and tumor growth. Additionally, in those tumors where loss of p16 occurs due to deletion of the chromosome 9p21 locus, adjacent genes such as *MTAP* and the interferon α and β cluster may be also affected and lost [98,99]. Previous studies have suggested that melanomas with low *MTAP* expression have decreased cGAS-STING signaling [100], a pathway strongly involved in SASP expression though NF-$\kappa\beta$ regulation [101–103]. Additionally, codeletion of *CDKN2A* and the interferon α and β cluster has been linked with decreased expression of immune cell genes in melanoma tumors [104]. Therefore, it is likely that multiple mechanisms exist in tumors with loss of 9p21 to suppress SASP gene expression and to modulate the tumor microenvironment.

How is immunological surveillance initiated on the tumor microenvironment? A large amount of evidence suggests that intratumoral senescence induction is critical for activation of the immune system and clearance of cancer cells [83,86,105,106], and abrogation of senescence-inducing pathways, mainly p16 and p21, have been shown to be critical to promote immune-checkpoint inhibitors resistance [107,108]. Thus, which is the main requirement for intratumoral immune system activation: senescence induction or high p16 expression? This is a complicated question since most senescent cells upregulate p16 as a manner to maintain the cell cycle arrest [26,109], and early abrogation of p16 activity overcomes senescence in vitro and in vivo [42,43,93,94], thus likely reducing the intratumoral senescence burden. In this regard, our previous data demonstrate that although etoposide can induce senescence in melanoma cells with stable p16 knockdown, these cells fail to increase the expression of *IL6* and *CXCL8* [18]. Moreover, although tumors with low p16 expression show a significant decrease in SASP factors, we did not observe differences in the amount of intratumoral senescent cells [18]. Interestingly, Novais et al. also found that suppression of p16 in a model of intervertebral disc regeneration decreases the SASP without altering the onset of senescence [110]. Altogether, these data suggest that p16 may regulate the tumor microenvironment and by extension intratumor immunity independently of senescence induction. This indeed, may explain previous observations where overexpression of p16 induces a cell cycle arrest without the SASP [50].

Importantly, the lack of SASP expression observed in cells with p16 suppression indicates that induction of senescence as a mechanism to abrogate malignant proliferation may be a suitable and safe therapy for cancers with null or very low p16 expression. More experiments analyzing the immune response landscape as well as the mechanisms by which p16 decreases SASP are needed to understand this observation and to develop better treatments for the ~50% of human cancers with decreased p16 expression [12].

5. Conclusions

In conclusion, suppression of p16 in tumor cells decreases the expression of interleukins, chemokines and other factors belonging to the SASP that in turn may remodel the tumor microenvironment, thereby impairing immunological surveillance (Figure 1). The observation that suppression of p16 decreases SASP gene expression is sustained by different studies in multiple models including cancer and fibrosis. Additionally, there are multiple high-throughput correlative studies in different diseases suggesting that low *CDKN2A* activity correlates with a decreased number and activity of intratumoral immune cells. Altogether, this suggests that p16 suppression is not only a cell cycle regulator but also a regulator of tissue homeostasis. More research is needed to understand whether this is a direct or indirect effect and whether this is due to canonical or non-canonical p16 roles. This is imperative since currently there are not pharmacological treatments specifically for p16-non canonical pathways [47].

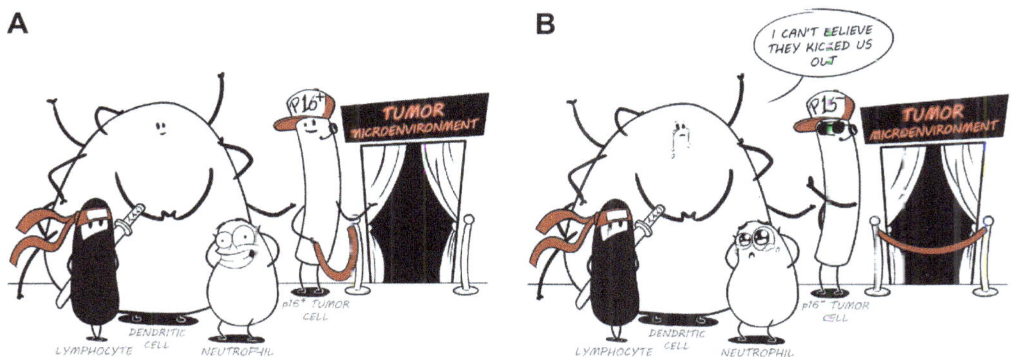

Figure 1. Cartoon representing the hypothesis discussed in this review: the possible role of p16 in regulation of immunological surveillance. (**A**) Represents a p16 positive tumor where cells of the immune system are invited into the tumor microenvironment and hence tumor immunosurveillance is fostered. (**B**) Represents a p16-null tumor which correlates with decreased number and activity of immune cells, thus impairing immunosurveillance and promoting tumor growth.

Funding: This work was supported by grants from the National Institutes of Health (F31CA250366 to K.E.L. and R37CA240625 to K.M.A.).

Institutional Review Board Statement: Not applicable.

Informed Consent Statement: Not applicable.

Data Availability Statement: Not applicable.

Acknowledgments: We thank Francisco J. Vazquez-Moreno for his beautiful cartoon in Figure 1.

Conflicts of Interest: The authors declare no conflict of interest.

References

1. Decker, W.K.; da Silva, R.F.; Sanabria, M.H.; Angelo, L.S.; Guimaraes, F.; Burt, B.M.; Kheradmand, F.; Paust, S. Cancer Immunotherapy: Historical Perspective of a Clinical Revolution and Emerging Preclinical Animal Models. *Front. Immunol.* **2017**, *8*, 829. [CrossRef]
2. Vaddepally, R.K.; Kharel, P.; Pandey, R.; Garje, R.; Chandra, A.B. Review of Indications of FDA-Approved Immune Checkpoint Inhibitors per NCCN Guidelines with the Level of Evidence. *Cancers* **2020**, *12*, 738. [CrossRef]
3. Aird, K.M.; Zhang, R. Nucleotide metabolism, oncogene-induced senescence and cancer. *Cancer Lett.* **2015**, *356*, 204–210. [CrossRef]
4. Kuilman, T.; Peeper, D.S. Senescence-messaging secretome: SMS-ing cellular stress. *Nat. Rev. Cancer* **2009**, *9*, 81–94. [CrossRef]
5. Coppe, J.P.; Patil, C.K.; Rodier, F.; Sun, Y.; Munoz, D.P.; Goldstein, J.; Nelson, P.S.; Desprez, P.Y.; Campisi, J. Senescence-associated secretory phenotypes reveal cell-nonautonomous functions of oncogenic RAS and the p53 tumor suppressor. *PLoS Biol.* **2008**, *6*, 2853–2868. [CrossRef]
6. Zhang, H.; Pan, K.H.; Cohen, S.N. Senescence-specific gene expression fingerprints reveal cell-type-dependent physical clustering of up-regulated chromosomal loci. *Proc. Natl. Acad. Sci. USA* **2003**, *100*, 3251–3256. [CrossRef]
7. Bavik, C.; Coleman, I.; Dean, J.P.; Knudsen, B.; Plymate, S.; Nelson, P.S. The gene expression program of prostate fibroblast senescence modulates neoplastic epithelial cell proliferation through paracrine mechanisms. *Cancer Res.* **2006**, *66*, 794–802. [CrossRef] [PubMed]
8. Wajapeyee, N.; Serra, R.W.; Zhu, X.; Mahalingam, M.; Green, M.R. Oncogenic BRAF induces senescence and apoptosis through pathways mediated by the secreted protein IGFBP7. *Cell* **2008**, *132*, 363–374. [CrossRef] [PubMed]
9. Kuilman, T.; Michaloglou, C.; Vredeveld, L.C.; Douma, S.; van Doorn, R.; Desmet, C.J.; Aarden, L.A.; Mooi, W.J.; Peeper, D.S. Oncogene-induced senescence relayed by an interleukin-dependent inflammatory network. *Cell* **2008**, *133*, 1019–1031. [CrossRef]
10. Acosta, J.C.; O'Loghlen, A.; Banito, A.; Guijarro, M.V.; Augert, A.; Raguz, S.; Fumagalli, M.; Da Costa, M.; Brown, C.; Popov, N.; et al. Chemokine signaling via the CXCR2 receptor reinforces senescence. *Cell* **2008**, *133*, 1006–1018. [CrossRef] [PubMed]
11. Campisi, J. Aging, cellular senescence, and cancer. *Annu. Rev. Physiol.* **2013**, *75*, 685–705. [CrossRef] [PubMed]
12. Gonzalez, S.; Serrano, M. A new mechanism of inactivation of the INK4/ARF locus. *Cell Cycle* **2006**, *5*, 1382–1384. [CrossRef] [PubMed]
13. Sharpless, N.E.; Sherr, C.J. Forging a signature of in vivo senescence. *Nat. Rev. Cancer* **2015**, *15*, 397–408. [CrossRef] [PubMed]
14. Balli, D.; Rech, A.J.; Stanger, B.Z.; Vonderheide, R.H. Immune Cytolytic Activity Stratifies Molecular Subsets of Human Pancreatic Cancer. *Clin. Cancer Res.* **2017**, *23*, 3129–3138. [CrossRef] [PubMed]
15. Wartenberg, M.; Cibin, S.; Zlobec, I.; Vassella, E.; Eppenberger-Castori, S.; Terracciano, L.; Eichmann, M.D.; Worni, M.; Gloor, B.; Perren, A.; et al. Integrated Genomic and Immunophenotypic Classification of Pancreatic Cancer Reveals Three Distinct Subtypes with Prognostic/Predictive Significance. *Clin. Cancer Res.* **2018**, *24*, 4444–4454. [CrossRef] [PubMed]
16. Morrison, C.; Pabla, S.; Conroy, J.M.; Nesline, M.K.; Glenn, S.T.; Dressman, D.; Papanicolau-Sengos, A.; Burgher, B.; Andreas, J.; Giamo, V.; et al. Predicting response to checkpoint inhibitors in melanoma beyond PD-L1 and mutational burden. *J. Immunother. Cancer* **2018**, *6*, 32. [CrossRef]
17. Chikenji, T.S.; Saito, Y.; Konari, N.; Nakano, M.; Mizue, Y.; Otani, M.; Fujimiya, M. p16(INK4A)-expressing mesenchymal stromal cells restore the senescence-clearance-regeneration sequence that is impaired in chronic muscle inflammation. *EBioMedicine* **2019**, *44*, 86–97. [CrossRef]
18. Buj, R.; Leon, K.E.; Anguelov, M.A.; Aird, K.M. Suppression of p16 alleviates the senescence-associated secretory phenotype. *Aging* **2021**, *13*, 3290. [CrossRef]
19. Ozenne, P.; Eymin, B.; Brambilla, E.; Gazzeri, S. The ARF tumor suppressor: Structure, functions and status in cancer. *Int. J. Cancer* **2010**, *127*, 2239–2247. [CrossRef]
20. Sherr, C.J. The INK4a/ARF network in tumour suppression. *Nat. Rev. Mol. Cell Biol.* **2001**, *2*, 731–737. [CrossRef]
21. Esteller, M.; Corn, P.G.; Baylin, S.B.; Herman, J.G. A gene hypermethylation profile of human cancer. *Cancer Res.* **2001**, *61*, 3225–3229. [PubMed]
22. Ruas, M.; Brookes, S.; McDonald, N.Q.; Peters, G. Functional evaluation of tumour-specific variants of p16INK4a/CDKN2A: Correlation with protein structure information. *Oncogene* **1999**, *18*, 5423–5434. [CrossRef]

23. Inoue, K.; Fry, E.A. Aberrant expression of p16(INK4a) in human cancers—A new biomarker? *Cancer Rep. Rev.* **2018**, *2*. [CrossRef] [PubMed]

24. Quelle, D.E.; Cheng, M.; Ashmun, R.A.; Sherr, C.J. Cancer-associated mutations at the INK4a locus cancel cell cycle arrest by p16INK4a but not by the alternative reading frame protein p19ARF. *Proc. Natl. Acad. Sci. USA* **1997**, *94*, 669–673. [CrossRef]

25. Barnum, K.J.; O'Connell, M.J. Cell cycle regulation by checkpoints. *Methods Mol. Biol.* **2014**, *1170*, 29–40. [CrossRef]

26. Serrano, M.; Lin, A.W.; McCurrach, M.E.; Beach, D.; Lowe, S.W. Oncogenic ras provokes premature cell senescence associated with accumulation of p53 and p16INK4a. *Cell* **1997**, *88*, 593–602. [CrossRef]

27. Liu, J.Y.; Souroullas, G.P.; Diekman, B.O.; Krishnamurthy, J.; Hall, B.M.; Sorrentino, J.A.; Parker, J.S. Sessions, G.A.; Gudkov, A.V.; Sharpless, N.E. Cells exhibiting strong p16 (INK4a) promoter activation in vivo display features of senescence. *Proc. Natl. Acad. Sci. USA* **2019**, *116*, 2603–2611. [CrossRef]

28. Baker, D.J.; Childs, B.G.; Durik, M.; Wijers, M.E.; Sieben, C.J.; Zhong, J.; Saltness, R.A.; Jeganathan, K.B.; Verzosa, G.C.; Pezeshki, A.; et al. Naturally occurring p16(Ink4a)-positive cells shorten healthy lifespan. *Nature* **2016**, *530*, 184–189. [CrossRef]

29. Gil, J.; Peters, G. Regulation of the INK4b-ARF-INK4a tumour suppressor locus: All for one or one for all. *Nat. Rev. Mol. Cell Biol.* **2006**, *7*, 667–677. [CrossRef]

30. Itahana, K.; Campisi, J.; Dimri, G.P. Mechanisms of cellular senescence in human and mouse cells. *Biogerontology* **2004**, *5*, 1–10. [CrossRef]

31. Bulavin, D.V.; Phillips, C.; Nannenga, B.; Timofeev, O.; Donehower, L.A.; Anderson, C.W.; Appella, E.; Fornace, A.J., Jr. Inactivation of the Wip1 phosphatase inhibits mammary tumorigenesis through p38 MAPK-mediated activation of the p16(Ink4a)-p19(Arf) pathway. *Nat. Genet.* **2004**, *36*, 343–350. [CrossRef]

32. Deng, Q.; Liao, R.; Wu, B.L.; Sun, P. High intensity ras signaling induces premature senescence by activating p38 pathway in primary human fibroblasts. *J. Biol. Chem.* **2004**, *279*, 1050–1059. [CrossRef]

33. Lin, A.W.; Barradas, M.; Stone, J.C.; van Aelst, L.; Serrano, M.; Lowe, S.W. Premature senescence involving p53 and p16 is activated in response to constitutive MEK/MAPK mitogenic signaling. *Genes Dev.* **1998**, *12*, 3008–3019. [CrossRef] [PubMed]

34. Wang, W.; Chen, J.X.; Liao, R.; Deng, Q.; Zhou, J.J.; Huang, S.; Sun, P. Sequential activation of the MEK-extracellular signal-regulated kinase and MKK3/6-p38 mitogen-activated protein kinase pathways mediates oncogenic ras-induced premature senescence. *Mol. Cell. Biol.* **2002**, *22*, 3389–3403. [CrossRef] [PubMed]

35. Bracken, A.P.; Kleine-Kohlbrecher, D.; Dietrich, N.; Pasini, D.; Gargiulo, G.; Beekman, C.; Theilgaard-Monch, K.; Minucci, S.; Porse, B.T.; Marine, J.C.; et al. The Polycomb group proteins bind throughout the INK4A-ARF locus and are disassociated in senescent cells. *Genes Dev.* **2007**, *21*, 525–530. [CrossRef] [PubMed]

36. Dietrich, N.; Bracken, A.P.; Trinh, E.; Schjerling, C.K.; Koseki, H.; Rappsilber, J.; Helin, K.; Hansen, K.H. Bypass of senescence by the polycomb group protein CBX8 through direct binding to the INK4A-ARF locus. *EMBO J.* **2007**, *26*, 1637–1648. [CrossRef] [PubMed]

37. Ito, T.; Teo, Y.V.; Evans, S.A.; Neretti, N.; Sedivy, J.M. Regulation of Cellular Senescence by Polycomb Chromatin Modifiers through Distinct DNA Damage- and Histone Methylation-Dependent Pathways. *Cell Rep.* **2018**, *22*, 3480–3492. [CrossRef] [PubMed]

38. Nakade, K.; Pan, J.; Yamasaki, T.; Murata, T.; Wasylyk, B.; Yokoyama, K.K. JDP2 (Jun Dimerization Protein 2)-deficient mouse embryonic fibroblasts are resistant to replicative senescence. *J. Biol. Chem.* **2009**, *284*, 10808–10817. [CrossRef]

39. Ohtani, N.; Zebedee, Z.; Huot, T.J.; Stinson, J.A.; Sugimoto, M.; Ohashi, Y.; Sharrocks, A.D.; Peters, G.; Hara, E. Opposing effects of Ets and Id proteins on p16INK4a expression during cellular senescence. *Nature* **2001**, *409*, 1067–1070. [CrossRef] [PubMed]

40. Passegue, E.; Wagner, E.F. JunB suppresses cell proliferation by transcriptional activation of p16 INK4a) expression. *EMBO J.* **2000**, *19*, 2969–2979. [CrossRef]

41. Swarbrick, A.; Roy, E.; Allen, T.; Bishop, J.M. Id1 cooperates with oncogenic Ras to induce metastatic mammary carcinoma by subversion of the cellular senescence response. *Proc. Natl. Acad. Sci. USA* **2008**, *105*, 5402–5407. [CrossRef]

42. Buj, R.; Chen, C.W.; Dahl, E.S.; Leon, K.E.; Kuskovsky, R.; Maglakelidze, N.; Navaratnarajah, M.; Zhang, G.; Doan, M.T.; Jiang, H.; et al. Suppression of p16 Induces mTORC1-Mediated Nucleotide Metabolic Reprogramming. *Cell Rep.* **2019**, *28*, 1971–1980.e1978. [CrossRef]

43. Damsky, W.; Micevic, G.; Meeth, K.; Muthusamy, V.; Curley, D.P.; Santhanakrishnan, M.; Erdelyi, I.; Flatt, J.T.; Huang, L.; Theodosakis, N.; et al. mTORC1 activation blocks BrafV600E-induced growth arrest but is insufficient for melanoma formation. *Cancer Cell* **2015**, *27*, 41–56. [CrossRef]

44. Dankort, D.; Filenova, E.; Collado, M.; Serrano, M.; Jones, K.; McMahon, M. A new mouse model to explore the initiation, progression, and therapy of BRAFV600E-induced lung tumors. *Genes Dev.* **2007**, *21*, 379–384. [CrossRef]

45. Goel, V.K.; Ibrahim, N.; Jiang, G.; Singhal, M.; Fee, S.; Flotte, T.; Westmoreland, S.; Haluska, F.S.; Hinds, P.W.; Haluska, F.G. Melanocytic nevus-like hyperplasia and melanoma in transgenic BRAFV600E mice. *Oncogene* **2009**, *28*, 2289–2298. [CrossRef]

46. Haferkamp, S.; Becker, T.M.; Scurr, L.L.; Kefford, R.F.; Rizos, H. p16INK4a-induced senescence is disabled by melanoma-associated mutations. *Aging Cell* **2008**, *7*, 733–745. [CrossRef]

47. Buj, R.; Aird, K.M. p16: Cycling off the beaten path. *Mol. Cell. Oncol.* **2019**, *6*, e1677140. [CrossRef]

48. Coppe, J.P.; Desprez, P.Y.; Krtolica, A.; Campisi, J. The senescence-associated secretory phenotype: The dark side of tumor suppression. *Annu. Rev. Pathol.* **2010**, *5*, 99–118. [CrossRef] [PubMed]

49. Gonzalez-Meljem, J.M.; Apps, J.R.; Fraser, H.C.; Martinez-Barbera, J.P. Paracrine roles of cellular senescence in promoting tumourigenesis. *Br. J. Cancer* **2018**, *118*, 1283–1288. [CrossRef] [PubMed]

50. Coppe, J.P.; Rodier, F.; Patil, C.K.; Freund, A.; Desprez, P.Y.; Campisi, J. Tumor suppressor and aging biomarker p16(INK4a) induces cellular senescence without the associated inflammatory secretory phenotype. *J. Biol. Chem.* **2011**, *286*, 36396–36403. [CrossRef] [PubMed]

51. Kang, C.; Xu, Q.; Martin, T.D.; Li, M.Z.; Demaria, M.; Aron, L.; Lu, T.; Yankner, B.A.; Campisi, J.; Elledge, S.J. The DNA damage response induces inflammation and senescence by inhibiting autophagy of GATA4. *Science* **2015**, *349*, aaa5612. [CrossRef]

52. Malaquin, N.; Olivier, M.A.; Martinez, A.; Nadeau, S.; Sawchyn, C.; Coppe, J.P.; Cardin, G.; Mallette, F.A.; Campisi, J.; Rodier, F. Non-canonical ATM/MRN activities temporally define the senescence secretory program. *EMBO Rep.* **2020**, *21*, e50718. [CrossRef] [PubMed]

53. Rodier, F.; Coppe, J.P.; Patil, C.K.; Hoeijmakers, W.A.; Munoz, D.P.; Raza, S.R.; Freund, A.; Campeau, E.; Davalos, A.R.; Campisi, J. Persistent DNA damage signalling triggers senescence-associated inflammatory cytokine secretion. *Nat. Cell. Biol.* **2009**, *11*, 973–979. [CrossRef] [PubMed]

54. Panda, A.C.; Abdelmohsen, K.; Gorospe, M. SASP regulation by noncoding RNA. *Mech. Ageing Dev.* **2017**, *168*, 37–43. [CrossRef] [PubMed]

55. Herranz, N.; Gallage, S.; Mellone, M.; Wuestefeld, T.; Klotz, S.; Hanley, C.J.; Raguz, S.; Acosta, J.C.; Innes, A.J.; Banito, A.; et al. mTOR regulates MAPKAPK2 translation to control the senescence-associated secretory phenotype. *Nat. Cell. Biol.* **2015**, *17*, 1205–1217. [CrossRef] [PubMed]

56. Laberge, R.M.; Sun, Y.; Orjalo, A.V.; Patil, C.K.; Freund, A.; Zhou, L.; Curran, S.C.; Davalos, A.R.; Wilson-Edell, K.A.; Liu, S.; et al. MTOR regulates the pro-tumorigenic senescence-associated secretory phenotype by promoting IL1A translation. *Nat. Cell. Biol.* **2015**, *17*, 1049–1061. [CrossRef]

57. Chandra, T.; Ewels, P.A.; Schoenfelder, S.; Furlan-Magaril, M.; Wingett, S.W.; Kirschner, K.; Thuret, J.Y.; Andrews, S.; Fraser, P.; Reik, W. Global reorganization of the nuclear landscape in senescent cells. *Cell Rep.* **2015**, *10*, 471–483. [CrossRef]

58. Criscione, S.W.; De Cecco, M.; Siranosian, B.; Zhang, Y.; Kreiling, J.A.; Sedivy, J.M.; Neretti, N. Reorganization of chromosome architecture in replicative cellular senescence. *Sci. Adv.* **2016**, *2*, e1500882. [CrossRef]

59. Narita, M.; Nunez, S.; Heard, E.; Narita, M.; Lin, A.W.; Hearn, S.A.; Spector, D.L.; Hannon, G.J.; Lowe, S.W. Rb-mediated heterochromatin formation and silencing of E2F target genes during cellular senescence. *Cell* **2003**, *113*, 703–716. [CrossRef]

60. Zhang, R.; Poustovoitov, M.V.; Ye, X.; Santos, H.A.; Chen, W.; Daganzo, S.M.; Erzberger, J.P.; Serebriiskii, I.G.; Canutescu, A.A.; Dunbrack, R.L.; et al. Formation of MacroH2A-containing senescence-associated heterochromatin foci and senescence driven by ASF1a and HIRA. *Dev. Cell* **2005**, *8*, 19–30. [CrossRef]

61. Sakaki, M.; Ebihara, Y.; Okamura, K.; Nakabayashi, K.; Igarashi, A.; Matsumoto, K.; Hata, K.; Kobayashi, Y.; Maehara, K. Potential roles of DNA methylation in the initiation and establishment of replicative senescence revealed by array-based methylome and transcriptome analyses. *PLoS ONE* **2017**, *12*, e0171431. [CrossRef]

62. Zhang, R.; Chen, W.; Adams, P.D. Molecular dissection of formation of senescence-associated heterochromatin foci. *Mol. Cell. Biol.* **2007**, *27*, 2343–2358. [CrossRef]

63. Cruickshanks, H.A.; McBryan, T.; Nelson, D.M.; Vanderkraats, N.D.; Shah, P.P.; van Tuyn, J.; Singh Rai, T.; Brock, C.; Donahue, G.; Dunican, D.S.; et al. Senescent cells harbour features of the cancer epigenome. *Nat. Cell. Biol.* **2013**, *15*, 1495–1506. [CrossRef]

64. Parry, A.J.; Hoare, M.; Bihary, D.; Hansel-Hertsch, R.; Smith, S.; Tomimatsu, K.; Mannion, E.; Smith, A.; D'Santos, P.; Russell, I.A.; et al. NOTCH-mediated non-cell autonomous regulation of chromatin structure during senescence. *Nat. Commun.* **2018**, *9*, 1840. [CrossRef]

65. Aird, K.M.; Iwasaki, O.; Kossenkov, A.V.; Tanizawa, H.; Fatkhutdinov, N.; Bitler, B.G.; Le, L.; Alicea, G.; Yang, T.L.; Johnson, F.B.; et al. HMGB2 orchestrates the chromatin landscape of senescence-associated secretory phenotype gene loci. *J. Cell Biol.* **2016**, *215*, 325–334. [CrossRef] [PubMed]

66. Tasdemir, N.; Banito, A.; Roe, J.S.; Alonso-Curbelo, D.; Camiolo, M.; Tschaharganeh, D.F.; Huang, C.H.; Aksoy, O.; Bolden, J.E.; Chen, C.C.; et al. BRD4 Connects Enhancer Remodeling to Senescence Immune Surveillance. *Cancer Discov.* **2016**, *6*, 612–629. [CrossRef] [PubMed]

67. Capell, B.C.; Drake, A.M.; Zhu, J.; Shah, P.P.; Dou, Z.; Dorsey, J.; Simola, D.F.; Donahue, G.; Sammons, M.; Rai, T.S.; et al. MLL1 is essential for the senescence-associated secretory phenotype. *Genes Dev.* **2016**, *30*, 321–336. [CrossRef] [PubMed]

68. Chen, H.; Ruiz, P.D.; McKimpson, W.M.; Novikov, L.; Kitsis, R.N.; Gamble, M.J. MacroH2A1 and ATM Play Opposing Roles in Paracrine Senescence and the Senescence-Associated Secretory Phenotype. *Mol. Cell* **2015**, *59*, 719–731. [CrossRef] [PubMed]

69. Hayakawa, T.; Iwai, M.; Aoki, S.; Takimoto, K.; Maruyama, M.; Maruyama, W.; Motoyama, N. SIRT1 suppresses the senescence-associated secretory phenotype through epigenetic gene regulation. *PLoS ONE* **2015**, *10*, e0116480. [CrossRef]

70. Leon, K.E.; Buj, R.; Lesko, E.; Dahl, E.S.; Chen, C.-W.; Imamura, Y.; Kossenkov, A.V.; Hobbs, R.P.; Arid, K.M. DOT1L modulates the senescence-associated secretory phenotype through epigenetic regulation of IL1A. *bioRxiv Prepr. Serv. Biol.* **2020**. [CrossRef]

71. Campisi, J. Cellular senescence as a tumor-suppressor mechanism. *Trends Cell Biol.* **2001**, *11*, S27–S31. [CrossRef]

72. Dimri, G.P. What has senescence got to do with cancer? *Cancer Cell* **2005**, *7*, 505–512. [CrossRef]

73. Krtolica, A.; Parrinello, S.; Lockett, S.; Desprez, P.Y.; Campisi, J. Senescent fibroblasts promote epithelial cell growth and tumorigenesis: A link between cancer and aging. *Proc. Natl. Acad. Sci. USA* **2001**, *98*, 12072–12077. [CrossRef] [PubMed]

74. Bhatia, B.; Multani, A.S.; Patrawala, L.; Chen, X.; Calhoun-Davis, T.; Zhou, J.; Schroeder, L.; Schneider-Broussard, R.; Shen, J.; Pathak, S.; et al. Evidence that senescent human prostate epithelial cells enhance tumorigenicity: Cell fusion as a potential mechanism and inhibition by p16INK4a and hTERT. *Int. J. Cancer* **2008**, *122*, 1483–1495. [CrossRef]

75. Coppe, J.P.; Kauser, K.; Campisi, J.; Beausejour, C.M. Secretion of vascular endothelial growth factor by primary human fibroblasts at senescence. *J. Biol. Chem.* **2006**, *281*, 29568–29574. [CrossRef] [PubMed]

76. Hoare, M.; Ito, Y.; Kang, T.W.; Weekes, M.P.; Matheson, N.J.; Patten, D.A.; Shetty, S.; Parry, A.J.; Menon, S.; Salama, R.; et al. NOTCH1 mediates a switch between two distinct secretomes during senescence. *Nat. Cell Biol.* **2016**, *18*, 979–992. [CrossRef] [PubMed]

77. Basisty, N.; Kale, A.; Jeon, O.H.; Kuehnemann, C.; Payne, T.; Rao, C.; Holtz, A.; Shah, S.; Sharma, V.; Ferrucci, L.; et al. A proteomic atlas of senescence-associated secretomes for aging biomarker development. *PLoS Biol.* **2020**, *18*, e3000599. [CrossRef]

78. Rao, S.G.; Jackson, J.G. SASP: Tumor Suppressor or Promoter? Yes! *Trends Cancer* **2016**, *2*, 676–687. [CrossRef] [PubMed]

79. Hernandez-Segura, A.; de Jong, T.V.; Melov, S.; Guryev, V.; Campisi, J.; Demaria, M. Unmasking Transcriptional Heterogeneity in Senescent Cells. *Curr. Biol.* **2017**, *27*, 2652–2660.e2654. [CrossRef]

80. Krizhanovsky, V.; Yon, M.; Dickins, R.A.; Hearn, S.; Simon, J.; Miething, C.; Yee, H.; Zender, L.; Lowe, S.W. Senescence of activated stellate cells limits liver fibrosis. *Cell* **2008**, *134*, 657–667. [CrossRef]

81. Jun, J.I.; Lau, L.F. The matricellular protein CCN1 induces fibroblast senescence and restricts fibrosis in cutaneous wound healing. *Nat. Cell Biol.* **2010**, *12*, 676–685. [CrossRef]

82. Pitiyage, G.N.; Slijepcevic, P.; Gabrani, A.; Chianea, Y.G.; Lim, K.P.; Prime, S.S.; Tilakaratne, W.M.; Fortune, F.; Parkinson E.K. Senescent mesenchymal cells accumulate in human fibrosis by a telomere-independent mechanism and ameliorate fibrosis through matrix metalloproteinases. *J. Pathol.* **2011**, *223*, 604–617. [CrossRef] [PubMed]

83. Kang, T.W.; Yevsa, T.; Woller, N.; Hoenicke, L.; Wuestefeld, T.; Dauch, D.; Hohmeyer, A.; Gereke, M.; Rudalska, R.; Potapova, A.; et al. Senescence surveillance of pre-malignant hepatocytes limits liver cancer development. *Nature* **2011**, *479*, 547–551. [CrossRef]

84. Georgilis, A.; Gil, J. Controlling secretion to limit chemoresistance. *Genes Dev.* **2016**, *30*, 1791–1792. [CrossRef] [PubMed]

85. Ruhland, M.K.; Loza, A.J.; Capietto, A.H.; Luo, X.; Knolhoff, B.L.; Flanagan, K.C.; Belt, B.A.; Alspach, E.; Leahy, K.; Luo, J.; et al. Stromal senescence establishes an immunosuppressive microenvironment that drives tumorigenesis. *Nat. Commun.* **2016**, *7*, 11762. [CrossRef] [PubMed]

86. Toso, A.; Revandkar, A.; Di Mitri, D.; Guccini, I.; Proietti, M.; Sarti, M.; Pinton, S.; Zhang, J.; Kalathur, M.; Civenni, G.; et al. Enhancing chemotherapy efficacy in Pten-deficient prostate tumors by activating the senescence-associated antitumor immunity. *Cell Rep.* **2014**, *9*, 75–89. [CrossRef]

87. Teo, Y.V.; Rattanavirotkul, N.; Olova, N.; Salzano, A.; Quintanilla, A.; Tarrats, N.; Kiourtis, C.; Muller, M.; Green, A.R.; Adams, P.D.; et al. Notch Signaling Mediates Secondary Senescence. *Cell Rep.* **2019**, *27*, 997–1007.e1005. [CrossRef] [PubMed]

88. Zelenay, S.; van der Veen, A.G.; Bottcher, J.P.; Snelgrove, K.J.; Rogers, N.; Acton, S.E.; Chakravarty, F.; Girotti, M.R.; Marais, R.; Quezada, S.A.; et al. Cyclooxygenase-Dependent Tumor Growth through Evasion of Immunity. *Cell* **2015**, *162*, 1257–1270. [CrossRef] [PubMed]

89. Rooney, M.S.; Shukla, S.A.; Wu, C.J.; Getz, G.; Hacohen, N. Molecular and genetic properties of tumors associated with local immune cytolytic activity. *Cell* **2015**, *160*, 48–61. [CrossRef]

90. DeLeon, T.T.; Almquist, D.R.; Kipp, B.R.; Langlais, B.T.; Mangold, A.; Winters, J.L.; Kosiorek, H.E.; Joseph, R.W.; Dronca, R.S.; Block, M.S.; et al. Assessment of clinical outcomes with immune checkpoint inhibitor therapy in melanoma patients with CDKN2A and TP53 pathogenic mutations. *PLoS ONE* **2020**, *15*, e0230306. [CrossRef]

91. Helgadottir, H.; Ghiorzo, P.; van Doorn, R.; Puig, S.; Levin, M.; Kefford, R.; Lauss, M.; Queirolo, P.; Pastorino, L.; Kapiteijn, E.; et al. Efficacy of novel immunotherapy regimens in patients with metastatic melanoma with germline CDKN2A mutations. *J. Med. Genet.* **2020**, *57*, 316–321. [CrossRef]

92. Lv, F.; Wu, J.; Miao, D.; An, W.; Wang, Y. p16 deficiency promotes nonalcoholic steatohepatitis via regulation of hepatic oxidative stress. *Biochem. Biophys. Res. Commun.* **2017**, *486*, 264–269. [CrossRef] [PubMed]

93. Chin, L.; Pomerantz, J.; Polsky, D.; Jacobson, M.; Cohen, C.; Cordon-Cardo, C.; Horner, J.W., 2nd; DePinho, R.A. Cooperative effects of INK4a and ras in melanoma susceptibility in vivo. *Genes Dev.* **1997**, *11*, 2822–2834. [CrossRef] [PubMed]

94. Krimpenfort, P.; Quon, K.C.; Mooi, W.J.; Loonstra, A.; Berns, A. Loss of p16Ink4a confers susceptibility to metastatic melanoma in mice. *Nature* **2001**, *413*, 83–86. [CrossRef] [PubMed]

95. Che, H.; Li, J.; Li, Y.; Ma, C.; Liu, H.; Qin, J.; Dong, J.; Zhang, Z.; Xian, C.J.; Miao, D.; et al. p16 deficiency attenuates intervertebral disc degeneration by adjusting oxidative stress and nucleus pulposus cell cycle. *eLife* **2020**, *9*, e52570. [CrossRef] [PubMed]

96. Kyritsi, K.; Francis, H.; Zhou, T.; Ceci, L.; Wu, N.; Yang, Z.; Meng, F.; Chen, L.; Baiocchi, L.; Kundu D.; et al. Downregulation of p16 Decreases Biliary Damage and Liver Fibrosis in the Mdr2(/) Mouse Model of Primary Sclerosing Cholangitis. *Gene Expr.* **2020**, *20*, 89–103. [CrossRef] [PubMed]

97. Faget, D.V.; Ren, Q.; Stewart, S.A. Unmasking senescence: Context-dependent effects of SASP in cancer. *Nat. Rev. Cancer* **2019**, *19*, 439–453. [CrossRef]

98. Zhang, H.; Chen, Z.H.; Savarese, T.M. Codeletion of the genes for p16INK4, methylthioadenosine phosphorylase, interferon-alpha1, interferon-beta1, and other 9p21 markers in human malignant cell lines. *Cancer Genet Cytogenet.* **1996**, *86*, 22–28. [CrossRef]

99. Roy, D.M.; Walsh, L.A.; Desrichard, A.; Huse, J.T.; Wu, W.; Gao, J.; Bose, P.; Lee, W.; Chan, T.A. Integrated Genomics for Pinpointing Survival Loci within Arm-Level Somatic Copy Number Alterations. *Cancer Cell* **2016**, *29*, 737–750. [CrossRef]

100. Kim, H.; Kim, H.; Feng, Y.; Li, Y.; Tamiya, H.; Tocci, S.; Ronai, Z.A. PRMT5 control of cGAS/STING and NLRC5 pathways defines melanoma response to antitumor immunity. *Sci. Transl. Med.* **2020**, *12*, eaaz5683. [CrossRef] [PubMed]

101. Dou, Z.; Ghosh, K.; Vizioli, M.G.; Zhu, J.; Sen, P.; Wangensteen, K.J.; Simithy, J.; Lan, Y.; Lin, Y.; Zhou, Z.; et al. Cytoplasmic chromatin triggers inflammation in senescence and cancer. *Nature* **2017**, *550*, 402–406. [CrossRef]

102. Ivanov, A.; Pawlikowski, J.; Manoharan, I.; van Tuyn, J.; Nelson, D.M.; Rai, T.S.; Shah, P.P.; Hewitt, G.; Korolchuk, V.I.; Passos, J.F.; et al. Lysosome-mediated processing of chromatin in senescence. *J. Cell Biol.* **2013**, *202*, 129–143. [CrossRef] [PubMed]

103. Gluck, S.; Guey, B.; Gulen, M.F.; Wolter, K.; Kang, T.W.; Schmacke, N.A.; Bridgeman, A.; Rehwinkel, J.; Zender, L.; Ablasser, A. Innate immune sensing of cytosolic chromatin fragments through cGAS promotes senescence. *Nat. Cell Biol.* **2017**, *19*, 1061–1070. [CrossRef] [PubMed]

104. Linsley, P.S.; Speake, C.; Whalen, E.; Chaussabel, D. Copy number loss of the interferon gene cluster in melanomas is linked to reduced T cell infiltrate and poor patient prognosis. *PLoS ONE* **2014**, *9*, e109760. [CrossRef] [PubMed]

105. Xue, W.; Zender, L.; Miething, C.; Dickins, R.A.; Hernando, E.; Krizhanovsky, V.; Cordon-Cardo, C.; Lowe, S.W. Senescence and tumour clearance is triggered by p53 restoration in murine liver carcinomas. *Nature* **2007**, *445*, 656–660. [CrossRef]

106. Braumuller, H.; Wieder, T.; Brenner, E.; Assmann, S.; Hahn, M.; Alkhaled, M.; Schilbach, K.; Essmann, F.; Kneilling, M.; Griessinger, C.; et al. T-helper-1-cell cytokines drive cancer into senescence. *Nature* **2013**, *494*, 361–365. [CrossRef]

107. Brenner, E.; Schorg, B.F.; Ahmetlic, F.; Wieder, T.; Hilke, F.J.; Simon, N.; Schroeder, C.; Demidov, G.; Riedel, T.; Fehrenbacher, B.; et al. Cancer immune control needs senescence induction by interferon-dependent cell cycle regulator pathways in tumours. *Nat. Commun.* **2020**, *11*, 1335. [CrossRef]

108. Hilke, F.J.; Sinnberg, T.; Gschwind, A.; Niessner, H.; Demidov, G.; Amaral, T.; Ossowski, S.; Bonzheim, I.; Rocken, M.; Riess, O.; et al. Distinct Mutation Patterns Reveal Melanoma Subtypes and Influence Immunotherapy Response in Advanced Melanoma Patients. *Cancers* **2020**, *12*, 2359. [CrossRef]

109. Stein, G.H.; Drullinger, L.F.; Soulard, A.; Dulic, V. Differential roles for cyclin-dependent kinase inhibitors p21 and p16 in the mechanisms of senescence and differentiation in human fibroblasts. *Mol. Cell. Biol.* **1999**, *19*, 2109–2117. [CrossRef]

110. Novais, E.J.; Diekman, B.O.; Shapiro, I.M.; Risbud, M.V. p16(Ink4a) deletion in cells of the intervertebral disc affects their matrix homeostasis and senescence associated secretory phenotype without altering onset of senescence. *Matrix Biol.* **2019**, *82*, 54–70. [CrossRef]

Review

The Cellular Senescence Stress Response in Post-Mitotic Brain Cells: Cell Survival at the Expense of Tissue Degeneration

Eric Sah [1], Sudarshan Krishnamurthy [1,2], Mohamed Y. Ahmidouch [1,3], Gregory J. Gillispie [1,4], Carol Milligan [5] and Miranda E. Orr [1,4,6,*]

1 Department of Internal Medicine, Section on Gerontology and Geriatric Medicine, Wake Forest School of Medicine, Winston-Salem, NC 27157, USA; esah@wakehealth.edu (E.S.); sukrishr@wakehealth.edu (S.K.); ahmimy17@wfu.edu (M.Y.A.); ggillisp@wakehealth.edu (G.J.G.)
2 Bowman Gray Center for Medical Education, Wake Forest School of Medicine, Winston-Salem, NC 27101, USA
3 Departments of Biology and Chemistry, Wake Forest University, Winston-Salem, NC 27109, USA
4 Sticht Center for Healthy Aging and Alzheimer's Prevention, Wake Forest School of Medicine, Winston-Salem, NC 27157, USA
5 Department of Neurobiology and Anatomy, Wake Forest School of Medicine, Winston-Salem, NC 27157, USA; milligan@wakehealth.edu
6 Salisbury VA Medical Center, Salisbury, NC 28144, USA
* Correspondence: morr@wakehealth.edu

Abstract: In 1960, Rita Levi-Montalcini and Barbara Booker made an observation that transformed neuroscience: as neurons mature, they become apoptosis resistant. The following year Leonard Hayflick and Paul Moorhead described a stable replicative arrest of cells in vitro, termed "senescence". For nearly 60 years, the cell biology fields of neuroscience and senescence ran in parallel, each separately defining phenotypes and uncovering molecular mediators to explain the 1960s observations of their founding mothers and fathers, respectively. During this time neuroscientists have consistently observed the remarkable ability of neurons to survive. Despite residing in environments of chronic inflammation and degeneration, as occurs in numerous neurodegenerative diseases, often times the neurons with highest levels of pathology resist death. Similarly, cellular senescence (hereon referred to simply as "senescence") now is recognized as a complex stress response that culminates with a change in cell fate. Instead of reacting to cellular/DNA damage by proliferation or apoptosis, senescent cells survive in a stable cell cycle arrest. Senescent cells simultaneously contribute to chronic tissue degeneration by secreting deleterious molecules that negatively impact surrounding cells. These fields have finally collided. Neuroscientists have begun applying concepts of senescence to the brain, including post-mitotic cells. This initially presented conceptual challenges to senescence cell biologists. Nonetheless, efforts to understand senescence in the context of brain aging and neurodegenerative disease and injury emerged and are advancing the field. The present review uses pre-defined criteria to evaluate evidence for post-mitotic brain cell senescence. A closer interaction between neuro and senescent cell biologists has potential to advance both disciplines and explain fundamental questions that have plagued their fields for decades.

Keywords: cellular senescence; post-mitotic; neuronal senescence; Alzheimer's disease; biology of aging; neurodegeneration; brain; geroscience; amyotrophic lateral sclerosis; tauopathy

Citation: Sah, E.; Krishnamurthy, S.; Ahmidouch, M.Y.; Gillispie, G.J.; Milligan, C.; Orr, M.E. The Cellular Senescence Stress Response in Post-Mitotic Brain Cells: Cell Survival at the Expense of Tissue Degeneration. *Life* **2021**, *11*, 229. https://doi.org/10.3390/life11030229

Academic Editor: Markus Riessland

Received: 1 February 2021
Accepted: 2 March 2021
Published: 11 March 2021

Publisher's Note: MDPI stays neutral with regard to jurisdictional claims in published maps and institutional affiliations.

1. Introduction

Many debilitating diseases affecting our modern population have resulted from the deterioration of biological processes suited for a 40-year lifespan. Exceptional examples are neurodegenerative diseases. Age is the single greatest risk factor for the most common neurodegenerative diseases including Alzheimer's disease (AD), Parkinson's disease (PD), and amyotrophic lateral sclerosis (ALS). Sporadic neurodegenerative diseases, i.e., those not inherited, rarely affect adults before the age of 50 and are nearly absent in adults younger than 40-years-old. In this way, evolution strongly favored nervous system health. Indeed, adaptation accounts for the success of humans in our recent history, a behavioral flexibility dependent on the nervous system. For these reasons neurodegenerative diseases and central nervous system (CNS) injuries (stroke) are devastating because of the limited regenerative capacity of the mature tissues.

The complexity of the structure and function of the nervous system is achieved through extensive developmental processes and is maintained in part due to the resiliency of the cells to maintain function and resist activating cell death processes. This first evidence of this phenomenon was reported by Rita Levi-Montalcini and Barbara Booker in 1960 in their pioneering experiments demonstrating the critical role of nerve growth factor (NGF) in sympathetic neuron growth and survival [1]. Applying NGF antiserum to newborn mouse neurons resulted in a 97–99% cell loss; the same strategy only resulted in a 34% neuron loss in the adult mouse [1]. Less differentiated cells that fail to interact with their target appear to die by a morphological process (nuclear cell death) that we can now attribute to apoptosis [2]. Apoptosis is an effective and efficient mechanism of cell death involving robust activation of caspases. More mature neurons appear to die in a slower process that was termed cytoplasmic with prominent changes occurring in mitochondria, endoplasmic reticulum (ER), and lysosomes. Several forms of neuronal cell death have been described [3]. Reactivation of cell cycle in post-mitotic neurons has also been reported to be an initiating event for neuronal death (reviewed in [4]). The alterations in mitochondrial function, fission and fusion, ER stress, protein misfolding and aggregations, autophagy, and expression of proteins associated with cell cycle regulators are observed in neurodegenerative diseases and assumed to be processes associated with the neuron's cell death. However, in a post-mitotic system, cellular evolution may have favored a pro-survival response for difficult-to-replace cells in post-mitotic tissues that simultaneously prevented malignancy in mitotically competent cells–senescence.

In 1961, Leonard Hayflick and Paul Moorhead reported a dogma-shifting observation from their cell culture experiments. At that time, primary cells grown in culture were believed to be immortal with indefinite replicative potential. However, Hayflick and Moorhead reported cessation of growth and eventual loss of the lines routinely after ~50 passages or one year in culture. They referred to the phenomenon as "senescence at the cellular level" and hypothesized that it was attributed to intrinsic factors [5]. Subsequent studies revealed that human cells track their cell divisions. Cultured human fibroblasts replicate 50–80 times and then no longer divide, which is referred to as replicative senescence or the "Hayflick limit" [6,7]. By mathematical definition, a "limit" can be approached but not achieved, which is ironically fitting for this phenomenon-translating these in vitro observations to tissues and living organisms have been proposed by many, but a consensus definition has not been reached. Toward this end, we frame senescence as a complex stress response that culminates a change in cell fate.

Replicative senescence, as defined by Hayflick and Moorhead, has clearly defined underlying biology (telomere attrition) and functional phenotypes (inability to divide). Subsequent studies have identified exogenous factors that can cause cell cycle arrest through telomere-independent mechanisms. As the number of exogenous senescence-inducing factors has expanded, so has the number of unique cell types being interrogated. Numerous phenotypes have emerged as a result. Senescent cells have been identified using various morphology markers; gene, protein, metabolic changes; and functional readouts and have been a subject of earlier reviews [8–11]. A specific combination of phenotypes defining senescence currently does not exist [12]; however, most agree that it is a stress-induced change in cell fate which includes a stable cell cycle arrest and cell death resistance.

The identity of the parent cell type and upstream signals has consequences on the post-senescence phenotype [13,14]. The resulting heterogeneity has presented challenges for identifying, defining, and studying senescent cells in vivo and across disciplines. Where biologists agree is that interpreting the senescence phenotype requires integrating various lines of distinct evidence placed in appropriate context. This is especially true for post-mitotic tissues such as the brain. The intent of this literature review is not to list studies that used an umbrella term "senescence" to describe a physiological response. Rather, it is to critically evaluate results from reports on brain cell senescence using pre-defined senescence-defining criteria: proliferative/cell cycle arrest, apoptosis resistance, senescence-associated secretory phenotype (i.e., cytokines, chemokines, pathogenic proteins, exosomes, miRNA, enzymes, etc.), and senescence associated b-galactosidase activity (SA β-gal) (Figure 1). The International Cell Senescence Association (ICSA) recently provided a list of key cellular and molecular features of senescence [9]. They acknowledged that post-mitotic cells may develop features of senescence, however, their recommendations largely focused on dividing cells. Some of which have major limitation when evaluating brain tissue (i.e., lipofuscin and SA β-gal). Here we provide an overview of senescent phenotypes, as relevant to post-mitotic cells, the assays used to assess them, and the markers relevant to senescence, each in the context of brain cell biology. We then review different brain cell types using these criteria. Overall, the present contribution aims to provide an accessible summary on senescent post-mitotic brain cells with criteria and interpretations relevant to the neurobiology of aging and disease.

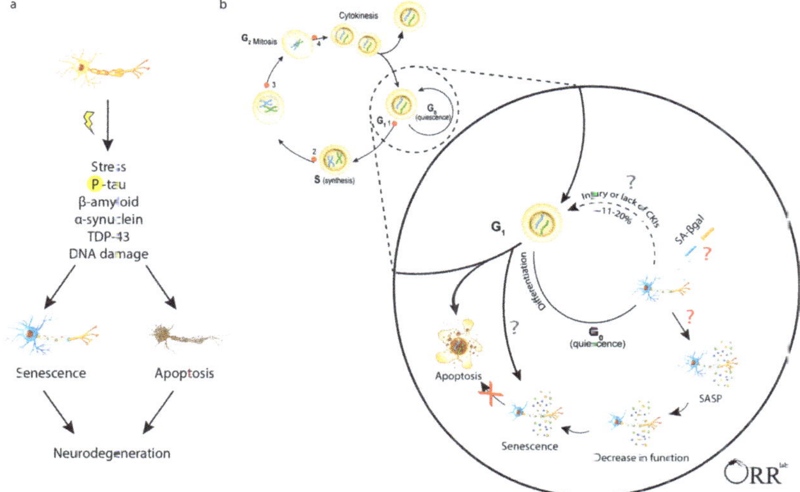

Figure 1. Senescence-associated mechanisms may confer exceptional resistance to cell death, but contribute to pathogenesis through inflammatory SASP. (**a**) The accumulation of stress, including protein aggregates and DNA damage, contributes to both senescence and apoptosis in post-mitotic cells. Protein aggregates including hyperphosphorylated tau, β-amyloid, and α-synuclein are seen in senescent cells, apoptotic cells, and patients with AD, PD, ALS and DLB. (**b**) Mitotic cells exit cell cycle and terminally differentiate into post-mitotic brain cells. Post-mitotic cell cycle re-entry can lead to cell death or senescence. In senescence, post-mitotic cells show stable cell cycle arrest, upregulate SCAPs, and release SASP. Neuronal SASP includes proinflammatory molecules and neurotoxic proteins. The SASP activates glia, drives inflammation, loss of neuronal connectivity, and perpetuates toxicity in a prion-like spread. SA β-gal can detect senescence, but it has a limited ability to discern between quiescent and senescent cells. AD: Alzheimer's disease; PD: Parkinson's disease; ALS: amyotrophic lateral sclerosis; DLB: dementia with Lewy bodies; CKI: cyclin-dependent kinases inhibitors; SCAPs: senescent cell anti-apoptotic pathways; SASP: senescence associated secretory phenotype.

2. Identifying Senescent Brain Cells

Heterogeneity of senescent cells has been revealed through transcriptomic profiling [14–16]. The senescence phenotype is guided by context differences in cell type, upstream stressor and environment. To yield a deeper understanding of cellular senescence signatures across distinct cell types in vivo, a careful examination of multiple key biomarkers is needed and has been the subject of many reviews, for example [17]. Below we provide an overview of this strategy, with specific focus on its utility to identifying senescent post mitotic brain cells.

2.1. Absence of Proliferation/Stable Cell Cycle Arrest

Various indicators of cell cycle arrest such as cell cycle inhibition and telomere attrition have been discussed for mitotically competent brain cells, for example [18]. Despite the persistent dogma that neurons permanently withdraw from the cell cycle upon terminal differentiation, numerous studies have demonstrated the expression of cell cycle proteins in post-mitotic neurons, that can give rise to dysfunctional hyperploid cells [19]. A decrease in telomere length is associated with cell division; notably telomere shortening has been observed in non-replicative neural brain populations in C57BL/6 mice in a cell cycle-independent manner [20]. Telomere shortening may occur in replication-independent scenarios in long-lived post-mitotic cells through oxidative stress and downregulation of telomeric factor *POT1* and shelterin subunit *TRF2* [21]. While markers of proliferation may not necessarily indicate mitotic competency in post-mitotic cells, they may reflect cell cycle re-entry out of G0 into G1 which could make them vulnerable to apoptosis or senescence. Cell cycle re-entry has been estimated to occur in ~11.5% of post-mitotic cortical neurons through DNA content variation and ~20% of post-mitotic neurons in AD through both DNA content variation and expression of cyclin B1 [22,23]. The data collectively indicate a link between aberrant neuronal cell cycle activity and neuronal dysfunction and disease. For example, multiple studies link AD associated Ab [24–26] and phosphorylated tau with aberrant cell cycle activity [27–29]. Cell cycle re-entry in the absence of AD pathology also has been described [30] and put forth as a potential novel therapeutic target for neurodegenerative diseases [31]. In this context, an open question remains whether proliferation markers may also apply to a pre-senescent phase in post-mitotic neurons. Co-expression of cell cycle mediators in post-mitotic cells, such as G1 proteins, in the absence of apoptotic markers (i.e., caspase 3) suggests an arrest of aberrant cell cycle activity, consistent with senescence. Alternatively, post-mitotic quiescent cells may more readily transition to senescence than mitotically competent cells, for example through changes in lysosomal activity [32]. Nonetheless, measuring molecular signatures of cell cycle activity may provide evidence for or against a senescence stress response in post-mitotic cells. Examples of studies looking at stable cell cycle arrest are discussed in Section 3.

2.2. Cell Death Resistance

Senescent cells display enhanced survival over their non-senescent counterparts by activating senescent cell anti-apoptotic pathways (SCAPs) [33,34]. Similarly, post-mitotic cells, including neurons, acquire a greater resistance to cell death as they mature [35,36]. A complete understanding of neuronal cell death resistance is not known; however, some pathways have been identified [37–40]. Similarly, while some SCAPs have been identified for senescent cells, this is a burgeoning research area. It is tempting to speculate that SCAP-mediated degeneration resistance may contribute to (or use similar mechanisms as) post-mitotic cell death resistance. In this way, identifying molecular regulators of cell death resistance in post-mitotic cells may apply to senescence, and vice versa. In response to injury, mitotically competent cells may proliferate; instead post-mitotic cell cycle reentry triggers degenerative processes [41]. In this review, we provide evidence that post-mitotic senescence in neural tissue may preserve cellular integrity by avoiding cell death [42]. Thus, cell cycle inhibitors such as INK4 cyclin-dependent kinases inhibitors (CKIs) (i.e., p16, p18, p19) and CK-interacting protein/kinase inhibitor protein CKIs (i.e., p21, p27, p57) may be protective and contribute to cell death resistance in post-mitotic cells [41]. Though a stable senescence cell cycle arrest may confer degeneration resistance to the affected post-mitotic cell, downstream consequences of its preserved survival may include neural network dysregulation and chronic inflammation through secreted factors.

2.3. Secretory Phenotype

Post-mitotic cells can produce a senescence-associated secretory phenotype (SASP) consistent with that of mitotically competent cells. For details on the definition, role, and common factors used to identify SASP in the brain, please refer to previous reviews on this topic [18,43,44]. Briefly, upregulation of NFκb activity and consequential production of canonical pro-inflammatory markers may occur in post-mitotic cells similar to that of dividing cells. Moreover, post-mitotic cells may produce unique SASP factors, like aggregation-prone proteins that impact protein homeostasis and drive neurodegenerative proteinopathies. For example, human postmortem brains from patients clinically diagnosed with AD, PD, and dementia with Lewy bodies (DLB) all show severe hyperphosphorylated microtubule-associated tau, β-amyloid, and α-synuclein loads, although topographical distribution of protein aggregates are different [45]. These protein aggregates may influence each other and synergistically promote the accumulation of one another [46,47]. Thus, SASP is an important component of post-mitotic senescence with implications for neurodegeneration.

2.4. Senescence Associated β-Galactosidase

In-depth details of the SA β-gal assay specific to brain tissue were described in our previous review [18]. Briefly, SA β-gal detects lysosomal-β-galactosidase activity at pH 6.0 [48]. While useful for distinguishing senescent cells in culture, it is detected in brain tissue independently of age or senescence [49]. The *GLB1* gene encodes for lysosomal beta-D-galactosidase and is the origin of the SA β-gal activity, but SA β-gal's cellular roles and mechanism in senescence are not fully understood [50]. In post-mitotic cells, interpretation of SA β-gal requires extra caution. Purkinje neurons in the cerebellum, CA2 neurons, and a subset of cortical neurons all display SA β-gal even in young mice [49]. A recent study demonstrated that lysosomal activity mediates the transition from deep quiescence to senescence [22]. Given that neurons are quiescent, assigning positive staining to senescence versus quiescence becomes especially challenging in static tissue. Even in vitro, SA β-gal positivity has been reported in neurons in the absence of other markers of senescence [51]. A more detailed review of SA β-gal used to identify senescence in post-mitotic cells is described in later sections.

2.5. Concluding Remarks on Identifying Senescent Cells

Evaluating senescence requires in-depth understanding of the cell type and markers of senescence. Post-mitotic cells have different characteristics from mitotically competent cells that should be considered when evaluating for senescence. In the following sections, we review studies reporting senescence in post-mitotic brain cells. We evaluate the methods used by key studies and compare them to our pre-defined criteria presented above and summarized in Figure 1.

3. Neurons

The adult human brain contains an estimated 86 billion neurons [52]. Barring neurodegenerative disease or brain trauma, nearly all cortical neurons (96–98%) remain alive during the lifespan. Their exceptional survival has been attributed to their restriction of apoptotic pathways, though the precise molecular details are not fully understood [16] An appreciation for dysfunctional, not missing, neurons has emerged over the past decade. For example, age-associated cognitive decline has been attributed to changes in neuronal chemistry, metabolism, and/or morphology, but not necessarily the progressive loss of neurons [53]. Re-evaluation of the literature and accumulating experimental evidence suggests that age- and disease-induced stressors on neurons initiates a neuronal senescence stress response as a means to avoid active degeneration and cellular loss. However, as we discuss, these neuronal structural and functional changes contribute to pathogenesis in neurodegenerative diseases [49,54–56]. For example, p16 and p21 expression has been reported in neurons and glial cells in postmortem motor and frontal association cortex of ALS patients [57], while microglia express p16, p53, and SASP in late-stage spinal cord of the ALS rat model [58] (please refer to mitotic review for microglia senescence [18]). As terminally differentiated cells arrested in G0, neurons either inherently fulfill one of the key defining features of senescence (near permanent cell cycle arrest); alternatively, they may arrest in G1 after cell cycle re-entry, which has been described in numerous degenerative diseases [22,23] (Figure 1). This phenomenon is not unique to neurons; a recent review provides a discussion on the topic of post-mitotic senescence across tissues [42]. Here we review supporting evidence that neurons, like mitotically competent cells, have the ability to mount a canonical senescence stress response.

3.1. Neuronal Senescence in Tauopathies and Peripheral Neuropathies

Senescent cell heterogeneity, in part, arises from differences in cell biology of the parent cell. Growing experimental evidence demonstrates that the phenotypic diversity of neuronal senescence reflects the heterogeneity of neuronal subpopulations. Historically, neurons have been classified by morphology, anatomical location and/or distinct shapes, and function that can be further classified by direction, action on other neurons, discharge patterns, and neurotransmitter utilization. Recent methodologies, in particular single nucleus transcriptomics, have provided an even deeper insight into neuronal heterogeneity [59]. For example, MAPT encodes the microtubule-associated protein tau proteins, which are often referred to as "neuron-specific" or "axon-specific" proteins. However, the diversity in tau proteins arise from extensive processing at the mRNA and protein levels. Six major tau protein isoforms are expressed in the adult brain which arise through alternative splicing; post-translational modifications further amplify the tau protein diversity by producing dozens of unique forms of tau protein that are differentially regulated and expressed based on developmental age and neuronal subtype [60,61]. Tau protein accumulation is the most common intraneuronal pathology among neurodegenerative diseases, though neuropathology and clinical presentations differ across diseases [62]. Among tauopathies, neurons containing neurofibrillary tangle (NFT) aggregates of heavily post-translationally modified tau are the closest correlate with neurodegeneration and dementia in AD, yet they are long-lived [63]. We recently determined that these neurons display a canonical senescence stress response [49]. Analyzing transcriptomic data from postmortem human brain provided the opportunity for a within-subjects comparison between neurons with or without NFTs. The transcriptomic and pathway analyses revealed expression patterns in NFT-bearing neurons consistent with senescence including upregulated anti-apoptotic/pro-survival pathways and concomitant inflammatory and secretory pathways. Using four independent tau transgenic mouse models, we found evidence for DNA damage; aberrant cellular respiration; karyomegaly; upregulation of cell cycle inhibitors, inflammation and inflammatory mediator NFκB. These phenotypes occurred concomitant with NFT formation and were reduced by genetically removing endogenous tau protein, indicating a molecular link between tau and neuronal senescence. Moreover, intermittent treatment with senolytics (dasatinib plus quercetin) caused ~35% reduction in NFTs that coincided with a reduction in senescence-associated gene signature (cell cycle inhibitors and inflammation). We did not observe neuronal senescence phenotypes in thalamic, midbrain, or cerebellar neurons. It remains unknown whether or not high expression of transgenic tau could ultimately drive neuronal senescence in these other neuronal subpopulations. However, work from other groups suggest that midbrain [56] and cerebellar neurons [54] may utilize other molecular mediators aside from MAPT/tau.

Overexpression of human non-mutated tau and its persistent phosphorylation also contributes to peripheral neuropathy and memory deficits [64]. Long-term and short-term memory were significantly impaired in female transgenic mice expressing all six human tau isoforms [64]. Peripheral neuropathy was evidenced as motor nerve conduction velocity (MNCV) slowing, paw tactile allodynia, paw heat hypoalgesia, and low paw density of intraepidermal nerve fibers in human tau mice compared to wild type mice [64]. Notably, neuronal senescence has been associated with cisplatin-induced peripheral neuropathy (CIPN). Primary DRG neurons treated with cisplatin upregulate SA β-gal activity and expression of *Mmp-9*, *Cdkn1a*, *Cdkn2a*, and display elevated translocation of HMGB1 compared to controls [65]. In a mouse model of CIPN, dorsal root ganglia (DRG) neuronal populations upregulated the DNA damage response pathway and *Cdkn1a* gene expression as determined by single-cell RNA-sequencing. Neuronal senescence was further verified by increased protein expression of p21, p-H2AX, NFκB-p65; SA β-gal; and lipofuscin granules [66]. Clearing p16 and/or p21 positive cells either pharmacologically with ABT263 or by utilizing suicide gene therapy (i.e., p16-3MR ganciclovir/herpes simplex virus thymidine kinase system) reversed CIPN as evidenced by improved mechanical and thermal thresholds [65]. Collectively these studies indicate senescence-associated neuronal dysfunction in the central and peripheral nervous system where tau may be linked.

3.2. Neuronal Senescence in Parkinson's Disease

The H1 MAPT haplotype [67] and single nucleotide polymorphisms in MAPT have been associated with age of onset and progression of PD [68]. Despite the strong genetic association with MAPT, tau pathology occurs only in about 50% of patients with PD [69]. Though cognitive dysfunction may occur, PD primarily affects motor behavior. Neurodegeneration in PD predominantly occurs in the substantia nigra where up to 70% of dopaminergic neurons can be lost in late disease stages [70,71]. These neurons express significantly lower levels of MAPT and tau protein than those in the cortex or hippocampus (~4-fold and ~6-fold difference, respectively) and do not develop tau pathology [72]. Instead, the hallmark protein deposit in PD is α-synuclein. Experiments in cell lines suggest that α-synuclein expression levels differentially regulate the cell cycle [73]; however, conclusive studies demonstrating α-synuclein-mediated neuronal senescence have not been reported. To date, the most comprehensive work on dopaminergic neuronal senescence involves special AT-rich sequence-binding protein 1 (SATB1) [56]. SATB1 functions as a transcription factor and chromatin architecture organizer [74]. SATB1 is overexpressed in various tumors and has been referred to as a T-cell-specific transcription factor given its importance in T cell development. A meta-analysis of genome-wide association studies comparing PD cases with controls identified SATB1 as a candidate risk gene [75]. Neurons in PD-vulnerable brain regions (e.g., substantia nigra pars compacta) display lower levels of SATB1 than neurons from the less susceptible ventral tegmental area [76]. Genetically reducing SATB1 in dopaminergic neurons drives a neuronal senescence response including elevated p21 protein expression, karyomegaly, SASP, and mitochondrial dysfunction [56]. SA β-gal and lipofuscin, hallmarks of mitotically competent senescence that also co-occur in neurons, were also observed. Mechanistically, SATB1 repressed dopaminergic neuron senescence by binding the regulatory region of CDKN1A. In the absence of SATB1, the CDKN1A encoded protein, p21, expression level increased to perpetuate the neuronal senescence stress response. When the authors reduced Cdkn1a in SATB1 knockout neurons, fewer senescent cells (as determined by SA β-gal) were observed, providing evidence for the mechanistic link between SATB1-p21 mediated neuronal senescence. Interestingly, reducing SATB1 in cortical neurons did not modulate Cdkn1a/p21 levels, which was attributed to a more open Cdkn1a locus in dopaminergic than cortical neurons. In contrast, tyrosine hydroxylase expressing neurons require Satb1 expression for their survival and will undergo neurodegeneration within three weeks of downregulated Satb1 [76]. This observation indicated that de-repression of Cdkn1a and concomitant increased p21 expression caused apoptosis and clearance by microglia. Evidence for this was observed by neuronal SASP production and concomitant microglia co-localization with tyrosine hydroxylase positive neurons. Follow-up experiments to deplete microglia after Satb1 reduction would conclusively demonstrate whether or not Cdkn1a-expressing dopaminergic neurons fulfill the criterion of apoptosis resistance in neuronal senescence. Indeed, elevated p21 expression induced apoptosis in vitro, indicating that it may not confer neuronal apoptosis resistance [77]. Nevertheless, the study by Riessland et al. determined a dopaminergic neuron-specific role of SATB1 in modulating Cdkn1a/p21 expression and downstream senescence phenotypes including karyomegaly, mitochondrial dysfunction, production of SASP, lysosomal dysfunction and presence of SA β-gal and lipofuscin [56].

In PD, disease-related stressors on neurons contribute to defects in several cellular systems ultimately involving alterations in Bcl-2 family signaling, JNK activation, p53 activation, expression of cell cycle regulators [78]. While many of these processes including those addressed above are thought to contribute to neuronal degeneration, some are also hypothesized to reflect survival-promoting mechanisms such as senescence. More recent studies focused on neuronal senescence in PD have revealed that overexpression of mutant p53, p21, or mutant Leucine-rich repeat kinase 2 (LRRK2) increased SA β-gal, and αSyn protein expression and fibril accumulation in vitro [77]. Transgenic mice expressing the same mutant $LRRK2_{G2019S}$ displayed elevated oligomeric αSyn, β-galactosidase and p21 expression. The increase in αSyn was due to impaired degradation, not increased transcription [77]. The results suggest that the $LRRK2_{G2019S}$ mutation may activate the p53-p21 senescence pathway, which is upstream of α-synuclein accumulation. While suggestive of senescence, the study did not evaluate cell cycle activity, apoptosis resistance or SASP production in the affected cells. Nonetheless, future studies to dissect if or how PD mutations may interact with *SATB1* would elucidate whether these pathways converge on a common senescence-associated pathway relevant to PD pathogenesis. An intellectual framework for proteinopathy-induced senescence in neurodegenerative diseases was first proposed in 2009 by Golde and Miller [79]. The idea warrants further studies as the emerging data from mechanistic studies that have directly tested this hypothesis (i.e., tauopathy, α-synucleinopathy, and β-amyloid) indicate that post-mitotic neurons are especially vulnerable to protein aggregation stress as highlighted in these aforementioned studies.

3.3. Neuronal Senescence in Aging

Senescent cells accumulate with advanced age even in the absence of disease. In 2012, Diana Jurk et al. evaluated neuronal senescence in naturally aged mice with or without increased DNA damage by genetically manipulating telomerase [54]. Age-associated DNA damage was associated with neuronal senescence in the brains of 32-month-old mice [54]. DNA damage foci, as determined by gH2A.X immunostaining, was elevated in cerebellar Purkinje and cortical neurons from 32-month-old mice compared to 4-month-old mice. These neurons also displayed activated p38 MAPK (phosphorylated at Thr^{180}/Tyr^{182}) indicative of DNA double strand breaks. Oxidative stress was assessed by visualizing cells with elevated lipid peroxidation product, 4-hydroxynonenal (4-HNE). Immunostaining with 4-HNE revealed cytoplasmic granular accumulation within the same subpopulations of cells. Similarly, these large neurons expressed higher levels of inflammatory protein IL-6 than other cell types. SA β-gal activity and lipofuscin (as measured by autofluorescence) showed similar overlapping patterns. Given the overlapping co-staining of multiple marker combinations, the authors hypothesized that DNA damage (gH2AX) increased with advanced age, which activated the DNA damage response (p-p38 MAPK) to induce a senescence-like pro-inflammatory (IL-6) and pro-oxidant phenotype (4-HNE) similar to mitotically competent cells (lipofuscin and SA β-gal). To begin evaluating mechanistic mediators of the senescence phenotype, they utilized transgenic mice with telomere dysfunction with or without *Cdkn1a*. Neurons from mice with telomere dysfunction (late generation telomerase knockout mice, F4 TERC-/-) displayed elevated levels of gH2AX, p-p38MAPK, 4HNE and IL6 compared to those with one functional copy of TERC. The genetic removal of *Cdkn1a* modulated these phenotypes in mice regardless of telomerase activity, however genotype and cell type specific phenotypes were observed. For example, in TERC wild type mice, the absence of p21 only significantly altered the 4HNE phenotype and only in Purkinje neurons (not cortical neurons). In contrast, in mice with telomere dysfunction, removing p21 did not modulate 4HNE. Instead, the absence of p21 significantly reduced gH2Ax and IL6 in Purkinje neurons and p-p38 and IL6 in cortical neurons. These results again highlight heterogeneity of the senescence stress response unique to different neuronal subpopulations. Nonetheless, removing p21 robustly reduced inflammation, as assessed through IL6, in both cellular populations to provide evidence that neuronal senescence may contribute to sterile inflammation with advanced age.

Insulin provides trophic support and drives excitatory signaling in neurons [80,81]. A loss of neuronal sensitivity to insulin, referred to as insulin resistance, coincides with their dysfunction and disease. The mechanisms driving insulin resistance in brain cells are not well understood, but risk factors include advanced age, obesity, peripheral insulin resistance, and metabolic dysfunction [82,83]. Recent studies in mice have demonstrated that brain insulin resistance induces neuronal senescence, which leads to synaptic dysfunction [55,84]. In these studies, insulin resistant neurons display several molecular, functional and morphological changes consistent with senescence [55]. Specifically, mice that developed spontaneous peripheral insulin resistance at either young (3-months-old) or old (24-months-old) age also displayed signs of brain insulin resistance (i.e., elevated insulin in the CSF, elevated pIRS1 (Ser^{307} and Ser^{612})], and senescence (i.e., neurite loss, elevated Cdkn1a and Cdkn2a and SA β-gal activity). This finding indicates that insulin resistance, like tau accumulation or loss of *SATB1*, may drive premature neuronal senescence in the absence of advanced age. The insulin resistant mice, regardless of age, behaved poorly on cognitive behavior tasks to indicate that neuronal insulin resistance/senescence co-occurred with poor brain function. Mechanistically, chronic insulin was shown to reduce hexokinase 2, impair glycolysis and increase levels of p25, a potent activator of both CDK5 and GSK3β. The simultaneous signals from CDK5 (neuronal cell death) and β-catenin (cell cycle re-entry) pushed neurons to enter a senescence-like state. A detailed signal transduction cascade was elucidated in vitro whereby insulin increased Ccnd1 and Cdkn2a expression, nuclear localization of β-catenin, cyclin D1 and p19ARF. The increase in p16INK4a and PML occurred later. Aberrant β-catenin also induced a parallel p53-p21 senescence pathway. The authors concluded that chronic insulin signaling induced a neuronal senescence phenotype through the over-stabilization and nuclear localization of β-catenin. Tau phosphorylation was not assessed, but given the increased activity of tau kinases Cdk5 and GSK3β and parallels with findings in Musi et al. [49], it is tempting to speculate that aberrant tau may also contribute to insulin resistance-mediated neuronal senescence.

3.4. General Considerations for Evaluating Neuronal Senescence

Observations across the aforementioned studies highlight the complexity of applying canonical senescence measures to post-mitotic cells. For example, we caution the use of lipofuscin and SA β-gal for neuronal senescence as these markers seemingly reflect shared phenotypes among neurons, across age and/or disease, that requires further investigation into their association with other senescence markers. The best example are cerebellar Purkinje neurons that display SA β-gal throughout the lifespan [49]. Jurk et al. noted, "the frequencies of neurons showing multiple markers of a senescent phenotype are very substantial, going well beyond 20% in Purkinje cells already in young mice brains" [54]. A key readout for this conclusion was SA β-gal staining. Given the early stages of defining neuronal senescence in vivo, it remains unknown whether SA β-gal positivity truly reflects senescence in Purkinje neurons, which could become senescent in early life due to their high energetic and metabolic demands. Other studies have demonstrated that cerebellar Purkinje neurons can survive and function as polyploid cells [85]. Neuronal polyploidy suggests that DNA replication occurred, but that neuronal mitosis stalled. Indeed, hyperploid neurons have been reported in preclinical and mild stages of AD as evidenced by immunofluorescence and slide-based cytometry methods cross-validated by chromogenic in situ hybridization [86]. The neurons avoid apoptosis, upregulate several cell cycle mediators and survive months in the adult mouse brain, which meets several criteria of a senescent cell. Importantly, cerebellar Purkinje neurons are indispensable for motor movement control. Notably, gait speed, coordination, and balance are significant predictors of mortality [87,88]. It is tempting to speculate that senescence of these neurons may contribute to the overall decline in health and increased mortality with advanced age. Alternatively, the physiological function of these neurons may require signaling through cellular and molecular pathways resulting in phenotypes typically attributed to senescence. For example, we routinely observe neuronal lipofuscin throughout the lifespan, though it notably increases with age; similarly, we observed high levels of SA β-gal activity in these same neuronal populations throughout the mouse lifespan [49]. Moreno-Blas et al. also proposed that SA β-gal may not be a reliable marker of senescence by itself [89]. Despite cortical neurons expressing senescence-associated phenotypes such as p21, γH2AX, ruptures of DNA, lipofuscin, SASP, and irregular nuclear morphology, they observed normal nuclear morophology in some neurons with high SA β-gal [89]. Instead, their data suggested that autophagy impairment/dysfunction, perhaps through lysosomal fusion with autophagosomes, critically contributed to the neuronal transition from quiescence to senescence, similar to that reported by [32]. Since SA β-gal positivity overlaps with lysosomal dysfunction, it may be useful to narrow down potential senescent cell candidates; however, as indicated by Moreno-Blas [89] and several studies reviewed here (i.e., [49,51,54]) it cannot be used in isolation. Similarly, neuronal lipofuscin staining was first reported in children in 1903 and has been later confirmed in several studies where it occurs in at least 20% of neurons by 9-years-old [90]. These aggregates of oxidation products of lipids, proteins, and metals autofluoresce non-specifically bind antibodies which can complicate interpretations of immunofluorescence assays and thus requires multiple controls. The pigment granules change with aging by increasing progressively in size, as well as their subcellular localization thus appropriate age-matched negative controls and antibody controls are necessary to interpret results. Within the aging field, the increased rate of lipofuscin formation and accumulation is considered a hallmark of both replicative and stress-induced senescence [91,92] and methods for its specific staining (i.e., Sudan Black B) are increasingly used to detect senescence in vitro and in vivo [93]. It is our opinion that at this time both lipofuscin and SA β-gal require further investigation before using them as decisive markers for neuronal senescence.

3.5. Concluding Remarks

Differentiated neurons are remarkably apoptosis resistant, but their vulnerability to excitotoxicity increases with age [94]. Neurons inherently lack the option to divide, but they upregulate cell cycle proteins in response to stress. The inability to replace these critical cells indispensable for maintaining life may have placed strong evolutionary pressure to favor stress-induced senescence over apoptosis. In this way, neuronal survival would be maintained though the number of dysfunctional cells would increase with advancing age. Indeed, this is what is observed in the human brain [52,53]. As the burgeoning field of neuronal senescence advances, we expect that the next wave of studies will reveal additional molecular regulators, clarify pathways previously identified, and differentiate between shared pathways and neuron subtype specific mechanisms. Additionally, with the increasing use of single cell technologies, we anticipate an increased ability to identify, track and study senescence with greater clarity on the phenotype(s) and how they change across the lifespan and in disease.

4. Astrocytes

Astrocytes are an abundant and heterogenous cell population within the central nervous system (CNS). They comprise 20–40% of the total glial cell population in the brain, depending on region, developmental stage, and species [95–97]. Along with oligodendrocytes, astrocytes originate from the neural tube [98]. Astrocytes differentiate from the glial progenitor cells proliferating in the forebrain subventricular zone as they migrate outwards to other regions of the brain [99]. The majority of astrocytes are considered post-mitotic, and in the absence of pathology or disease, they display low rates of turnover and proliferation [100].

Astrocytes vary in function and morphology. Distinct types, including radial astrocytes, fibrous astrocytes, and protoplasmic astrocytes have been elucidated within the CNS based on structure, distribution, and function, as well as their expression level of the different isoforms and splice variants of the intermediate filament protein glial fibrillary acidic protein (GFAP) [101,102]. Astrocytes have been implicated in maintaining water and ionic homeostasis, providing metabolic and structural support to neurons, and regulating the blood–brain barrier (BBB) [102,103]. They also cooperate with microglia to control local neuroinflammation and neuronal restoration following damage to the CNS. Similar to microglia, astrocytes prune synapses and remove cellular debris within the synapse in healthy and diseased brains [104,105]. Genes crucial for astrocyte function such as Excitatory Amino Acid Transporters 1 (*EAAT1*) and 2 (*EAAT2*), potassium transporter *Kir4*, and water transporter *AQP4* involved in glutamate, glutamine, potassium, and water homeostasis in the brain have shown to be downregulated when astrocytes become senescent [106]. Thus, their change in function associated with senescence can lead to detrimental effects including the onset of various neurodegenerative pathologies [103,107–110].

Astrocyte senescence is often wrongly conflated with astrogliosis or astrocyte reactivity. Reactive astrogliosis involves structural changes to the astrocytes alongside cellular proliferation and migration [100,109]. Reactive astrocytes, also known as A1 cells, have been shown to be induced by activated neuroinflammatory microglia through the secretion of Il-1α, TNFα, and C1q cytokines. Upregulated expression of GFAP is a known marker of reactive astrocytes, and its levels are also increased during aging [111,112]. In contrast, radiation-induced senescent astrocytes demonstrated a downregulation of GFAP [113]. Reactive A1s lose their ability to promote neuronal survival, outgrowth, synaptogenesis, and phagocytosis and induce death of neurons and oligodendrocytes [114]. A1s have also been shown to be present in the brains in many neurodegenerative disorders, including AD, PD, and Huntington's disease [114,115]. The benefit of astrogliosis and subsequent scar formation is the protection of the surrounding neurons and tissue and restriction of inflammation and pathology. However, dysfunction in reactive astrocytes can lead to neuronal dysfunction, and eventually degeneration that can contribute to various CNS disorders. Many of these features are similar with a senescence phenotype, including morphology changes and secretion of pro-inflammatory molecules.

Astrocytes undergo a senescence-like stress response, which has been referred to as "astrosenescence" and described as a functional change from neurosupportive to neuroinflammatory [116]. Oxidative stress, exhaustive replication, inhibition of proteasomes, and an increase in glucose concentration elicit an astrocyte response consistent with senescence, in vivo and in vitro (reviewed: [116,117]). For example, replicative senescent primary human fetal [118] and rat [119] astrocytes displayed an arrest of growth and cell cycle progression; the human fetal astrocytes also upregulated gene expression of *TP53* and *CDKN1A*. Astrocytes do not express TERT [120] and replicative senescence was not avoided with telomerase reverse transcriptase (hTERT) expression [118], indicating that telomere-length independent mechanisms govern replicative senescence in astrocytes. Inhibiting p53 function with human papillomavirus type 16 E6, however, delayed the onset of senescence, implying a p53-dependent mechanism of replicative senescence in astrocytes [118]. Increased SA β-Gal activity, marked by staining kits, was also observed in many of these studies [113,117,121,122]. Strengths and weaknesses of using this method for labeling brain cells have been discussed [18].

Radiation cancer therapy has potential to induce senescence [123]. The effect of radiation therapy on astrocytes in vivo was examined by evaluating human brain from individuals receiving cranial radiation cancer therapy [113]. Senescent cells were identified with immunohistochemical labeling of p16, heterochromatin protein Hp1γ, and expression of Δ133p53, an inhibitory isoform of p53. Elevated p16 and Hp1γ largely co-localized with astrocytes in patient brains that had received radiation, but not in control tissue. Expression of Δ133p53 was primarily in astrocytes, and its role in senescence was explored in vitro. They found that these irradiated astrocytes in vitro had diminished Δ133p53, and developed a phenotype associated with other senescent cells, such as increased SA β-Gal activity, p16, and IL-6. However, restoration of Δ133p53 expression inhibited and prevented further senescence, promoted DNA repair, and prevented astrocyte-mediated neuroinflammation and neurotoxicity [113]. Collectively, this study [113] and others [106,124] have characterized the radiation-induced senescence phenotype in astrocytes to include decreased proliferation and increased SA β-Gal activity, along with typical increased expression of p53, p21, and p16, which were analyzed using Western Blot [113,121].

Senescent astrocytes downregulate genes associated with activation, including GFAP and genes involved in the processing and presentation of antigens by major histocompatibility complex class II proteins, while upregulating pro-inflammatory genes [121]. Increased expression of p16, p21, p53, and MMP3 have also been associated with astrocytes undergoing senescence and those isolated from aged brains [125]. The downregulation of genes associated with development and differentiation, coinciding with the upregulation of pro-inflammatory genes, manifest as functional changes (i.e., inflammatory stress response). This may perpetuate a pro-inflammatory feedback loop that is stably maintained by senescence-associated changes in gene expression and transcript processing [126].

Astrocyte senescence increases with age in the human brain and in AD [127,128] and PD [129]. The consequences of astrocyte senescence are myriad. Functionally, astrocytes communicate with nearby neurons and the surrounding vasculature to clear disease-specific protein aggregates, including β-amyloid, the accumulation of which has been linked to the progression of AD [96,130]. The release of SASP factors by senescent astrocytes including IL-6, IL-8, MMP3, MMP10, and TIM2 were found to contribute to β-amyloid accumulation, phosphorylation of tau protein, and an increase in NFTs [125,131]. An increased risk for PD has been linked to contact with the herbicide paraquat (PQ), which an environmental neurotoxin. Complementary in vivo and in vitro approaches were used to evaluate mouse and human astrocyte responses to PQ [129]. PQ-treated astrocytes developed several features consistent with senescence, including upregulated *Cdkn2a*/p16. Importantly, senescent cell removal improved neurogenesis in the subventricular zone, reduced neuronal loss and rescued motor function deficits in PQ-treated mice [129]. Collectively their results highlight astrocyte senescence as a mechanism of PQ-associated neuropathology and brain dysfunction, and represents an appealing therapeutic target for the treatment of PD.

Concluding Remarks

"Astrosenescence" is a complex and heterogeneous process that necessitates evaluating astrocyte structure, distribution, function and molecular expression profiles. Measuring the expression level of GFAP [113,125] can help differentiate whether upregulated pro-inflammatory cytokines and chemokines expression reflect astrogliosis or astrosenescence [125,131]. The most consistently shared features across senescent astrocytes were arrest of growth and cell cycle progression, increased expression of p53 and p21, and p16 [113,117,121] and some evidence of increased SA β-Gal activity [117]. Collectively the studies reviewed here indicated that functional changes associated with senescent astrocytes contribute to chronic neurodegenerative diseases and may propagate inflammation and induce senescence to surrounding cells [132,133]. Targeting them for removal represents an opportunity to intervene in neurodegenerative diseases.

5. Endothelial Cells

Endothelial cells form a single layer of cells called endothelium that line the blood vessels of the circulatory system. They have an array of functions in vascular homeostasis such as regulating blood flow, immune cell recruitment, maintaining blood vessel tone, and hormone trafficking [134,135]. While endothelial cell function is heterogeneous and tissue-specific, several studies have demonstrated that endothelial cells can become senescent in adipose tissue, coronary arteries, and in the human umbilical cord using observations of morphology changes, SA β-gal activity, and SASP through DNA microarray [136–138]. While there is a great literature describing senescent endothelial cells throughout the body, the focus of this section turns to brain microvascular endothelial cells.

Brain endothelial cells are mostly post-mitotic with minimal proliferation [139–141]. They express a high density of tight junction and adherens junction proteins and high transendothelial electrical resistance [142–144]. Functionally, brain endothelial cells contribute to the BBB, regulate local cerebral blood flow as a part of the neurovascular unit (NVU), and thus have important implications for brain diseases [145–147]. The BBB is a highly selective semipermeable barrier with tight junctions that closely regulates the biochemical composition of the brain by restricting the free diffusion of nutrients, hormones, and pharmaceuticals [148]. The tight junctions force molecular traffic to take place through the endothelial membrane through sealing of the paracellular space and by establishing a polarized, transporting epithelial and endothelial phenotype [149]. During aging, endothelial cells experience senescence-associated stressors including oxidative stress, DNA damage accumulation, telomere shortening, increased NFκB signaling and decreased Sirt1 expression [136,150]. Recent studies suggest that brain endothelial cell senescence could contribute to BBB dysfunction though neurovascular uncoupling and reactive oxygen species [151]. Indeed, increased BBB permeability and vascular dysregulation have been observed in patients with early cognitive dysfunction, cerebral microvascular diseases, and AD [152–154]. However, the co-occurrence of senescent endothelial cells with aging and disease makes it difficult to discern whether they are upstream mediators or downstream consequences of diseases. BBB dysfunction has been observed in patients with AD, Multiple Sclerosis (MS), traumatic brain injury (TBI), and stroke, featuring overexpression of MMP-2 and MMP-9 [155–159]. Molecular cascades such as activation of MMP's have been suggested to induce senescence [160]. Thus, it is possible that brain insult leading to BBB dysfunction causes senescence as well.

There have been several recent studies specifically examining cerebrovascular endothelial senescent cells induced by external stimuli and natural aging. In one study, rat primary cerebromicrovascular endothelial cells were delivered 2–8 Gy of γ-irradiation using a 137Cs gamma irradiator [161]. After irradiation, cerebromicrovascular endothelial cells' DNA damage was examined using the Comet Assay with alkaline single-cell gel electrophoresis followed by fluorescent imaging of the nuclei Comet Assay visualizes the amount of DNA which leaves the nucleus as a marker for DNA strand breaks [162]. Irradiation caused DNA fragments to migrate out of the nuclei, indicating increased DNA damage [161]. SA β-gal staining was positive in a dose-dependent manner, p16 and p53 upregulation was observed, and a SASP was seen with upregulation of *IL-6, IL-1α, GM-CSF, G-CSF, MIP-1α, MCP-1*, eotaxin, and *IL-1β* via RT-PCR to verify senescence [161]. In another study discerning cerebromicrovascular endothelial cell senescence, 3-month-old and 28-month-old C57BL/6 mice gene expression profiles were compared by single-cell RNA sequencing [163]. The mean expression of senescence core genes (*Cdkn2a, Bmi1, Trp53, Hmga1, Chek1, Chek2, Prodh, Tnfrsf10b, Cdkn1a, Dao*), senescence effector genes (*Ppp1ca, Ahcy, Brf1, Map2k3, Map2k6, Smurf2, Tgfblil, Srsf1, Angptl2*), and SASP genes (*Ccl2, Ccl24, Ccl3, Ccl5, Ctnnb1, Cxcl1, Cxcl10, Cxcl12, Cxcl2, Cxcl16, Hgf, Hmgb1, Icam1, Igfbp2, Igfbp3, Igfbp4, Igfbp6, Igfbp7, Il15, Il18, Il1α, Il1β, Il2, Il6, Mif, Mmp12, Mmp13, Mmp14, Pgf, Plat, Timp2, Serpine1, Ccl3, Ccl4, Ang, Csf2, Kitl, Serpine2, Tnfrsf1a, Hzfi, Nrg1, Ereg, Areg*) were used to calculate a running enrichment score [163]. Higher senescence gene enrichment scores were found in brains from aged versus young mice [163] These SASP factors have potential to promote neuroinflammation and affect BBB integrity [164]. Future studies may further clarify the downstream impact of senescent endothelial cells on their neighboring environment [165].

Other studies also highlight the difficulty of disentangling cause and effect of brain cell senescence, aging and disease. Emerging evidence suggests aberrant angiogenesis, and potentially endothelial senescence, may occur as bystander effects of other cell's SASP. Studies using the rTg4510 mouse model of tauopathy have revealed an increased number of blood vessels and concomitant upregulation of angiogenesis-related genes such as *Vegfa*, *Serpine1*, and *Plau* [166]. Confocal imaging demonstrated aberrant vasculature near neurons with tau-containing NFTs which display a senescence-like phenotype (please refer to Section 3: Neurons) [49]. Together, these studies suggest that factors secreted by senescent NFT-containing neurons may negatively impact surrounding cells, which could drive aberrant angiogenesis. Alternatively, aberrant cerebrovasculature could be upstream of tau accumulation and contribute to NFT formation. To translate these studies to human clinical conditions, postmortem human AD brains with tau pathology were investigated for cerebrovascular senescence [167]. Cerebral microvessels were isolated from 16 subjects with a Braak NFT score of V/VI (B3) and 12 subjects with a Braak NFT score of 0/I/II (e.g., high neuropathology versus low neuropathology). Upregulation of senescence was inferred by elevated expression of *Serpine1*, *Cxcl8*, *Cxcl1*, *Cxcl2*, *Csf2*, and *Cdkn1a*; however, other markers of senescence were not evaluated [167]. Whether tauopathy causes endothelial senescence and induces a leaky BBB and/or endothelial senescence affects the vascular microenvironment will require further investigation [168].

Concluding Remarks

Most of the aforementioned studies examined brain endothelial cell senescence by analyzing expression of senescence-associated genes [161,163,167]. Some studies also examined SASP genes [163,167]. Future studies are needed to evaluate cell cycle arrest, SCAPs, DNA-damage responses, resistance to apoptosis to define and validate senescence in brain endothelial cells [169,170]. Of interest for future studies will be determining brain region-specific differences in endothelial senescence and to better identify their mechanistic impact on the neighboring cells and environment.

6. Oligodendrocytes

Oligodendrocytes (OLs) are derived from oligodendrocyte precursor cells (OPCs) in a highly regulated process [171]. OPCs differentiate into pre-OLs, and later into mature, myelinating OLs in the presence of differentiation-promoting transcription factors [172]. The primary role of mature OLs includes myelination of neuronal axons in the CNS. Additionally, OLs play a role in providing metabolic support to myelinated axons, especially in axons that spike at high frequencies [173]. OLs have also been implicated in information processing, and defects in OL maturation are linked with behavioral abnormalities [173,174]. OLs are highly vulnerable to oxidative stress and mitochondrial injury, and OL loss occurs upon exposure to inflammatory cytokines [171,175,176]. OLs are also highly susceptible to accumulation of DNA damage during normal aging and have been indicated as a potential upstream cause of cellular aging leading to neurodegeneration, illustrated by the involvement of myelin in several neurodegenerative disorders [175,176]. DNA damage is a known mediator of senescence suggesting a potential relationship between senescence of oligodendrocytes and neurodegenerative disorders. Senescent OLs could result in defective myelination as seen in several neurodegenerative disorders [177–179]. For instance, loss of OLs can lead to demyelination as seen in MS.

Only a few studies exist that try to validate senescence in OLs. A rodent model with a novel senescence marker utilizing the p16 promoter, ZsGreen, crossed with the established APP/PS1 AD model was used to look for senescence in different cell types, including OLs [180]. OPCs showed upregulation of p21, p16, and SA-β-gal activity [180]. However, no senescence was observed in OLs (immunohistochemically stained with OL marker, CNP, and ZsGreen p16 senescence reporter) while OPCs were senescent and unable to differentiate into OLs [180]. It is likely that increased susceptibility of OPCs to the microenvironment increase the incidence of senescence in these cells compared to OLs. It may be possible that senescence in OLs occurs through p16-independent mechanisms. For example, a recent study reported an age-associated increase in p16-positive oligodendrocytes, but they were not cleared using senolytic approaches [181]. Brain cell type specific responses to senolytic clearance [181] highlights the heterogeneity of senescent cells even when in the same tissue, which may (in part) reflect cell type diversity in complex tissues such as the brain [59,182,183].

In a study of white matter lesions of frozen postmortem human brain tissue from patients who were over 65 years old, OLs exhibited elevated SA-β-gal [184]. Immunohisto-chemistry was used to double-label white matter tissue with SA-β-gal to identify cell types [i.e., astrocytes (GFAP+), microglia (CD68+), and oligodendrocytes (OSP+)]. Additionally, OLs also showed increased levels of 8-OHdG, a marker for oxidative stress, but did not display high levels of p16 [184]. Comparison of mRNA using qRT-PCR revealed a 1.5-fold increase in *TP53*, *H2AX*, and *CDKN1B* [184]. *CDKN1B* encodes p27kip1 and its upregulation results in the induction of a senescent phenotype [185]. Elevated *H2AX* and *TP53* are indicative of increased DNA damage and are also suggestive of a senescent phenotype. However, to confirm true senescence in these OLs, additional inspection of (1) SASP factors, (2) resistance to apoptosis through SCAPs, and (3) the presence of proliferation markers would be beneficial.

Concluding Remarks

Although there are several studies that examine OPC senescence (refer to our other review on senescence in mitotically active brain cells for literature regarding OPC senescence) in multiple disease processes, there are limited data regarding the senescence of OLs in natural aging and other disease models. Limited studies mentioned suggest that SA-β-gal and gene expression analysis may be used to see if OLs have a senescent, but the results are inconsistent. For example, SA-β-gal activity was seen while high levels of p16 was not observed. Further study is required to establish the true senescence status of these cells and their potential role in aging and disease.

7. Summary

Cellular senescence has been best studied and characterized in mitotically competent cells [180]. However, most cells in the brain including neurons, astrocytes, endothelial cells, and oligodendrocytes have very low or no cell turnovers and show mostly post-mitotic phenotypes. Most post-mitotic brain cells that survive brain development will remain throughout the lifespan. While this feature historically precluded the study of senescence in the brain due to the early definition requiring mitotic competency, brain cell types are highly susceptible to acquiring a lifetime worth of damage known to drive senescence (Table 1). These include oxidative stress, DNA damage, and protein accumulation, which impact cell cycle and secretory phenotypes. While senescent cells continue to survive due to their apoptosis resistance, they tend to partially lose (or change) their function and increase expression of pro-inflammatory molecules. SASP from senescent cells can affect the microenvironment in the brain by its paracrine effect, causing other neighboring cells regardless of cell types to go senescent [133]. In the brain, these dysregulations manifest as an increase in neuroinflammation, increased BBB permeability, loss of neuronal synapses, demyelination, and dysregulated metabolism [179-186]. Collectively, these features have been associated with impaired cognition, and clearance of senescent cells as a therapeutic strategy has shown to reduce pathology, inflammation, and neuronal dysfunction [49,65,181,187].

Table 1. Biomarkers previously described to verify senescent cells.

Cell Type	Biomarkers	Reference
Neurons	SA-β-gal	[54], [55], [56], [65], [66], [77], [84], [89]
	H2ax	[49], [66], [89]
	Comet Assay	[89]
	Nuclear morphology	[89]
	Cytosolic HMGB1	[65]
	Telomere-associated DNA damage foci	[84]
	p53	[77]
	p21	[49], [54], [55], [56], [57], [65], [66], [77], [89]
	p16	[49], [55], [65], [84]
	SASP	[49], [54], [56], [65], [66], [89]
	Lipofuscin	[54], [56], [66], [89]
	p38MAPK	[54]
Astrocytes	SA-β-gal	[106], [113], [118], [119], [121], [122], [124], [125], [126], [127], [129]
	Heterochromatin protein Hp1γ	[113]
	Senescence-associated heterochromatin foci	[121]
	BrdU	[118], [129]
	53BP1 foci	[129]
	Lamins	[106], [122], [129]
	GADD45A	[122]
	p53	[118], [121], [122], [124], [125]
	p21	[118], [121], [122], [124], [125]
	p16	[106], [113], [118], [121], [126], [127], [129]
	SASP	[106], [113], [124], [125], [126], [127], [129]
	p38MAPK	[127]
Brain Endothelial Cells	Senescence-associated genes	[161], [163], [167]
	Comet Assay	[161]
	SASP	[161], [167]
Oligodendrocytes	SA-β-gal	[184]

Characteristics needed to positively identify senescent cells that were described in our previous review [18] can be translated from mitotically competent cells to post-mitotic cells, but with caution. For example, absence of proliferation using proliferative markers (i.e., BrdU, Ki-67) should rarely be considered to characterize post-mitotic senescence as most post-mitotic cells show a lack of proliferation and turnover. Instead, presence of these markers may indicate aberrant cell cycle activity consistent with either a fate of apoptosis or, potentially, pre-senescence. In neurons, SA β-gal staining was positive in even young Purkinje neurons. SA β-gal activity may be observed due to metabolic demands rather than as a marker for senescence. Astrocytes present their own challenge as reactive astrogliosis show senescence-like phenotype with changes in cell shape and secretion of pro-inflammatory molecules. Endothelial cells were evaluated as senescent mainly with gene expression analysis of senescence-associated genes, and there is a lack of verification with DNA damage, cell cycle arrest machinery, and resistance to apoptosis. OLs also showed elevated SA-β-gal activity but no activation of p16, which also questions the validity of the SA-β-gal assay in the brain. Although there is a lack of studies evaluating senescence in OLs, those reviewed here overall did not utilize all of our predefined criteria, similar to studies on dividing cells mentioned in our previous review [18]. Post-mitotic cells discussed in this review highlight the need to critically look at multiple markers of senescence.

Although senescence was initially, and exclusively, studied in mitotic cells, the literature reviewed here (Table 1) provides evidence that post-mitotic cells also undergo senescence as a complex stress response. The emergence of this new field, senescence in the brain, requires clarity of defining features. While tempting to label cells as "senescent" in many of these studies, a thorough evaluation of cell biology placed in the context of the cellular environment must also be considered. As a result, the field of neuroscience has pressured senescence biologists to clarify definitions and labels. As neuroscientists, we are in the early stages of applying methodologies and principles form senescence biology to the brain. In return, neuroscientists have over 60 years of lessons and principles of exceptional resistance to cell death to share with senescence biologists. A closer interaction and sharing of concepts between neuroscience and senescent cell biologists will propel both fields. As these efforts progress, we will continue to clarify definitions and revisit interpretations from the foundational studies reviewed here.

Author Contributions: Conceptualization, C.M. and M.E.O.; writing—original draft preparation, Abstract and introduction: E.S., C.M. and M.E.O.; Identifying senescent brain cells: E.S., G.J.G. and M.E.O.; neurons: C.M. and M.E.O.; astrocytes: M.Y.A.; endothelial cells: E.S.; oligodendrocytes: S.K.; Summary: E.S and M.E.O.; Writing—review and editing, E.S., S.K., M.Y.A., G.J.G., C.M. and M.E.O.; visualization, E.S., G.J.G. and M.E.O.; supervision, M.E.O.; funding acquisition. M.E.O. All authors have read and agreed to the published version of the manuscript.

Funding: G.J.G. is supported by the Wake Forest T32AG033534; MEO is funded by Veterans Affairs, grant number IK2BX003804; NIH/NIA, grant number R01AG058293; New Vision Research, Charleston Conference for Alzheimer's disease; the Cure Alzheimer's Fund; the Alzheimer's disease Drug Discovery Foundation.

Institutional Review Board Statement: Not applicable.

Informed Consent Statement: Not applicable.

Data Availability Statement: Not applicable.

Acknowledgments: The authors would like to thank Anna Zaia Rodrigues, Emma Bennett, Olivia Posey, and Alyssa Hampton for insightful comments. We would also like to thank Timothy Orr for graphical illustration support.

Conflicts of Interest: The funders had no role in the design of the study; in the collection, analyses, or interpretation of data; in the writing of the manuscript, or in the decision to publish the results.

References

1. Levi-Montalcini, R.; Booker, B. Destruction of the Sympathetic Ganglia in Mammals by an Antiserum to a Nerve-Growth Protein. *Proc. Natl. Acad. Sci. USA* **1960**, *46*, 384–391. [CrossRef]
2. Oppenheim, R.; Milligan, C. Programmed Cell Death. In *Neuroscience and Biobehavioral Psychology*; Elsevier: New York, NY, USA, 2017. [CrossRef]
3. Fricker, M.; Tolkovsky, A.M.; Borutaite, V.; Coleman, M.; Brown, G.C. Neuronal Cell Death. *Physiol. Rev.* **2018**, *98*, 813–880. [CrossRef] [PubMed]
4. Greene, L.A.; Liu, D.X.; Troy, C.M.; Biswas, S.C. Cell cycle molecules define a pathway required for neuron death in development and disease. *Biochim. Biophys. Acta* **2007**, *1772*, 392–401. [CrossRef] [PubMed]
5. Hayflick, L.; Moorhead, P.S. The serial cultivation of human diploid cell strains. *Exp. Cell Res.* **1961**, *25*, 585–621. [CrossRef]
6. Hayflick, L. The Limited in Vitro Lifetime of Human Diploid Cell Strains. *Exp. Cell Res.* **1965**, *37*, 614–636. [CrossRef]

7. Liu, J.; Wang, L.; Wang, Z.; Liu, J.P. Roles of Telomere Biology in Cell Senescence, Replicative and Chronological Ageing. *Cells* **2019**, *8*, 54. [CrossRef]

8. Hernandez-Segura, A.; Nehme, J.; Demaria, M. Hallmarks of Cellular Senescence. *Trends Cell Biol.* **2018**, *28*, 436–453. [CrossRef] [PubMed]

9. Gorgoulis, V.; Adams, P.D.; Alimonti, A.; Bennett, D.C.; Bischof, O.; Bishop, C.; Campisi, J.; Collado, M.; Evangelou, K.; Ferbeyre, G.; et al. Cellular Senescence: Defining a Path Forward. *Cell* **2019**, *179*, 813–827. [CrossRef]

10. Matjusaitis, M.; Chin, G.; Sarnoski, E.A.; Stolzing, A. Biomarkers to identify and isolate senescent cells. *Ageing Res. Rev.* **2016**, *29*, 1–12. [CrossRef]

11. Dodig, S.; Cepelak, I.; Pavic, I. Hallmarks of senescence and aging. *Biochem. Med.* **2019**, *29*, 030501. [CrossRef]

12. Roy, A.L.; Sierra, F.; Howcroft, K.; Singer, D.S.; Sharpless, N.; Hodes, R.J.; Wilder, E.L.; Anderson, J.M. A Blueprint for Characterizing Senescence. *Cell* **2020**, *183*, 1143–1146. [CrossRef] [PubMed]

13. Basisty, N.; Kale, A.; Jeon, O.H.; Kuehnemann, C.; Payne, T.; Rao, C.; Holtz, A.; Shah, S.; Sharma, V.; Ferrucci, L.; et al. A proteomic atlas of senescence-associated secretomes for aging biomarker development. *PLoS Biol.* **2020**, *18*, e3000599. [CrossRef] [PubMed]

14. Hernandez-Segura, A.; de Jong, T.V.; Melov, S.; Guryev, V.; Campisi, J.; Demaria, M. Unmasking Transcriptional Heterogeneity in Senescent Cells. *Curr. Biol.* **2017**, *27*, 2652–2660 e2654. [CrossRef]

15. Casella, G.; Munk, R.; Kim, K.M.; Piao, Y.; De, S.; Abdelmohsen, K.; Gorospe, M. Transcriptome signature of cellular senescence. *Nucleic Acids Res.* **2019**, *47*, 7294–7305. [CrossRef] [PubMed]

16. Wiley, C.D.; Flynn, J.M.; Morrissey, C.; Lebofsky, R.; Shuga, J.; Dong, X.; Unger, M.A.; Vijg, J.; Melov, S.; Campisi, J. Analysis of individual cells identifies cell-to-cell variability following induction of cellular senescence. *Aging Cell* **2017**, *16*, 1043–1050. [CrossRef] [PubMed]

17. Herranz, N.; Gil, J. Mechanisms and functions of cellular senescence. *J. Clin. Investig.* **2018**, *128*, 1238–1246. [CrossRef]

18. Gillispie, G.J.; Sah, E.; Krishnamurthy, S.; Ahmidouch, M.Y.; Zhang, B.; Orr, M.E. Evidence of the Cellular Senescence Stress Response in Mitotically Active Brain Cells-Implications for Cancer and Neurodegeneration. *Life* **2021**, *11*, 153. [CrossRef]

19. Barrio-Alonso, E.; Fontana, B.; Valero, M.; Frade, J.M. Pathological Aspects of Neuronal Hyperploidization in Alzheimer's Disease Evidenced by Computer Simulation. *Front. Genet.* **2020**, *11*, 287. [CrossRef] [PubMed]

20. Ain, Q.; Schmeer, C.; Penndorf, D.; Fischer, M.; Bondeva, T.; Forster, M.; Haenold, R.; Witte, O.W.; Kretz, A. Cell cycle-dependent and -independent telomere shortening accompanies murine brain aging. *Aging* **2018**, *10*, 3397–3420. [CrossRef] [PubMed]

21. Jacome Burbano, M.S.; Gilson, E. Long-lived post-mitotic cell aging: Is a telomere clock at play? *Mech. Ageing Dev.* **2020**, *189*, 111256. [CrossRef]

22. Mosch, B.; Morawski, M.; Mittag, A.; Lenz, D.; Tarnok, A.; Arendt, T. Aneuploidy and DNA replication in the normal human brain and Alzheimer's disease. *J. Neurosci.* **2007**, *27*, 6859–6867. [CrossRef]

23. Fischer, H.G.; Morawski, M.; Bruckner, M.K.; Mittag, A.; Tarnok, A.; Arendt, T. Changes in neuronal DNA content variation in the human brain during aging. *Aging Cell* **2012**, *11*, 628–633. [CrossRef] [PubMed]

24. Varvel, N.H.; Bhaskar, K.; Patil, A.R.; Pimplikar, S.W.; Herrup, K.; Lamb, B.T. Abeta oligomers induce neuronal cell cycle events in Alzheimer's disease. *J. Neurosci.* **2008**, *28*, 10786–10793. [CrossRef]

25. Modi, P.K.; Jaiswal, S.; Sharma, P. Regulation of Neuronal Cell Cycle and Apoptosis by MicroRNA 34a. *Mol. Cell. Biol.* **2016**, *36*, 84–94. [CrossRef]

26. Chauhan, M.; Modi, P.K.; Sharma, P. Aberrant activation of neuronal cell cycle caused by dysregulation of ubiquitin ligase Itch results in neurodegeneration. *Cell Death Dis.* **2020**, *11*, 441. [CrossRef] [PubMed]

27. Hradek, A.C.; Lee, H.P.; Siedlak, S.L.; Torres, S.L.; Jung, W.; Han, A.H.; Lee, H.G. Distinct chronology of neuronal cell cycle re-entry and tau pathology in the 3xTg-AD mouse model and Alzheimer's disease patients. *J. Alzheimers Dis.* **2015**, *43*, 57–65. [CrossRef] [PubMed]

28. Andorfer, C.; Acker, C.M.; Kress, Y.; Hof, P.R.; Duff, K.; Davies, P. Cell-cycle reentry and cell death in transgenic mice expressing nonmutant human tau isoforms. *J. Neurosci.* **2005**, *25*, 5446–5454. [CrossRef]

29. Delobel, P.; Lavenir, I.; Ghetti, B.; Holzer, M.; Goedert, M. Cell-cycle markers in a transgenic mouse model of human tauopathy: Increased levels of cyclin-dependent kinase inhibitors p21Cip1 and p27Kip1. *Am. J. Pathol.* **2006**, *168*, 878–887. [CrossRef]

30. Barrio-Alonso, E.; Hernandez-Vivanco, A.; Walton, C.C.; Perea, G.; Frade, J.M. Cell cycle reentry triggers hyperploidization and synaptic dysfunction followed by delayed cell death in differentiated cortical neurons. *Sci. Rep.* **2018**, *8*, 14316. [CrossRef] [PubMed]

31. Marlier, Q.; D'Aes, T.; Verteneuil, S.; Vandenbosch, R.; Malgrange, B. Core cell cycle machinery is crucially involved in both life and death of post-mitotic neurons. *Cell. Mol. Life Sci.* **2020**, *77*, 4553–4571. [CrossRef]

32. Fujimaki, K.; Li, R.; Chen, H.; Della Croce, K.; Zhang, H.H.; Xing, J.; Bai, F.; Yao, G. Graded regulation of cellular quiescence depth between proliferation and senescence by a lysosomal dimmer switch. *Proc. Natl. Acad. Sci. USA* **2019**, *116*, 22624–22634. [CrossRef]

33. Zhu, Y.I.; Tchkonia, T.; Pirtskhalava, T.; Gower, A.C.; Ding, H.; Giorgadze, N.; Palmer, A.K.; Ikeno, Y.; Hubbard, G.B.; Lenburg, M. The Achilles' heel of senescent cells: From transcriptome to senolytic drugs. *Aging Cell* **2015**, *14*, 644–658. [CrossRef] [PubMed]

34. Soto-Gamez, A.; Quax, W.J.; Demaria, M. Regulation of Survival Networks in Senescent Cells: From Mechanisms to Interventions. *J. Mol. Biol.* **2019**, *431*, 2629–2643. [CrossRef]

35. Kole, A.J.; Annis, R.P.; Deshmukh, M. Mature neurons: Equipped for survival. *Cell Death Dis.* **2013**, *4*, e689. [CrossRef]

36. Oppenheim, R.W. Cell death during development of the nervous system. *Annu. Rev. Neurosci.* **1991**, *14*, 453–501. [CrossRef] [PubMed]
37. Wright, K.M.; Smith, M.I.; Farrag, L.; Deshmukh, M. Chromatin modification of Apaf-1 restricts the apoptotic pathway in mature neurons. *J. Cell Biol.* **2007**, *179*, 825–832. [CrossRef]
38. Donovan, M.; Cotter, T.G. Caspase-independent photoreceptor apoptosis in vivo and differential expression of apoptotic protease activating factor-1 and caspase-3 during retinal development. *Cell Death Differ.* **2002**, *9*, 1220–1231. [CrossRef] [PubMed]
39. Yakovlev, A.G.; Ota, K.; Wang, G.; Movsesyan, V.; Bao, W.L.; Yoshihara, K.; Faden, A.I. Differential expression of apoptotic protease-activating factor-1 and caspase-3 genes and susceptibility to apoptosis during brain development and after traumatic brain injury. *J. Neurosci.* **2001**, *21*, 7439–7446. [CrossRef]
40. Annis, R.P.; Swahari, V.; Nakamura, A.; Xie, A.X.; Hammond, S.M.; Deshmukh, M. Mature neurons dynamically restrict apoptosis via redundant premitochondrial brakes. *FEBS J.* **2016**, *283*, 4569–4582. [CrossRef] [PubMed]
41. Xia, P.; Liu, Y.; Chen, J.; Cheng, Z. Cell Cycle Proteins as Key Regulators of Postmitotic Cell Death. *Yale J. Biol. Med.* **2019**, *92*, 641–650. [PubMed]
42. Sapieha, P.; Mallette, F.A. Cellular Senescence in Postmitotic Cells: Beyond Growth Arrest. *Trends Cell Biol.* **2018**, *28*, 595–607. [CrossRef] [PubMed]
43. Chinta, S.J.; Woods, G.; Rane, A.; Demaria, M.; Campisi, J.; Andersen, J.K. Cellular senescence and the aging brain. *Exp. Gerontol.* **2015**, *68*, 3–7. [CrossRef]
44. Ruben, L.N. Recombinant DNA produced human IL-2, injected in vivo, will substitute for carrier priming of helper function in the South African clawed toad, Xenopus laevis. *Immunol. Lett.* **1986**, *13*, 227–230. [CrossRef]
45. Walker, L.; McAleese, K.E.; Thomas, A.J.; Johnson, M.; Martin-Ruiz, C.; Parker, C.; Colloby, S.J.; Jellinger, K.; Attems, J. Neuropathologically mixed Alzheimer's and Lewy body disease: Burden of pathological protein aggregates differs between clinical phenotypes. *Acta Neuropathol.* **2015**, *129*, 729–748. [CrossRef]
46. Outeiro, T.F.; Koss, D.J.; Erskine, D.; Walker, L.; Kurzawa-Akanbi, M.; Burn, D.; Donaghy, P.; Morris, C.; Taylor, J.P.; Thomas, A.; et al. Dementia with Lewy bodies: An update and outlook. *Mol. Neurodegener.* **2019**, *14*, 5. [CrossRef]
47. Clinton, L.K.; Blurton-Jones, M.; Myczek, K.; Trojanowski, J.Q.; LaFerla, F.M. Synergistic Interactions between Abeta, tau, and alpha-synuclein: Acceleration of neuropathology and cognitive decline. *J. Neurosci.* **2010**, *30*, 7281–7289. [CrossRef] [PubMed]
48. Dimri, G.P.; Lee, X.; Basile, G.; Acosta, M.; Scott, G.; Roskelley, C.; Medrano, E.E.; Linskens, M.; Rubelj, I.; Pereira-Smith, O.; et al. A biomarker that identifies senescent human cells in culture and in aging skin in vivo. *Proc. Natl. Acad. Sci. USA* **1995**, *92*, 9363–9367. [CrossRef]
49. Musi, N.; Valentine, J.M.; Sickora, K.R.; Baeuerle, E.; Thompson, C.S.; Shen, Q.; Orr, M.E. Tau protein aggregation is associated with cellular senescence in the brain. *Aging Cell* **2018**, *17*, e12840. [CrossRef]
50. Lee, B.Y.; Han, J.A.; Im, J.S.; Morrone, A.; Johung, K.; Goodwin, E.C.; Kleijer, W.J.; DiMaio, D.; Hwang, E.S. Senescence-associated beta-galactosidase is lysosomal beta-galactosidase. *Aging Cell* **2006**, *5*, 187–195. [CrossRef] [PubMed]
51. Piechota, M.; Sunderland, P.; Wysocka, A.; Nalberczak, M.; Sliwinska, M.A.; Radwanska, K.; Sikora, E. Is senescence-associated beta-galactosidase a marker of neuronal senescence? *Oncotarget* **2016**, *7*, 81099–81109. [CrossRef] [PubMed]
52. Herculano-Houzel, S. The remarkable, yet not extraordinary, human brain as a scaled-up primate brain and its associated cost. *Proc. Natl. Acad. Sci. USA* **2012**, *109* (Suppl. 1), 10661–10668. [CrossRef]
53. von Bartheld, C.S. Myths and truths about the cellular composition of the human brain: A review of influential concepts. *J. Chem. Neuroanat.* **2018**, *93*, 2–15. [CrossRef] [PubMed]
54. Jurk, D.; Wang, C.; Miwa, S.; Maddick, M.; Korolchuk, V.; Tsolou, A.; Gonos, E.S.; Thrasivoulou, C.; Saffrey, M.J.; Cameron, K.; et al. Postmitotic neurons develop a p21-dependent senescence-like phenotype driven by a DNA damage response. *Aging Cell* **2012**, *11*, 996–1004. [CrossRef]
55. Chow, H.M.; Shi, M.; Cheng, A.; Gao, Y.; Chen, G.; Song, X.; So, R.W.L.; Zhang, J.; Herrup, K. Age-related hyperinsulinemia leads to insulin resistance in neurons and cell-cycle-induced senescence. *Nat. Neurosci.* **2019**, *22*, 1806–1819. [CrossRef] [PubMed]
56. Riessland, M.; Kolisnyk, B.; Kim, T.W.; Cheng, J.; Ni, J.; Pearson, J.A.; Park, E.J.; Dam, K.; Acehan, D.; Ramos-Espiritu, L.S.; et al. Loss of SATB1 Induces p21-Dependent Cellular Senescence in Post-mitotic Dopaminergic Neurons. *Cell Stem Cell* **2019**, *25*, 514–530.e8. [CrossRef]
57. Vazquez-Villasenor, I.; Garwood, C.J.; Heath, P.R.; Simpson, J.E.; Ince, P.G.; Wharton, S.B. Expression of p16 and p21 in the frontal association cortex of ALS/MND brains suggests neuronal cell cycle dysregulation and astrocyte senescence in early stages of the disease. *Neuropathol. Appl. Neurobiol.* **2020**, *46*, 171–185. [CrossRef] [PubMed]
58. Trias, E.; Beilby, P.R.; Kovacs, M.; Ibarburu, S.; Varela, V.; Barreto-Nunez, R.; Bradford, S.C.; Beckman, J.S.; Barbeito, L. Emergence of Microglia Bearing Senescence Markers During Paralysis Progression in a Rat Model of Inherited ALS. *Front. Aging Neurosci.* **2019**, *11*, 42. [CrossRef] [PubMed]
59. Lake, B.B.; Ai, R.; Kaeser, G.E.; Salathia, N.S.; Yung, Y.C.; Liu, R.; Wildberg, A.; Gao, D.; Fung, H.L.; Chen, S.; et al. Neuronal subtypes and diversity revealed by single-nucleus RNA sequencing of the human brain. *Science* **2016**, *352*, 1586–1590. [CrossRef] [PubMed]
60. Andreadis, A.; Brown, W.M.; Kosik, K.S. Structure and novel exons of the human tau gene. *Biochemistry* **1992**, *31*, 10626–10633. [CrossRef]

61. Goedert, M.; Spillantini, M.G.; Jakes, R.; Rutherford, D.; Crowther, R.A. Multiple isoforms of human microtubule-associated protein tau: Sequences and localization in neurofibrillary tangles of Alzheimer's disease. *Neuron* **1989**, *3*, 519–526. [CrossRef]

62. Orr, M.E.; Sullivan, A.C.; Frost, B. A Brief Overview of Tauopathy: Causes, Consequences, and Therapeutic Strategies. *Trends Pharmacol. Sci.* **2017**, *38*, 637–648. [CrossRef]

63. de Calignon, A.; Spires-Jones, T.L.; Pitstick, R.; Carlson, G.A.; Hyman, B.T. Tangle-bearing neurons survive despite disruption of membrane integrity in a mouse model of tauopathy. *J. Neuropathol. Exp. Neurol.* **2009**, *68*, 757–761. [CrossRef]

64. Marquez, A.; Guernsey, L.S.; Frizzi, K.E.; Cundiff, M.; Constantino, I.; Muttalib, N.; Arenas, F.; Zhou, X.; Lim, S.H.; Ferdousi, M.; et al. Tau associated peripheral and central neurodegeneration: Identification of an early imaging marker for tauopathy. *Neurobiol. Dis.* **2021**, *151*, 105273. [CrossRef]

65. Acklin, S.; Zhang, M.; Du, W.; Zhao, X.; Plotkin, M.; Chang, J.; Campisi, J.; Zhou, D.; Xia, F. Depletion of senescent-like neuronal cells alleviates cisplatin-induced peripheral neuropathy in mice. *Sci. Rep.* **2020**, *10*, 14170. [CrossRef] [PubMed]

66. Calls, A.; Torres-Espin, A.; Navarro, X.; Yuste, V.J.; Udina, E.; Bruna, J. Cisplatin-induced peripheral neuropathy is associated with neuronal senescence-like response. *Neuro Oncol.* **2021**, *23*, 88–99. [CrossRef] [PubMed]

67. Ezquerra, M.; Pastor, P.; Gaig, C.; Vidal-Taboada, J.M.; Cruchaga, C.; Munoz, E.; Marti, M.J.; Valldeoriola, F.; Aguilar, M.; Calopa, M.; et al. Different MAPT haplotypes are associated with Parkinson's disease and progressive supranuclear palsy. *Neurobiol. Aging* **2011**, *32*, 547.e11–547.e16. [CrossRef]

68. Davis, A.A.; Andruska, K.M.; Benitez, B.A.; Racette, B.A.; Perlmutter, J.S.; Cruchaga, C. Variants in GBA, SNCA, and MAPT influence Parkinson disease risk, age at onset, and progression. *Neurobiol. Aging* **2016**, *37*, 209.e201–209.e207. [CrossRef]

69. Zhang, X.; Gao, F.; Wang, D.; Li, C.; Fu, Y.; He, W.; Zhang, J. Tau Pathology in Parkinson's Disease. *Front. Neurol.* **2018**, *9*, 809. [CrossRef] [PubMed]

70. Lang, A.E.; Lozano, A.M. Parkinson's disease. First of two parts. *N. Engl. J. Med.* **1998**, *339*, 1044–1053. [CrossRef]

71. Dauer, W.; Przedborski, S. Parkinson's disease: Mechanisms and models. *Neuron* **2003**, *39*, 889–909. [CrossRef]

72. Santa-Maria, I.; Hernandez, F.; Smith, M.A.; Perry, G.; Avila, J.; Moreno, F.J. Neurotoxic dopamine quinone facilitates the assembly of tau into fibrillar polymers. *Mol. Cell. Biochem.* **2005**, *278*, 203–212. [CrossRef] [PubMed]

73. Rodriguez-Losada, N.; de la Rosa, J.; Larriva, M.; Wendelbo, R.; Aguirre, J.A.; Castresana, J.S.; Ballaz, S.J. Overexpression of alpha-synuclein promotes both cell proliferation and cell toxicity in human SH-SY5Y neuroblastoma cells. *J. Adv. Res.* **2020**, *23*, 37–45. [CrossRef]

74. Yasui, D.; Miyano, M.; Cai, S.; Varga-Weisz, P.; Kohwi-Shigematsu, T. SATB1 targets chromatin remodelling to regulate genes over long distances. *Nature* **2002**, *419*, 641–645. [CrossRef] [PubMed]

75. Chang, D.; Nalls, M.A.; Hallgrimsdottir, I.B.; Hunkapiller, J.; van der Brug, M.; Cai, F.; International Parkinson's Disease Genomics, C.; andMe Research, T.; Kerchner, G.A.; Ayalon, G.; et al. A meta-analysis of genome-wide association studies identifies 17 new Parkinson's disease risk loci. *Nat. Genet.* **2017**, *49*, 1511–1516. [CrossRef] [PubMed]

76. Brichta, L.; Shin, W.; Jackson-Lewis, V.; Blesa, J.; Yap, E.L.; Walker, Z.; Zhang, J.; Roussarie, J.P.; Alvarez, M.J.; Califano, A.; et al. Identification of neurodegenerative factors using translatome-regulatory network analysis. *Nat. Neurosci.* **2015**, *18*, 1325–1333. [CrossRef] [PubMed]

77. Ho, D.H.; Seol, W.; Son, I. Upregulation of the p53–p21 pathway by G2019S LRRK2 contributes to the cellular senescence and accumulation of alpha-synuclein. *Cell Cycle* **2019**, *18*, 467–475. [CrossRef]

78. Levy, O.A.; Malagelada, C.; Greene, L.A. Cell death pathways in Parkinson's disease: Proximal triggers, distal effectors, and final steps. *Apoptosis* **2009**, *14*, 478–500. [CrossRef]

79. Golde, T.E.; Miller, V.M. Proteinopathy-induced neuronal senescence: A hypothesis for brain failure in Alzheimer's and other neurodegenerative diseases. *Alzheimers Res. Ther.* **2009**, *1*, 5. [CrossRef]

80. Gralle, M. The neuronal insulin receptor in its environment. *J. Neurochem.* **2017**, *140*, 359–367. [CrossRef]

81. Liang, H.; Nie, J.; Van Skike, C.E.; Valentine, J.M.; Orr, M.E. Mammalian Target of Rapamycin at the Crossroad Between Alzheimer's Disease and Diabetes. *Adv. Exp. Med. Biol.* **2019**, *1128*, 185–225. [CrossRef]

82. Kullmann, S.; Heni, M.; Hallschmid, M.; Fritsche, A.; Preissl, H.; Haring, H.U. Brain Insulin Resistance at the Crossroads of Metabolic and Cognitive Disorders in Humans. *Physiol. Rev.* **2016**, *96*, 1169–1209. [CrossRef]

83. Orr, M.E.; Salinas, A.; Buffenstein, R.; Oddo, S. Mammalian target of rapamycin hyperactivity mediates the detrimental effects of a high sucrose diet on Alzheimer's disease pathology. *Neurobiol. Aging* **2014**, *35*, 1233–1242. [CrossRef]

84. Ogrodnik, M.; Zhu, Y.; Langhi, L.G.P.; Tchkonia, T.; Kruger, P.; Fielder, E.; Victorelli, S.; Ruswhandi, R.A.; Giorgadze, N.; Pirtskhalava, T.; et al. Obesity-Induced Cellular Senescence Drives Anxiety and Impairs Neurogenesis. *Cell Metab.* **2019**, *29*, 1061–1077 e1068. [CrossRef]

85. Yang, Y.; Herrup, K. Loss of neuronal cell cycle control in ataxia-telangiectasia: A unified disease mechanism. *J. Neurosci.* **2005**, *25*, 2522–2529. [CrossRef]

86. Arendt, T.; Bruckner, M.K.; Mosch, B.; Losche, A. Selective cell death of hyperploid neurons in Alzheimer's disease. *Am. J. Pathol.* **2010**, *177*, 15–20. [CrossRef]

87. Toots, A.; Rosendahl, E.; Lundin-Olsson, L.; Nordstrom, P.; Gustafson, Y.; Littbrand, H. Usual gait speed independently predicts mortality in very old people: A population-based study. *J. Am. Med. Dir. Assoc.* **2013**, *14*, 529.e1–529.e6. [CrossRef]

88. Blain, H.; Carriere, I.; Sourial, N.; Berard, C.; Favier, F.; Colvez, A.; Bergman, H. Balance and walking speed predict subsequent 8-year mortality independently of current and intermediate events in well-functioning women aged 75 years and older. *J. Nutr. Health Aging* **2010**, *14*, 595–600. [CrossRef]

89. Moreno-Blas, D.; Gorostieta-Salas, E.; Pommer-Alba, A.; Mucino-Hernandez, G.; Geronimo-Olvera, C.; Maciel-Baron, L.A.; Konigsberg, M.; Massieu, L.; Castro-Obregon, S. Cortical neurons develop a senescence-like phenotype promoted by dysfunctional autophagy. *Aging* **2019**, *11*, 6175–6198. [CrossRef]

90. Goyal, V.K. Lipofuscin pigment accumulation in human brain during aging. *Exp. Gerontol.* **1982**, *17*, 481–487. [CrossRef]

91. Sitte, N.; Merker, K.; Grune, T.; von Zglinicki, T. Lipofuscin accumulation in proliferating fibroblasts in vitro: An indicator of oxidative stress. *Exp. Gerontol.* **2001**, *36*, 475–486. [CrossRef]

92. Georgakopoulou, E.A.; Tsimaratou, K.; Evangelou, K.; Fernandez Marcos, P.J.; Zoumpourlis, V.; Trougakos, I.P.; Kletsas, D.; Bartek, J.; Serrano, M.; Gorgoulis, V.G. Specific lipofuscin staining as a novel biomarker to detect replicative and stress-induced senescence. A method applicable in cryo-preserved and archival tissues. *Aging* **2013**, *5*, 37–50. [CrossRef]

93. Evangelou, K.; Lougiakis, N.; Rizou, S.V.; Kotsinas, A.; Kletsas, D.; Munoz-Espin, D.; Kastrinakis, N.G.; Pouli, N.; Marakos, P.; Townsend, P.; et al. Robust, universal biomarker assay to detect senescent cells in biological specimens. *Aging Cell* **2017**, *16*, 192–197. [CrossRef]

94. Calvo, M.; Sanz-Blasco, S.; Caballero, E.; Villalobos, C.; Nunez, L. Susceptibility to excitotoxicity in aged hippocampal cultures and neuroprotection by non-steroidal anti-inflammatory drugs: Role of mitochondrial calcium. *J. Neurochem.* **2015**, *132*, 403–417. [CrossRef] [PubMed]

95. von Bartheld, C.S.; Bahney, J.; Herculano-Houzel, S. The search for true numbers of neurons and glial cells in the human brain: A review of 150 years of cell counting. *J. Comp. Neurol.* **2016**, *524*, 3865–3895. [CrossRef]

96. Kim, Y.; Park, J.; Choi, Y.K. The Role of Astrocytes in the Central Nervous System Focused on BK Channel and Heme Oxygenase Metabolites: A Review. *Antioxidants* **2019**, *8*, 121. [CrossRef]

97. Herculano-Houzel, S. The glia/neuron ratio: How it varies uniformly across brain structures and species and what that means for brain physiology and evolution. *Glia* **2014**, *62*, 1377–1391. [CrossRef] [PubMed]

98. Stiles, J.; Jernigan, T.L. The basics of brain development. *Neuropsychol. Rev.* **2010**, *20*, 327–348. [CrossRef]

99. Cayre, M.; Canoll, P.; Goldman, J.E. Cell migration in the normal and pathological postnatal mammalian brain. *Prog. Neurobiol.* **2009**, *88*, 41–63. [CrossRef]

100. Sofroniew, M.V.; Vinters, H.V. Astrocytes: Biology and pathology. *Acta Neuropathol.* **2010**, *119*, 7–35. [CrossRef]

101. Contet, C.; Goulding, S.P.; Kuljis, D.A.; Barth, A.L. BK Channels in the Central Nervous System. *Int. Rev. Neurobiol.* **2016**, *128*, 281–342. [CrossRef] [PubMed]

102. Tabata, H. Diverse subtypes of astrocytes and their development during corticogenesis. *Front. Neurosci.* **2015**, *9*, 114. [CrossRef]

103. de Majo, M.; Koontz, M.; Rowitch, D.; Ullian, E.M. An update on human astrocytes and their role in development and disease. *Glia* **2020**, *68*, 685–704. [CrossRef]

104. Jung, Y.J.; Chung, W.S. Phagocytic Roles of Glial Cells in Healthy and Diseased Brains. *Biomol. Ther.* **2018**, *26*, 350–357. [CrossRef] [PubMed]

105. Gomez-Arboledas, A.; Davila, J.C.; Sanchez-Mejias, E.; Navarro, V.; Nunez-Diaz, C.; Sanchez-Varo, R.; Sanchez-Mico, M.V.; Trujillo-Estrada, L.; Fernandez-Valenzuela, J.J.; Vizuete, M.; et al. Phagocytic clearance of presynaptic dystrophies by reactive astrocytes in Alzheimer's disease. *Glia* **2018**, *66*, 637–653. [CrossRef]

106. Limbad, C.; Oron, T.R.; Alimirah, F.; Davalos, A.R.; Tracy, T.E.; Gan, L.; Desprez, P.Y.; Campisi, J. Astrocyte senescence promotes glutamate toxicity in cortical neurons. *PLoS ONE* **2020**, *15*, e0227887. [CrossRef] [PubMed]

107. Chung, W.S.; Allen, N.J.; Eroglu, C. Astrocytes Control Synapse Formation, Function, and Elimination. *Cold Spring Harb. Perspect. Biol.* **2015**, *7*, a020370. [CrossRef]

108. Barres, B.A. The mystery and magic of glia: A perspective on their roles in health and disease. *Neuron* **2008**, *60*, 430–440. [CrossRef]

109. Phatnani, H.; Maniatis, T. Astrocytes in neurodegenerative disease. *Cold Spring Harb. Perspect. Biol.* **2015**, *7*. [CrossRef] [PubMed]

110. Munger, E.L.; Edler, M.K.; Hopkins, W.D.; Ely, J.J.; Erwin, J.M.; Perl, D.P.; Mufson, E.J.; Hof, P.R.; Sherwood, C.C.; Raghanti, M.A. Astrocytic changes with aging and Alzheimer's disease-type pathology in chimpanzees. *J. Comp. Neurol.* **2019**, *527*, 1179–1195. [CrossRef]

111. Boisvert, M.M.; Erikson, G.A.; Shokhirev, M.N.; Allen, N.J. The Aging Astrocyte Transcriptome from Multiple Regions of the Mouse Brain. *Cell Rep.* **2018**, *22*, 269–285. [CrossRef]

112. Zamanian, J.L.; Xu, L.; Foo, L.C.; Nouri, N.; Zhou, L.; Giffard, R.G.; Barres, B.A. Genomic analysis of reactive astrogliosis. *J. Neurosci.* **2012**, *32*, 6391–6410. [CrossRef] [PubMed]

113. Turnquist, C.; Beck, J.A.; Horikawa, I.; Obiorah, I.E.; Von Muhlinen, N.; Vojtesek, B.; Lane, D.P.; Grunseich, C.; Chahine, J.J.; Ames, H.M.; et al. Radiation-induced astrocyte senescence is rescued by Delta133p53. *Neuro Oncol.* **2019**, *21*, 474–485. [CrossRef] [PubMed]

114. Liddelow, S.A.; Guttenplan, K.A.; Clarke, L.E.; Bennett, F.C.; Bohlen, C.J.; Schirmer, L.; Bennett, M.L.; Munch, A.E.; Chung, W.S.; Peterson, T.C.; et al. Neurotoxic reactive astrocytes are induced by activated microglia. *Nature* **2017**, *541*, 481–487. [CrossRef]

115. Li, K.; Li, J.; Zheng, J.; Qin, S. Reactive Astrocytes in Neurodegenerative Diseases. *Aging Dis.* **2019**, *10*, 664–675. [CrossRef] [PubMed]

116. Cohen, J.; Torres, C. Astrocyte senescence: Evidence and significance. *Aging Cell* **2019**, *18*, e12937. [CrossRef] [PubMed]

117. Han, X.; Zhang, T.; Liu, H.; Mi, Y.; Gou, X. Astrocyte Senescence and Alzheimer's Disease: A Review. *Front. Aging Neurosci.* **2020**, *12*, 148. [CrossRef]

118. Evans, R.J.; Wyllie, F.S.; Wynford-Thomas, D.; Kipling, D.; Jones, C.J. A P53-dependent, telomere-independent proliferative life span barrier in human astrocytes consistent with the molecular genetics of glioma development. *Cancer Res.* **2003**, *63*, 4854–4861.

119. Pertusa, M.; García-Matas, S.; Rodríguez-Farré, E.; Sanfeliu, C.; Cristòfol, R. Astrocytes aged in vitro show a decreased neuroprotective capacity. *J. Neurochem.* **2007**, *101*, 794–805. [CrossRef] [PubMed]

120. Spilsbury, A.; Miwa, S.; Attems, J.; Saretzki, G. The role of telomerase protein TERT in Alzheimer's disease and in tau-related pathology in vitro. *J. Neurosci.* **2015**, *35*, 1659–1674. [CrossRef]

121. Bitto, A.; Sell, C.; Crowe, E.; Lorenzini, A.; Malaguti, M.; Hrelia, S.; Torres, C. Stress-induced senescence in human and rodent astrocytes. *Exp. Cell Res.* **2010**, *316*, 2961–2968. [CrossRef]

122. Görg, B.; Karababa, A.; Shafigullina, A.; Bidmon, H.J.; Häussinger, D. Ammonia-induced senescence in cultured rat astrocytes and in human cerebral cortex in hepatic encephalopathy. *Glia* **2015**, *63*, 37–50. [CrossRef] [PubMed]

123. Eriksson, D.; Stigbrand, T. Radiation-induced cell death mechanisms. *Tumour. Biol.* **2010**, *31*, 363–372. [CrossRef] [PubMed]

124. Turnquist, C.; Horikawa, I.; Foran, E.; Major, E.O.; Vojtesek, B.; Lane, D.P.; Lu, X.; Harris, B.T.; Harris, C.C. p53 isoforms regulate astrocyte-mediated neuroprotection and neurodegeneration. *Cell Death Differ.* **2016**, *23*, 1515–1528. [CrossRef] [PubMed]

125. Simmnacher, K.; Krach, F.; Schneider, Y.; Alecu, J.E.; Mautner, L.; Klein, P.; Roybon, L.; Prots, I.; Xiang, W.; Winner, B. Unique signatures of stress-induced senescent human astrocytes. *Exp. Neurol.* **2020**, *334*, 113466. [CrossRef]

126. Lye, J.J.; Latorre, E.; Lee, B.P.; Bandinelli, S.; Holley, J.E.; Gutowski, N.J.; Ferrucci, L.; Harries, L.W. Astrocyte senescence may drive alterations in GFAPα, CDKN2A p14(ARF), and TAU3 transcript expression and contribute to cognitive decline. *Geroscience* **2019**, *41*, 561–573. [CrossRef] [PubMed]

127. Bhat, R.; Crowe, E.P.; Bitto, A.; Moh, M.; Katsetos, C.D.; Garcia, F.U.; Johnson, F.B.; Trojanowski, J.Q.; Sell, C.; Torres, C. Astrocyte senescence as a component of Alzheimer's disease. *PLoS ONE* **2012**, *7*, e45069. [CrossRef]

128. Salminen, A.; Ojala, J.; Kaarniranta, K.; Haapasalo, A.; Hiltunen, M.; Soininen, H. Astrocytes in the aging brain express characteristics of senescence-associated secretory phenotype. *Eur. J. Neurosci.* **2011**, *34*, 3–11. [CrossRef] [PubMed]

129. Chinta, S.J.; Woods, G.; Demaria, M.; Rane, A.; Zou, Y.; McQuade, A.; Rajagopalan, S.; Limbad, C.; Madden, D.T.; Campisi, J.; et al. Cellular Senescence Is Induced by the Environmental Neurotoxin Paraquat and Contributes to Neuropathology Linked to Parkinson's Disease. *Cell Rep.* **2018**, *22*, 930–940. [CrossRef]

130. Thal, D.R.; Ronisz, A.; Tousseyn, T.; Rijal Upadhaya, A.; Balakrishnan, K.; Vandenberghe, R.; Vandenbulcke, M.; von Arnim, C.A.F.; Otto, M.; Beach, T.G.; et al. Different aspects of Alzheimer's disease-related amyloid beta-peptide pathology and their relationship to amyloid positron emission tomography imaging and dementia. *Acta Neuropathol. Commun.* **2019**, *7*, 178. [CrossRef]

131. Fakhoury, M. Microglia and Astrocytes in Alzheimer's Disease: Implications for Therapy. *Curr. Neuropharmacol.* **2018**, *16*, 508–518. [CrossRef]

132. Csipo, T.; Lipecz, A.; Ashpole, N.M.; Balasubramanian, P.; Tarantini, S. Astrocyte senescence contributes to cognitive decline. *Geroscience* **2020**, *42*, 51–55. [CrossRef]

133. Acosta, J.C.; Banito, A.; Wuestefeld, T.; Georgilis, A.; Janich, P.; Morton, J.P.; Athineos, D.; Kang, T.W.; Lasitschka, F.; Andrulis, M.; et al. A complex secretory program orchestrated by the inflammasome controls paracrine senescence. *Nat. Cell Biol.* **2013**, *15*, 978–990. [CrossRef] [PubMed]

134. Rajendran, P.; Rengarajan, T.; Thangavel, J.; Nishigaki, Y.; Sakthisekaran, D.; Sethi, G.; Nishigaki, I. The vascular endothelium and human diseases. *Int. J. Biol. Sci.* **2013**, *9*, 1057–1069. [CrossRef]

135. Greenwood, J.; Heasman, S.J.; Alvarez, J.I.; Prat, A.; Lyck, R.; Engelhardt, B. Review: Leucocyte-endothelial cell crosstalk at the blood-brain barrier: A prerequisite for successful immune cell entry to the brain. *Neuropathol. Appl. Neurobiol.* **2011**, *37*, 24–39. [CrossRef]

136. Minamino, T.; Miyauchi, H.; Yoshida, T.; Ishida, Y.; Yoshida, H.; Komuro, I. Endothelial cell senescence in human atherosclerosis: Role of telomere in endothelial dysfunction. *Circulation* **2002**, *105*, 1541–1544. [CrossRef]

137. Barinda, A.J.; Ikeda, K.; Nugroho, D.B.; Wardhana, D.A.; Sasaki, N.; Honda, S.; Urata, R.; Matoba, S.; Hirata, K.I.; Emoto, N. Endothelial progeria induces adipose tissue senescence and impairs insulin sensitivity through senescence associated secretory phenotype. *Nat. Commun.* **2020**, *11*, 481. [CrossRef] [PubMed]

138. Hayashi, T.; Matsui-Hirai, H.; Miyazaki-Akita, A.; Fukatsu, A.; Funami, J.; Ding, Q.F.; Kamalanathan, S.; Hattori, Y.; Ignarro, L.J.; Iguchi, A. Endothelial cellular senescence is inhibited by nitric oxide: Implications in atherosclerosis associated with menopause and diabetes. *Proc. Natl. Acad. Sci. USA* **2006**, *103*, 17018–17023. [CrossRef]

139. Ekstrand, J.; Hellsten, J.; Tingstrom, A. Environmental enrichment, exercise and corticosterone affect endothelial cell proliferation in adult rat hippocampus and prefrontal cortex. *Neurosci. Lett.* **2008**, *442*, 203–207. [CrossRef]

140. Hobson, B.; Denekamp, J. Endothelial proliferation in tumours and normal tissues: Continuous labelling studies. *Br. J. Cancer* **1984**, *49*, 405–413. [CrossRef]

141. Wong, A.D.; Ye, M.; Levy, A.; Rothstein, J.; Bergles, D.; Searson, P.C. The blood-brain barrier: An engineering perspective. *Front Neuroeng.* **2013**, *6*, 7. [CrossRef] [PubMed]

142. Stamatovic, S.M.; Keep, R.F.; Andjelkovic, A.V. Brain endothelial cell-cell junctions: How to "open" the blood brain barrier. *Curr. Neuropharmacol.* **2008**, *6*, 179–192. [CrossRef] [PubMed]

143. Katt, M.E.; Xu, Z.S.; Gerecht, S.; Searson, P.C. Human Brain Microvascular Endothelial Cells Derived from the BC1 iPS Cell Line Exhibit a Blood-Brain Barrier Phenotype. *PLoS ONE* **2016**, *11*, e0152105. [CrossRef] [PubMed]
144. Rosas-Hernandez, H.; Cuevas, E.; Lantz, S.M.; Paule, M.G.; Ali, S.F. Isolation and Culture of Brain Microvascular Endothelial Cells for In Vitro Blood-Brain Barrier Studies. *Methods Mol. Biol.* **2018**, *1727*, 315–331. [CrossRef] [PubMed]
145. Song, H.W.; Foreman, K.L.; Gastfriend, B.D.; Kuo, J.S.; Palecek, S.P.; Shusta, E.V. Transcriptomic comparison of human and mouse brain microvessels. *Sci. Rep.* **2020**, *10*, 12358. [CrossRef]
146. Daneman, R.; Prat, A. The blood-brain barrier. *Cold Spring Harb. Perspect. Biol.* **2015**, *7*, a020412. [CrossRef]
147. Guerra, G.; Lucariello, A.; Perna, A.; Botta, L.; De Luca, A.; Moccia, F. The Role of Endothelial Ca(2−) Signaling in Neurovascular Coupling: A View from the Lumen. *Int. J. Mol. Sci.* **2018**, *19*, 938. [CrossRef]
148. Pardridge, W.M. The Isolated Brain Microvessel: A Versatile Experimental Model of the Blood-Brain Barrier. *Front. Physiol.* **2020**, *11*, 398. [CrossRef]
149. Bauer, H.C.; Krizbai, I.A.; Bauer, H.; Traweger, A. "You Shall Not Pass"—Tight junctions of the blood brain barrier. *Front. Neurosci.* **2014**, *8*, 392. [CrossRef]
150. Soares, J.P.; Cortinhas, A.; Bento, T.; Leitao, J.C.; Collins, A.R.; Gaivao, I.; Mota, M.P. Aging and DNA damage in humans: A meta-analysis study. *Aging* **2014**, *6*, 432–439. [CrossRef] [PubMed]
151. Tarantini, S.; Tran, C.H.T.; Gordon, G.R.; Ungvari, Z.; Csiszar, A. Impaired neurovascular coupling in aging and Alzheimer's disease: Contribution of astrocyte dysfunction and endothelial impairment to cognitive decline. *Exp. Gerontol.* **2017**, *94*, 52–58. [CrossRef]
152. Nation, D.A.; Sweeney, M.D.; Montagne, A.; Sagare, A.P.; D'Orazio, L.M.; Pachicano, M.; Sepehrband, F.; Nelson, A.R.; Buennagel, D.P.; Harrington, M.G.; et al. Blood-brain barrier breakdown is an early biomarker of human cognitive dysfunction. *Nat. Med.* **2019**, *25*, 270–276. [CrossRef]
153. Farrall, A.J.; Wardlaw, J.M. Blood-brain barrier: Ageing and microvascular disease-systematic review and meta-analysis. *Neurobiol. Aging* **2009**, *30*, 337–352. [CrossRef] [PubMed]
154. Machida, T.; Dohgu, S.; Takata, F.; Matsumoto, J.; Kimura, I.; Koga, M.; Nakamoto, K.; Yamauchi, A.; Kataoka, Y. Role of thrombin-PAR1-PKCtheta/delta axis in brain pericytes in thrombin-induced MMP-9 production and blood-brain barrier dysfunction in vitro. *Neuroscience* **2017**, *350*, 146–157. [CrossRef]
155. Brouns, R.; Wauters, A.; De Surgeloose, D.; Marien, P.; De Deyn, P.P. Biochemical markers for blood-brain barrier dysfunction in acute ischemic stroke correlate with evolution and outcome. *Eur. Neurol.* **2011**, *65*, 23–31. [CrossRef]
156. Zhang, S.; An, Q.; Wang, T.; Gao, S.; Zhou, G. Autophagy- and MMP-2/9-mediated Reduction and Redistribution of ZO-1 Contribute to Hyperglycemia-increased Blood-Brain Barrier Permeability During Early Reperfusion in Stroke. *Neuroscience* **2018**, *377*, 126–137. [CrossRef] [PubMed]
157. Guilfoyle, M.R.; Carpenter, K.L.; Helmy, A.; Pickard, J.D.; Menon, D.K.; Hutchinson, P.J. Matrix Metalloproteinase Expression in Contusional Traumatic Brain Injury: A Paired Microdialysis Study. *J. Neurotrauma* **2015**, *32*, 1553–1559. [CrossRef] [PubMed]
158. Fainardi, E.; Castellazzi, M.; Bellini, T.; Manfrinato, M.C.; Baldi, E.; Casetta, I.; Paolino, E.; Granieri, E.; Dallocchio, F. Cerebrospinal fluid and serum levels and intrathecal production of active matrix metalloproteinase-9 (MMP-9) as markers of disease activity in patients with multiple sclerosis. *Mult. Scler.* **2006**, *12*, 294–301. [CrossRef]
159. Montagne, A.; Zhao, Z.; Zlokovic, B.V. Alzheimer's disease: A matter of blood-brain barrier dysfunction? *J. Exp. Med.* **2017** *214*, 3151–3169. [CrossRef]
160. Rao, J.S.; Bhoopathi, P.; Chetty, C.; Gujrati, M.; Lakka, S.S. MMP-9 short interfering RNA induced senescence resulting in inhibition of medulloblastoma growth via p16(INK4a) and mitogen-activated protein kinase pathway. *Cancer Res.* **2007**, *67*, 4956–4964. [CrossRef] [PubMed]
161. Ungvari, Z.; Podlutsky, A.; Sosnowska, D.; Tucsek, Z.; Toth, P.; Deak, F.; Gautam, T.; Csiszar, A.; Sonntag, W.E. Ionizing radiation promotes the acquisition of a senescence-associated secretory phenotype and impairs angiogenic capacity in cerebromicrovascular endothelial cells: Role of increased DNA damage and decreased DNA repair capacity in microvascular radiosensitivity. *J. Gerontol. A Biol. Sci. Med. Sci.* **2013**, *68*, 1443–1457. [CrossRef]
162. Collins, A.R. The comet assay for DNA damage and repair: Principles, applications, and limitations. *Mol. Biotechnol.* **2004**, *26*, 249–261. [CrossRef]
163. Kiss, T.; Nyul-Toth, A.; Balasubramanian, P.; Tarantini, S.; Ahire, C.; DelFavero, J.; Yabluchanskiy, A.; Csipo, T.; Farkas, E.; Wiley, G.; et al. Single-cell RNA sequencing identifies senescent cerebromicrovascular endothelial cells in the aged mouse brain. *Geroscience* **2020**, *42*, 429–444. [CrossRef]
164. Yamazaki, Y.; Baker, D.J.; Tachibana, M.; Liu, C.C.; van Deursen, J.M.; Brott, T.G.; Bu, G.; Kanekiyo, T. Vascular Cell Senescence Contributes to Blood-Brain Barrier Breakdown. *Stroke* **2016**, *47*, 1068–1077. [CrossRef] [PubMed]
165. Graves, S.I.; Baker, D.J. Implicating endothelial cell senescence to dysfunction in the ageing and diseased brain. *Basic Clin. Pharmacol. Toxicol.* **2020**, *127*, 102–110. [CrossRef] [PubMed]
166. Bennett, R.E.; Robbins, A.B.; Hu, M.; Cao, X.; Betensky, R.A.; Clark, T.; Das, S.; Hyman, B.T. Tau induces blood vessel abnormalities and angiogenesis-related gene expression in P301L transgenic mice and human Alzheimer's disease. *Proc. Natl. Acad. Sci. USA* **2018**, *115*, E1289–E1298. [CrossRef]
167. Bryant, A.G.; Hu, M.; Carlyle, B.C.; Arnold, S.E.; Frosch, M.P.; Das, S.; Hyman, B.T.; Bennett, R.E. Cerebrovascular Senescence Is Associated With Tau Pathology in Alzheimer's Disease. *Front. Neurol.* **2020**, *11*, 575953. [CrossRef] [PubMed]

168. Michalicova, A.; Majerova, P.; Kovac, A. Tau Protein and Its Role in Blood-Brain Barrier Dysfunction. *Front. Mol. Neurosci.* **2020**, *13*, 570045. [CrossRef]
169. Minamino, T.; Miyauchi, H.; Yoshida, T.; Tateno, K.; Kunieda, T.; Komuro, I. Vascular cell senescence and vascular aging. *J. Mol. Cell. Cardiol.* **2004**, *36*, 175–183. [CrossRef]
170. Spazzafumo, L.; Mensa, E.; Matacchione, G.; Galeazzi, T.; Zampini, L.; Recchioni, R.; Marcheselli, F.; Prattichizzo, F.; Testa, R.; Antonicelli, R.; et al. Age-related modulation of plasmatic beta-Galactosidase activity in healthy subjects and in patients affected by T2DM. *Oncotarget* **2017**, *8*, 93338–93348. [CrossRef]
171. Bradl, M.; Lassmann, H. Oligodendrocytes: Biology and pathology. *Acta Neuropathol.* **2010**, *119*, 37–53. [CrossRef]
172. van Tilborg, E.; de Theije, C.G.M.; van Hal, M.; Wagenaar, N.; de Vries, L.S.; Benders, M.J.; Rowitch, D.H.; Nijboer, C.H. Origin and dynamics of oligodendrocytes in the developing brain: Implications for perinatal white matter injury. *Glia* **2018**, *66*, 221–238. [CrossRef]
173. Moore, S.; Meschkat, M.; Ruhwedel, T.; Trevisiol, A.; Tzvetanova, I.D.; Battefeld, A.; Kusch, K.; Kole, M.H.P.; Strenzke, N.; Möbius, W.; et al. A role of oligodendrocytes in information processing. *Nat. Commun.* **2020**, *11*, 5497. [CrossRef] [PubMed]
174. Nave, K.A.; Ehrenreich, H. Myelination and oligodendrocyte functions in psychiatric diseases. *JAMA Psychiatry* **2014**, *71*, 582–584. [CrossRef]
175. Tse, K.H.; Herrup, K. DNA damage in the oligodendrocyte lineage and its role in brain aging. *Mech. Ageing Dev.* **2017**, *161*, 37–50. [CrossRef]
176. Giacci, M.K.; Bartlett, C.A.; Smith, N.M.; Iyer, K.S.; Toomey, L.M.; Jiang, H.; Guagliardo, P.; Kilburn, M.R.; Fitzgerald, M. Oligodendroglia Are Particularly Vulnerable to Oxidative Damage after Neurotrauma In Vivo. *J. Neurosci.* **2018**, *38*, 6491–6504. [CrossRef]
177. Chen, J.H.; Hales, C.N.; Ozanne, S.E. DNA damage, cellular senescence and organismal ageing: Causal or correlative? *Nucleic Acids Res.* **2007**, *35*, 7417–7428. [CrossRef] [PubMed]
178. Ettle, B.; Schlachetzki, J.C.M.; Winkler, J. Oligodendroglia and Myelin in Neurodegenerative Diseases: More Than Just Bystanders? *Mol. Neurobiol.* **2016**, *53*, 3046–3062. [CrossRef]
179. Kritsilis, M.; Rizou, S.; Koutsoudaki, P.N.; Evangelou, K.; Gorgoulis, V.G.; Papadopoulos, D. Ageing, Cellular Senescence and Neurodegenerative Disease. *Int. J. Mol. Sci.* **2018**, *19*, 2937. [CrossRef] [PubMed]
180. Zhang, P.; Kishimoto, Y.; Grammatikakis, I.; Gottimukkala, K.; Cutler, R.G.; Zhang, S.; Abdelmohsen, K.; Bohr, V.A.; Misra Sen, J.; Gorospe, M.; et al. Senolytic therapy alleviates Abeta-associated oligodendrocyte progenitor cell senescence and cognitive deficits in an Alzheimer's disease model. *Nat. Neurosci.* **2019**, *22*, 719–728. [CrossRef]
181. Ogrodnik, M.; Evans, S.A.; Fielder, E.; Victorelli, S.; Kruger, P.; Salmonowicz, H.; Weigand, B.M.; Patel, A.D.; Pirtskhalava, T.; Inman, C.L.; et al. Whole-body senescent cell clearance alleviates age-related brain inflammation and cognitive impairment in mice. *Aging Cell* **2021**, e13296. [CrossRef]
182. Werkman, I.L.; Lentferink, D.H.; Baron, W. Macroglial diversity: White and grey areas and relevance to remyelination. *Cell. Mol. Life Sci.* **2021**, *78*, 143–171. [CrossRef] [PubMed]
183. Masuda, T.; Sankowski, R.; Staszewski, O.; Prinz, M. Microglia Heterogeneity in the Single-Cell Era. *Cell Rep.* **2020**, *30*, 1271–1281. [CrossRef]
184. Al-Mashhadi, S.; Simpson, J.E.; Heath, P.R.; Dickman, M.; Forster, G.; Matthews, F.E.; Brayne, C.; Ince, P.G.; Wharton, S.B.; Medical Research Council Cognitive, F.; et al. Oxidative Glial Cell Damage Associated with White Matter Lesions in the Aging Human Brain. *Brain Pathol.* **2015**, *25*, 565–574. [CrossRef] [PubMed]
185. Pruitt, S.C.; Freeland, A.; Rusiniak, M.E.; Kunnev, D.; Cady, G.K. Cdkn1b overexpression in adult mice alters the balance between genome and tissue ageing. *Nat. Commun.* **2013**, *4*, 2626. [CrossRef]
186. Martinez-Cue, C.; Rueda, N. Cellular Senescence in Neurodegenerative Diseases. *Front. Cell. Neurosci.* **2020**, *14*, 16. [CrossRef] [PubMed]
187. Bussian, T.J.; Aziz, A.; Meyer, C.F.; Swenson, B.L.; van Deursen, J.M.; Baker, D.J. Clearance of senescent glial cells prevents tau-dependent pathology and cognitive decline. *Nature* **2018**, *562*, 578–582. [CrossRef] [PubMed]

life

Review

Evidence of the Cellular Senescence Stress Response in Mitotically Active Brain Cells—Implications for Cancer and Neurodegeneration

Gregory J. Gillispie [1,2], Eric Sah [1], Sudarshan Krishnamurthy [1,3], Mohamed Y. Ahmidouch [1,4], Bin Zhang [5] and Miranda E. Orr [1,2,6,*]

[1] Section of Gerontology and Geriatric Medicine, Department of Internal Medicine, Wake Forest School of Medicine, Winston-Salem, NC 27157, USA; ggillisp@wakehealth.edu (G.J.G.); esah@wakehealth.edu (E.S.); sukrishn@wakehealth.edu (S.K.); ahmimy17@wfu.edu (M.Y.A.)

[2] Sticht Center for Healthy Aging and Alzheimer's Prevention, Wake Forest School of Medicine, Winston-Salem, NC 27157, USA

[3] Bowman Gray Center for Medical Education, Wake Forest School of Medicine, Winston-Salem, NC 27101, USA

[4] Wake Forest University, Winston-Salem, NC 27109, USA

[5] Department of Genetics and Genomic Sciences, Department of Pharmacological Sciences, Mount Sinai Center for Transformative Disease Modeling, Icahn Institute for Data Science and Genomic Technology, Icahn School of Medicine at Mount Sinai, New York, NY 10029, USA; bin.zhang@mssm.edu

[6] Salisbury VA Medical Center, Salisbury, NC 28144, USA

* Correspondence: morr@wakehealth.edu

check for **updates**

Citation: Gillispie, G.J.; Sah, E.; Krishnamurthy, S.; Ahmidouch, M.Y.; Zhang, B.; Orr, M.E. Evidence of the Cellular Senescence Stress Response in Mitotically Active Brain Cells—Implications for Cancer and Neurodegeneration. *Life* **2021**, *11*, 153. https://doi.org/10.3390/life11020153

Academic Editor: Markus Riessland

Received: 19 January 2021
Accepted: 9 February 2021
Published: 17 February 2021

Publisher's Note: MDPI stays neutral with regard to jurisdictional claims in published maps and institutional affiliations.

Abstract: Cellular stress responses influence cell fate decisions. Apoptosis and proliferation represent opposing reactions to cellular stress or damage and may influence distinct health outcomes. Clinical and epidemiological studies consistently report inverse comorbidities between age-associated neurodegenerative diseases and cancer. This review discusses how one particular stress response, cellular senescence, may contribute to this inverse correlation. In mitotically competent cells, senescence is favorable over uncontrolled proliferation, i.e., cancer. However, senescent cells notoriously secrete deleterious molecules that drive disease, dysfunction and degeneration in surrounding tissue. In recent years, senescent cells have emerged as unexpected mediators of neurodegenerative diseases. The present review uses pre-defined criteria to evaluate evidence of cellular senescence in mitotically competent brain cells, highlights the discovery of novel molecular regulators and discusses how this single cell fate decision impacts cancer and degeneration in the brain. We also underscore methodological considerations required to appropriately evaluate the cellular senescence stress response in the brain.

Keywords: cellular senescence; Alzheimer's disease; biology of aging; neurodegeneration; brain; geroscience; senolytics; tauopathy; cancer; stress response

1. Introduction

The risk of both neurodegenerative disease and cancer increases with advanced age due to increased damage accumulation and decreased repair capabilities; yet the relative odds of developing one or the other are inversely correlated [1–5]. Molecular profiling studies have identified disrupted genes, proteins, and signaling pathways shared by neurodegenerative diseases and cancer, but in opposing directions. For example, p53 is upregulated in Alzheimer's disease (AD), Parkinson's disease, and Huntington's disease, but downregulated in many cancers (reviewed [6]). Similarly mutations of the Parkin gene (PARK2) have been shown to simultaneously contribute both to Parkinson's disease and tumor suppression [7]. A recent study performed transcriptomic analyses of four different tissues from four different species at ages across their lifespan [8]. Across samples, the largest number of shared risk single nucleotide polymorphisms (SNPs) were in the genomic locus containing the long non-coding RNA ANRIL which modulates many cell cycle regulators including *CDKN2A/B* [9], which codes for p16^INK4A (hereon referred to as p16), one of the best characterized mediators of cellular senescence [10,11]. Notably, SNPs in this locus were identified in the brain, as well as other tissues analyzed [8]. These results point toward aberrant cell cycle, and in particular senescence, as a key age-associated molecular pathway worth further study.

Cellular senescence has emerged as a hallmark biological process that promotes aging (reviewed [12]). The pillars of aging, including cellular senescence, are highly interconnected and do not occur in isolation [13]. For example, epigenetic changes, telomere attrition, DNA damage and mitochondrial dysfunction all may induce cellular senescence, which then contributes to dysfunctional nutrient signaling and proteostasis. Consequences of cellular senescence include stem cell exhaustion and chronic inflammation. Thus, cellular senescence represents an intersection of aging hallmarks [13]. While best studied as an anti-cancer stress response, recent studies highlight its pro-degenerative role in AD and tauopathies [11,14,15]. As such, cellular senescence may contribute to the inverse correlation between the risk for developing neurodegeneration and that for cancer.

Bulk tissue analyses, while informative at a macroscopic level, may not capture important changes occurring in single cells. Senescent cell abundance increases with aging, but the relative contribution to a tissue is relatively low and may be missed in bulk analyses [16] (reviewed [11]). Several laboratories are using single cell technologies to assign cell type specificity to tissue-level observations [17], but to date these analyses have not included senescent cells in the brain. To maximize generalization and interpretation across studies, in this review we only evaluate studies which investigated cellular senescence with cell type specificity, and not bulk analyses. We specifically focus on mitotically competent brain cells; due to space considerations, the topic of postmitotic brain cellular senescence is reviewed in a separate manuscript (Sah et al., *Life*, in review). The present compilation provides evidence on conditions in which cellular senescence may benefit (anti-cancer) or negatively impact (neurodegeneration) brain health. In doing so, this review explores how the cellular senescence stress response may simultaneously distinguish and connect AD and cancer risk.

2. Identifying Senescent Brain Cells

The identities of the parent cell type and upstream senescence-inducing stressors have consequences on the post-senescence phenotype [18,19]. The resulting heterogeneity has presented challenges for identifying, defining, and studying senescent cells in vivo and across disciplines. Senescent cells have been identified using various morphology markers; gene, protein, epigenetic, metabolic changes; and functional readouts and have been a subject of earlier reviews [20–24]. Transcriptomic analyses in particular show significant promise for identifying senescent cells. They have been utilized to study characteristics of senescence common between different cells-of-origin and modes of senescence induction and may help identify senescence markers which are more powerful than current traditional senescent markers such as p16 [25]. However, a specific combination of phenotypes defining cellular senescence currently does not exist [26]. Where biologists agree is that interpreting the senescence phenotype requires integrating various lines of distinct evidence placed in appropriate context [21]. This is especially true for postmitotic tissues such as the brain. We reviewed the literature using pre-defined senescence-defining criteria: proliferative/cell cycle arrest, apoptosis resistance, senescence-associated secretory phenotype, and senescence-associated β-galactosidase activity (Figure 1). As with measuring other biological processes, the interpretation of results requires integrating several lines of evidence. We explain the bases of these chosen criteria in this section.

2.1. Absence of Proliferation/Stable Cell Cycle Arrest

Replicative senescence, as observed by Hayflick and Moorhead [27], may be fundamentally restricted to mitotically competent cells. Brain cells which are susceptible to this fate include neural stem cells (NSCs), neural progenitor cells (NPCs), oligodendrocyte progenitor cells (OPCs), and microglia [28–30]. A subpopulation of astrocytes may be capable of cell division and trans-differentiating into neurons [31], however to what extent this happens in vivo and whether it occurs enough to induce replicative senescence in these rare cells is unclear. Of note, postmitotic cells may also undergo cellular senescence [32], including neurons [15,33–35]; a review of postmitotic senescent brain cells will be available in a separate manuscript (Sah et al., *Life*, in review). A wide range of methods have been validated to measure cell proliferation both in vitro and in vivo, including cell cycle specific markers, BrdU incorporation, ploidy, and the quantity and size of stem-cell containing non-adherent cultures (e.g., neurospheres or brain organoids). A reduction in proliferation at the tissue or cell population level (in situ tissue analyses or in vitro cell culture, respectively) may indicate that a portion of cells have undergone senescence or a general slowing of the cell cycle. To discern these differences and assign a senescence arrest, evaluating single cells is necessary.

Mechanistically, replicative senescence is achieved through intrinsically mediated (cell autonomous) telomere attrition [27] and has been the focus of several recent reviews [21,36,37]. Briefly, telomeres progressively shorten with age due to successive cell divisions and reduced telomerase activity. When chromosomes reach a critical truncated length, cells cease to replicate [36,37]. Average or relative telomere length can be measured using polymerase chain reaction (PCR), terminal restriction fragment (TRF) analysis, single-telomere length (STELA) analysis, and several different fluorescent in situ hybridization (FISH) methods [38]. Quantifying the shortest telomeres, as opposed to average length, may be more beneficial depending on the application and can be accomplished using the telomere shortest length assay (TeSLA) [38,39]. Similarly, telomerase activity can be quantified with a variety of measurement techniques including, most commonly, the PCR-based telomeric repeat amplification protocol (TRAP) [40]. A number of methods have also recently been developed, improving upon the convenience, throughput, sensitivity, and reliability of TRAP [41–43]. These methods provide objective readouts when evaluating potential mechanistic mediators of replicative senescence.

The most broadly accepted and universal phenotype among senescent cells is the change in cell fate that accompanies cell cycle arrest. This change in cell fate goes beyond a reduction in proliferation and is distinct from differentiation. While a reduction in cell proliferation can be reversible and attributable to many non-senescent causes, true senescence involves an irreversible cell cycle arrest. Senescent cells will not re-enter the cell cycle if signaled to do so, if the original stressor is removed, or even if reprogrammed with Yamanaka factors [44–47]. The stressors, pathways, and cell cycle inhibitors involved have been reviewed [45,48–50]. Many senescent cells accomplish cell cycle inhibition by upregulating p16 and/or p21^{CIP1} (hereon referred to as p21); many other cell cycle inhibitors have been reported as well [45]. Based on the prevalence of upregulated p16 across studies, reporter mice using the p16 promoter have been utilized for many of the in vivo experiments (reviewed [51]). A caveat to this in vivo work is its dependence on p16-mediated senescence, overlooking other pathways. Nonetheless, these results have guided the field to date. In summary, a stable cell cycle arrest indicative of cellular senescence includes the evaluation of cell proliferation; expression of cell cycle inhibitors, including p16, p21 and p19^{INK4d} (hereon referred to as p19); telomere attrition; and inability to re-enter the cell cycle when provided relevant stimuli. Additional morphological and functional changes accompanying senescence may include increased cell size; altered nuclei that are either enlarged (karyomegaly) or syncytia (multinucleated) [15,52,53].

2.2. Apoptosis Resistance

The enhanced survival of senescent cells depends on the activation of senescent cell anti-apoptotic pathways (SCAPs) [54,55]. The SCAP molecular pathways which have been identified thus far include BCL-2/BCL-XL, PI3K/AKT, p53/p21/serpines, dependence receptors/tyrosine kinases, HIF-1α and the unfolded protein response. The unfolded protein response in particular may be especially important for proteinopathies which cause many neurodegenerative dementing diseases. The senescence-associated phenotype can be identified by the expression of these SCAP pathways and/or by their resilience when exposed to apoptosis-inducing factors. Senescent cells often utilize multiple, redundant (although not all) SCAPs to survive in an otherwise toxic microenvironment. As such, targeting/evaluating multiple pathways is critical when identifying apoptosis resistance. Additionally, senescent cells should not express markers of apoptosis (i.e., TUNEL and caspase-3 negative).

2.3. Secretory Phenotye

The senescence-associated secretory phenotype (SASP) represents a deleterious effect of chronic senescent cell survival (reviewed [56,57]). Like other stressed cells, senescent cells secrete molecules to communicate to neighboring cells. These include chemokines and cytokines which signal to immune cells to clear them. Other secreted molecules include extracellular remodeling factors, exosomes, miRNAs, growth factors, and proteases that alter the environment and may induce senescence in other cells. The SASP is beneficial for tissue remodeling and wound healing (reviewed [21,58,59]). However, as senescent cells accumulate with age and pathology, the SASP contributes not only to local, temporary inflammation, but also chronic, systemic inflammation. This inflammation, which may have other contributors in addition to senescent cells, has been shown to be disadvantageous for a number of age-related diseases, including those in the brain. SASP factors differ across parent cell type and even within the same cell type exposed to different stressors. The phenotypic diversity of SASP factors represents a major challenge for developing a unifying profile and has been the subject of numerous reviews [56–59]. Moreover, how the SASP may change within a senescent cell over time remains unknown. Nonetheless, some of the common factors used to identify SASP in the studies reviewed here include IL-1α, IL-6, IL-1β, TNF-α, and CXCL1, CCL4, CCL6 and TGF-β.

2.4. Senescence-Associated β-Galactosidase

Senescent cells in vitro retain lysosomal β-galactosidase activity at pH 6.0, which is referred to as senescence-associated β-galactosidase (SA β-gal) activity [60]. The mechanisms surrounding SA β-gal activity are not fully elucidated, although studies have determined that *GLB1*, the gene encoding lysosomal β-D-galactosidase, is required [61]. This finding indicates an absence of a unique senescence-specific enzyme. Instead SA β-gal activity may reflect a change in lysosomal content or function. Toward this end, a recent study determined that lysosomes mediate whether cells remain quiescent or become senescent. As lysosomal function decreased, cells progressively lost the ability to reverse out of quiescence and thus became senescent [62]. However, the relation to SA β-gal staining was not investigated, so it remains unknown whether it coincides with the transition from deep quiescence to senescence. Notably, pharmacologically targeting lysosomal β-galactosidase clears senescent cells as evidenced by reduced expression of cell cycle inhibitor genes associated with senescence [63]. While these studies provide evidence for the utility of the SA β-gal assay to identify senescent cells, appropriately measuring SA β-gal activity requires careful methodological attention.

SA β-gal evaluation is a histological assay that requires enzymatic activity of GLB1. [61]. Due to this need for enzymatic activity, ideal experiments necessitate fresh frozen tissue, yet oftentimes the use of archived frozen or fixed tissues are reported in the literature. Similarly, it is imperative that the precise pH is reported and serial sections are processed at both pH 6.0 and pH 4.0 to discern between global differences in lysosomal activity from senescence-specific differences. Appropriate controls should include young, age-matched disease-free and/or untreated tissues that are processed side-by-side with the experimental samples.

Given that lysosomal function differences between quiescent and senescent cells are on a continuum and subtle [62], SA β-gal cannot be used as a single surrogate marker. Additionally, not all senescent cells acquire the phenotype. We emphasize the importance of considering the cellular phenotype in entirety when evaluating the senescent phenotype [21]; we used the methodological criteria described above to evaluate SA β-gal interpretations throughout the review.

2.5. Concluding Remarks on Identifying Senescent Cells

Evaluating cellular senescence requires both an appreciation for the phenotypic complexity of cellular senescence, expertise in the parent cell type and an understanding of how they respond to stress. Beyond the pre-defined criteria used in this review, additional markers may be applied to supplement the characterization of senescent cells. Morphological observations such as an enlarged, flattened morphology and a disrupted nuclear membrane and metabolic dysregulation such as increased ROS and mitochondria dysfunction are also commonly employed, among others [24]. In the following sections, we review studies reporting senescent cells of mitotically competent brain cells using the pre-defined criteria presented above and summarized in Figure 1. While reviewing the literature we evaluated the methods used to define senescence and how these phenotypes were placed in the context of brain cell biology.

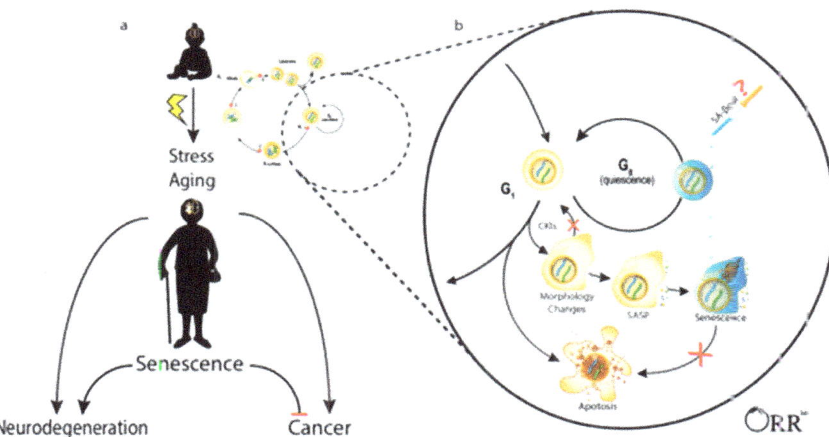

Figure 1. Cellular senescence in mitotically competent brain cells may impact risk for developing cancer and neurodegeneration. (**a**) The accumulation of cellular damage throughout the lifespan induces cell stress responses that impact cell fate decisions. Neurodegeneration increases with age and inversely correlates with risk of developing age-associated brain cancer. Cellular senescence may influence this clinical observation. (**b**) Mitotically competent brain cells may enter quiescence, a reversible G0 arrest. Damaged cells may undergo apoptosis, other cell death processes or senescence. In senescence, the cell permanently exits the cell cycle, upregulates cyclin-dependent kinase inhibitors (CKIs), exhibits morphological changes, acquires a senescence-associated secretory phenotype (SASP), and upregulates senescent cell anti-apoptotic pathways (SCAPs). Senescence-associated beta galactosidase (SA β-gal) staining can be used to identify some senescent cells but also labels some quiescent cells and requires careful methodological techniques and interpretations. For simplicity, senescence was illustrated at G1; however, cells may enter senescence arrest at G0 or G2 as well [64].

3. Neuronal Precursor Cells

Brain maturation continues after birth for three months in mice [65] and 20 years in humans [66] Even beyond that, the adult brain maintains populations of self-renewing, multipotent NSCs first identified in rats [67] and later in humans [68–70]. Adult neurogenesis occurs primarily in the subgranular zone (SGZ) in the hippocampal dentate gyrus (DG) and the subventricular zone (SVZ) around the lateral ventricles of the forebrain. In humans, the primary progenitor of the SVZ is a subpopulation of specialized, quiescent NSCs known as B cells which give rise to interneurons and OPCs. These are sometimes referred to as SVZ astrocytes due to their morphological structure and their co-expression of the glial fibrillary acidic protein (GFAP), Nestin and carbohydrate Lewis X (LeX) [71]. These specialized NSCs share some physical and molecular characteristics with typical astrocytes, including branched processes, intermediate filament bundles, cell bodies in the cytoplasm, and gap junctions [72]. Thus, precise phenotyping measures are required to identify neurogenic from non-neurogenic astrocytes. Some studies suggest that human NSCs populations persist and contribute to neurogenesis throughout adulthood [73–80]. Others have detected significant reductions during childhood, ranging from simply lower activity all the way to negligible or nonexistent in adults [81–84]. Thus, most studies reviewed in this section were performed in human cells in vitro or in mice.

In healthy adult mammalian tissue, NSCs exist primarily in a quiescent arrest [85]. During quiescence, cells lower their metabolic activity and cell division rate to minimize damage to DNA, proteins, and mitochondria which can lead to cancer, senescence, and exhaustion of the stem cell population [86,87]. Various environmental and behavioral activities have been shown to activate quiescent NSCs (e.g., exercise, sleep, learning) through diverse cell signaling pathways (e.g., neurotransmitters, Notch, neurotrophins, Wnt) [87]. Once activated, NSCs proliferate, migrate, and differentiate toward NPCs and OPCs and terminally differentiate into neurons, astrocytes, and oligodendrocytes [88,89]. Given that aberrant NSC activation may contribute to cancer, the ability for them to utilize the senescence stress response may benefit short-term health and survival.

The incidence of cancer, including in the brain, increases with age [90], and evidence suggests that NSCs are often the cells of origin for brain tumor formation [91,92]. Evolutionarily, the ability for proliferative cells to undergo senescence in response to cell damage provided an advantage over malignancy [93,94]. Some evidence suggests that patient survival rates are favorable when brain cancer cells become senescent [95]. Pilocytic astrocytoma is a slowly growing benign brain tumor derived from astrocytes and is the most common pediatric brain tumor. Genetic mutations in the proto-oncogene B-Raf gene, *BRAF*, cause pilocytic astrocytoma. Human fetal neurospheres transduced with a constitutively active form of *BRAF* were evaluated for senescence-like features. The mutation promoted colony formation and in early passages proliferation did not differ as assessed through BrdU incorporation. However, after five passages BrdU uptake was notably reduced in BRAF mutant cells compared to controls, and was shortly thereafter followed by proliferative arrest. The remaining cells appeared viable as per light microscopy and displayed a 35-fold increase in the percentage of cells with SA β-gal activity with elevated PAI-1 and p16 compared to controls [95]. The *BRAF* mutant NSCs displayed a progressive decrease in SOX2 expression that coincided with the increase in other senescence-associated protein upregulation, suggesting a loss of neural stemness as they progressed into senescence. Telomere length did not differ between NSCs indicating the senescence arrest was independent of telomere attrition. The research team confirmed elevated p16 in 86% (57/66) of patient-derived pilocytic astrocytoma tumors. Moreover, individuals whose tumors were negative for p16 (9 cases) had significantly shorter survival than p16 positive cases. Tumor suppressor TP53 (p53) was detected in 23% of pilocytic astrocytoma samples and significantly correlated with p16 in the cases that were scored. SA β-gal was also observed in >50% of primary tumor cells derived from patients (n = 2) at low passage number. Collectively, these data suggest that elevated p16, and presumably NSC senescence, is favorable for survival in patients with pilocytic astrocytoma.

Human NSCs may undergo cellular senescence in response to non-genetic carcinogens as well. For example, radiation decreased human NSC proliferation and caused cell cycle arrest as measured by BrdU uptake in vitro, acutely increased DNA damage repair (γH2AX levels returned to baseline by 2 hours) and apoptosis (PARP cleavage returned to baseline by 28 h) [96]. Cellular metabolism was also increased on a per cell basis, as measured by the XTT assay, and neuronal differentiation was impaired, suggesting the surviving cells were senescent rather than quiescent [96]. Irradiation also reduced proliferation of NSCs in the SGZ and SVZ of rats as measured by BrdU incorporation in vivo [97]. Neuronal differentiation was reduced 97% in vivo, but NSCs isolated from these rats were still able to differentiate towards neurons in vitro. Furthermore, non-irradiated NSCs also showed an 81% reduction in neuronal differentiation when transplanted into irradiated rats. Combined, these results suggest the effects of irradiation on NSCs may have been indirectly induced by the damaged microenvironment [97], thus highlighting the importance of healthy cells and environment; both of which are negatively impacted by senescence.

Canonical senescence phenotypes acquired by irradiated NSCs include cell cycle arrest and SA β-gal staining [98]. Unique features include an increased expression of astrocyte markers which has been observed in vitro [98] and in vivo [99]. This differs from astrocytes which downregulate GFAP when undergoing senescence [100]. Development of SASP has not reproducibly occurred in irradiated NSCs across studies. For example, Zou et al., did not detect SASP factors IL-6, IL-8, IL-α concomitant with cell cycle arrest and SA β-gal positive NSCs [98]. However, using a similar cell line, Schneider et al. reported radiation-induced NSC cell cycle arrest via p21 and p27^{KIP}, and Rb-dephosphorylation [99]. These NSCs also showed an enlarged, flattened morphology and an increase in SA β-gal staining. In contrast to Zou et al., they found an increased secretion of several cytokines including IL-6 suggestive of a canonical senescence phenotype. Interestingly, irradiated NSCs still lost their markers of stemness (i.e., Nestin and SOX2) even when researchers inhibited the astrocytic differentiation pathway [99], indicating the reduced proliferation of these cells was not simply due to their differentiation into astrocytes. Together, these studies indicate that cell cycle arrest and SA β-gal positivity are not sufficient to label NSCs as senescent; however, these features combined with elevated p21 and p27^{KIP} and morphological changes co-occur with SASP production and more complete senescent phenotype.

Senescence phenotypes in NSCs have been observed in response to chemical carcinogens as well. Independently, Dong et al. [101] using NSCs derived from the SVZ of mice and Daniele et al. [102] using human NSCs derived from embryonic stem cells, observed senescence-like phenotypes after treatment with hydroxyurea (HU). HU is a ribonucleotide reductase inhibitor in the "antimetabolite" family of chemotherapies known to induce DNA damage. Both mouse and human NSCs showed senescent-like phenotypes including reduced proliferation via neurosphere formation, p21, and p16, increased DNA damage via p53 and γH2AX, and an increase in SA β-gal staining. A reduction in apoptosis was also found in NSCs derived from rats [101] and the proinflammatory transcription factor NF-κB was found in human NSCs [102]. Collectively, these in vitro studies suggest that NSCs develop features of senescence in response to HU, although further investigation into the irreversibility of this state, a more in-depth characterization of their resistance to apoptosis, and investigation of SASP production downstream of NF-κB activation would strengthen this assertion.

As so far, the literature indicates that senescence may protect NSCs from becoming cancerous in response to known carcinogens [103]. A long term effect of senescent cell accumulation is chronic inflammation and degeneration. The brain is especially susceptible to age-associated neurodegenerative diseases (reviewed [104]). For example, AD onset prior to the age of 50-years-old is exceptionally rare, with the exception of familial cases. In contrast, 10% of adults aged 65-years-old and roughly 40% over the age of 80-years old have an AD diagnosis. Notably, aberrant NSC behavior has been reported in patients with AD [105,106]. To explore whether AD pathology is a cause or effect of NSC senescence, we evaluated literature from postmortem human brain studies as well as mechanistic studies in vitro.

AD is a progressive neurodegenerative disease characterized by the presence of amyloid-β (Aβ) plaques and tau-containing neurofibrillary tangles [107]. Evidence from neurosphere assays suggests that Aβ_{42} [108] and tau [109] overexpression negatively impact NSC function through senescence-associated pathways. For example, He et al. [108] investigated senescence in response to Aβ_{42} oligomers in vitro and in vivo. Neurospheres derived from mouse hippocampi were exposed to various concentrations of Aβ_{42} for up to 5 days and assessed for proliferation, differentiation, SA β-gal, and toxicity. Apoptosis was increased with 10 μM Aβ_{42}, but remained unchanged at lower concentrations. Aβ_{42} also reduced proliferation and increased p16 in a dose-dependent manner. In Aβ producing TgAPP/PS1 mice, the number of Nestin-positive NSCs double stained for p16 increased 1.3-fold and the number of SA β-gal positive cells in the dentate gyrus increased 1.7-fold. FPR2 (a G protein-coupled receptor which mediates the inflammatory response) and p38 MAPK (a stress response protein) were both implicated in the senescence induction of NSCs [108]. These in vivo data were largely driven by the use of p16 antibody staining and SA β-gal, both of which are somewhat unreliable due to lack of specificity (as discussed in [110] and [15], respectively). Nonetheless, the suggestive data are compelling for further investigation.

Tau protein accumulation is the most common intraneuronal pathology among neurodegenerative diseases [111,112]. We recently discovered that tau protein accumulation drives senescence in the brain [15], which was confirmed shortly thereafter by Bussian et al. [113]. Between the two independent studies, several senescent cell types were identified, but neither group looked at the role of tauopathy on NSC senescence. Nonetheless, evidence suggests that tau protein is critical for normal NSC behavior [109] and plays a role in abnormal activity in tauopathies [111]. We generated NSC-containing neurospheres from tau transgenic mice overexpressing either human frontotemporal dementia mutant tauP301L or human tau and compared phenotypes to those generated from wild type mice [109]. We found that mutant tauP301L overexpression resulted in abnormal proliferation and differentiation as indicated by MTT proliferation assays and filopodial/spine morphology measures, respectively. With repeated passaging, the non-adherent neurosphere culture proliferation rate significantly slowed and lost the ability to differentiate into brain cells. Instead, plating the Nestin-positive cells as adherent cultures revealed large, flattened cells with large cell somas containing numerous vacuoles and enlarged nuclei. These adherent cells remained in culture until finally discarding the culture plate after several weeks of attempts to expand the culture or differentiate into brain cells. While we did not examine cell cycle inhibitor expression or SA β-gal, the other defining features of cellular senescence were observed [109]. The Tau Consortium has generated at least 31 induced pluripotent stem cell (iPSCs) lines from 140 MAPT mutation/risk variant carriers and cognitively normal controls [114]. As the scientific community continues to utilize this resource, we anticipate more clarity surrounding the effects of aberrant tau on cellular senescence across brain cell types during early development and in disease.

Similarly, Voisin and colleagues [110] developed human iPSCs from a single patient with Huntington's disease, a rare inherited neurodegenerative disease arising from mutations in the *HTT* gene. In this study, HTT CAG repeat-corrected (C116) cells from the same donors were used as a disease-free control. Using qRT-PCR and a gene ontology approach, they found expression of FOXO3 and its transcriptional targets to antagonize p16 expression. Furthermore, they identified p16 and related genes as key mediators of senescence in Huntington's disease NSCs and medium spiny neurons. Cells derived from the Huntington's disease patient additionally showed decreased proliferation and elevated SA β-gal staining in vitro [110]. These results suggest that Huntington's disease mutations may negatively impact NSC function and perhaps contribute to disease through senescence-mediated pathogenesis; however, senescence phenotypes will need to be confirmed in postmortem human brain tissue from patients with Huntington's disease to validate these intriguing preclinical findings.

Over 200 disease causing mutations to *POLG*, a subunit of the mitochondrial DNA repair protein polymerase gamma (polγ), have been identified—many of which primarily affect the nervous systems. Using fibroblasts from two patients carrying *POLG* mutations, Liang et al. [115] generated iPSCs and then differentiated them into NSCs. While the iPSCs recapitulated some known *POLG* mutation phenotypes, the NSCs especially showed characteristic signs of mitochondrial dysfunction, including reduced energy production, abnormal mitochondrial volume, respiratory chain complex 1 loss, and increased ROS. Other evidence include upregulated mitochondrial uncoupler protein 2 (UCP2, which maintains mitochondrial membrane potential) and reduced phosphorylated Sirtuin 1 (SIRT1, a NAD-dependent protein deacetylase) [115]. In addition to mitochondrial dysfunction, these NSCs showed increased levels of SA β-gal staining and p16 expression suggestive of senescence. While the study did demonstrate SA β-gal staining and potentially a reduction in cell proliferation via p16, their characterization of senescence did not include the cell's resistance to apoptosis or a SASP. Instead the focus of the study was primarily on mitochondrial dysfunction given its strong link with *POLG* mutation disorders. Together these studies highlight the on-going research and approaches utilized by the neuroscience field to understand mechanisms surrounding age-associated neurodegenerative diseases. Cellular senescence is emerging as a potentially significant cell stress pathway worth continued investigation.

While the above studies provided examples of the effects of carcinogens, genetic or pathological variants on NSC function, other studies have investigated sporadic, age-associated phenotypes. Tissues accumulate senescent cells with advanced chronological age. Gao et al. [116] isolated SVZ neurospheres in vitro from 23-month-old mice and compared them to 3-week-old counterparts; an age where developmental neurogenesis has completed but the brain is still maturing. Neurosphere formation, proliferation, and neuronal differentiation were significantly lower in aged NSCs. Further, aged NSCs showed increased p16 expression and shortened telomeres via qRT-PCR. SA β-gal staining, apoptosis, and SASP were not investigated. These changes were partially alleviated by flavonoids of the plant Ribes meyeriy, [116] suggesting the drug's anti-aging and potentially senotherapeutic effects.

Using the senescence accelerated (SAMP8) mouse model at 2, 6, and 12 months of age, Hu et al. [117] showed an age-related increase in NSC senescent phenotypes. NSCs, labelled via Sox2 and either Nestin or GFAP, were found to decrease with age in the hippocampus. Furthermore, SA β-gal staining in the hippocampus increased at 6 months and again at 12 months. The percentage of senescent NSCs, labelled via Sox2 and p16, also increased with age. Isolated NSCs from these mice were less able to form neurospheres or differentiate into neurons. DNA damage-associated cell cycle arrest was evidenced by elevated γH2AX, p53, p16, and p21. Exogenously applied embryonic stem cell extracellular vesicles reduced many of these senescence-associated outcomes both in SAMP8 mice and in vitro. RNA sequencing analyses revealed that *Myt1*, a regulator of neurogenesis, was downregulated with passaging and age and upregulated with the extracellular vesicle treatment. Consistent with these observations, *Myt1* knockout mice developed many of the same senescence-associated phenotypes and were resistant to the beneficial effects of the extracellular vesicles [117]. These data suggest that *Myt1* may be a novel regulator of NSC senescence and that extracellular vesicles from embryonic stem cells have potential to treat age-related neurodegeneration.

A recent study by Xiao et al. [118] investigated senescence phenotypes in aged hypothalamic NSCs. Neurospheres derived from 18-month-old wild type mice were fewer, smaller, and had lower proliferation than those from 3-month-old mice. From these experiments, they identified the long non-coding RNA Hnscr as playing a role in NSC senescence. *Hnscr* null mice displayed similarities to wild type counterparts at young ages, but accelerated aging at 18 months, including reduced NSC proliferation and increased SA β-gal staining. Similar, but less striking, results were seen with viral suppression of *Hnscr* specifically in hypothalamus NSCs. RNA sequencing and bioinformatic approaches revealed cell senescence, apoptosis and inflammatory responses among the altered pathways.

In particular, p16 was upregulated as confirmed by qRT-PCR. Lastly, wild type mice treated with Theaflavin 3-Gallate, an Hnscr mimetic, had lower NSC senescence compared to vehicle treated controls. Outcome measures included SA β-gal staining and RNAseq data consistent with cell cycle arrest, apoptosis resistance, and a SASP suggest a *bona fide* senescence arrest. Collectively their data highlight Hnscr as a critical mediator of hypothalamic NSC senescence. Future studies are needed to determine whether this broadly applies to other NSC populations or SCs in general. Nonetheless, this study highlights a significant advance in senescence biology by evaluating brain cell populations.

Overall, these studies indicate that NSCs acquire senescence-like phenotypes in response to carcinogens and stressors known to promote cancer or neurodegenerative disease. Given the inverse relationship between the risk for developing AD and cancer, it is tempting to speculate that senescence may protect against cancer but promote neurodegeneration. However, further research is needed to draw more conclusive inferences. Through reviewing the literature, we identified important areas for clarifying "senescence" in NSCs. One is the difference between quiescence and senescence. Given that cell cycle arrest is a defining feature of both, the permanency of the arrest needs to be experimentally determined. A gold standard assay to confirm the cell cycle activity in NSCs is their ability to divide and give rise to multipotent clones when provided with cytokine growth factors in the culture, e.g., the neurosphere assay. While useful to identify proliferative arrest ex vivo; cell cycling is not easily discernable in postmortem tissue, especially since similar molecular pathways are used for both quiescent and senescence arrest. For example, the transcription factor Bmi-1 regulates NSC self-renewal by suppressing p16 and p19 [119–121]. p16 regulates whether DG NSCs exit quiescence and undergo neurogenesis in response to running [122] and genetically removing p16 lessens the age-associated neurogenesis decline in the SVZ, but not DG [123]. *Cdkn1a*, which codes for the cell cycle inhibitor p21, expression is required to appropriately regulate NSC proliferation/quiescence during early and middle age [124], and in response to ischemia [125]. These studies highlight the physiological roles of commonly used markers of cellular senescence (i.e., Bmi-1, p16, p19 and p21) in NSC maintenance and behavior in an age, stress and region-specific manner. Thus, their differential expression in histological tissues may represent a snapshot of regulated NSC quiescence, making it difficult to discriminate from senescence. Distinguishing between quiescent and senescent NSCs in particular remains a challenging roadblock to the field.

Other distinguishing markers can help differentiate senescence from quiescence or other causes of decreased neurogenesis. For example, senescent stem cells remain metabolically active whereas a reduction in metabolism is one of the key characteristics of quiescence [87,126,127]. While little is known about the secretome of quiescent cells, limited evidence suggests a potential overlap with the SASP including IL-1α, IL-6, IL-8, CCL4, and CCL6 [128,129]. Apoptosis resistance, to our knowledge, has not been investigated as a marker of NSC senescence. Nonetheless, NSCs naturally display a general resistance to apoptosis [130,131]. Whether or not these pathways change with age or in response to other stressors could be used to help distinguish between senescence and quiescence. A recent study identified a gradual entrance into senescence from quiescence regulated by lysosomal function [62]. While this may complicate the discernment between quiescence and senescence, it suggests that lysosomal content and activity may help decipher these cell states.

Concluding Remarks

In summary, true senescence may have been present in some of these studies, but most have only partially characterized the phenotype. Other studies have taken for granted senescence-related phenomena which occur in other cell types, but have not yet been demonstrated in NSCs. The Hayflick Limit and the SASP, for example, both represent opportunities for future experimentation in NSCs. In addition, while many studies have relied heavily upon SA β-gal staining, few have adhered to proper protocols (see Section 2.4.). Nonetheless, mechanistic studies utilizing brain NSCs are advancing our understanding of the molecular regulators of cellular senescence, including neurodegenerative disease-associated tau [109], *Foxo3* [110], *Myt1* [117], and lncRNA Hnscr [118]. Whether or not they are cell type specific remains unknown, but they highlight the utility and importance of evaluating senescence in brain cells. Future studies will be needed for more in-depth characterization of senescent NSCs, their effects in brain pathology, and interventions specifically targeting this cell population.

4. Oligodendrocyte Precursor Cells

OPCs form 5–8% of all cells in the adult brain [132]. Developmentally, OPCs arise from multiple progenitor cell pools in the developing spinal cord and forebrain (reviewed [133,134]). Postnatally, OPCs are predominantly found in the SVZ and migrate to white matter regions, where they proliferate and differentiate into oligodendrocytes and astrocytes [134,135]. OPC multipotency provides an opportunity for malignant gliomas [136] including oligodendroglial [137] and astrocytic tumors [138]. Physiologically, OPCs differentiate into oligodendrocytes throughout life and contribute to myelin turnover [139–142], adaptive myelination [141,143,144] and regenerative processes following demyelinating insults [132,145]. The ability of OPCs to respond appropriately to myelin maintenance or stress signals determines whether they positively impact myelination [132,139–142] or potentially divide uncontrollably and become cancerous [136–138]. While cellular senescence protects against the latter, evolutionary pressure for the senescence stress response did not account for the long lifespans of modern day humans. OPC senescence, thus, may contribute to neurodegenerative phenotypes in later life. Evidence for the tradeoff between OPC senescence, cancer and neurodegeneration is presented in this section.

When grown in optimal culture conditions, OPCs challenge the Hayflick limit and replicate indefinitely [146]. Prolonged culture results in elevated expression of many cell cycle inhibitors (Cip/Kip proteins: p21, p27 and p57, and INK4s: p18 and p19) in the absence of cell cycle arrest. This non-senescence profile may be attributed to significant upregulation of positive cell cycle regulators Cdk2, Cdk4 and cyclin D1, D3 and E [146]. These experiments suggest that a healthy brain environment could provide OPCs an opportunity for indefinite replication (e.g., could be beneficial for brain maintenance but contribute to cancer). However, suboptimal culture conditions including prolonged exposure to 15% fetal bovine serum, contact inhibition due to overgrowth/confluency, and low dose genotoxic drugs all produced senescence-like phenotypes. Culture shock conditions resulted in proliferative arrest, including a failure to incorporate BrdU, and flattened morphology [146]. Notably, elevated p16 was not detected in any culture conditions, but viral-mediated over-expression of p16 strongly inhibited BrdU incorporation [146]. While these results indicate that p16 activation can cause OPC proliferative arrest, OPCs do not innately upregulate p16 in response to many known senescence-inducing stressors. Additionally, p21 is required for OPC differentiation to oligodendrocytes [147], which presents challenges for interpreting OPC differentiation from senescence. Thus, two of the most commonly used markers of senescence, p16 and p19, require additional considerations and methodologies and highlight the importance of including measures of DNA damage response pathways, resistance to apoptosis, and the SASP when evaluating OPC senescence.

Age-associated deficits in remyelination often co-occur with a lack of OPCs in the lesion site (e.g., poor recruitment) and/or delayed or failed OPC differentiation. The successful execution of recruitment and differentiation requires a complex interaction between the OPCs and their environment. Thus, poor remyelination may reflect impaired OPCs and/or suboptimal environmental cues. Cell autonomous deficits in OPCs are inferred by increased DNA damage, upregulated *Cdkn2a*, and reduced cellular respiration in OPCs isolated from 20-24 month-old mice compared to 2–3 month-old mice [148]. Given the increase in DNA damage, as evidenced by single cell comet assays, poor OPC proliferation and migration may protect against the development of malignant gliomas. In this way, a senescence response may be beneficial over tumorigenicity. Similarly, cultured OPCs from aged mice were less proliferative as shown by decreased BrdU incorporation and cell cycle arrest [149]. Examining whether these cells are resistant to apoptosis and identifying SASP factors associated with the cellular environment would validate the senescence phenotype in OPCs, and provide much needed information regarding whether they acquire a toxic SASP or remain benignly arrested.

OPCs are present abundantly in grey and white matter, and are believed to survey their microenvironment, migrate to lesioned areas for remyelination through repopulation of oligodendrocytes, and respond to inflammatory cues [132,150]. Although the non-progenitor characteristics of OPCs are less understood, they respond to CNS injury, ischemia, and neurodegeneration, along with microglia [150,151]. Depletion of the endogenous OPC pool, reduced migration of OPCs in a pathological environment, and lack of differentiation of OPCs lead to failure of remyelination in brain lesions, suggesting a critical role for OPCs in neurodegenerative diseases [134]. Interestingly, in post-mortem human brain tissue of patients with AD, elevated levels of p21 were observed in OPCs surrounding Aβ plaques, but not in regions devoid of pathology [152]. The authors reported that >80% of large Aβ plaques (>50 μm) contained cells that co-expressed Olig2 and p21. Interestingly, brain samples from non-demented controls or with mild cognitive impairment (MCI) contained few Olig2 or NG2 positive cells regardless of p21 or p16. This observation suggests that the Aβ plaque environment induces OPCs to proliferate, migrate and/or hone to the plaque environment. Given that p21 is a marker of OPC differentiation, one possible explanation for elevated p21 expressing OPCs near plaques is that they are differentiating into oligodendrocytes, which may be delayed by the plaque environment as described with aging [153]. Similarly, multiple sclerosis brain lesions more often contain OPCs that have failed to differentiate rather than a lack of OPCs [154–156]. In rodent models of these conditions, increasing the number of OPCs does not improve remyelination [157], presumably due to the unfavorable environment. Thus, it is tempting to speculate that the increase in p21 expressing Olig2 cells near Aβ plaques [152] represents a proper OPC response to proliferate and migrate to the lesion, but the unfavorable plaque environment has delayed or prevented their differentiation into mature oligodendrocytes. In this way, identifying the Aβ plaque-associated molecule(s) that impair OPC differentiation could provide an opportunity for therapeutic intervention. Across studies, additional markers are needed to distinguish whether p21 expressing OPCs are actively differentiating into oligodendrocytes (albeit slowly) or senescent.

Mouse models of Aβ plaque accumulation support the notion that OPCs hone to the plaque environment. The 3xTg-AD mouse model develops Aβ and tau pathology characteristic of patients with AD [158]. Hypertrophic OPCs were observed surrounding and infiltrating Aβ plaques in brains of 3xTgAD mice, which again suggests that OPCs hone to regions of Aβ pathology [159]. Similarly, TgAPP/PS1, mimicked the findings of human brain whereby increased *Cdkn2a* mRNA expression, SA β-gal, and IL-6 levels (SASP factor) were found in OPCs surrounding Aβ plaques to suggest that extracellular protein accumulation may negatively impact OPC function [152]. A novel ZsGreen senescence marker utilizing the p16 promoter was crossed to TgAPP/PS1 mice. The TgAPP/PS1 × ZsGreen mice displayed elevated reporter expression compared to wild type ZsGreen mice indicating significant upregulation of p16 [152]. Treatment with senolytics, Dasatinib and Quercetin [54], alleviated senescence phenotypes, reduced the presence of IL-1β, IL-6 and TNF-α (SASP factors), ameliorated cognitive deficits, and reduced neuroinflammation in OPCs of TgAPP/PS1 mice [152]. Although senescent cells were cleared with administration of senolytics, Aβ load remained unchanged in these mice [152], which may suggest that senescent cells alters cognition independent of Aβ. Moreover, these results may suggest that senolytics cleared the negative regulator of OPC differentiation thus allowing them to develop into mature oligodendrocytes. In this case, other senescent cells may be negatively impacting OPC function through their SASP. Indeed other senescent cell types have been identified in AD including neurons with neurofibrillary tangle pathology [15]. Interestingly, enhanced OPC function has been reported in mice expressing mutant human tau that drives senescence in many other brain cell types including neurons and microglia [15,113,160]. The authors conclude that damaged axons, possibly from senescent tau-containing neurons, promotes OPC differentiation. It is tempting to speculate that neuronal SASP may be responsible for this altered OPC behavior. Indeed, our analyses of transcriptomic differences between neurons with or without NFTs indicate that NFT-bearing senescent neurons [161] differentially express growth factors that may influence OPC behavior.

OPC senescence has been suggested to contribute to disease progression in rodent models of progressive multiple sclerosis (PMS) [162]. High-mobility group box 1 (HMGB1) was identified as a component of the secretome of senescent NPCs [162]. Moreover, HMGB1 induced OPC senescence, as defined by upregulated gene expression of *Cdkn2a*, *Mmp-2*, and *Igfbp2*, and impaired the ability to OPCs to differentiate into oligodendrocytes [162]. The authors interpret that treatment of NPCs with rapamycin, an mTOR inhibitor, reversed the senescence phenotype as evidenced by a decrease in *Cdkn2a* mRNA and protein expression, and decreased SA β-gal intensity [162]. OPCs cultured in media isolated from these rapamycin-treated NPCs were found to differentiate into oligodendrocytes at a higher rate than OPCs cultured in PMS NPC media without rapamycin treatment, further suggesting a role played by senescent NPC environment in successful differentiation of OPCs [162]. Given that senescent cells are stably arrested, we question whether the NPCs in the PMS model are instead in deep quiescence and perhaps not fully senescent [62]. Nonetheless, the beneficial effects of rapamycin on OPC function warrant further investigation for demyelinating diseases.

Concluding Remarks

OPC multipotency offers the brain regenerative capacity throughout life, but at the cost of cancer development. OPCs migrate to lesions, proliferate, and repopulate the area with oligodendrocytes, in an effort to remyelinate and repair. Due to their ability to replicate indefinitely in culture, cellular senescence may play an important protective anti-cancer role in vivo. While OPC senescence-associated phenotypes have been identified in vitro, the in vivo data require further investigation. In particular, the most compelling study in AD used p21 as a marker of OPC senescence [152]. However, given that p21 upregulation is required for OPC differentiation, we cannot discern if these OPCs are experiencing delayed differentiation or if they became senescent and can no longer differentiate. Many of the studies suggest that OPCs hone to sites of injury but fail to differentiate; determining whether these cells represent senescent OPCs requires further investigation. Though several studies have demonstrated senescence-like phenotypes in OPCs, we could not find evidence for SASP or SCAPs. These areas are of primary interest as they will inform on whether future therapeutic approaches should focus on the use of senolytics to treat OPC dysfunction.

5. Microglia

Microglia are known as the primary immune cell of the central nervous system, consisting of 0.5–16.6% of all brain cells in humans [163]. While neurons, astrocytes, and oligodendrocytes originate from the neural tube, microglia are derived from yolk-sac macrophage precursors and migrate to the brain before the blood-brain barrier forms [164,165]. Microglia mostly display postmitotic phenotypes with an estimated turnover rate of 0.08% per day in healthy human cortex with ^{14}C birth dating indicating an average age of 4.2 years [166]. However, they readily proliferate locally in response to injuries [167–169]. Recent studies have also shown that there are latent microglia progenitors that can proliferate and differentiate into microglia in the microglia-depleted brain [170].

Microglia serve a dual function of maintaining the health of the central nervous system and acting as brain macrophages with immune surveillance by monitoring their cellular environment and promoting repair. For example, they remove dying neurons [171], protein aggregates such as Aβ [172] and tau [173], and tissue debris, as well as support synaptic pruning [172]. Through their release of anti-inflammatory cytokines and trophic factors such as IL-4, IL-10, IL-13, TGF-β, they contribute to neuronal and oligodendrocyte survival [174,175]. However, they also can be activated to secrete pro-inflammatory cytokines, TNF-α, IL-1β, IL-6, MIP-1α, reactive oxygen species, and nitric oxide [176,177]. Opposite to anti-tumorigenic pro-inflammatory cytokines, anti-inflammatory cytokines and trophic factor secretion by microglia are known to be pro-tumorigenic as they promote an immunosuppression of the tumor microenvironment, tissue repair, and angiogenesis that enable tumor invasion and survival [178]. Brain cancers such as gliomas are infiltrated with glioma associated microglia (GAM) that are recruited and reprogrammed by glioma cells [179,180]. These microglia increase glioma growth due to their decreased tumor sensing and immune response, while releasing mitogens and invasion promoting factors [181,182]. Given the role of microglia in neuroprotective and immune-modulatory functions, aberrant microglial function has also been linked to neurological diseases such as AD and PD [165].

Numerous studies have reported that microglia can become senescent and/or dystrophic. In the microglia literature, "dystrophic" and "senescent" are often used interchangeably [183]. However, we note that this is distinct from the cellular senescence criteria we use to evaluate the literature. Specifically, dystrophic microglia show morphological changes such as altered cytoplasmic structures and iron overload, but do not necessarily display the other criteria (i.e., shortened telomeres, cell-cycle arrest, lack of proliferation, resistance to apoptosis, and SASP). Activated microglia secrete proinflammatory cytokines and reactive oxygen species similar to senescent microglia. The key difference between activated and senescent microglia is that senescent microglia are unable to proliferate. Here we review microglia senescence, focusing on the pre-defined phenotypes outlined in Section 2.

Activated microglia are those in a hypersensitive state with exaggerated inflammatory responses. Activating stimuli include natural aging, infections, and other systemic inflammatory episodes [184]. For example, aging microglia display impaired neuroprotective abilities including low motility, reduced phagocytic capacity, and the secretion of pro-inflammatory molecules and reactive oxygen species. Repeated intraperitoneal injection of lipopolysaccharide (LPS) induces similar effects of systemic inflammation both in vivo and in vitro. In response to these stressors, activated microglia secrete proinflammatory molecules and reactive oxygen species. To identify different types of activated microglia including the traditionally defined M1 (pro-inflammatory) and M2 (anti-inflammatory) states, mRNA expression or protein levels of inflammatory markers such as *TNF-α*, *IL-1β*, and *TLR2*, and histological analysis of morphology markers such as *Iba-1*, *CD68*, and *CD11b* have been commonly examined [177,185–187]. Single-cell RNA sequencing has recently been used identify a more broadly defined activation response microglia (ARM)/disease-associated microglia (DAM) to examine specific microglial activation gene signatures including *APOE*, *CST3*, and *CD74* [188]. In contrast, various distinctive measures have been used as a biomarker for microglia senescence including detection of senescence-associated heterochromatin foci (SAHF), accumulation of lipofuscin, SA β-gal activity, secretion of matrix metalloproteinase-1 (MMP-1), nitrotyrosine, and other pro-inflammatory molecules, loss of lamin B1 expression, and upregulated p53, p21, and p16 markers [164,189–192]. Altered autophagy and impaired mitochondrial functions were also reported [176,192]. Thus, it is important to not only examine the morphology and secretion of pro-inflammatory molecules of microglia, but also to examine other markers of senescence to distinguish senescent microglia from activated microglia.

Studies in vitro have revealed that microglia can acquire various senescence-like phenotypes. Primary microglia obtained from an amyotrophic lateral sclerosis (ALS) rat model expressing SOD1^{G93A} developed flat morphology, SA β-gal, and increased expressions of p16, p53, and MMP-1 at 12 days in vitro compared to 2 days in vitro. [191]. A similar experiment looked at microglia cells isolated from neonatal mice and compared the cultured cells at 2, 10, and 16 days in vitro [193]. In comparison to culture day 2, day 16 cells revealed altered morphology, increased matrix metalloproteinase (MMP-2) activation,

NF-κB activation, miR-146a expression, SA β-gal activity and decreased autophagy and expression of Toll-like receptor TLR-2 and TLR-4 [193]. Cell culture studies have demonstrated that telomere shortening contributed to microglia senescence. Reduced expression of murine *Tert* and telomere-associated genes were observed in activated primary murine microglia from C57Bl6 mice using mRNA isolation and qRT-PCR [194]. The transcriptomic changes associated with telomerase activity is believed to contribute to microglia senescence as reduced *Tert* mRNA expression is also observed in both ischemia and AD-like pathology in vivo [194]. Experiments in vitro with the BV2 microglia cell line have shown that LPS stimuli repeated 6 times every 48 hours induces acute neuroinflammation, drives heterochromatic foci, increases p53 expression, and increases SA β-gal staining [195]. Thus, chronic inflammation has been suggested as an intrinsic and extrinsic inducer of microglia senescent phenotypes [193,195].

Interestingly, human microglia isolated from Braak stage III or greater AD neuropathology also revealed a higher quantity of microglia with short telomeres compared to that of non-demented individuals using the same assays [196], which provides evidence that microglia from human AD brain with advanced tau neuropathology may be susceptible to cellular senescence. Notably, tau accumulation induces a microglia senescence-like phenotype in transgenic mice [113]. Isolated microglia from 6-month-old TgPS19 transgenic mice that express high levels of mutant human tauP301S showed increased expression of cell cycle regulators p16, p19, and p21, and pro-inflammatory IL-6 compared to control mice through fluorescence-activated cell sorting (FACS) and qRT-PCR analysis [113]. Although this study did not evaluate telomere length, together with the human brain data [196] and other studies on tau-associated senescence [15], the data suggest that neuronal tau pathology may induce a senescence phenotype in microglia.

A relationship between telomere length and senescence has been demonstrated with natural aging as well. Microglia isolated from 30-month-old mice displayed significantly shorter telomeres via telomere flow-FISH and decreased telomerase activity via TRAP compared to the younger controls [196]. Compared to natural aging microglia, LPS injection can generate a model for microglial senescence relatively quickly. Mice from several different studies display strong pro-inflammatory responses to LPS injection, such as large increases in *Tnfα*, Il-6, and Il-1β expression in both a neurodegenerative mouse model of ME7 prion diseases and in middle-aged mice [197,198]. While these pro-inflammatory markers are consistent with SASP, this study [197] did not address other features of senescence such as loss of proliferation, DNA damage, aberrant expression of cell cycle inhibitors, and resistance to apoptosis. Notably, not all markers of senescence fully translate to in vivo studies. An absence of p21/p53 pathway activation, lack of telomere shortening, and increased number of Ki-67 markers were reported for microglia isolated from 24-month-old mouse brains [199]. The authors concluded that the aged microglia may be dysfunctional, but that they do not display a senescent profile [199]. This striking discrepancy between widely accepted in vitro senescent markers and in vivo studies requires better approaches for understanding the differences between induced senescence in vitro and aging in vivo. Furthermore, despite how inflammation can trigger a senescence-like profile in microglial cells, the discrepancy in results between the natural aging and primed microglia studies should perhaps urge researchers to reexamine primed microglia as a model for cellular senescence beyond their pro-inflammatory responses.

Concluding Remarks

Altogether, these studies verified microglial senescence-like phenotypes through observations of morphology [191,193,199], DNA damage [194-196,199,200], lack of proliferation [199], cell-cycle arrest machinery [113,191,195,199], SA β-gal [191,193,195,199], and pro-inflammatory SASP [113,184,193,199]. It is important to note that microglial senescence is complex and heterogeneous. In order to highlight microglial cells as senescent and not simply dysfunctional, we conclude that telomere shortening, activation of cell-cycle arrest machinery, lack of proliferation markers, resistance to apoptosis, and pro-inflammatory

SASP markers are crucial to verify microglial senescent state. The commonly used SA β-gal assay may provide additional insight into microglia lysosomal activity, but cannot be used as a single surrogate to identify senescent cells. In this regard, of the reviewed studies, the study by Stojiljkovic et al. [199] met the predefined criteria outlined in the introduction. As the microglia senescence field moves forward, we recognize that clarity on definitions is needed. Oftentimes, dystrophic and senescent are used interchangeably, but do not imply biological cellular senescence as described in this review. Moreover, activated microglia display many characteristics of senescence. Careful evaluation of the full senescence phenotype is needed to make the distinction.

6. Summary

The accumulation of senescent cells has been reported across tissues in model organisms. Several recent studies have implicated senescent cell accumulation with neurodegenerative diseases [15,34,113,152,201]. Given that age increases both the incidence of neurodegenerative diseases and senescent cell accumulation, determining directionality and causality in the human brain remains unclear [14]. The articles reviewed in this manuscript support the possibility that the cellular senescence stress response may prevent the development of brain cancer. In this way, cells that undergo senescence may evade cancer, but drive neurodegeneration in later life. Thus, the cellular senescence response may contribute to the predisposition of individuals towards either cancer or neurodegeneration. This relationship, however, is complicated by a multitude of factors which have been identified to concurrently increase the risk of neurodegeneration, senescence and cancer. Senescence itself may also increase the incidence of cancer via the SASP [202].

Several concurrent characteristics are needed to positively identify senescent cells. These include proliferative arrest, a permanent change in cell fate, the SASP, a resistance to apoptosis, and possible staining with SA β-gal. Evaluating these criteria in vivo and in situ presents challenges, especially in the human brain. As such, most studies evaluated in this literature review did not measure or meet all of our predefined criteria for defining cellular senescence. While most senescent phenotypes were observed in the brain cell types reviewed, they were rarely observed simultaneously in a single study. By only partially characterizing the phenotype, the results leave room for alternative explanations beyond senescence. A reduction in proliferation and increase in p16 or p21 could indicate quiescent, not senescent NSCs. In particular, whether senescent NSCs exhibit SCAPs has not yet been investigated and their secretory phenotypes have been rarely characterized. Features associated with the senescent phenotype of OPCs are relatively less known. While it has been established that the surrounding environment strongly influence OPCs, further study of the in vivo senescent phenotype is required to reveal SASP or SCAPs associated with OPCs. Current literature utilizes p21 as a biomarker for the senescent phenotype. However, the upregulation of p21 in the differentiation process of OPCs confounds the use of this marker in senescence. In microglia, the SASP was a primary method for identifying senescent cells, but activated microglia also secrete pro-inflammatory molecules and reactive oxygen species. In general, a lack of other methodologies were used to verify that microglia cells were senescent and not simply activated in these studies. Future studies should further verify a resistance to apoptosis in senescent microglia and differentiate between activated, dystrophic, and senescent microglia via proliferative markers.

Staining with SA β-gal is perhaps the most prominent method for identifying senescent brain cells. However, the mechanistic link between SA β-gal staining and senescence remains unclear. The strong staining of quiescent postmitotic neurons additionally calls into question the utility of the method especially in the brain [15,152]. Additionally, many studies discussed in this review used archived tissue and did not utilize proper methods including insufficient controls, replicates, and poor reporting of pH. Lastly, the difficulties associated with dual labeling with SA β-gal makes it difficult to discern which cell type is being observed in vivo, meaning most evidence of cell-specific SA β-gal staining comes from in vitro studies.

Great advances have been made in the past few years regarding the presence of senescent cells in vivo. The use of rodent models carrying reporter transgenes have provided conclusive evidence that senescent cells accumulate with age across tissues. As scientists work to translate these findings to human tissue, there is a need to develop techniques and methodologies to identify, track and study senescent cells at the single cell resolution similar to what has been accomplished in mice. With the advancement of single cell transcriptomics and high-resolution digital profiling, we are on the precipice of developing similar data for human tissues [17,26]. Such data coupled with advanced systems biology approaches [203–206] provide an opportunity to identify robust multianalyte patterns across cells simultaneously and at the single cell resolution. As emphasized throughout this review, no single marker can identify senescent cells. Multianalyte transcriptomic and proteomic data therefore will provide the opportunity to confidently identify senescent cells. Directionality, magnitude and concurrence of senescence-associated pathways will be important in data interpretation. For example, many of the cell types here caution against the utility of using gold standard p16 and p21, especially in isolation, given their critical importance for brain cell function (i.e., NSC quiescence, OPC differentiation, microglial activation). However, a co-occurrence of upregulated cell cycle inhibitors, SCAPs and SASP with concomitant reduction in cell proliferation, differentiation and lysosomal pathways would provide stronger evidence that a cell has entered senescence. We expect comprehensive bioinformatic analyses of these multianalyte measures will also reveal cell-specific SCAPs and SASP with detailed signaling circuits that can be used for drug development. Given that clinical trials for the treatment of AD are underway (NCT04063124 and NCT04685590), results from high resolution studies will be critical in advancing this therapeutic approach.

In conclusion, though additional work is needed to confidently identify, profile and study senescent cells in the brain, this area of research holds great promise. We have highlighted several studies which, while focused on a specific cell type, have broader impacts for senescence in the rest of the body. Novel mechanisms regulating senescence have been explored in brain cells [110,117,118] and senescence in the brain has been shown to have major effects on the entire organism [207]. Future studies focused on senescence in the brain will continue to advance the fields of biology of aging, neuroscience and cancer; and potentially lead to interventions that broadly impact human health and longevity.

Author Contributions: Conceptualization, G.J.G. and M.E.O.; writing—original draft preparation, Abstract and introduction: E.S., B.Z. and M.E.O.; Identifying senescent brain cells: G.J.G. and M.E.O.; NSCs: G.J.G. and M.Y.A.; OPCs: S.K.; microglia: E.S.; Summary: B.Z. and G.J.G.; writing—review and editing, G.J.G., E.S., S.K., M.Y.A., B.Z. and M.E.O.; visualization, M.E.O.; supervision, M.E.O.; funding acquisition, M.E.O. All authors have read and agreed to the published version of the manuscript.

Funding: G.J.G. is funded by the Wake Forest T32, grant number 5T32AG033534; M.E.O. is funded by Veterans Affairs, grant number IK2BX003804; NIH/NIA, grant number F01AG068293; New Vision Research, Charleston Conference for Alzheimer's disease; the Cure Alzheimer's Fund; the Alzheimer's disease Drug Discovery Foundation; B.Z. is funded by NIH/NIA, grant numbers U01AG046170, RF1AG054014, RF1AG057440, R01AG057907, U01AG052411 and R01AG068030

Institutional Review Board Statement: Not applicable.

Informed Consent Statement: Not applicable.

Data Availability Statement: Not applicable.

Acknowledgments: The authors would like to thank Anna Zaia Rodrigues, Valentina Garbarino, and Emma Bennett for insightful comments and edits and Timothy Orr for graphical illustration support.

Conflicts of Interest: The funders had no role in the design of the study; in the collection, analyses, or interpretation of data; in the writing of the manuscript, or in the decision to publish the results.

References

1. Driver, J.A.; Beiser, A.; Au, R.; Kreger, B.E.; Splansky, G.L.; Kurth, T.; Kiel, D.P.; Lu, K.P.; Seshadri, S.; Wolf, P.A. Inverse association between cancer and Alzheimer's disease: Results from the Framingham Heart Study. *BMJ* **2012**, *344*, e1442. [CrossRef] [PubMed]
2. Roe, C.M.; Fitzpatrick, A.L.; Xiong, C.; Sieh, W.; Kuller, L.; Miller, J.P.; Williams, M.M.; Kopan, R.; Behrens, M.I.; Morris, J.C. Cancer linked to Alzheimer disease but not vascular dementia. *Neurology* **2009**, *74*, 106–112. [CrossRef] [PubMed]
3. Tabarés-Seisdedos, R.; Dumont, N.; Baudot, A.; Valderas, J.M.; Climent, J.; Valencia, A.; Crespo-Facorro, B.; Vieta, E.; Gómez-Beneyto, M.; Martínez, S.; et al. No paradox, no progress: Inverse cancer comorbidity in people with other complex diseases. *Lancet Oncol.* **2011**, *12*, 604–608. [CrossRef]
4. Driver, J.A.; Logroscino, G.; Buring, J.E.; Gaziano, J.M.; Kurth, T. A prospective cohort study of cancer incidence following the diagnosis of Parkinson's disease. *Cancer Epidemiol. Biomark. Prev.* **2007**, *16*, 1260–1265. [CrossRef]
5. Shi, H.B.; Tang, B.; Liu, Y.W.; Wang, X.F.; Chen, G.J. Alzheimer disease and cancer risk: A meta-analysis. *J. Cancer Res. Clin. Oncol.* **2015**, *141*, 485–494. [CrossRef]
6. Houck, A.L.; Seddighi, S.; Driver, J.A. At the Crossroads Between Neurodegeneration and Cancer: A Review of Overlapping Biology and Its Implications. *Curr. Aging Sci.* **2019**, *11*, 77–89. [CrossRef]
7. Liu, J.; Zhang, C.; Hu, W.; Feng, Z. Parkinson's disease-associated protein Parkin: An unusual player in cancer. *Cancer Commun. (Lond.)* **2018**, *38*, 40. [CrossRef]
8. Irizar, P.A.; Schäuble, S.; Esser, D.; Groth, M.; Frahm, C.; Priebe, S.; Baumgart, M.; Hartmann, N.; Marthandan, S.; Menzel, U.; et al. Transcriptomic alterations during ageing reflect the shift from cancer to degenerative diseases in the elderly. *Nat. Commun.* **2018**, *9*, 327. [CrossRef] [PubMed]
9. Congrains, A.; Kamide, K.; Ohishi, M.; Rakugi, H. ANRIL: Molecular Mechanisms and Implications in Human Health. *Int. J. Mol. Sci.* **2013**, *14*, 1278–1292. [CrossRef]
10. Rayess, H.; Wang, M.B.; Srivatsan, E.S. Cellular senescence and tumor suppressor gene p16. *Int. J. Cancer* **2011**, *130*, 1715–1725. [CrossRef] [PubMed]
11. He, S.; Sharpless, N.E. Senescence in Health and Disease. *Cell* **2017**, *169*, 1000–1011. [CrossRef] [PubMed]
12. López-Otín, C.; Blasco, M.A.; Partridge, L.; Serrano, M.; Kroemer, G. The Hallmarks of Aging. *Cell* **2013**, *153*, 1194–1217. [CrossRef]
13. McHugh, D.; Gil, J. Senescence and aging: Causes, consequences, and therapeutic avenues. *J. Cell Biol.* **2017**, *217*, 65–77. [CrossRef]
14. Saez-Atienzar, S.; Masliah, E. Cellular senescence and Alzheimer disease: The egg and the chicken scenario. *Nat. Rev. Neurosci.* **2020**, *21*, 433–444. [CrossRef] [PubMed]
15. Musi, N.; Valentine, J.M.; Sickora, K.R.; Baeuerle, E.; Thompson, C.S.; Shen, Q.; Orr, M.E. Tau protein aggregation is associated with cellular senescence in the brain. *Aging Cell* **2018**, *17*, e12840. [CrossRef]
16. Justice, J.N.; Gregory, M.H.; Tchkonia, T.; Lebrasseur, N.K.; Kirkland, J.L.; Kritchevsky, S.B.; Nicklas, B.J. Cellular Senescence Biomarker p16INK4a+ Cell Burden in Thigh Adipose is Associated With Poor Physical Function in Older Women. *J. Gerontol. Ser. A Boil. Sci. Med Sci.* **2017**, *73*, 939–945. [CrossRef] [PubMed]
17. Uyar, B.; Palmer, D.; Kowald, A.; Escobar, H.M.; Barrantes, I.; Möller, S.; Akalin, A.; Fuellen, G. Single-cell analyses of aging, inflammation and senescence. *Ageing Res. Rev.* **2020**, *64*, 101156. [CrossRef]
18. Hernandez-Segura, A.; De Jong, T.V.; Melov, S.; Guryev, V.; Campisi, J.; DeMaria, M. Unmasking Transcriptional Heterogeneity in Senescent Cells. *Curr. Biol.* **2017**, *27*, 2652–2660.e4. [CrossRef]
19. Selbach, M.; Schwanhäusser, B.; Thierfelder, N.; Fang, Z.; Khanin, R.; Rajewsky, N. Widespread changes in protein synthesis induced by microRNAs. *Nat. Cell Biol.* **2008**, *455*, 58–63. [CrossRef]
20. Hernandez-Segura, A.; Nehme, J.; DeMaria, M. Hallmarks of Cellular Senescence. *Trends Cell Biol.* **2018**, *28*, 436–453. [CrossRef]
21. Gorgoulis, V.; Adams, P.D.; Alimonti, A.; Bennett, D.C.; Bischof, O.; Bishop, C.; Campisi, J.; Collado, M.; Evangelou, K.; Ferbeyre, G.; et al. Cellular Senescence: Defining a Path Forward. *Cell* **2019**, *179*, 813–827. [CrossRef]
22. Matjusaitis, M.; Chin, G.; Sarnoski, E.A.; Stolzing, A. Biomarkers to identify and isolate senescent cells. *Ageing Res. Rev.* **2016**, *29*, 1–12. [CrossRef]
23. Dodig, S.; Čepelak, I.; Pavić, I. Hallmarks of senescence and aging. *Biochem. Med.* **2019**, *29*, 483–497. [CrossRef]
24. González-Gualda, E.; Baker, A.G.; Fruk, L.; Muñoz-Espín, D. A guide to assessing cellular senescence in vitro and in vivo. *FEBS J.* **2021**, *288*, 56–80. [CrossRef]
25. Casella, G.; Munk, R.; Kim, K.M.; Piao, Y.; De, S.; Abdelmohsen, K.; Gorospe, M. Transcriptome signature of cellular senescence. *Nucleic Acids Res.* **2019**, *47*, 7294–7305. [CrossRef]
26. Roy, A.L.; Sierra, F.; Howcroft, K.; Singer, D.S.; Sharpless, N.; Hodes, R.J.; Wilder, E.L.; Anderson, J.M. A Blueprint for Characterizing Senescence. *Cell* **2020**, *183*, 1143–1146. [CrossRef]
27. Hayflick, L.; Moorhead, P. The serial cultivation of human diploid cell strains. *Exp. Cell Res.* **1961**, *25*, 585–621. [CrossRef]
28. Homem, C.C.F.; Repic, M.; Knoblich, J.A. Proliferation control in neural stem and progenitor cells. *Nat. Rev. Neurosci.* **2015**, *16*, 647–659. [CrossRef] [PubMed]
29. Tomita, K.-I.; Aida, J.; Izumiyama-Shimomura, N.; Nakamura, K.-I.; Ishikawa, N.; Matsuda, Y.; Arai, T.; Ishiwata, T.; Kumasaka, T.; Takahashi-Fujigasaki, J.; et al. Changes in telomere length with aging in human neurons and glial cells revealed by quantitative fluorescence in situ hybridization analysis. *Geriatr. Gerontol. Int.* **2018**, *18*, 1507–1512. [CrossRef] [PubMed]

30. Minamino, T.; Miyauchi, H.; Yoshida, T.; Komuro, I. Endothelial cell senescence in human atherosclerosis: Role of telomeres in endothelial dysfunction. *J. Cardiol.* **2003**, *41*, 1541–1544. [CrossRef] [PubMed]

31. Qian, H.; Kang, X.; Hu, J.; Zhang, D.; Liang, Z.; Meng, F.; Zhang, X.; Xue, Y.; Maimon, R.; Dowdy, S.F.; et al. Reversing a model of Parkinson's disease with in situ converted nigral neurons. *Nat. Cell Biol.* **2020**, *582*, 550–556. [CrossRef]

32. Sapieha, P.; Mallette, F.A. Cellular Senescence in Postmitotic Cells: Beyond Growth Arrest. *Trends Cell Biol.* **2018**, *28*, 595–607. [CrossRef] [PubMed]

33. Jurk, D.; Wang, C.; Miwa, S.; Maddick, M.; Korolchuk, V.; Tsolou, A.; Gonos, E.S.; Thrasivoulou, C.; Saffrey, M.J.; Cameron, K.; et al. Postmitotic neurons develop a p21-dependent senescence-like phenotype driven by a DNA damage response. *Aging Cell* **2012**, *11*, 996–1004. [CrossRef] [PubMed]

34. Riessland, M.; Kolisnyk, B.; Kim, T.W.; Cheng, J.; Ni, J.; Pearson, J.A.; Park, E.J.; Dam, K.; Acehan, D.; Ramos-Espiritu, L.S.; et al. Loss of SATB1 Induces p21-Dependent Cellular Senescence in Post-mitotic Dopaminergic Neurons. *Cell Stem Cell* **2019**, *25*, 514–530.e8. [CrossRef] [PubMed]

35. Chow, H.-M.; Shi, M.; Cheng, A.; Gao, Y.; Chen, G.; Song, X.; So, R.W.L.; Zhang, J.; Herrup, K. Age-related hyperinsulinemia leads to insulin resistance in neurons and cell-cycle-induced senescence. *Nat. Neurosci.* **2019**, *22*, 1806–1819. [CrossRef]

36. Turner, K.J.; Vasu, V.; Griffin, D.K. Telomere Biology and Human Phenotype. *Cells* **2019**, *8*, 73. [CrossRef]

37. Zhu, Y.; Liu, X.; Ding, X.; Wang, F.; Geng, X. Telomere and its role in the aging pathways: Telomere shortening, cell senescence and mitochondria dysfunction. *Biogerontology* **2019**, *20*, 1–16. [CrossRef] [PubMed]

38. Lai, T.-P.; Wright, W.E.; Shay, J.W. Comparison of telomere length measurement methods. *Philos. Trans. R. Soc. B Biol. Sci.* **2018**, *373*, 20160451. [CrossRef]

39. Lai, T.-P.; Zhang, N.; Noh, J.; Mender, I.; Tedone, E.; Huang, E.; Wright, W.E.; Danuser, G.; Shay, J.W. A method for measuring the distribution of the shortest telomeres in cells and tissues. *Nat. Commun.* **2017**, *8*, 1–14. [CrossRef]

40. Zhang, X.; Lou, X.; Xia, F. Advances in the detection of telomerase activity using isothermal amplification. *Theranostics* **2017**, *7*, 1847–1862. [CrossRef]

41. Zheng, K.-W.; Liu, C.; Meng, Q.; Hao, Y.-H.; Zheng, J.-P.; Li, W.; Tan, Z. One-Step High-Throughput Telomerase Activity Measurement of Cell Populations, Single Cells, and Single-Enzyme Complexes. *ACS Omega* **2020**, *5*, 24666–24673. [CrossRef]

42. Liu, Y.; Li, S.; Zhang, L.; Zhao, Q.; Li, N.; Wu, Y. Catalytic Hairpin Assembly-Assisted Rolling Circle Amplification for High-Sensitive Telomerase Activity Detection. *ACS Omega* **2020**, *5*, 11836–11841. [CrossRef]

43. Díaz-Cartagena, D.C.; Hernández-Cancel, G.; Bracho-Rincón, D.P.; González-Feliciano, J.A. Cunci, L.; González, C.I.; Cabrera, C.R. Label-Free Telomerase Activity Detection via Electrochemical Impedance Spectroscopy. *ACS Omega* **2019**, *4*, 16724–16732. [CrossRef]

44. Takahashi, A.; Ohtani, N.; Yamakoshi, K.; Iida, S.-I.; Tahara, H.; Nakayama, K.; Nakayama, K.I.; Ide, T.; Saya, H.; Hara, E. Mitogenic signalling and the p16INK4a–Rb pathway cooperate to enforce irreversible cellular senescence. *Nat. Cell Biol.* **2006**, *8*, 1291–1297. [CrossRef]

45. Childs, B.G.; Gluscevic, M.; Baker, D.J.; Laberge, R.-M.; Marquess, D.; Dananberg, J.; Van Deursen, J.M. Senescent cells: An emerging target for diseases of ageing. *Nat. Rev. Drug Discov.* **2017**, *16*, 718–735. [CrossRef] [PubMed]

46. Banito, A.; Rashid, S.T.; Acosta, J.C.; De Li, S.; Pereira, C.F.; Geti, I.; Pinho, S.; Silva, J.C.; Azuara, V.; Walsh, M.; et al. Senescence impairs successful reprogramming to pluripotent stem cells. *Genes Dev.* **2009**, *23*, 2134–2139. [CrossRef] [PubMed]

47. Mertens, J.; Paquola, A.C.; Ku, M.; Hatch, E.M.; Böhnke, L.; Ladjevardi, S.; McGrath, S.; Campbell, B.; Lee, H.; Herdy, J.R.; et al. Directly Reprogrammed Human Neurons Retain Aging-Associated Transcriptomic Signatures and Reveal Age-Related Nucleocytoplasmic Defects. *Cell Stem Cell* **2015**, *17*, 705–718. [CrossRef]

48. van Deursen, J.M. The role of senescent cells in ageing. *Nature* **2014**, *509*, 439–446. [CrossRef] [PubMed]

49. Childs, B.G.; Durik, M.; Baker, D.J.; Van Deursen, J.M. Cellular senescence in aging and age-related disease: From mechanisms to therapy. *Nat. Med.* **2015**, *21*, 1424–1435. [CrossRef]

50. Muñoz-Espín, D.; Serrano, M. Cellular senescence: From physiology to pathology. *Nat. Rev. Mol. Cell Biol.* **2014**, *15*, 482–496. [CrossRef]

51. Fan, Y.; Cheng, J.; Zeng, H.; Shao, L. Senescent Cell Depletion through Targeting BCL-Family Proteins and Mitochondria. *Front. Physiol.* **2020**, *11*, 593630. [CrossRef]

52. Ogrodnik, M.; Salmonowicz, H.; Gladyshev, V.N. Integrating cellular senescence with the concept of damage accumulation in aging: Relevance for clearance of senescent cells. *Aging Cell* **2019**, *18*, e12841. [CrossRef] [PubMed]

53. Herranz, N.; Gil, J. Mechanisms and functions of cellular senescence. *J. Clin. Investig.* **2018**, *128*, 1238–1246. [CrossRef]

54. Zhu, Y.; Tchkonia, T.; Pirtskhalava, T.; Gower, A.C.; Ding, H.; Giorgadze, N.; Palmer, A.K.; Ikeno, Y.; Hubbard, G.B.; Lenburg, M.; et al. The Achilles' heel of senescent cells: From transcriptome to senolytic drugs. *Aging Cell* **2015**, *14*, 644–658. [CrossRef]

55. Soto-Gamez, A.; Quax, W.J.; DeMaria, M. Regulation of Survival Networks in Senescent Cells: From Mechanisms to Interventions. *J. Mol. Biol.* **2019**, *431*, 2629–2643. [CrossRef] [PubMed]

56. Rao, S.G.; Jackson, J.G. SASP: Tumor Suppressor or Promoter? Yes! *Trends Cancer* **2016**, *2*, 676–687. [CrossRef] [PubMed]

57. Birch, J.; Gil, J. Senescence and the SASP: Many therapeutic avenues. *Genes Dev.* **2020**, *34*, 1565–1576. [CrossRef] [PubMed]

58. Lopes-Paciencia, S.; Saint-Germain, E.; Rowell, M.-C.; Ruiz, A.F.; Kalegari, P.; Ferbeyre, G. The senescence-associated secretory phenotype and its regulation. *Cytokine* **2019**, *117*, 15–22. [CrossRef] [PubMed]

59. Schafer, M.J.; Haak, A.J.; Tschumperlin, D.J.; Lebrasseur, N.K. Targeting Senescent Cells in Fibrosis: Pathology, Paradox, and Practical Considerations. *Curr. Rheumatol. Rep.* **2018**, *20*, 3. [CrossRef]

60. Dimri, G.P.; Lee, X.; Basile, G.; Acosta, M.; Scott, G.; Roskelley, C.; Medrano, E.E.; Linskens, M.; Rubelj, I.; Pereira-Smith, O.; et al. A biomarker that identifies senescent human cells in culture and in aging skin in vivo. *Proc. Natl. Acad. Sci. USA* **1995**, *92*, 9363–9367. [CrossRef] [PubMed]

61. Lee, B.Y.; Han, J.A.; Im, J.S.; Morrone, A.; Johung, K.; Goodwin, E.C.; Kleijer, W.J.; DiMaio, D.; Hwang, E.S. Senescence-associated beta-galactosidase is lysosomal beta-galactosidase. *Aging Cell* **2006**, *5*, 187–195. [CrossRef]

62. Fujimaki, K.; Li, R.; Chen, H.; Della Croce, K.; Zhang, H.H.; Xing, J.; Bai, F.; Yao, G. Graded regulation of cellular quiescence depth between proliferation and senescence by a lysosomal dimmer switch. *Proc. Natl. Acad. Sci. USA* **2019**, *116*, 22624–22634. [CrossRef]

63. Cai, Y.; Zhou, H.; Zhu, Y.; Sun, Q.; Ji, Y.; Xue, A.; Wang, Y.; Chen, W.; Yu, X.; Wang, L.; et al. Elimination of senescent cells by beta-galactosidase-targeted prodrug attenuates inflammation and restores physical function in aged mice. *Cell Res.* **2020**, *30*, 574–589. [CrossRef] [PubMed]

64. Mao, Z.; Ke, Z.; Gorbunova, V.; Seluanov, A. Replicatively senescent cells are arrested in G1 and G2 phases. *Aging* **2012**, *4*, 431–435. [CrossRef] [PubMed]

65. Hammelrath, L.; Škokić, S.; Khmelinskii, A.; Hess, A.; Van Der Knaap, N.; Staring, M.; Lelieveldt, B.P.; Wiedermann, D.; Hoehn, M. Morphological maturation of the mouse brain: An in vivo MRI and histology investigation. *NeuroImage* **2016**, *125*, 144–152. [CrossRef]

66. Semple, B.D.; Blomgren, K.; Gimlin, K.; Ferriero, D.M.; Noble-Haeusslein, L.J. Brain development in rodents and humans: Identifying benchmarks of maturation and vulnerability to injury across species. *Prog. Neurobiol.* **2013**, *106-107*, 1–16. [CrossRef]

67. Altman, J. Are new neurons formed in the brains of adult mammals? *Science* **1962**, *135*, 1127–1128. [CrossRef] [PubMed]

68. Sanai, N.; Tramontin, A.D.; Quiñones-Hinojosa, A.; Barbaro, N.M.; Gupta, N.; Kunwar, S.; Lawton, M.T.; McDermott, M.W.; Parsa, A.T.; Verdugo, J.M.-G.; et al. Unique astrocyte ribbon in adult human brain contains neural stem cells but lacks chain migration. *Nat. Cell Biol.* **2004**, *427*, 740–744. [CrossRef]

69. Temple, S. Division and differentiation of isolated CNS blast cells in microculture. *Nat. Cell Biol.* **1989**, *340*, 471–473. [CrossRef] [PubMed]

70. Reynolds, B.; Weiss, S. Generation of neurons and astrocytes from isolated cells of the adult mammalian central nervous system. *Science* **1992**, *255*, 1707–1710. [CrossRef]

71. Gonzalez-Perez, O.; Quiñones-Hinojosa, A. Astrocytes as neural stem cells in the adult brain. *J. Stem Cells* **2012**, *7*, 181–188.

72. Doetsch, F.; Caillé, I.; Lim, D.A.; García-Verdugo, J.M.; Alvarez-Buylla, A. Subventricular Zone Astrocytes Are Neural Stem Cells in the Adult Mammalian Brain. *Cell* **1999**, *97*, 703–716. [CrossRef]

73. Knoth, R.; Singec, I.; Ditter, M.; Pantazis, G.; Capetian, P.; Meyer, R.P.; Horvat, V.; Volk, B.; Kempermann, G. Murine Features of Neurogenesis in the Human Hippocampus across the Lifespan from 0 to 100 Years. *PLoS ONE* **2010**, *5*, e8809. [CrossRef] [PubMed]

74. Spalding, K.L.; Bergmann, O.; Alkass, K.; Bernard, S.; Salehpour, M.; Huttner, H.B.; Boström, E.; Westerlund, I.; Vial, C.; Buchholz, B.A.; et al. Dynamics of Hippocampal Neurogenesis in Adult Humans. *Cell* **2013**, *153*, 1219–1227. [CrossRef]

75. Eriksson, P.S.; Perfilieva, E.; Björk-Eriksson, T.; Alborn, A.-M.; Nordborg, C.; Peterson, D.A.; Gage, F.H. Neurogenesis in the adult human hippocampus. *Nat. Med.* **1998**, *4*, 1313–1317. [CrossRef]

76. Moreno-Jiménez, E.P.; Flor-García, M.; Terreros-Roncal, J.; Rábano, A.; Cafini, F.; Pallas-Bazarra, N.; Ávila, J.; Llorens-Martín, M. Adult hippocampal neurogenesis is abundant in neurologically healthy subjects and drops sharply in patients with Alzheimer's disease. *Nat. Med.* **2019**, *25*, 554–560. [CrossRef] [PubMed]

77. Tobin, M.K.; Musaraca, K.; Disouky, A.; Shetti, A.; Bheri, A.; Honer, W.G.; Kim, N.; Dawe, R.J.; Bennett, D.A.; Arfanakis, K.; et al. Human Hippocampal Neurogenesis Persists in Aged Adults and Alzheimer's Disease Patients. *Cell Stem Cell* **2019**, *24*, 974–982.e3. [CrossRef]

78. Nait-Oumesmar, B.; Picard-Riera, N.; Kerninon, C.; Decker, L.; Seilhean, D.; Höglinger, G.U.; Hirsch, E.C.; Reynolds, R.; Evercooren, A.B.-V. Activation of the subventricular zone in multiple sclerosis: Evidence for early glial progenitors. *Proc. Natl. Acad. Sci. USA* **2007**, *104*, 4694–4699. [CrossRef] [PubMed]

79. Donega, V.; Burm, S.M.; Van Strien, M.E.; Van Bodegraven, E.J.; Paliukhovich, I.; Geut, H.; Van De Berg, W.D.J.; Li, K.W.; Smit, A.B.; Basak, O.; et al. Transcriptome and proteome profiling of neural stem cells from the human subventricular zone in Parkinson's disease. *Acta Neuropathol. Commun.* **2019**, *7*, 84. [CrossRef]

80. Boldrini, M.; Fulmore, C.A.; Tartt, A.N.; Simeon, L.R.; Pavlova, I.; Poposka, V.; Rosoklija, G.B.; Stankov, A.; Arango, V.; Dwork, A.J.; et al. Human Hippocampal Neurogenesis Persists throughout Aging. *Cell Stem Cell* **2018**, *22*, 589–599.e5. [CrossRef]

81. Sorrells, S.F.; Paredes, M.F.; Cebrian-Silla, A.; Sandoval, K.; Qi, D.; Kelley, K.W.; James, D.; Mayer, S.; Chang, J.; Auguste, K.I.; et al. Human hippocampal neurogenesis drops sharply in children to undetectable levels in adults. *Nature* **2018**, *555*, 377–381. [CrossRef] [PubMed]

82. Dennis, C.V.; Suh, L.S.; Rodriguez, M.L.; Kril, J.J.; Sutherland, G.T. Human adult neurogenesis across the ages: An immunohistochemical study. *Neuropathol. Appl. Neurobiol.* **2016**, *42*, 621–638. [CrossRef]

83. Sanai, N.; Nguyen, T.; Ihrie, R.A.; Mirzadeh, Z.; Tsai, H.-H.; Wong, M.; Gupta, N.; Berger, M.S.; Huang, E.; Garcia-Verdugo, J.-M.; et al. Corridors of migrating neurons in the human brain and their decline during infancy. *Nat. Cell Biol.* **2011**, *478*, 382–386. [CrossRef]

84. Mathews, K.J.; Allen, K.M.; Boerrigter, D.; Ball, H.; Weickert, C.S.; Double, K.L. Evidence for reduced neurogenesis in the aging human hippocampus despite stable stem cell markers. *Aging Cell* **2017**, *16*, 1195–1199. [CrossRef] [PubMed]

85. Morshead, C.M.; Reynolds, B.A.; Craig, C.G.; McBurney, M.W.; Staines, W.A.; Morassutti, D.; Weiss, S.; Van Der Kooy, D. Neural stem cells in the adult mammalian forebrain: A relatively quiescent subpopulation of subependymal cells. *Neuron* **1994**, *13*, 1071–1082. [CrossRef]

86. Cheung, T.H.; Rando, T.A. Molecular regulation of stem cell quiescence. *Nat. Rev. Mol. Cell Biol.* **2013**, *14*, 329–340. [CrossRef] [PubMed]

87. Urbán, N.; Blomfield, I.M.; Guillemot, F. Quiescence of Adult Mammalian Neural Stem Cells: A Highly Regulated Rest. *Neuron* **2019**, *104*, 834–848. [CrossRef] [PubMed]

88. Obernier, K.; Alvarez-Buylla, A. Neural stem cells: Origin, heterogeneity and regulation in the adult mammalian brain. *Development* **2019**, *146*, dev156059. [CrossRef]

89. Nicaise, A.M.; Willis, C.M.; Crocker, S.J.; Pluchino, S. Stem Cells of the Aging Brain. *Front. Aging Neurosci.* **2020**, *12*, 247. [CrossRef]

90. Patel, A.P.; Fisher, J.L.; Nichols, E.; Abd-Allah, F.; Abdela, J.; Abdelalim, A.; Abraha, H.; Agius, D.; Alahdab, F.; Alam, T.; et al. Global, regional, and national burden of brain and other CNS cancer, 1990-2016: A systematic analysis for the Global Burden of Disease Study. *Lancet Neurol.* **2019**, *18*, 376–393. [CrossRef]

91. Azzarelli, R.; Simons, B.D.; Philpott, A. The developmental origin of brain tumours: A cellular and molecular framework. *Development* **2018**, *145*, dev162693. [CrossRef] [PubMed]

92. Lee, J.H.; Lee, J.E.; Kahng, J.Y.; Kim, S.H.; Park, J.S.; Yoon, S.J.; Um, J.-Y.; Kim, W.K.; Lee, J.-K.; Park, J.; et al. Human glioblastoma arises from subventricular zone cells with low-level driver mutations. *Nature* **2018**, *560*, 243–247. [CrossRef]

93. Di Micco, R.; Fumagalli, M.; Cicalese, A.; Piccinin, S.; Gasparini, P.; Luise, C.; Schurra, C.; Garre', M.; Nuciforo, P.G.; Bensimon, A.; et al. Oncogene-induced senescence is a DNA damage response triggered by DNA hyper-replication. *Nat. Cell. Biol.* **2006**, *444*, 638–642. [CrossRef] [PubMed]

94. Kuilman, T.; Michaloglou, C.; Vredeveld, L.C.; Douma, S.; Van Doorn, R.; Desmet, C.J.; Aarden, L.A.; Mooi, W.J.; Peeper, D.S. Oncogene-Induced Senescence Relayed by an Interleukin-Dependent Inflammatory Network. *Cell* **2008**, *133*, 1019–1031. [CrossRef] [PubMed]

95. Raabe, E.H.; Lim, K.S.; Kim, J.M.; Meeker, A.; Mao, X.-G.; Nikkhah, G.; Maciaczyk, J.; Kahlert, U.; Jain, D.; Bar, E.; et al. BRAF Activation Induces Transformation and Then Senescence in Human Neural Stem Cells: A Pilocytic Astrocytoma Model. *Clin. Cancer Res.* **2011**, *17*, 3590–3599. [CrossRef] [PubMed]

96. Acharya, M.M.; Lan, M.L.; Kan, V.H.; Patel, N.H.; Giedzinski, E.; Tseng, B.P.; Limoli, C.L. Consequences of ionizing radiation-induced damage in human neural stem cells. *Free. Radic. Biol. Med.* **2010**, *49*, 1846–1855. [CrossRef]

97. Monje, M.L.; Mizumatsu, S.; Fike, J.R.; Palmer, T.D. Irradiation induces neural precursor-cell dysfunction. *Nat. Med.* **2002**, *8*, 955–962. [CrossRef]

98. Zou, Y.; Zhang, N.; Ellerby, L.M.; Davalos, A.R.; Zeng, X.; Campisi, J.; Desprez, P.-Y. Responses of human embryonic stem cells and their differentiated progeny to ionizing radiation. *Biochem. Biophys. Res. Commun.* **2012**, *426*, 100–105. [CrossRef]

99. Schneider, L.; Pellegatta, S.; Favaro, R.; Pisati, F.; Roncaglia, P.; Testa, G.; Nicolis, S.K.; Finocchiaro, G.; Di Fagagna, F.D. DNA Damage in Mammalian Neural Stem Cells Leads to Astrocytic Differentiation Mediated by BMP2 Signaling through JAK-STAT. *Stem Cell Rep.* **2013**, *1*, 123–138. [CrossRef]

100. Bitto, A.; Sell, C.; Crowe, E.; Lorenzini, A.; Malaguti, M.; Hrelia, S.; Torres, C. Stress-induced senescence in human and rodent astrocytes. *Exp. Cell Res.* **2010**, *316*, 2961–2968. [CrossRef]

101. Dong, C.-M.; Wang, X.-L.; Wang, G.-M.; Zhang, W.-J.; Zhu, L.; Gao, S.; Yang, D.-J.; Qin, Y.; Liang, Q.-J.; Chen, Y.-L.; et al. A stress-induced cellular aging model with postnatal neural stem cells. *Cell Death Dis.* **2014**, *5*, e1116. [CrossRef] [PubMed]

102. Daniele, S.; Da Pozzo, E.; Iofrida, C.; Martini, C. Human Neural Stem Cell Aging Is Counteracted by α-Glycerylphosphorylethanolamine. *ACS Chem. Neurosci.* **2016**, *7*, 952–963. [CrossRef]

103. Campisi, J. Aging, Cellular Senescence, and Cancer. *Annu. Rev. Physiol.* **2013**, *75*, 685–705. [CrossRef] [PubMed]

104. Hou, Y.; Dan, X.; Babbar, M.; Wei, Y.; Hasselbalch, S.G.; Croteau, D.L.; Bohr, V.A. Ageing as a risk factor for neurodegenerative disease. *Nat. Rev. Neurol.* **2019**, *15*, 565–581. [CrossRef] [PubMed]

105. Houben, S.; Leroy, K.; Ando, K.; Yilmaz, Z.; Widomski, C.; Buée, L.; Brion, J.-P. Genetic ablation of tau in postnatal neurons rescues decreased adult hippocampal neurogenesis in a tauopathy model. *Neurobiol. Dis.* **2019**, *127*, 131–141. [CrossRef]

106. Komuro, Y.; Xu, G.; Bhaskar, K.; Lamb, B.T. Human tau expression reduces adult neurogenesis in a mouse model of tauopathy. *Neurobiol. Aging* **2015**, *36*, 2034–2042. [CrossRef] [PubMed]

107. Jack, C.R.; Bennett, D.A.; Blennow, K.; Carrillo, M.C.; Dunn, B.; Haeberlein, S.B.; Holtzman, D.M.; Jagust, W.; Jessen, F.; Karlawish, J.; et al. NIA-AA Research Framework: Toward a biological definition of Alzheimer's disease. *Alzheimers Dement.* **2018**, *14*, 535–562. [CrossRef]

108. He, N.; Jin, W.L.; Lok, K.H.; Wang, Y.; Yin, M.; Wang, Z.J. Amyloid-beta(1-42) oligomer accelerates senescence in adult hippocampal neural stem/progenitor cells via formylpeptide receptor. *Cell Death Dis.* **2013**, *4*, e924. [CrossRef]

109. Orr, M.E.; Pitstick, R.; Canine, B.; Ashe, K.H.; Carlson, G.A. Genotype-Specific Differences between Mouse CNS Stem Cell Lines Expressing Frontotemporal Dementia Mutant or Wild Type Human Tau. *PLoS ONE* **2012**, *7*, e39328. [CrossRef]

110. Voisin, J.; Farina, F.; Naphade, S.; Fontaine, M.; Tshilenge, K.T.; Aguirre, C.G.; Lopez-Ramirez, A.; Dancourt, J.; Ginisty, A.; Nair, S.S.; et al. FOXO3 targets are reprogrammed as Huntington's disease neural cells and striatal neurons face senescence with p16(INK4a) increase. *Aging Cell* **2020**, *19*, e13226. [CrossRef]

111. Orr, M.E.; Sullivan, A.C.; Frost, B. A Brief Overview of Tauopathy: Causes, Consequences, and Therapeutic Strategies. *Trends Pharmacol. Sci.* **2017**, *38*, 637–648. [CrossRef]

112. Strang, K.H.; Golde, T.E.; Giasson, B.I. MAPT mutations, tauopathy, and mechanisms of neurodegeneration. *Lab. Investig.* **2019**, *99*, 912–928. [CrossRef] [PubMed]

113. Bussian, T.J.; Aziz, A.; Meyer, C.F.; Swenson, B.L.; Van Deursen, J.M.; Baker, D.J. Clearance of senescent glial cells prevents tau-dependent pathology and cognitive decline. *Nat. Cell Biol.* **2018**, *562*, 578–582. [CrossRef] [PubMed]

114. Karch, C.M.; Kao, A.W.; Karydas, A.; Onanuga, K.; Martinez, R.; Argouarch, A.; Wang, C.; Huang, C.; Sohn, P.D.; Bowles, K.R.; et al. A Comprehensive Resource for Induced Pluripotent Stem Cells from Patients with Primary Tauopathies. *Stem Cell Rep.* **2019**, *13*, 939–955. [CrossRef]

115. Liang, K.X.; Kristiansen, C.K.; Mostafavi, S.; Vatne, G.H.; Zantingh, G.A.; Kianian, A.; Tzoulis, C.; Høyland, L.E.; Ziegler, M.; Perez, R.M.; et al. Disease-specific phenotypes in iPSC -derived neural stem cells with POLG mutations. *EMBO Mol. Med.* **2020**, *12*, 12146. [CrossRef]

116. Gao, J.; Wu, Y.; He, D.; Zhu, X.; Li, H.; Liu, H.; Liu, H. Anti-aging effects of Ribes meyeri anthocyanins on neural stem cells and aging mice. *Aging* **2020**, *12*, 17738–17753. [CrossRef]

117. Hu, G.; Xia, Y.; Chen, B.; Zhang, J.; Gong, L.; Chen, Y.; Li, Q.; Wang, Y.; Deng, Z. ESC-sEVs Rejuvenate Aging Hippocampal NSCs by Transferring SMADs to Regulate the MYT1-Egln3-Sirt1 Axis. *Mol. Ther.* **2021**, *29*, 103–120. [CrossRef]

118. Xiao, Y.Z.; Yang, M.; Xiao, Y.; Guo, Q.; Huang, Y.; Li, C.J.; Cai, D.; Luo, X.H. Reducing Hypothalamic Stem Cell Senescence Protects against Aging-Associated Physiological Decline. *Cell Metab.* **2020**, *31*, 534–548.e5. [CrossRef] [PubMed]

119. Molofsky, A.V.; Pardal, R.; Iwashita, T.; Park, I.-K.; Clarke, M.F.; Morrison, S.J. Bmi-1 dependence distinguishes neural stem cell self-renewal from progenitor proliferation. *Nat. Cell Biol.* **2003**, *425*, 962–967. [CrossRef]

120. Molofsky, A.V.; He, S.; Bydon, M.; Morrison, S.J.; Pardal, R. Bmi-1 promotes neural stem cell self-renewal and neural development but not mouse growth and survival by repressing the p16Ink4a and p19Arf senescence pathways. *Genes Dev.* **2005**, *19*, 1432–1437. [CrossRef]

121. Bruggeman, S.W.; Valk-Lingbeek, M.E.; Van Der Stoop, P.P.; Jacobs, J.J.; Kieboom, K.; Tanger, E.; Hulsman, D.; Leung, C.; Arsenijevic, Y.; Marino, S.; et al. Ink4a and Arf differentially affect cell proliferation and neural stem cell self-renewal in Bmi1-deficient mice. *Genes Dev.* **2005**, *19*, 1438–1443. [CrossRef]

122. Micheli, L.; D'Andrea, G.; Ceccarelli, M.; Ferri, A.; Scardigli, R.; Tirone, F. p16Ink4a Prevents the Activation of Aged Quiescent Dentate Gyrus Stem Cells by Physical Exercise. *Front. Cell. Neurosci.* **2019**, *13*, 10. [CrossRef] [PubMed]

123. Molofsky, A.V.; Slutsky, S.G.; Joseph, N.M.; He, S.; Pardal, R.; Krishnamurthy, J.; Sharpless, N.E.; Morrison, S.J. Increasing p16INK4a expression decreases forebrain progenitors and neurogenesis during ageing. *Nat. Cell Biol.* **2006**, *443*, 448–452. [CrossRef]

124. Kippin, T.E.; Martens, D.J.; Van Der Kooy, D. p21 loss compromises the relative quiescence of forebrain stem cell proliferation leading to exhaustion of their proliferation capacity. *Genes Dev.* **2005**, *19*, 756–767. [CrossRef]

125. Qiu, J.; Takagi, Y.; Harada, J.; Rodrigues, N.; Moskowitz, M.A.; Scadden, D.T.; Cheng, T. Regenerative Response in Ischemic Brain Restricted by p21cip1/waf1. *J. Exp. Med.* **2004**, *199*, 937–945. [CrossRef]

126. Wiley, C.D.; Campisi, J. From Ancient Pathways to Aging Cells—Connecting Metabolism and Cellular Senescence. *Cell Metab.* **2016**, *23*, 1013–1021. [CrossRef]

127. Terzi, M.Y.; Izmirli, M.; Gogebakan, B. The cell fate: Senescence or quiescence. *Mol. Biol. Rep.* **2016**, *43*, 1213–1220. [CrossRef] [PubMed]

128. Anwar, T.; Sen, B.; Aggarwal, S.; Nath, R.; Pathak, N.; Katoch, A.; Aiyaz, M.; Trehanpati, N.; Khosla, S.; Ramakrishna, G. Differentially regulated gene expression in quiescence versus senescence and identification of ARID5A as a quiescence associated marker. *J. Cell. Physiol.* **2018**, *233*, 3695–3712. [CrossRef] [PubMed]

129. Salmenperä, P.; Karhemo, P.-R.; Räsänen, K.; Laakkonen, P.; Vaheri, A. Fibroblast spheroids as a model to study sustained fibroblast quiescence and their crosstalk with tumor cells. *Exp. Cell Res.* **2016**, *345*, 17–24. [CrossRef]

130. Tomé, M.; Tchorz, J.; Gassmann, M.; Bettler, B. Constitutive activation of Notch2 signalling confers chemoresistance to neural stem cells via transactivation of fibroblast growth factor receptor-1. *Stem Cell Res.* **2019**, *35*, 101390. [CrossRef]

131. Yadirgi, G.; Leinster, V.; Acquati, S.; Bhagat, H.; Shakhova, O.; Marino, S. Conditional Activation of Bmi1 Expression Regulates Self-renewal, Apoptosis, and Differentiation of Neural Stem/Progenitor Cells In Vitro and In Vivo. *Stem Cells* **2011**, *29*, 700–712. [CrossRef] [PubMed]

132. Levine, J.M.; Reynolds, R.; Fawcett, J.W. The oligodendrocyte precursor cell in health and disease. *Trends Neurosci.* **2001**, *24*, 39–47. [CrossRef]

133. Rowitch, D.H. Glial specification in the vertebrate neural tube. *Nat. Rev. Neurosci.* **2004**, *5*, 409–419. [CrossRef]

134. Gallo, V.; Deneen, B. Glial Development: The Crossroads of Regeneration and Repair in the CNS. *Neuron* **2014**, *83*, 283–308. [CrossRef] [PubMed]

135. van Tilborg, E.; de Theije, C.G.; van Hal, M.; Wagenaar, N.; de Vries, L.S.; Benders, M.J.; Rowitch, D.H.; Nijboer, C.H. Origin and dynamics of oligodendrocytes in the developing brain: Implications for perinatal white matter injury. *Glia* **2018**, *66*, 221–238. [CrossRef] [PubMed]

136. Galvao, R.P.; Kasina, A.; McNeill, R.S.; Harbin, J.E.; Foreman, O.; Verhaak, R.G.W.; Nishiyama, A.; Miller, C.R.; Zong, H. Transformation of quiescent adult oligodendrocyte precursor cells into malignant glioma through a multistep reactivation process. *Proc. Natl. Acad. Sci. USA* **2014**, *111*, E4214–E4223. [CrossRef] [PubMed]

137. Lindberg, N.; Kastemar, M.; Olofsson, T.; Smits, A.; Uhrbom, L. Oligodendrocyte progenitor cells can act as cell of origin for experimental glioma. *Oncogene* **2009**, *28*, 2266–2275. [CrossRef] [PubMed]

138. Lindberg, N.; Jiang, Y.; Xie, Y.; Bolouri, H.; Kastemar, M.; Olofsson, T.; Holland, E.C.; Uhrbom, L. Oncogenic Signaling Is Dominant to Cell of Origin and Dictates Astrocytic or Oligodendroglial Tumor Development from Oligodendrocyte Precursor Cells. *J. Neurosci.* **2014**, *34*, 14644–14651. [CrossRef]

139. Young, K.M.; Psachoulia, K.; Tripathi, R.B.; Dunn, S.-J.; Cossell, L.; Attwell, D.; Tohyama, K.; Richardson, W.D. Oligodendrocyte Dynamics in the Healthy Adult CNS: Evidence for Myelin Remodeling. *Neuron* **2013**, *77*, 873–885. [CrossRef] [PubMed]

140. Tripathi, R.B.; Jackiewicz, M.; McKenzie, I.A.; Kougioumtzidou, E.; Grist, M.; Richardson, W.D. Remarkable Stability of Myelinating Oligodendrocytes in Mice. *Cell Rep.* **2017**, *21*, 316–323. [CrossRef] [PubMed]

141. Hughes, E.G.; Orthmann-Murphy, J.L.; Langseth, A.J.; Bergles, D.E. Myelin remodeling through experience-dependent oligodendrogenesis in the adult somatosensory cortex. *Nat. Neurosci.* **2018**, *21*, 696–706. [CrossRef] [PubMed]

142. Hill, R.A.; Li, A.M.; Grutzendler, J. Lifelong cortical myelin plasticity and age-related degeneration in the live mammalian brain. *Nat. Neurosci.* **2018**, *21*, 683–695. [CrossRef]

143. Mitew, S.; Gobius, I.; Fenlon, L.R.; McDougall, S.J.; Hawkes, D.; Xing, Y.L.; Bujalka, H.; Gundlach, A.L.; Richards, L.J.; Kilpatrick, T.J.; et al. Pharmacogenetic stimulation of neuronal activity increases myelination in an axon-specific manner. *Nat. Commun.* **2018**, *9*, 1–16. [CrossRef]

144. Gibson, E.M.; Purger, D.; Mount, C.W.; Goldstein, A.K.; Lin, G.L.; Wood, L.S.; Inema, I.; Miller, S.E.; Bieri, G.; Zuchero, J.B.; et al. Neuronal Activity Promotes Oligodendrogenesis and Adaptive Myelination in the Mammalian Brain. *Science* **2014**, *344*, 1252304. [CrossRef]

145. Zawadzka, M.; Rivers, L.E.; Fancy, S.P.; Zhao, C.; Tripathi, R.; Jamen, F.; Young, K.; Goncharevich, A.; Pohl, H.; Rizzi, M.; et al. CNS-Resident Glial Progenitor/Stem Cells Produce Schwann Cells as well as Oligodendrocytes during Repair of CNS Demyelination. *Cell Stem Cell* **2010**, *6*, 578–590. [CrossRef]

146. Tang, D.G.; Tokumoto, Y.M.; Apperly, J.A.; Lloyd, A.C.; Raff, M.C. Lack of Replicative Senescence in Cultured Rat Oligodendrocyte Precursor Cells. *Science* **2001**, *291*, 868–871. [CrossRef]

147. Zezula, J.; Casaccia-Bonnefil, P.; Ezhevsky, S.A.; Osterhout, D.J.; Levine, J.M.; Dowdy, S.F.; Chao, M.V.; Koff, A. p21 cip1 is required for the differentiation of oligodendrocytes independently of cell cycle withdrawal. *EMBO Rep.* **2001**, *2*, 27–34. [CrossRef]

148. Neumann, B.; Baror, R.; Zhao, C.; Segel, M.; Dietmann, S.; Rawji, K.S.; Foerster, S.; McClain, C.R.; Chalut, K.; Van Wijngaarden, P.; et al. Metformin Restores CNS Remyelination Capacity by Rejuvenating Aged Stem Cells. *Cell Stem Cell* **2019**, *25*, 473–485.e8. [CrossRef] [PubMed]

149. Kujuro, Y.; Suzuki, N.; Kondo, T. Esophageal cancer-related gene 4 is a secreted inducer of cell senescence expressed by aged CNS precursor cells. *Proc. Natl. Acad. Sci. USA* **2010**, *107*, 8259–8264. [CrossRef] [PubMed]

150. Kirby, L.; Jin, J.; Cardona, J.G.; Smith, M.D.; Martin, K.A.; Wang, J.; Strasburger, H.; Herbst, L.; Alexis, M.; Karnell, J.; et al. Oligodendrocyte precursor cells present antigen and are cytotoxic targets in inflammatory demyelination. *Nat. Commun.* **2019**, *10*, 1–20. [CrossRef]

151. Fernandez-Castaneda, A.; Gaultier, A. Adult oligodendrocyte progenitor cells—Multifaceted regulators of the CNS in health and disease. *Brain Behav. Immun.* **2016**, *57*, 1–7. [CrossRef]

152. Zhang, P.; Kishimoto, Y.; Grammatikakis, I.; Gottimukkala, K.; Cutler, R.G.; Zhang, S.; Abdelmohsen, K.; Bohr, V.A.; Sen, J.M.; Gorospe, M.; et al. Senolytic therapy alleviates Abeta-associated oligodendrocyte progenitor cell senescence and cognitive deficits in an Alzheimer's disease model. *Nat. Neurosci.* **2019**, *22*, 719–728. [CrossRef]

153. Sim, F.J.; Zhao, C.; Penderis, J.; Franklin, R.J.M. The Age-Related Decrease in CNS Remyelination Efficiency Is Attributable to an Impairment of Both Oligodendrocyte Progenitor Recruitment and Differentiation. *J. Neurosci.* **2002**, *22*, 2451–2459. [CrossRef] [PubMed]

154. Boyd, A.; Zhang, H.; Williams, A. Insufficient OPC migration into demyelinated lesions is a cause of poor remyelination in MS and mouse models. *Acta Neuropathol.* **2013**, *125*, 841–859. [CrossRef]

155. Chang, A.; Tourtellotte, W.W.; Rudick, R.; Trapp, B.D. Premyelinating Oligodendrocytes in Chronic Lesions of Multiple Sclerosis. *N. Engl. J. Med.* **2002**, *346*, 165–173. [CrossRef] [PubMed]

156. Kuhlmann, T.; Miron, V.; Cuo, Q.; Wegner, C.; Antel, J.; Bruck, W. Differentiation block of oligodendroglial progenitor cells as a cause for remyelination failure in chronic multiple sclerosis. *Brain* **2008**, *131*, 1749–1758. [CrossRef]

157. Woodruff, R.H.; Franklin, R.J.M. Demyelination and remyelination of the caudal cerebellar peduncle of adult rats following stereotaxic injections of lysolecithin, ethidium bromide, and complement/anti-galactocerebroside: A comparative study. *Glia* **1999**, *25*, 216–228. [CrossRef]

158. Oddo, S.; Caccamo, A.; Shepherd, J.D.; Murphy, M.P.; Golde, T.E.; Kayed, R.; Metherate, R.; Mattson, M.P.; Akbari, Y.; LaFerla, F.M. Triple-transgenic model of Alzheimer's disease with plaques and tangles: Intracellular Abeta and synaptic dysfunction. *Neuron* **2003**, *39*, 409–421. [CrossRef]

159. Vanzulli, I.; Papanikolaou, M.; De-La-Rocha, I.C.; Pieropan, F.; Rivera, A.D.; Gomez-Nicola, D.; Verkhratsky, A.; Rodríguez, J.J.; Butt, A.M. Disruption of oligodendrocyte progenitor cells is an early sign of pathology in the triple transgenic mouse model of Alzheimer's disease. *Neurobiol. Aging* **2020**, *94*, 130–139. [CrossRef]

160. Ossola, B.; Zhao, C.; Compston, A.; Pluchino, S.; Franklin, R.J.M.; Spillantini, M.G. Neuronal expression of pathological tau accelerates oligodendrocyte progenitor cell differentiation. *Glia* **2015**, *64*, 457–471. [CrossRef]

161. Dunckley, T.; Beach, T.G.; Ramsey, K.E.; Grover, A.; Mastroeni, D.; Walker, D.G.; LaFleur, B.J.; Coon, K.D.; Brown, K.M.; Caselli, R.; et al. Gene expression correlates of neurofibrillary tangles in Alzheimer's disease. *Neurobiol. Aging* **2006**, *27*, 1359–1371. [CrossRef] [PubMed]

162. Nicaise, A.M.; Wagstaff, L.J.; Willis, C.M.; Paisie, C.; Chandok, H.; Robson, P.; Fossati, V.; Williams, A.; Crocker, S.J. Cellular senescence in progenitor cells contributes to diminished remyelination potential in progressive multiple sclerosis. *Proc. Natl. Acad. Sci. USA* **2019**, *116*, 9030–9039. [CrossRef] [PubMed]

163. Lawson, L.J.; Perry, V.H.; Gordon, S. Turnover of resident microglia in the normal adult mouse brain. *Neuroscience* **1992**, *48*, 405–415. [CrossRef]

164. Angelova, D.M.; Brown, D.R. Microglia and the aging brain: Are senescent microglia the key to neurodegeneration? *J. Neurochem.* **2019**, *151*, 676–688. [CrossRef]

165. Ginhoux, F.; Prinz, M. Origin of Microglia: Current Concepts and Past Controversies. *Cold Spring Harb. Perspect. Biol.* **2015**, *7*, a020537. [CrossRef]

166. Réu, P.; Khosravi, A.; Bernard, S.; Mold, J.E.; Salehpour, M.; Alkass, K.; Perl, S.; Tisdale, J.; Possnert, G.; Druid, H.; et al. The Lifespan and Turnover of Microglia in the Human Brain. *Cell Rep.* **2017**, *20*, 779–784. [CrossRef]

167. Sierra, A.; Abiega, O.; Shahraz, A.; Neumann, H. Janus-faced microglia: Beneficial and detrimental consequences of microglial phagocytosis. *Front. Cell. Neurosci.* **2013**, *7*, 6. [CrossRef]

168. Ginhoux, F.; Greter, M.; Leboeuf, M.; Nandi, S.; See, P.; Gokhan, S.; Mehler, M.F.; Conway, S.J.; Ng, L.G.; Stanley, E.R.; et al. Fate Mapping Analysis Reveals That Adult Microglia Derive from Primitive Macrophages. *Science* **2010**, *330*, 841–845. [CrossRef]

169. Schulz, C.; Perdiguero, E.G.; Chorro, L.; Szabo-Rogers, H.; Cagnard, N.; Kierdorf, K.; Prinz, M.; Wu, B.; Jacobsen, S.E.W.; Pollard, J.W.; et al. A Lineage of Myeloid Cells Independent of Myb and Hematopoietic Stem Cells. *Science* **2012**, *336*, 86–90. [CrossRef]

170. Elmore, M.R.P.; Najafi, A.R.; Koike, M.A.; Dagher, N.N.; Spangenberg, E.E.; Rice, R.A.; Kitazawa, M.; Matusow, B.; Nguyen, H.; West, B.L.; et al. Colony-Stimulating Factor 1 Receptor Signaling Is Necessary for Microglia Viability, Unmasking a Microglia Progenitor Cell in the Adult Brain. *Neuron* **2014**, *82*, 380–397. [CrossRef]

171. Nomura, K.; Vilalta, A.; Allendorf, D.H.; Hornik, T.C.; Brown, G.C. Activated Microglia Desialylate and Phagocytose Cells via Neuraminidase, Galectin-3, and Mer Tyrosine Kinase. *J. Immunol.* **2017**, *198*, 4792–4801. [CrossRef]

172. Tejera, D.; Mercan, D.; Sanchez-Caro, J.M.; Hanan, M.; Greenberg, D.; Soreq, H.; Latz, E.; Golenbock, D.; Heneka, M.T. Systemic inflammation impairs microglial Abeta clearance through NLRP3 inflammasome. *EMBO J.* **2019**, *38*, e101064. [CrossRef] [PubMed]

173. Luo, W.; Liu, W.; Hu, X.; Hanna, M.; Caravaca, A.; Paul, S.M. Microglial internalization and degradation of pathological tau is enhanced by an anti-tau monoclonal antibody. *Sci. Rep.* **2015**, *5*, 11161. [CrossRef]

174. Tang, Y.; Le, W. Differential Roles of M1 and M2 Microglia in Neurodegenerative Diseases. *Mol. Neurobiol.* **2016**, *53*, 1181–1194. [CrossRef]

175. Kaur, C.; Rathnasamy, G.; Ling, E.-A. Biology of Microglia in the Developing Brain. *J. Neuropathol. Exp. Neurol.* **2017**, *76*, 736–753. [CrossRef]

176. Koellhoffer, E.C.; McCullough, L.D.; Ritzel, R.M. Old Maids: Aging and Its Impact on Microglia Function. *Int. J. Mol. Sci.* **2017**, *18*, 769. [CrossRef]

177. Afridi, R.; Lee, W.-H.; Suk, K. Microglia Gone Awry: Linking Immunometabolism to Neurodegeneration. *Front. Cell. Neurosci.* **2020**, *14*, 246. [CrossRef] [PubMed]

178. Wu, S.Y.; Watabe, K. The roles of microglia/macrophages in tumor progression of brain cancer and metastatic disease. *Front. Biosci. (Landmark Ed.)* **2017**, *22*, 1805–1829. [CrossRef]

179. Gutmann, D.H.; Kettenmann, H. Microglia/Brain Macrophages as Central Drivers of Brain Tumor Pathobiology. *Neuron* **2019**, *104*, 442–449. [CrossRef]

180. Abels, E.R.; Maas, S.L.; Nieland, L.; Wei, Z.; Cheah, P.S.; Tai, E.; Kolsteeg, C.-J.; Dusoswa, S.A.; Ting, D.T.; Hickman, S.; et al. Glioblastoma-Associated Microglia Reprogramming Is Mediated by Functional Transfer of Extracellular miR-21. *Cell Rep.* **2019**, *28*, 3105–3119.e7. [CrossRef] [PubMed]

181. Maas, S.L.N.; Abels, E.R.; Van De Haar, L.L.; Zhang, X.; Morsett, L.; Sil, S.; Guedes, J.; Sen, P.; Prabhakar, S.; Hickman, S.E.; et al. Glioblastoma hijacks microglial gene expression to support tumor growth. *J. Neuroinflammation* **2020**, *17*, 120. [CrossRef] [PubMed]

182. Dumas, A.A.; Pomella, N.; Rosser, G.; Guglielmi, L.; Vinel, C.; Millner, T.O.; Rees, J.; Aley, N.; Sheer, D.; Wei, J.; et al. Microglia promote glioblastoma via mTOR-mediated immunosuppression of the tumour microenvironment. *EMBO J.* **2020**, *39*, e103790. [CrossRef]

183. Streit, W.J.; Xue, Q.S.; Tischer, J.; Bechmann, I. Microglial pathology. *Acta Neuropathol. Commun.* **2014**, *2*, 142. [CrossRef] [PubMed]

184. Hoeijmakers, L.; Heinen, Y.; Van Dam, A.-M.; Lucassen, P.J.; Korosi, A. Microglial Priming and Alzheimer's Disease: A Possible Role for (Early) Immune Challenges and Epigenetics? *Front. Hum. Neurosci.* **2016**, *10*, 398. [CrossRef]

185. Hoogland, I.C.; Houbolt, C.; van Westerloo, D.J.; van Gool, W.A.; van de Beek, D. Systemic inflammation and microglial activation: Systematic review of animal experiments. *J. Neuroinflammation* **2015**, *12*, 114. [CrossRef]

186. Morgan, J.T.; Chana, G.; Pardo, C.A.; Achim, C.; Semendeferi, K.; Buckwalter, J.; Courchesne, E.; Everall, I.P. Microglial Activation and Increased Microglial Density Observed in the Dorsolateral Prefrontal Cortex in Autism. *Biol. Psychiatry* **2010**, *68*, 368–376. [CrossRef]

187. Hendrickx, D.A.; Van Eden, C.G.; Schuurman, K.G.; Hamann, J.; Huitinga, I. Staining of HLA-DR, Iba1 and CD68 in human microglia reveals partially overlapping expression depending on cellular morphology and pathology. *J. Neuroimmunol.* **2017**, *309*, 12–22. [CrossRef]

188. Thrupp, N.; Frigerio, C.S.; Wolfs, L.; Skene, N.G.; Fattorelli, N.; Poovathingal, S.; Fourne, Y.; Matthews, P.M.; Theys, T.; Mancuso, R.; et al. Single-Nucleus RNA-Seq Is Not Suitable for Detection of Microglial Activation Genes in Humans. *Cell Rep.* **2020**, *32*, 108189. [CrossRef]

189. Aird, K.M.; Zhang, R. Detection of Senescence-Associated Heterochromatin Foci (SAHF). *Methods Mol. Biol.* **2013**, *965*, 185–196. [CrossRef]

190. Sierra, A.; Gottfried-Blackmore, A.C.; McEwen, B.S.; Bulloch, K. Microglia derived from aging mice exhibit an altered inflammatory profile. *Glia* **2007**, *55*, 412–424. [CrossRef]

191. Trias, E.; Beilby, P.R.; Kovacs, M.; Ibarburu, S.; Varela, V.; Barreto-Núñez, R.; Bradford, S.C.; Beckman, J.S.; Barbeito, L. Emergence of Microglia Bearing Senescence Markers During Paralysis Progression in a Rat Model of Inherited ALS. *Front. Aging Neurosci.* **2019**, *11*, 42. [CrossRef]

192. Chen, N.C.; Partridge, A.T.; Sell, C.; Torres, C.; Martín-García, J. Fate of microglia during HIV-1 infection: From activation to senescence? *Glia* **2017**, *65*, 431–446. [CrossRef]

193. Caldeira, C.; Oliveira, A.F.; Ecunha, C.; Vaz, A.R.; Falcão, A.S.; Efernandes, A.; Ebrites, D. Microglia change from a reactive to an age-like phenotype with the time in culture. *Front. Cell. Neurosci.* **2014**, *8*, 152. [CrossRef]

194. Kronenberg, G.; Uhlemann, R.; Schöner, J.; Wegner, S.; Boujon, V.; Deigendesch, N.; Endres, M.; Gertz, K. Repression of telomere-associated genes by microglia activation in neuropsychiatric disease. *Eur. Arch. Psychiatry Clin. Neurosci.* **2017**, *267*, 473–477. [CrossRef]

195. Yu, H.-M.; Zhao, Y.-M.; Luo, X.-G.; Feng, Y.; Ren, Y.; Shang, H.; He, Z.-Y.; Chen, S.-D.; Wang, X.-Y. Repeated Lipopolysaccharide Stimulation Induces Cellular Senescence in BV2 Cells. *Neuroimmunomodulation* **2012**, *19*, 131–136. [CrossRef]

196. Flanary, B.E.; Sammons, N.W.; Nguyen, C.; Walker, D.; Streit, W.J. Evidence That Aging And Amyloid Promote Microglial Cell Senescence. *Rejuvenation Res.* **2007**, *10*, 61–74. [CrossRef]

197. Nikodemova, M.; Small, A.L.; Kimyon, R.S.; Watters, J.J. Age-dependent differences in microglial responses to systemic inflammation are evident as early as middle age. *Physiol. Genom.* **2016**, *48*, 336–344. [CrossRef]

198. Niraula, A.; Sheridan, J.F.; Godbout, J.P. Microglia Priming with Aging and Stress. *Neuropsychopharmacology* **2016**, *42*, 318–333. [CrossRef]

199. Stojiljkovic, M.R.; Ain, Q.; Bondeva, T.; Heller, R.; Schmeer, C.; Witte, O.W. Phenotypic and functional differences between senescent and aged murine microglia. *Neurobiol. Aging* **2019**, *74*, 56–69. [CrossRef]

200. Scheffold, A.; Holtman, I.R.; Dieni, S.; Brouwer, N.; Katz, S.-F.; Jebaraj, B.M.C.; Kahle, P.J.; Hengerer, B.; Lechel, A.; Stilgenbauer, S.; et al. Telomere shortening leads to an acceleration of synucleinopathy and impaired microglia response in a genetic mouse model. *Acta Neuropathol. Commun.* **2016**, *4*, 87. [CrossRef]

201. Chinta, S.J.; Woods, G.; DeMaria, M.; Rane, A.; Zou, Y.; McQuade, A.; Rajagopalan, S.; Limbad, C.; Madden, D.T.; Campisi, J.; et al. Cellular Senescence Is Induced by the Environmental Neurotoxin Paraquat and Contributes to Neuropathology Linked to Parkinson's Disease. *Cell Rep.* **2018**, *22*, 930–940. [CrossRef] [PubMed]

202. Shen, W.H.; Liu, L.; Zeng, S. Senescence and Cancer. *Cancer Transl. Med.* **2018**, *4*, 70–74. [CrossRef]

203. Zhang, B.; Gaiteri, C.; Bodea, L.-G.; Wang, Z.; McElwee, J.; Podtelezhnikov, A.A.; Zhang, C.; Xie, T.; Tran, L.; Dobrin, R.; et al. Integrated Systems Approach Identifies Genetic Nodes and Networks in Late-Onset Alzheimer's Disease. *Cell* **2013**, *153*, 707–720. [CrossRef]

204. Huang, F.; Wang, M.; Liu, R.; Wang, J.Z.; Schadt, E.; Haroutunian, V.; Katsel, P.; Zhang, B.; Wang, X. CDT2-controlled cell cycle reentry regulates the pathogenesis of Alzheimer's disease. *Alzheimers Dement.* **2019**, *15*, 217–231. [CrossRef] [PubMed]

205. Wang, M.; Li, A.; Sekiya, M.; Beckmann, N.D.; Quan, X.; Schrode, N.; Fernando, M.B.; Yu, A.; Zhu, L.; Cao, J.; et al. Transformative Network Modeling of Multi-omics Data Reveals Detailed Circuits, Key Regulators, and Potential Therapeutics for Alzheimer's Disease. *Neuron* **2021**, *109*, 257–272.e14. [CrossRef]

206. Neff, R.A.; Wang, M.; Vatansever, S.; Guo, L.; Ming, C.; Wang, Q.; Wang, E.; Horgusluoglu-Moloch, E.; Song, W.-M.; Li, A.; et al. Molecular subtyping of Alzheimer's disease using RNA sequencing data reveals novel mechanisms and targets. *Sci. Adv.* **2021**, *7*, eabb5398. [CrossRef]
207. Zhang, Y.; Kim, M.S.; Jia, B.; Yan, J.; Zuniga-Hertz, J.P.; Han, C.; Cai, D. Hypothalamic stem cells control ageing speed partly through exosomal miRNAs. *Nat. Cell Biol.* **2017**, *548*, 52–57. [CrossRef] [PubMed]

Opinion

Secreted Protein Acidic and Rich in Cysteine as A Regeneration Factor: Beyond the Tissue Repair

Abdelaziz Ghanemi [1,2], Mayumi Yoshioka [2] and Jonny St-Amand [1,2,*]

[1] Department of Molecular Medicine, Faculty of Medicine, Laval University, Québec, QC G1V 0A6, Canada; abdelaziz.ghanemi@crchudequebec.ulaval.ca

[2] Functional Genomics Laboratory, Endocrinology and Nephrology Axis, CHU de Québec-Université Laval Research Center, Québec, QC G1V 4G2, Canada; mayumi.yoshioka@crchudequebec.ulaval.ca

* Correspondence: jonny.st-amand@crchudequebec.ulaval.ca; Tel.: + 1-(418)-525-4444 (ext. 46448); Fax: +1-(418)-654-2298

Abstract: Diverse pathologies (inflammation, tissues injuries, cancer, etc.) and physiological conditions (obesity, physical activity, etc.) induce the expression/secretion of the matricellular protein, secrete protein acidic and rich in cysteine (SPARC). SPARC contributes to the creation of an environment that is suitable for tissue regeneration through a variety of roles, including metabolic homeostasis, inflammation reduction, extracellular matrix remodeling and collagen maturation. Such a homeostatic environment optimizes tissue regeneration and improves tissues' repair ability. These properties that SPARC has within the regeneration contexts could have a variety of applications, such as in obesity, cancer, sarcopenia, diabetes and bioengineering.

Keywords: secreted protein acidic and rich in cysteine; regeneration; homeostasis

Citation: Ghanemi, A.; Yoshioka, M.; St-Amand, J. Secreted Protein Acidic and Rich in Cysteine as A Regeneration Factor: Beyond the Tissue Repair. *Life* **2021**, *11*, 38. https://doi.org/10.3390/life11010038

Received: 7 December 2020
Accepted: 7 January 2021
Published: 8 January 2021

Publisher's Note: MDPI stays neutral with regard to jurisdictional claims in published maps and institutional affiliations.

Tissue regeneration is a vital process allowing organisms to overcome biological disturbances and adapt to changes and physiological development via the renewal, growth and restoration of diverse cells and tissues. The regeneration ability changes throughout the lifespan, which leads to diverse tissue malfunctions and diseases [1]. The regenerative process could be either normal or limited (abnormal) depending on the biological environment. Indeed, under healthy environmental conditions (stem cells growth ratio [2], growth factors [3], hormones [4,5], pH [6,7], etc.), the regenerative processes are optimized. They allow for regular tissue development and adaptation to the corresponding biological functions. However, under physiological (ageing [8,9]) or pathological (cancer [10], obesity [11], inflammation [12], etc.) conditions, or when impacted by disturbing stimuli or exogenous factors (such as radiations [13]), tissues' regeneration ability and functions could be impaired. To overcome this "negative" regeneration environment, the organism has a variety of tools to compensate or reduce the intensity or the impacts. These correcting or counteracting mechanisms are mediated through what could be considered regeneration factors. Among these molecules, secreted protein acidic and rich in cysteine (SPARC) has a variety of roles and implications. One of the SPARC properties is its ability to optimize the regeneration environment with an improved cellular regenerative capacity from different perspectives (metabolics, tissue repair, oxidation, inflammation, cancer, etc.), as illustrated below.

SPARC, also known as BM-40 or osteonectin (32 kDa [14]), is a matricellular (extracellular matrix-associated) protein. Unlike its name (osteonectin) might suggest. SPARC expression is not limited to bones, but this glycoprotein is also present in diverse tissues including non-mineralized tissues, in platelets [15] and in muscles [16]. Such wide distribution correlates with SPARC roles during embryogenesis [17] as well as during tissue repair, cell turnover, cellular differentiation and remodeling [18–22], which are key steps in tissue regeneration. Therefore, SPARC expression or levels increase following injuries such as myocardial injury [23], myopathies [24] and in situations (either physiological or pathological) where tissues undergo

95

changes (repair, renewal and remodeling) such as during obesity [18,25], skeletal muscle regeneration [26], cancer [27], systemic sclerosis, hepatic fibrosis [28] and physical exercise. Indeed, SPARC/*Sparc* expression increases in the skeletal muscle during training [29], as well as following electrical pulse stimulation in muscle cells (considered to be the in vitro equivalent of exercise) [30]. Such situations do represent a disturbance of the homeostasis that leads to a "negative" regeneration environment. Therefore, biological processes that overcome such a homeostatic disturbance, restore a suitable environment for regeneration and rescue the affected tissues to allow better developmental patters are required. Interestingly, the situations in which SPARC is overexpressed are mainly those requiring regeneration, either to repair tissues (injury) or adapt to tissue changes (obesity, exercised muscle, etc.). These specific patterns highlight SPARC as a regenerative factor. In addition, the importance of the extracellular matrix in regeneration suggests close interactions between SPARC, the extracellular matrix [31] and matricellular protein components such as thrombospondin-2 [32] during the regeneration process.

Tissue regeneration is a process that requires the implication of numerous cellular organelles and the use of energy. Thus, regeneration has metabolic and biochemical needs to which the cellular machinery has to adapt [33]. In this context, SPARC has been shown to be implicated in a variety of metabolic functions, such as glucose tolerance improvement [34], while it is also required for both glucose homeostasis maintenance and insulin secretion [35]. In the skeletal muscle, SPARC also seems to act towards improved metabolic properties and functions [18,24], including mitochondrial functions [30,36,37], which is of interest knowing the importance of the mitochondria during regeneration [38, 39]. Importantly, our latest study suggests that exercise-induced muscle phenotype changes are SPARC-dependent [40]. These SPARC properties are also completed by their important roles in energy balance and storage. For instance, SPARC inhibits adipogenesis [41] and its inactivation leads to an enhancement of high-fat diet-induced obesity [42]. These patterns correlate with the role of SPARC in brown adipocyte activation and lipid usage in white adipocytes [43]. Such energy metabolism effects—in addition to optimizing the regeneration (synchronization)—also lead to increased energy usage, thus reducing the risk of obesity through increased energy expenditure. This represents another illustration of how SPARC counteracts the "negative" regeneration environment, since obesity itself represents a status of impaired regeneration [44]. Indeed, during obesity, many factors lead to such a "negative" regeneration environment due to all the conditions induced by or associated with obesity, such as inflammation, insulin resistance, metabolic disorders [45, 46] and even stem cell changes [47,48], that impact regeneration. SPARC is extremely important for bone formation, remodeling and regeneration [14,32,49–51]. This is important as well, not only for the structural homeostasis, but also for both locomotion and, most importantly, the energy metabolism. Indeed, the skeletal muscle that governs most of the energy expenditure [52] is supported by the skeleton with which it forms the locomotor (musculoskeletal) system. Therefore, the good metabolic and contractile function (strength) of muscles would require homeostatic skeleton development due to the close ties between both bones and skeletal muscles, including synchronized development [53].

Furthermore, in addition to such metabolic implications, SPARC is also involved in other growth and homeostasis-related patterns, including cancer homeostasis. SPARC is overexpressed during cancer [27] and has been reported to have anti-cancer properties [54,55]. SPARC has also been shown to have interesting roles within the inflammatory processes [56,57]. It has anti-inflammatory properties [56] and can, for instance, protect from adverse cardiac inflammation during viral myocarditis [58]. These properties of controlling cancer and inflammation development would impact the microenvironment, contributing to an improved homeostasis. Moreover, SPARC is required for the immune system functions [59], which is relevant, for instance, during immune-modulatory therapy to support the regeneration of injured muscles [60] and muscle healing [61]. Importantly, more roles are yet to be explored in terms of SPARC contribution at the physiological levels, such as in cardiomyocyte contraction [23]. This cardiac role would improve the blood circulation for diverse cells, which are vital for tissue regeneration). In addition, the therapeutic practice of cardiac regeneration [2,62] could benefit from SPARC properties in cardiac regeneration [19,63] as well. Beyond the cardiac properties, SPARC has roles in the cardiovascular properties, as suggested by its production by both bone-marrow-derived cells during myocardial fibrosis (in left ventricular pressure overload) [64] and pericytes, with a possible role in postinfarct healing [65], which is supported by the possible classification of SPARC as a marker for vascular complications in pre-diabetics [66].

All these highlighted properties point to SPARC as a regeneration factor. It not only has significant roles in tissue repair or development but contributes directly and indirectly to generating a "positive" biological environment that optimizes regeneration, as summarized in the graphical abstract. Moreover, other factors that work towards reducing the regeneration ability, such as ageing [1,67] and oxidative stress [68,69], are also counteracted—at least indirectly—by SPARC effects. For instance, SPARC-induced increased muscle functions (including via interactions with actin in skeletal muscle [24]) and metabolism would increase the antioxidant effect induced by exercise [70]. This contributes to the improvement in the regeneration environment by decreasing the oxidative stress. Furthermore, an improved muscular function (including during exercise) would lead to reducing the accumulation of the lactic acid and, therefore, better control of the pH, which both impacts muscle fatigue [71,72] and represents another important factor for different cellular functions [73], including those related to regeneration [74,75]. In addition, ageing-induced collagen loss [76] would be counteracted via the roles of SPARC in collagen properties [77–80]. Moreover, many SPARC effects counteract ageing impacts. In this context, ageing is a factor that decreases the regeneration ability [67], and with which we see an increased risk of obesity [81], sarcopenia [82,83], osteoporosis [84], etc. This points to SPARC not only as a regeneration factor that counteracts the ageing-related decrease in regeneration ability, but also as a factor with key roles against ageing-induced conditions that lead to health problems including sarcopenia, obesity (a health problem that could increase with the ongoing COVID-19 crisis [85]) and osteoporosis, through metabolic, structural and functional roles, and the impacts SPARC has on the corresponding tissues and organs (muscles, adipose tissue, bone, etc.). Therefore, SPARC remains worth exploring in the ageing process and geriatric research. These examples represent additional illustrations of SPARC's contribution to creating the optimal environment for regeneration, and further point to it as a regenerative factor.

These patterns show complimentary roles in terms of the implications of SPARC in tissue repair, and the diverse metabolic and homeostatic effects it mediates [86]. Importantly, the fact that SPARC is overexpressed during pathological situations such as obesity and cancer, as well as during physical activity (physiological adaptation), further indicates that it could represent feedback. Rather than a damaging factor, SPARC would aim to counteract/correct the negative impacts induced by the pathological situations such as inflammation and tissue damage through properties including regeneration ability, as illustrated for the skeletal muscle [87]. Indeed, conditions (pathological and physiological) that lead to impaired regeneration by creating a negative environment are the same conditions under which SPARC overexpression has been reported. Such overexpressed SPARC improves the regeneration ability and reduces the negative environment by inducing functional and metabolic enhancement at different tissues. These actions reverse, correct or reduce the impacts those initial conditions had on regeneration, which will lead to a SPARC-induced corrected regeneration ability.

This paper presents SPARC as a promising therapeutic tool in a variety of health conditions, ranging from metabolism and inflammation to obesity and sarcopenia. Importantly, SPARC could also be an option in the area of tissue engineering based on its involvement in and impacts on the regenerative processes, especially with the known implications of SPARC in the functions of stem cells [88,89], as well as other types of cells such as erythroid progenitors [90]. Thus, SPARC-related pathways also represent a potential pharmacological target to optimize therapies in regenerative medicine as an adjuvant to optimize the regeneration environment of the targeted tissues and organs.

Author Contributions: A.G. drafted the manuscript; A.G., M.Y. and J.S.-A. critically revised the manuscript. All authors have read and agreed to the published version of the manuscript.

Funding: This research received no external funding.

Institutional Review Board Statement: Not applicable.

Informed Consent Statement: Not applicable.

Data Availability Statement: Not applicable.

Acknowledgments: Abdelaziz Ghanemi received a merit scholarship for foreign students from the Ministry of Education and Higher Education of Quebec, Canada, The Fonds de recherche du Québec—Nature et technologies (FRQNT) is responsible for managing the program (Bourses d'excellence pour étudiants étrangers du Ministère de l'Éducation et de l'Enseignement supérieur du Québec, Le Fonds de recherche du Québec—Nature et technologies (FRQNT) est responsable de la gestion du programme). The graphical abstract was created using images from: http://smart.servier.com. Servier Medical Art by Servier is licensed under a Creative Commons Attribution 3.0 Unported License.

Conflicts of Interest: The authors declare no conflict of interest.

References

1. Yun, M.H. Changes in Regenerative Capacity through Lifespan. *Int. J. Mol. Sci.* **2015**, *16*, 25392–25432. [CrossRef] [PubMed]
2. Zhang, Y.; Mignone, J.; MacLellan, W.R. Cardiac Regeneration and Stem Cells. *Physiol. Rev.* **2015**, *95*, 1189–1204. [CrossRef] [PubMed]
3. Sarker, M.D.; Naghieh, S.; McInnes, A.D.; Schreyer, D.J.; Chen, X.B. Regeneration of peripheral nerves by nerve guidance conduits: Influence of design, biopolymers, cells, growth factors, and physical stimuli. *Prog. Neurobiol.* **2018**, *171*, 125–150. [CrossRef] [PubMed]
4. Easterling, M.R.; Engbrecht, K.M.; Crespi, E.J. Endocrine Regulation of Epimorphic Regeneration. *Endocrinology* **2019**, *160*, 2969–2980. [CrossRef] [PubMed]
5. Wojda, S.J.; Donahue, S.W. Parathyroid hormone for bone regeneration. *J. Orthop. Res.* **2018**, *36*, 2586–2594. [CrossRef]
6. Shen, Y.; Liu, W.C.; Lin, K.; Pan, H.; Darvell, B.; Peng, S.; Wen, C.; Deng, L.; Lu, W.W.; Chang, J. Interfacial pH: A Critical Factor for Osteoporotic Bone Regeneration. *Langmuir* **2011**, *27*, 2701–2708. [CrossRef]
7. Shen, Y.; Liu, W.; Wen, C.; Pan, H.; Wang, T.; Darvell, B.W.; Lu, W.W.; Huang, W. Bone regeneration: Importance of local pH—Strontium-doped borosilicate scaffold. *J. Mater. Chem.* **2012**, *22*, 8662–8670. [CrossRef]
8. Pibiri, M. Liver regeneration in aged mice: New insights. *Aging* **2018**, *10*, 1801–1824. [CrossRef]
9. Varela-Eirin, M.; Loureiro, J.; Fonseca, E.; Corrochano, S.; Caeiro, J.R.; Collado, M.; Mayan, M.D. Cartilage regeneration and ageing: Targeting cellular plasticity in osteoarthritis. *Ageing Res. Rev.* **2018**, *42*, 56–71. [CrossRef]
10. Li, H.; Ye, Z.-H. Microenvironment of liver regeneration in liver cancer. *Chin. J. Integr. Med.* **2017**, *23*, 555–560. [CrossRef]
11. Ambrosi, T.H.; Scialdone, A.; Graja, A.; Gohlke, S.; Jank, A.M.; Bocian, C.; Woelk, L.; Fan, H.; Logan, D.W.; Schürmann, A.; et al. Adipocyte Accumulation in the Bone Marrow during Obesity and Aging Impairs Stem Cell-Based Hematopoietic and Bone Regeneration. *Cell. Stem. Cell.* **2017**, *20*, 771–784.e6. [CrossRef]
12. Cooke, J.P. Inflammation and Its Role in Regeneration and Repair. *Circ. Res.* **2019**, *124*, 1166–1168. [CrossRef] [PubMed]
13. Korolev, Y.N.; Geniatulina, M.S.; Mikhailik, L.V.; Nikulina, L.A. Intracellular regeneration of adrenocorticocytes in response to the prophylactic application of low-intensity electromagnetic radiation under the conditions of radiation (an experimental study). *Vopr. Kurortol. Fizioter. Lech. Fiz. Kul'tury* **2019**, *96*, 43–49. [CrossRef] [PubMed]
14. Rosset, E.M.; Bradshaw, A.D. SPARC/osteonectin in mineralized tissue. *Matrix Biol.* **2016**, *54*, 78–87. [CrossRef] [PubMed]
15. Maillard, C.; Malaval, L.; Delmas, P. Immunological screening of SPARC/Osteonectin in nonmineralized tissues. *Bone* **1992**, *13*, 257–264. [CrossRef]
16. Melouane, A.; Ghanemi, A.; Aubé, S.; Yoshioka, M.; St-Amand, J. Differential gene expression analysis in ageing muscle and drug discovery perspectives. *Ageing Res. Rev.* **2018**, *41*, 53–63. [CrossRef]
17. Motamed, K. SPARC (osteonectin/BM-40). *Int. J. Biochem. Cell Biol.* **1999**, *31*, 1363–1366. [CrossRef]
18. Ghanemi, A.; Melouane, A.; Yoshioka, M.; St-Amand, J. Secreted protein acidic and rich in cysteine and bioenergetics: Extracellular matrix, adipocytes remodeling and skeletal muscle metabolism. *Int. J. Biochem. Cell Biol.* **2019**, *117*, 105627. [CrossRef]
19. Bradshaw, A.D. The role of secreted protein acidic and rich in cysteine (SPARC) in cardiac repair and fibrosis: Does expression of SPARC by macrophages influence outcomes? *J. Mol. Cell. Cardiol.* **2016**, *93*, 156–161. [CrossRef]
20. McCurdy, S.; Baicu, C.F.; Heymans, S.; Bradshaw, A.D. Cardiac extracellular matrix remodeling: Fibrillar collagens and Secreted Protein Acidic and Rich in Cysteine (SPARC). *J. Mol. Cell. Cardiol.* **2010**, *48*, 544–549. [CrossRef]

21. Yan, Q.; Sage, E.H. SPARC, a Matricellular Glycoprotein with Important Biological Functions. *J. Histochem. Cytochem.* **1999**, *47*, 1495–1505. [CrossRef]

22. Bradshaw, A.D.; Sage, E.H. SPARC, a matricellular protein that functions in cellular differentiation and tissue response to injury. *J. Clin. Investig.* **2001**, *107*, 1049–1054. [CrossRef] [PubMed]

23. Deckx, S.; Johnson, D.M.; Rienks, M.; Carai, P.; Van Deel, E.; Van der Velden, J.; Sipido, K.R.; Heymans, S.; Papageorgiou, A.P. Extracellular SPARC increases cardiomyocyte contraction during health and disease. *PLoS ONE* **2019**, *14*, e0209534. [CrossRef]

24. Jørgensen, L.H.; Jepsen, P.L.; Boysen, A.; Dalgaard, L.B.; Hvid, L.G.; Ørtenblad, N.; Ravn, D.; Selathurai, J.; Møller-Jensen, J.; Lochmuller, H.; et al. SPARC Interacts with Actin in Skeletal Muscle In Vitro and In Vivo. *Am. J. Pathol.* **2017**, *187*, 457–474. [CrossRef]

25. Shen, Y.; Zhao, Y.; Yuan, L.; Yi, W.; Zhao, R.; Yi, Q.; Yong, T. SPARC is over-expressed in adipose tissues of diet-induced obese rats and causes insulin resistance in 3T3-L1 adipocytes. *Acta Histochem.* **2014**, *116*, 158–166. [CrossRef] [PubMed]

26. Petersson, S.J.; Jørgensen, L.H.; Andersen, D.C.; Nørgaard, R.C.; Jensen, C.H.; Schroeder, H.D. SPARC is up-regulated during skeletal muscle regeneration and inhibits myoblast differentiation. *Histol. Histopathol.* **2013**, *28*, 1451–1460. [CrossRef]

27. Ghanemi, A.; Yoshioka, M.; St-Amand, J. Secreted protein acidic and rich in cysteine and cancer: A homeostatic hormone? *Cytokine* **2020**, *127*, 154996. [CrossRef]

28. Scavelli, K.; Chatterjee, A.; Rhee, U.J. Secreted Protein Acidic and Rich in Cysteine in Ocular Tissue. *J. Ocul. Pharmacol. Ther.* **2015**, *31*, 396–405. [CrossRef]

29. Ghanemi, A.; Melouane, A.; Yoshioka, M.; St-Amand, J. Exercise and High-Fat Diet in Obesity: Functional Genomics Perspectives of Two Energy Homeostasis Pillars. *Genes* **2020**, *11*, 875. [CrossRef]

30. Melouane, A.; Yoshioka, M.; Kanzaki, M.; St-Amand, J. Sparc, an EPS-induced gene, modulates the extracellular matrix and mitochondrial function via ILK/AMPK pathways in C_2C_{12} cells. *Life Sci.* **2019**, *229*, 277–287. [CrossRef]

31. Niklason, L. Understanding the Extracellular Matrix to Enhance Stem Cell-Based Tissue Regeneration. *Cell. Stem. Cell.* **2018**, *22*, 302–305. [CrossRef] [PubMed]

32. Delany, A.M.; Hankenson, K. Thrombospondin-2 and SPARC/osteonectin are critical regulators of bone remodeling. *J. Cell Commun. Signal.* **2009**, *3*, 227–238. [CrossRef] [PubMed]

33. Caldez, M.J.; Van Hul, N.; Koh, H.W.; Teo, X.Q.; Fan, J.J.; Tan, P.Y.; Dewhurst, M.R.; Too, P.G.; Talib, S.Z.A.; Chiang, B.E.; et al. Metabolic Remodeling during Liver Regeneration. *Dev. Cell* **2018**, *47*, 425–438.e5. [CrossRef] [PubMed]

34. Aoi, W.; Hirano, N.; Lassiter, D.G.; Björnholm, M.; Chibalin, A.V.; Sakuma, K.; Tanimura, Y.; Mizushima, K.; Takagi, T.; Naito, Y.; et al. Secreted protein acidic and rich in cysteine (SPARC) improves glucose tolerance via AMP-activated protein kinase activation. *FASEB J.* **2019**, *33*, 10551–10562. [CrossRef]

35. Atorrasagasti, C.; Onorato, A.; Gimeno, M.L.; Andreone, L.; Garcia, M.; Malvicini, M.; Fiore, E.; Bayo, J.; Perone, M.J.; Mazzolini, G. SPARC is required for the maintenance of glucose homeostasis and insulin secretion in mice. *Clin. Sci.* **2019**, *133*, 351–365. [CrossRef]

36. Melouane, A.; Carbonell, A.; Yoshioka, M.; Puymirat, J.; St-Amand, J. Implication of SPARC in the modulation of the extracellular matrix and mitochondrial function in muscle cells. *PLoS ONE* **2018**, *13*, e0192714. [CrossRef]

37. Melouane, A.; Yoshioka, M.; St-Amand, J. Extracellular matrix/mitochondria pathway: A novel potential target for sarcopenia. *Mitochondrion* **2020**, *50*, 63–70. [CrossRef]

38. Smith, G.M.; Gallo, G. The role of mitochondria in axon development and regeneration. *Dev. Neurobiol.* **2018**, *78*, 221–237. [CrossRef]

39. Han, S.M.; Baig, H.S.; Hammarlund, M. Mitochondria Localize to Injured Axons to Support Regeneration. *Neuron* **2016**, *92*, 1308–1323. [CrossRef]

40. Ghanemi, A.; Melouane, A.; Yoshioka, M.; St-Amand, J. Exercise Training of Secreted Protein Acidic and Rich in Cysteine (*Sparc*) KO Mice Suggests That Exercise-Induced Muscle Phenotype Changes Are SPARC-Dependent. *Appl. Sci.* **2020**, *10*, 9108. [CrossRef]

41. Nie, J.; Sage, E.H. SPARC inhibits adipogenesis by its enhancement of beta-catenin signaling. *J. Biol. Chem.* **2009**, *284*, 1279–1290. [CrossRef] [PubMed]

42. Nie, J.; Bradshaw, A.D.; Delany, A.M.; Sage, E.H. Inactivation of SPARC enhances high-fat diet-induced obesity in mice. *Connect. Tissue Res.* **2011**, *52*, 99–108. [CrossRef] [PubMed]

43. Mukherjee, S.; Choi, M.J.; Kim, S.W.; Yun, J.-W. Secreted protein acidic and rich in cysteine (SPARC) regulates thermogenesis in white and brown adipocytes. *Mol. Cell. Endocrinol.* **2020**, *506*, 110757. [CrossRef] [PubMed]

44. Ghanemi, A.; Yoshioka, M.; St-Amand, J. Regeneration during Obesity: An Impaired Homeostasis. *Animals* **2020**, *10*, 2344. [CrossRef] [PubMed]

45. Ghanemi, A.; Yoshioka, M.; St-Amand, J. Broken Energy Homeostasis and Obesity Pathogenesis: The Surrounding Concepts. *J. Clin. Med.* **2018**, *7*, 453. [CrossRef] [PubMed]

46. Ghanemi, A.; St-Amand, J. Redefining obesity toward classifying as a disease. *Eur. J. Intern. Med.* **2018**, *55*, 20–22. [CrossRef]

47. Hermetet, F.; Buffière, A.; Aznague, A.; De Barros, J.-P.P.; Bastie, J.-N.; Delva, L.; Quéré, R. High-fat diet disturbs lipid raft/TGF-β signaling-mediated maintenance of hematopoietic stem cells in mouse bone marrow. *Nat. Commun.* **2019**, *10*, 1–11. [CrossRef]

48. Wilkinson, A.C.; Yamazaki, S. The hematopoietic stem cell diet. *Int. J. Hematol.* **2018**, *107*, 634–641. [CrossRef]

49. Ribeiro, N.; Sousa, S.R.; Brekken, R.A.; Monteiro, F.J. Role of SPARC in bone remodeling and cancer-related bone metastasis. *J. Cell Biochem.* **2014**, *115*, 17–26. [CrossRef]

50. Baloğlu, M.; Özkorkmaz, E.G. Biochemical and immunohistochemical investigations on bone formation and remodelling in ovariectomised rats with tamoxifen citrate administration. *Folia Morphol.* **2019**, *78*, 789–797. [CrossRef]

51. Delany, A.M.; McMahon, D.J.; Powell, J.S.; Greenberg, D.A.; Kurland, E.S. Osteonectin/SPARC polymorphisms in Caucasian men with idiopathic osteoporosis. *Osteoporos. Int.* **2007**, *19*, 969–978. [CrossRef]

52. Zurlo, F.; Larson, K.; Bogardus, C.; Ravussin, E. Skeletal muscle metabolism is a major determinant of resting energy expenditure. *J. Clin. Investig.* **1990**, *86*, 1423–1427. [CrossRef]

53. Di Girolamo, D.J.; Kiel, D.P.; Esser, K.A. Bone and Skeletal Muscle: Neighbors with Close Ties. *J. Bone Miner. Res.* **2013**, *28*, 1509–1518. [CrossRef] [PubMed]

54. Nagaraju, G.P.; Sharma, D. Anti-cancer role of SPARC, an inhibitor of adipogenesis. *Cancer Treat. Rev.* **2011**, *37*, 559–566. [CrossRef] [PubMed]

55. Chlenski, A.; Guerrero, L.J.; Peddinti, R.; Spitz, J.A.; Leonhardt, P.T.; Yang, Q.; Tian, Y.; Salwen, H.R.; Cohn, S.L.; Yang, Q. Anti-angiogenic SPARC peptides inhibit progression of neuroblastoma tumors. *Mol. Cancer* **2010**, *9*, 138. [CrossRef] [PubMed]

56. Ghanemi, A.; Yoshioka, M.; St-Amand, J. Secreted protein acidic and rich in cysteine and inflammation: Another homeostatic property? *Cytokine* **2020**, *133*, 155179. [CrossRef]

57. Sangaletti, S.; Tripodo, C.; Cappetti, B.; Casalini, P.; Chiodoni, C.; Piconese, S.; Santangelo, A.; Parenza, M.; Arioli, I.; Miotti, S.; et al. SPARC Oppositely Regulates Inflammation and Fibrosis in Bleomycin-Induced Lung Damage. *Am. J. Pathol.* **2011**, *179*, 3000–3010. [CrossRef]

58. Rienks, M.; Carai, P.; van Teeffelen, J.; Eskens, B.; Verhesen, W.; Hemmeryckx, B.; Johnson, D.M.; van Leeuwen, R.; Jones, E.A.; Heymans, S.; et al. SPARC preserves endothelial glycocalyx integrity, and protects against adverse cardiac inflammation and injury during viral myocarditis. *Matrix Biol.* **2018**, *74*, 21–34. [CrossRef]

59. Rempel, S.A.; Hawley, R.C.; Gutiérrez, J.A.; Mouzon, E.; Bobbitt, K.R.; Lemke, N.; Schultz, C.R.; Schultz, L.R.; Golembieski, W.; Koblinski, J.; et al. Splenic and immune alterations of the Sparc-null mouse accompany a lack of immune response. *Genes Immun.* **2007**, *8*, 262–274. [CrossRef]

60. Sass, F.A.; Fuchs, M.; Pumberger, M.; Geissler, S.; Duda, G.N.; Perka, C.; Schmidt-Bleek, K. Immunology Guides Skeletal Muscle Regeneration. *Int. J. Mol. Sci.* **2018**, *19*, 835. [CrossRef]

61. Muire, P.J.; Mangum, L.H.; Wenke, J.C. Time Course of Immune Response and Immunomodulation During Normal and Delayed Healing of Musculoskeletal Wounds. *Front. Immunol.* **2020**, *11*. [CrossRef]

62. Ahuja, P.; Sdek, P.; MacLellan, W.R. Cardiac Myocyte Cell Cycle Control in Development, Disease, and Regeneration. *Physiol. Rev.* **2007**, *87*, 521–544. [CrossRef] [PubMed]

63. Stary, M.; Pasteiner, W.; Summer, A.; Hrdina, A.; Eger, A.; Weitzer, G. Parietal endoderm secreted SPARC promotes early cardiomyogenesis in vitro. *Exp. Cell Res.* **2005**, *310*, 331–343. [CrossRef] [PubMed]

64. Riley, H.J.; Kelly, R.R.; Van Laer, A.O.; Neff, L.S.; Dasgupta, S.; Baicu, C.F.; McDonald, L.T.; LaRue, A.C.; Zile, M.R.; Bradshaw, A.D. SPARC Production by Bone Marrow-Derived Cells Contributes to Myocardial Fibrosis in Pressure-Overload. *Am. J. Physiol. Heart Circ. Physiol.* **2020**. [CrossRef]

65. Avolio, E.; Mangialardi, G.; Slater, S.C.; Alvino, V.V.; Gu, Y.; Cathery, W.; Beltrami, A.P.; Katare, R.; Heesom, K.; Caputo, M.; et al. Secreted Protein Acidic and Cysteine Rich Matricellular Protein Is Enriched in the Bioactive Fraction of the Human Vascular Pericyte Secretome. *Antioxid. Redox Signal.* **2020**. [CrossRef]

66. Benincasa, G.; Franzese, M.; Schiano, C.; Marfella, R.; Miceli, M.; Infante, T.; Sardu, C.; Zanfardino, M.; Affinito, O.; Mansueto, G.; et al. DNA methylation profiling of CD04(+)/CD08(+) T cells reveals pathogenic mechanisms in increasing hyperglycemia: PIRAMIDE pilot study. *Annu. Med. Surg.* **2020**, *60*, 218–226. [CrossRef]

67. West, M.D.; Sternberg, H.; Labat, I.; Janus, J.; Chapman, K.B.; Malik, N.N.; De Grey, A.D.; LaRocca, D. Toward a unified theory of aging and regeneration. *Regen. Med.* **2019**, *14*, 867–886. [CrossRef]

68. Chen, F.; Liu, Y.; Wong, N.-K.; Xiao, J.; So, K.-F. Oxidative Stress in Stem Cell Aging. *Cell Transplant.* **2017**, *26*, 1483–1495. [CrossRef]

69. Wang, J.; Wang, H. Oxidative Stress in Pancreatic Beta Cell Regeneration. *Oxidative Med. Cell. Longev.* **2017**, *2017*, 1930261. [CrossRef]

70. Brioche, T.; Lemoine-Morel, S. Oxidative Stress, Sarcopenia, Antioxidant Strategies and Exercise: Molecular Aspects. *Curr. Pharm. Des.* **2016**, *22*, 2664–2678. [CrossRef]

71. Hultman, E.; Spriet, L.L.; Söderlund, K. Biochemistry of muscle fatigue. *Biomed. Biochim. Acta* **1986**, *45*, S97–S106. [PubMed]

72. Sahlin, K. Muscle fatigue and lactic acid accumulation. *Acta Physiol. Scand. Suppl.* **1986**, *556*, 83–91. [PubMed]

73. Capuano, P.; Capasso, G. The importance of intracellular pH in the regulation of cell function. *G Ital. Nfrol.* **2003**, *20*, 139–150.

74. Lönnqvist, S.; Emanuelsson, P.; Kratz, G. Influence of acidic pH on keratinocyte function and re-epithelialisation of human in vitro wounds. *J. Plast. Surg. Hand Surg.* **2015**, *49*, 346–352. [CrossRef] [PubMed]

75. Kruse, C.R.; Singh, M.; Targosinski, S.; Sinha, I.; Sørensen, J.A.; Eriksson, E.; Nuutila, K. The effect of pH on cell viability, cell migration, cell proliferation, wound closure, and wound reepithelialization: In vitro and in vivo study. *Wound Repair Regen.* **2017**, *25*, 260–269. [CrossRef] [PubMed]

76. Shuster, S. Osteoporosis, like skin ageing, is caused by collagen loss which is reversible. *J. R. Soc. Med.* **2020**, *113*, 158–160. [CrossRef] [PubMed]

77. Ramu, I.; Buchholz, S.M.; Patzak, M.S.; Goetze, R.G.; Singh, S.K.; Richards, F.M.; Jodrell, D.I.; Sipos, B.; Ströbel, P.; Ellenrieder, V.; et al. SPARC dependent collagen deposition and gemcitabine delivery in a genetically engineered mouse model of pancreas cancer. *EBioMedicine* **2019**, *48*, 161–168. [CrossRef]
78. Bradshaw, A.D.; Puolakkainen, P.; Wight, T.N.; Helene, S.E.; Dasgupta, J.; Davidson, J.M. SPARC-Null Mice Display Abnormalities in the Dermis Characterized by Decreased Collagen Fibril Diameter and Reduced Tensile Strength. *J. Investig. Dermatol.* **2003**, *120*, 949–955. [CrossRef]
79. Francki, A.; Bradshaw, A.D.; Bassuk, J.A.; Howe, C.C.; Couser, W.G.; Sage, E.H. SPARC regulates the expression of collagen type I and transforming growth factor-beta1 in mesangial cells. *J. Biol. Chem.* **1999**, *274*, 32145–32152. [CrossRef]
80. Rentz, T.J.; Poobalarahi, F.; Bornstein, P.; Sage, E.H.; Bradshaw, A.D. SPARC regulates processing of procollagen I and collagen fibrillogenesis in dermal fibroblasts. *J. Biol. Chem.* **2007**, *282*, 22062–22071. [CrossRef]
81. Jura, M.; Kozak, L.P. Obesity and related consequences to ageing. *Age (Dordrecht)* **2016**, *38*, 1–18. [CrossRef] [PubMed]
82. Marzetti, E.; Calvani, R.; Tosato, M.; Cesari, M.; Di Bari, M.; Cherubini, A.; Collamati, A.; D'Angelo, E.; Pahor, M.; Bernabei, R.; et al. Sarcopenia: An overview. *Aging Clin. Exp. Res.* **2017**, *29*, 11–17. [CrossRef] [PubMed]
83. Melouane, A.; Ghanemi, A.; Yoshioka, M.; St-Amand, J. Functional genomics applications and therapeutic implications in sarcopenia. *Mutat. Res.* **2019**, *781*, 175–185. [CrossRef]
84. Aspray, T.J.; Hill, T.R. Osteoporosis and the Ageing Skeleton. *Subcell. Biochem.* **2019**, *91*, 453–476. [CrossRef] [PubMed]
85. Ghanemi, A.; Yoshioka, M.; St-Amand, J. Will an obesity pandemic replace the coronavirus disease-2019 (COVID-19) pandemic? *Med. Hypotheses* **2020**, *144*, 110042. [CrossRef]
86. Ghanemi, A.; Yoshioka, M.; St-Amand, J. Secreted Protein Acidic and Rich in Cysteine: Metabolic and Homeostatic Properties beyond the Extracellular Matrix Structure. *Appl. Sci.* **2020**, *10*, 2388. [CrossRef]
87. Jørgensen, L.H.; Petersson, S.J.; Sellathurai, J.; Andersen, D.C.; Thayssen, S.; Sant, D.J.; Jensen, C.H.; Schroeder, H.D. Secreted Protein Acidic and Rich in Cysteine (SPARC) in Human Skeletal Muscle. *J. Histochem. Cytochem.* **2008**, *57*, 29–39. [CrossRef]
88. Cheng, L.; Sun, X.; Guo, J.; Lu, G. SPARC support the expansion of cord blood stem cells in vitro. *Cel. Res.* **2003**, *18*, S49. [CrossRef]
89. Zhu, J.; Wang, L.-Y.; Li, C.-Y.; Wu, J.-Y.; Zhang, Y.-T.; Pang, K.-P.; Wei, Y.; Du, L.-Q.; Liu, M.; Wu, X.-Y. SPARC promotes self-renewal of limbal epithelial stem cells and ocular surface restoration through JNK and p38-MAPK signaling pathways. *Stem Cells* **2019**, *38*, 134–145. [CrossRef]
90. Luo, Z.; Luo, P.; Yu, Y.; Zhao, Q.; Zhao, X.; Cheng, L. SPARC promotes the development of erythroid progenitors. *Exp. Hematol.* **2012**, *40*, 828–836. [CrossRef]

MDPI
St. Alban-Anlage 66
4052 Basel
Switzerland
Tel. +41 61 683 77 34
Fax +41 61 302 89 18
www.mdpi.com

Life Editorial Office
E-mail: life@mdpi.com
www.mdpi.com/journal/life

AF291708

THE HISTORY OF
ANCIENT GREECE

THE HISTORY OF ANCIENT GREECE

A military and political account of the ancient Greeks including the Persian Wars, the
Battle of Marathon and the campaigns of Alexander the Great and his conquest of Asia

NIGEL RODGERS

LORENZ BOOKS

CONTENTS

INTRODUCTION

Our world began with Greece – the world, that is, of Western civilization. The ancient Greeks were the first in so many things that we reveal our debt to them almost every time we open our mouths. Anarchy, astronomy, athletics, ballet, biography, biology, comedy, democracy, diplomacy, drama, ecology, economics, eroticism, history, marathon, maths, music, oligarchy, philosophy, physics, poetry, strategy tactics, technology, theatre, tragedy, tyranny, zoology – all are Greek words. They describe concepts or activities that, while they were not always invented by Greeks, were developed by them into forms we now recognize. Without the ancient Greeks, our modern world would not exist.

Greek civilization reached its zenith in classical Athens, a very small city by modern standards. It did so quickly – in less than 200 years (*c.*500–300BC) – but the impact has lasted millennia. Nothing like this intensity of experimentation had ever been seen before. Arguably, nothing that the world has seen since has wholly matched it either.

The Greeks were not, of course, the first civilization. They borrowed from many older, more sophisticated peoples around them. They took their alphabet from the Phoenicians (in modern Lebanon), and were influenced by Egyptian sculpture and architecture.

But the Greeks used their alphabet to write poetry, plays, history and philosophy, which the Phoenicians had never attempted. Greeks breathed life into the imposing but static statues of Egypt to create the first wholly naturalistic, life-like sculptures. These images still form our ideals of the perfect body. Anyone today who works out in a gym (which comes from *gymnasion,* another Greek word) is paying unconscious homage to the Greek ideal of bodily beauty.

The Greeks were the first to think about and question almost everything: politics, religion, the physical world and the moral one. They used Babylonian star books to speculate and calculate daringly (and often accurately) about the universe: one Greek, Aristarchus of Samos, even suggesting that the Earth went around the Sun – a shocking idea at that time and for the next 2,000 years.

Discarding religious or political authority, they were perhaps the most dauntless innovators in human history. Ancient Greece can often seem startlingly modern when compared both to rival cultures and to many later civilizations. This resemblance is not superficial. In some ways, they *were* modern.

From the Greeks' adventurous questioning arose modern science and democracy. These formidable twins now

Above: Standing high above the sea, the temple to Poseidon at Sunium, built c.*440BC, signalled to ships that they were nearing Athens, the Greek city where democracy and classical art both reached their zenith.*

Left: The theatre was among the most typical Greek buildings. The theatre at Epidaurus, dating from the 4th century BC, *is remarkably well preserved with still marvellous acoustics.*

Left: The Greeks spread from their original homeland in the Greek peninsula and islands, first across the Aegean, then around the Mediterranean and Black Sea, and finally across Asia, founding independent cities as they went.

shape the world. Anyone concerned with the way in which the *contemporary* world runs needs to know something about the ancient Greeks.

The dilemmas that the ancient Greeks were the first to discuss and confront 2,500 years ago – Is democracy always best? Should every citizen's voice be counted equally? – concern us still.

THE FIRST DEMOCRACIES

The Greeks created the world's first democracies, now the officially preferred form of government the world over. Democracy meant rule by all the people, a genuine revolution in human history. There was little in the world of the Mediterranean and Middle East *c.*600BC, where absolute monarchies and priestly hierarchies were the norm, to suggest that this was about to happen.

But in Greece, hereditary monarchy was dying out or had already been replaced by aristocracies or tyrants (meaning unconstitutional rulers with broad popular support, not necessarily dictators), while there was never a separate priesthood. Even more crucial perhaps, in the Classical Age (*c.*500–*c.*300BC) Greece

was never unified into a large centrally controlled state. Instead, it remained divided into scores of fiercely independent city-states (as we translate the world *polis*, although 'citizen-state' might be more accurate).

This was partly due to geography. Lacking large river valleys or plains, Greece is divided by mountains into small valleys, which inherently encouraged

Below: Delphi was the holiest Greek shrine, considered the centre of the world and sacred to the god Apollo. Its oracle, always deliberately enigmatic, was revered by all Greeks.

such individualism. But the results made Greek civilization, as it developed focused on each *polis*, crucially different from all earlier civilizations. No one had really done politics (our word comes from *polis*) before the Greeks took to arguing and experimenting, at times violently, about the best forms of government.

LABORATORIES FOR HUMANKIND

To ancient Greeks, modern democracies, where every few years we elect MPs or congressmen to a distant capital, would not seem truly democratic at all. At best Greeks would call them elected oligarchies. A Greek democracy demanded much more from its citizens: fighting as soldiers or sailors; serving on the huge juries or councils or as annual officials, often chosen by lot. Every citizen could vote and speak in the Assembly. This directly proposed policies and elected generals, punishing the latter if they failed. This was democracy in its most direct and most radical form.

Pericles, the supreme democratic statesman of Athens, declared in his funeral oration in 431BC: "Everyone is interested not only in their own affairs but in public affairs too, and is well informed about politics. We think that the man who minds only his own business has no business in the city. We Athenians personally decide or discuss policies." (The Greek word *idiot* meant someone who chose a private life – an idiotic choice for Greeks.) With life lived at such intensity, Greek cities became laboratories for humankind.

The Victorian philosopher J.S. Mill declared that the Battle of Marathon in 490BC – when Athens first defeated Persia – was of greater importance in *British* history than the Battle of Hastings. Without that victory, Athenian democracy might not have survived. And without democracy, Athens' open society could not have prospered, encouraging theatre, philosophy, science, history, architecture, art and drama.

Below: The Charioteer of Delphi, *dating from* c.470BC, *superbly exemplifies early classical sculpture. One of the very few bronze statues to survive, it was originally painted with lifelike colours.*

Above: Far from being always made of cold white marble, most Greek statues glowed with vivid colour, as shown in this kore, *a statue of a young girl of the 6th century* BC.

Greek democracy had, however, definite limits: women were excluded from public life – as in almost all cultures and countries before the 20th century – and slavery was an essential part of Greek life. But almost every civilization, at the time, and for long after, relied on slaves, and slavery in Athens remained mostly domestic and small-scale.

A worse criticism of the Greeks would be their endless wars. Just enough city-states united to repel the Persian invasions of 490–478BC, but that unity proved unique. The Greeks reverted to fighting each other, often calling in outside powers – even Persia – to help. Such disunity led to their conquest, first by the Macedonians, then by the Romans. It was the dark side of the Greeks' pursuit of individual excellence or perfection.

THE PURSUIT OF PERFECTION

The Greeks were intensely competitive as individuals, striving 'always to be best'. (These words of Homer, the greatest Greek poet, about his hero Achilles, inspired Alexander the Great when conquering the Persian Empire.) Greeks were driven by two concepts: love of honour (*philotimaea*) and desire for *areté* (excellence, goodness, perfection, a term applicable to anything from athletics to philosophy). But the pursuit of excellence was not restricted to individuals.

It was the Athenian people, the *demos*, who supported the building of the great temples of the Acropolis – they voted on it in 447BC after impassioned debate. The Athenian people attended the plays of the great dramatists Aeschylus, Sophocles and Euripides. The Athenian experience of the 5th century BC shows that democracy does not have to mean dumbing down to the lowest intellectual level: it can mean raising up a whole city to unprecedented cultural and political heights. *Areté* was not detached from life but was pursued in the dusty clamour of the streets, even by philosophers such as Socrates.

VIBRANT CLASSICISM

For many people ancient Greece can seem a cold, remote world of smooth marble statues and pristine white temples, lacking human interest. Such images mislead. Far from being cold and passionless, the Greeks burned and quivered with passions and desires – personal, political and intellectual – that often led to disaster, not perfection. Greek statues were once brilliantly painted with vivid colours, eyes and hair in particular giving a startlingly lifelike appearance. Even the Elgin Marbles once blazed with colour. Buildings, too, were painted in ways that might seem garish today. Time has stripped the paint and gilding off Greek remains, leaving them more sober-looking than their creators intended.

Few original Greek statues survive – a few in marble, even fewer bronzes. Too often we have only some mediocre copies of the great originals, made for the Romans. We have lost even more Greek painting. What has survived, however, reveals that the Greeks again pioneered a naturalistic art. The far more numerous extant vases, often magnificent artworks, tell us much about Greek daily life.

THE EXPANSION OF GREECE

Today the Greeks are confined to Greece proper and Cyprus. But in antiquity they spread – first all around the Mediterranean as far as Marseilles, and then across western and central Asia. This second expansion occurred under Alexander the Great (336–323BC). A supreme military genius, Alexander overthrew the Persian Empire and founded many cities that helped to Hellenize (the Greeks called themselves Hellenes) western Asia. By 200BC, a traveller could go from southern France all the way to the borders of India speaking only Greek, at least in the cities.

This expanded Greek world fell in the end mostly to Rome, which by 30BC had conquered the Greeks, often with a greedy brutality. But the Romans, if lacking Greek brilliance, became superb preservers and transmitters of Greek culture (though not Greek democracy) across Western Europe. The resulting Graeco-Roman civilization is the bedrock of the modern world. The Greeks are not just ancient history buried in museums: their ideas, arguments, ambitions and culture have helped shape Western history since the Renaissance and still underly much of the modern world. To understand ourselves today, we need to look back to the Greeks.

Above: The influence of Greek art has persisted down the centuries. The front of the British Museum, London, built by Robert Smirke in 1823–46, consciously echoes Greek models.

Below: Among the most famous of Greeks, Alexander the Great of Macedonia (reigned 336–323BC) overthrew the Greeks' old enemy, the immense Persian empire, but also restricted their cities' cherished independence.

TIMELINE

Above: The Parthenon, the supreme Greek temple, still rises above Athens despite the vicissitudes of 2,500 years.

Greece has one of the longest histories in the world. The origins of Classical Greece, that supremely accomplished civilization of the 5th and 4th centuries BC, lie in misty prehistory. Archaeology and legend are at first our sole guides, for history proper begins only c.550BC. (All dates before 550BC are approximate.) Around 2000BC, as Europe's first civilization emerged on Crete, ancestors of the Greeks appeared in the Greek peninsula and archipelago. By 1200BC they had created the Mycenaean civilization, a dynamic, resplendent culture that later generations peopled with heroes, unaware that a Dark Age, illiterate and impoverished, lay between them. The Greek age proper began with the first Olympic Games, traditionally held in 776BC. From then on Greek life accelerated, reaching its undoubted climax between 500 and 300BC, the age of the Persian Wars, Pericles, Socrates, the great dramatists and generals. But Greek history continued to develop long after Alexander the Great's death in 323BC, becoming linked with that of Rome, its political conqueror but cultural captive. Even the fall of Rome in AD476 merely eclipsed Greek brilliance, which re-emerged in the 15th century to bedazzle the Renaissance.

3000–700BC

3000BC Beginnings of Minoan civilization.
2000BC Building of first palaces in Crete; destruction of Bronze Age Lerna.
1700BC Destruction of first palaces in Crete; start of Second Palatial Period.
1600BC Beginnings of Mycenaean civilization in mainland Greece.
1570BC Palaces in Crete rebuilt after earthquake; Minoan culture at its zenith.
1550BC First shaft-graves at Mycenae.
1500–1470BC Volcanic eruption on Thira devastates Minoan civilization.
1450BC Mycenaeans occupy Knossos.
1380BC Final destruction of Knossos; Mycenaean trade and influence spread.
1300–1250BC Building of Treasury of Atreus, Lion Gate at Mycenae.
1287BC Battle of Cadesh between Egypt and Hittites.
1200BC Destruction of palace at Pylos.
1190BC Traditional date of Trojan War; Egypt repels the Sea Peoples.
1150BC Final collapse of Mycenaean civilization; start of Greek Dark Ages.
1050BC Dorian migrations into Greece; Ionian migration to Asia Minor.
900–800BC Rise of aristocracies in Greece.
776BC First Olympiad (Olympic Games).
760–730BC Homer composes *The Iliad* and *The Odyssey*; adaptation of Phoenician alphabet by Greeks.
753BC **Founding of city of Rome.**
750BC Foundation of Cumae in Italy, first Greek colony in west.
735BC Foundation of the first Greek colonies in Sicily at Naxos (Catania) and Syracuse.
730–710BC Sparta's first conquest of Messenia.
700BC Hesiod writes *Work and Days* and *Theognis*; according to tradition Deioces founds Median kingdom; introduction of hoplite-style fighting.

Pages 14–17: The coins of ancient Greece were some of the very first ever to be minted. They were often brilliantly decorative, and they remain among the most beautiful and well designed that the world has seen.

699–500BC

682BC List of annual *archons* at Athens begins; Gyges seizes Lydian throne.
669BC Sparta defeated by Argos under King Pheidion.
660BC Lycurgan reforms in Sparta; it crushes Messenian revolt.
***c.*650BC** Rise of tyrants across Greece.
632BC Cylon tries to seize power in Athens; first colony in Libya at Cyrene.
620BC Dracon's Law Code published in Athens; foundation of Byzantium on Bosphorus, Naucratis in Egypt.
612BC Fall of Nineveh, Assyrian capital; first Black Sea colonies (Istrus, Olbia, etc).
***c.*600BC** Thrasybulus tyrant of Miletus; Ionian Enlightenment; first triremes; Sappho, Pittacus and Alcaeus in Lesbos.
594BC Legislation of Solon in Athens.
589BC Foundation of Acragas in Sicily.
585BC Eclipse of sun, predicted by Thales, halts battle between Media and Lydia.
561BC Pisistratus seizes power in Athens; Croesus becomes king of Lydia.
550BC Achaemenid Empire of Persia founded by Cyrus the Great; Sparta forms the Peloponnesian League.
546BC Cyrus conquers Lydia; Sparta defeats Argos, annexes Thyreatis.
538BC Cyrus captures Babylon; he liberates the Jews.
527BC Pisistratus dies; succeeded by sons Hippias and Hipparchus.
514BC Harmodius and Aristogeiton kill Hipparchus; Persian expedition crosses the Danube.
510BC Hippias expelled from Athens with Spartan help.
508BC Cleisthenes' reforms in Athens.
507BC The Spartan invasion under Cleomones is repelled; Athens defeats Boeotians and Chalcidians, and gains Chalcidian territory.

499–450BC

499BC Outbreak of Ionian Revolt.
494BC Defeat of Ionians at Lade by Persia; fall of Miletus; Sparta defeats Argos.
490BC Athenians defeat Persian invasion at Marathon.
487BC First recorded use of ostracism at Athens; *archons* appointed by lot.
486BC Death of Darius the Great of Persia: accession of Xerxes.
483BC New silver lode found at Laurium: Themistocles wins debate on building fleet; Persians dig canal through Mt Athos.
481BC League of Corinth formed to resist Persia under Spartan leadership.
480BC Second Persian invasion under command of Xerxes; August battles of Thermopylae and Artemisium; Athens occupied by the Persians; September Battle of Salamis: Persian fleet destroyed; Carthaginian attack defeated at Himera.
479BC Persians under Mardonius defeated at Plataea by Spartan-led army; Persian fleet defeated at Mycale; revolt of Ionia.
478BC Sparta withdraws from Greek alliance; formation of the League of Delos.
474BC Greeks defeat Etruscans at Cumae.
470BC Birth of Socrates.
467BC Battle of Eurymedon: last Persian fleet destroyed by Athens.
464BC Earthquake at Sparta; *helot* revolt.
463BC Cimon leads Athenian force to help suppress Messenians.
462BC Democratic reforms of Ephialtes.
460BC Outbreak of war between Sparta and Athens; pay for jurors introduced.
459BC Athens sends fleet to help Egyptian revolt against Persia.
458BC *Zeugitae* admitted to *archonship*; building of Long Walls of Athens.
457BC Athens conquers Boeotia.
454BC Loss of Egyptian expedition; Confederacy Treasury moved to Athens.

449–415BC

449BC Peace of Callias with Persia; Athens invites the Greeks to restore her temples.
447BC Parthenon begun; Athens loses Boeotia.
446BC 30 Years' Peace with Sparta (actually to 431BC).
443BC Ostracism of Thucydides, son of Melesias, confirms Pericles' supremacy.
438BC Gold and ivory giant statue of Athena set up in Parthenon.
436BC Foundation of Amphipolis by Athens.
432BC Defensive alliance of Athens with Corcyra; Megarian Decree passed.
431BC Outbreak of Peloponnesian War; invasion of Attica by Peloponnesian army.
430BC Outbreak of plague devastates Athens; Pericles tried and fined.
429BC Pericles reinstated and dies; birth of Plato.
427BC Revolt of Lesbos crushed: debate on how to treat prisoners in Athens.
425BC Athenians occupy Pylos and capture Spartans on Sphacteria: Sparta sues unsuccessfully for peace.
424BC Battle of Delium: Athenians defeated in Boeotia; loss of Amphipolis to Spartan general Brasidas leads to banishment of Thucydides; Congress of Gela in Sicily: Hermocrates propounds 'Monroe Doctrine' for Sicily.
422BC Battle of Amphipolis: deaths of Cleon and Brasidas.
421BC Peace of Nicias.
420BC Alcibiades dominates Assembly.
418BC Battle of Mantinea: Athens and Argos defeated by Sparta.
416BC Athens captures and sacks Melos: 'Melian Debate'.
415BC Mutilation of Herms: Syracusan expedition sails; Alcibiades, recalled to face trial, escapes to Sparta.

414–390BC

414BC Sparta reopens war with Athens; sends Gylippus to help Syracusans.
413BC Spartans occupy fort of Deceleia; Demosthenes sent to Syracuse with reinforcements; great battle in Syracuse harbour; disastrous loss of expedition.
412BC Revolt of Athenian allies; Treaty of Miletus between Sparta and Persia.
411BC Oligarchic revolution at Athens; moderate oligarchy proposed.
410BC Battle of Cyzicus leads to restoration of full democracy in Athens.
409BC Carthage invades Greek Sicily.
408BC Athenians under Alcibiades recapture Byzantium and Chalcedon.
407BC Alcibiades returns to Athens; Prince Cyrus comes down to the Aegean.
406BC Alcibiades leaves Athens after defeat at Notion; Battle of Arginusae; Acragas besieged by Carthaginians; death of playwright Euripides.
405BC Lysander, Spartan *navarch*, defeats Athenian fleet at Aegospotami; end of Athenian power, and blockade of Athens; Dionysius I becomes tyrant of Syracuse.
404BC Surrender of Athens; Long Walls pulled down; dictatorship of the 30.
403BC Spartan garrison on Acropolis; Thrasybulus seizes Piraeus; restoration of democracy, and general amnesty.
401BC 'March of the 10,000' behind Cyrus into Persian Empire; Cyrus killed at Cunaxa; Xenophon leads Greeks home.
399BC Trial and execution of Socrates.
397BC Dionysius I captures Motya in Sicily; Sparta makes truce with Persia.
395BC Athens rebuilding Long Walls.
394BC Thebes, Corinth and Athens Army beaten by Spartans at Battle of Corinth.
393BC Athens completes her Long Walls.
390BC Iphicrates defeats Spartans with light-armed *peltasts*

389–340BC

386BC The King's Peace: Sparta abandons Ionians in return for Persian support.
382BC Spartans seize Cadmaea (citadel) of Thebes and install pro-Spartan oligarchy.
379BC Spartans expelled from Cadmaea; revolution in Thebes led by Epaminondas.
378BC Athens forms 2nd Confederacy.
376BC Timotheus defeats Spartan fleet at Naxos; Mausolus *satrap* of Caria; Jason of Pherae establishes rule in Thessaly.
371BC Thebes routs Spartans at Leuctra but is checked by Jason.
370BC Assassination of Jason of Pherae; Epaminondas marches into Peloponnese.
369BC Foundation of Messene and liberation of *helots* by Thebans.
367BC Death of Dionysius I of Syracuse, succeeded by Dionysius II.
362BC Epaminondas killed at Mantinea; 'Revolt of the Satraps' against Persian king.
359BC Accession of Philip II of Macedon; defeats invading tribes; accession of Artaxerxes III, dynamic Persian king.
357BC Philip captures Amphipolis and marries Olympias; Dion returns to Sicily and 'liberates' Syracuse; start of the Athens' War of the Allies.
356BC Philomelus of Phocis seizes Delphi, starting Sacred War; Philip captures Potidaea; birth of Alexander.
354BC Murder of Dion in Syracuse; Athens makes peace with allies. Onomarchus of Phocis defeats Philip;
353BC Philip captures Methone; death of Mausolus of Halicarnassus.
352BC Philip defeats Phocians and becomes *tagus* (ruler) of Thessaly.
347BC Death of Plato.
346BC Peace of Philocrates ends Sacred War: Philip as protector of Delphi.
340BC Philip attacks Byzantium; Alexander left as regent of Macedonia.

339–327BC

338BC Battle of Chaeronea: Theban and Athenian armies defeated by Philip; death of Isocrates, Athenian orator; murder of Artaxerxes III by Bagoas, his Vizier.
337BC Council of Corinth elects Philip General of the Greeks for anti-Persian crusade; death of Timoleon in Sicily.
336BC Macedonian advance guard sent to Asia; Philip II murdered; accession of Alexander III; swift descent on Greece, where he is proclaimed General; accession of Darius III in Persia.
335BC Alexander campaigns in Thrace; Alexander's destruction of Thebes; Aristotle begins teaching at Athens.
334BC Alexander crosses into Asia; defeats Persians at Battle of Granicus; liberates Ionia; sieges of Miletus and Halicarnassus.
333BC Alexander cuts Gordian Knot; death of Greek mercenary general Memnon; change in Persian tactics; Battle of Issus (November): Alexander routs Persians; Darius flees east.
332BC Siege of Tyre (January–July); siege of Gaza; Alexander enters Egypt.
331BC Foundation of city of Alexandria; trip to consult oracle at Siwah; Alexander routs Persians at Gaugamela; enters Babylon and Susa; Sparta defeated by Macedonia.
330BC Alexander burns Persepolis; Darius murdered by Bessus; Alexander executes Philotas and Parmenion.
329BC Alexander crosses Hindu Kush to Bactria; crosses River Oxus; founds Alexandria Eschate; winters in Bactria.
328BC Campaign against Spitamenes; Alexander quarrels with and kills Cleitus.
327BC Capture of Sogdian Rock; Alexander marries Roxane; tries to introduce Persian *proskynesis*; Pages' Conspiracy – Callisthenes executed; enters India at the end of the year.

326–278BC

326BC Alexander defeats Porus at Hydaspes; conquers Punjab; advance to River Beas, where troops mutiny; forced to turn back.
325BC Conquest of the Malli: Alexander almost fatally wounded; he sails down Indus to Ocean; marches through Gedrosian Desert with many fatalities; voyage of Nearchus along coast.
324BC Execution of corrupt *satraps*; Susa weddings of Persians and Macedonians; mutiny of discharged veterans at Opis; Exiles Decree at Olympia; death of Hephaistion (October).
323BC Alexander enters Babylon; Greek cities hail him as a god; Alexander dies on 10 June; wars of the Diadochi (successors) and Lamian War of the Greeks against Macedonia; revolt of colonists in Bactria.
322BC Ptolemy gains control of Egypt: Lamian War ends in Greek defeat by Antipater; deaths of Demosthenes and Aristotle; Athenian democracy curtailed.
312BC Seleucus I takes over eastern *satrapies*; founds Seleucia-on-the-Tigris.
303BC Seleucus cedes Indian territories in return for 500 elephants.
301BC Battle of Ipsus: death of Antigonus I and division of world into four kingdoms.
300BC Zeno sets up Stoic School in Athens; Seleucus founds Antioch in Syria.
297BC Pyrrhus I king of Epirus (to 272BC).
295BC Library at Alexandria founded.
283BC Death of Ptolemy I, founder of Ptolemaic dynasty, in Egypt.
280BC Seleucus I defeats and kills Lysimachus, ending his kingdom, then murdered himself; Antiochus I succeeds him (to 261).
280–275BC Pyrrhus fights Romans in Italy without success; foundation of Achaean League in southern Greece.
279BC Building of Pharos at Alexandria.

274–200BC

274–232BC Reign of Ashoka, first Buddhist emperor, in India.

270BC Hieron emerges as saviour of Syracuse, assuming crown as Hieron II.

263–41BC Eumenes I of Pergamum starts to assert independence from Seleucids.

264BC First Punic War between Carthage and Rome starts.

*c.***255BC** Bactria breaks away from Seleucid control, followed by Parthia.

245–213BC Aratus dominates Achaean League, seizes Corinth (243BC).

244BC Agis IV (to 241) tries to introduce radical reforms at Sparta and is killed.

241BC End of First Punic War: Rome takes over most of Sicily.

235BC Cleomenes III King of Sparta (to 222BC) introduces radical reforms before being defeated; Euthydemnus I seizes power in Bactria and expands kingdom.

230BC Attalus I defeats Gauls.

229BC Athens 'buys out' Macedonian garrison, in effect becoming neutral.

228BC Rome makes Illyria protectorate.

223BC Antiochus III the Great succeeds to Seleucid throne (to 187BC).

221BC Philip V succeeds to Macedonian throne (to 179BC); Ptolemy IV defeats Antiochus III at Raphia 218.

218–202BC Second Punic War.

217BC Peace Conference at Naupactus: warning of the 'shadow of Rome'.

216BC Romans defeated at Cannae.

215BC Alliance of Philip and Carthage leads to First Macedonian War (to 205BC); Syracuse switches support to Hannibal.

212BC Antiochus starts eastern campaigns; fall of Syracuse to Romans: Archimedes killed in the fighting.

202BC Hannibal finally defeated at Zama.

200BC Egypt defeated by Antiochus III at Ionion, loses southern Syria/Palestine.

199–70BC

197BC Macedonia defeated at Battle of Cyroscephalae.

196BC Flaminius declares 'Liberty for Greeks' at Corinth.

192BC Start of 'First Syrian' War of Rome against Antiochus III.

190BC Antiochus defeated at Magnesia.

188BC Treaty of Apamea: Antiochus loses all land west of Taurus mountains.

171–138BC Mithradates I of Parthia expands kingdom at Seleucid expense.

171–168BC Third Macedonian War.

170BC Eucratides I rules united Indo-Greek kingdom (to 155BC).

168BC Perseus of Macedonia defeated at Pydna; Antiochus IV forced to abandon conquest of Egypt; his intervention in Jerusalem sparks Maccabee revolt.

166BC Romans enslave 150,000 Epirotes, make Delos a free port; slave trade booms.

155–130BC Menander, king of huge Indo-Greek kingdom reaching the Ganges, possibly converts to Buddhism.

146BC Romans sack Corinth; make Achaea and Macedonia Roman provinces.

133BC Attalus III of Pergamum dies leaving kingdom to Rome.

130BC Romans crush Aristonicus' Utopian revolt and make Asia Minor a Roman province; Parthians capture Babylonia.

122BC Gauis Gracchus gives tax-farming rights for Asia and Greece to his allies in Rome, leading to ruinous extortion.

120BC Mithradates VI king of Pontus.

88BC Mithradates VI overruns Rome's eastern provinces offering 'liberty' to the Greeks; massacre of 80,000 Italians on Delos and other islands.

86–85BC Sulla sacks Athens.

73BC Lucullus defeats Mithradates and drives him out of Pontus to Crimea, where he commits suicide.

69BC–AD1462

66–63BC Pompey reorganizes the east: Syria and Bithynia-Pontus become provinces Judaea and others client states.

51BC Parthians invade Syria; Cleopatra becomes co-monarch of Egypt.

48BC Battle of Pharsalus: Pompey flees to Egypt and is killed; Caesar following, meets Cleopatra and has affair.

47BC Caesar leaves Egypt; Cleopatra later follows him to Rome; Cicero in retirement summarizes Greek philosophy.

44BC Caesar assassinated

42BC Battle of Philippi: Cassius and Brutus defeated by Antony and Octavian; Antony takes eastern empire; winters in Athens.

41BC Antony meets Cleopatra at Tarsus.

36BC Antony invades Parthia but is forced to retreat.

31BC Battle of Actium: Octavian's forces defeat Antony and Cleopatra

30BC Antony and Cleopatra commit suicide: Egypt annexed to Roman Empire by Octavian; Kushans (Scythians) overrun last Indo-Greek kingdom

AD14 Death of Augustus (Octavian), first Roman emperor; succeeded by Tiberius.

AD66 Nero visits Greece, wins all prizes at the Olympics and declares 'liberty' for Greek cities.

AD124 Hadrian visits Athens and makes it head of Panhellenic League, completes Temple of Olympian Zeus.

AD393 Last Olympic Games held.

AD397–8 Visigoths ravage Greece.

AD529 Justinian closes Academy and other philosophy schools in Athens

1438 Council of Florence; Bessarion stays on in Italy, rekindling knowledge of Greek.

1453 Fall of Constantinople to Ottoman Turks; some Greek scholars flee to Italy.

1462 Platonist Academy founded in Florence to study Greek philosophy.

PART ONE

CLASSICAL GREECE

The Parthenon, most famous of all Greek temples, rises supremely above the city of Athens Epitomizing classical Greece, it was built by the world's first true democracy. Greek democracy had emerged only around 500BC and reached its final form just before the Parthenon was built. This is no coincidence: the Greeks passionately sought perfection in their politics *and* in their art. Their achievements in both mark the start of Western history. To this day, men and women fight and die for democracy (a Greek word) The history of classical Greece is, to a large extent, the history of the first democracies.

Democracy was not attained or defended just with words or ideas, much though Greeks loved both Greeks had to fight for their freedom against internal and external enemies. The greatest external enemy was the Persian Empire. Its rise and fall roughly coincided with Athens' democracy. In the great conflict of the Persian Wars (499–478BC), the fragmented Greek states (around 1,000 scattered from Spain to Cyprus) united to repel the Persian superpower. The major battles of those wars – Marathon, Thermopylae, Salamis and Plataea – are among the pivotal contests in human history.

But the Greeks never united again. Instead, they fought each other unceasingly, until finally they fell to other powers: first Macedonia, then Rome. But in their centuries of independence, the Greeks experimented constantly, trying out every form of government: democracy, monarchy, dictatorship, oligarchy, even communism (on the Aeolian Islands, near Sicily). In doing so, they invented philosophy and history besides politics. Reading Greek history entails far more than reading about ancient ruins: it means rediscovering our political and intellectual origins.

Left: The buildings on the Athenian Acropolis, such as the Erechtheum, still proclaim the achievements of classical Greece at its democratic zenith.

CHAPTER I

THE GREEK AWAKENING

c.2000–500BC

Greek history starts far earlier than the Greeks themselves suspected (or anyone else before recent archaeological discoveries). People who were possibly the Greeks' ancestors appeared in the peninsula around 2000BC. At about the same time, and perhaps not by coincidence, the sophisticated Minoan civilization arose in Crete and the Cyclades. While the Minoans were almost certainly not Greek, the Mycenaeans, who took over Crete and were influenced by its culture, definitely were.

Heroic myths cluster around the Mycenaean palaces of the late Bronze Age (1600–1200BC). This suggests that there is a kernel of truth to most Greek legends, although they should not be regarded as history. Greeks of the historical period proper (from c.550BC) mostly believed their legends, especially when related by Homer, but they had only the haziest ideas of their actual prehistory. In particular, they were unaware that between them and their legendary ancestors lay the chasm of the Greek Dark Ages. When Greek life revived after 800BC, it took very different forms from Mycenaean Greece.

One factor remained constant, however: the mountainous landscape, with valleys often isolated from each other, meant that Greece remained divided politically into many small, even tiny states. This helped to give Greek life its passionate intensity centred on the *polis*, the citizen-state.

Left: The huge grave shafts of Mycenae – "rich in gold" according to Homer – are the first Greek buildings, dating to c.1550BC.

MINOAN CRETE
THE FIRST CIVILIZATION 2000–1100 BC

Above: This vivid fresco of children boxing was found on Thira, the Aegean island devastated by the volcanic eruption in c.1500 BC that is thought to have half-wrecked Minoan civilization.

Below: The grand staircase at Cnossus Palace, the largest of the Cretan palaces, reveals the surprisingly modern-looking ingenuity of Minoan architects.

Europe's first civilization arose in Crete and the Aegean islands during the Bronze Age. It is called Minoan after its legendary king Minos. According to the Athenian historian Thucydides, Minos was the first to rule the seas and islands. Thucydides was long thought to be just repeating myths about the king of Cnossus, who reputedly exacted tribute from Athens: 14 girls and boys to be fed to the Minotaur, the bull-headed monstrous offspring of Minos' wife, Queen Pasiphae, and a bull. But, since Arthur Evans began excavations at Cnossus in 1900, controversially restoring many of the ruins, Thucydides has been proved partly right: there was a great civilization in Crete with outposts across the Aegean.

Cretan civilization started modestly in the 3rd millennium. Around 2000 BC the first palaces – multi-storeyed buildings centred on courtyards – were built at Cnossus, Phaestos and Mallia. They were influenced by Egyptian and Near Eastern models, for the Cretans were by then sailing south and east, bringing home ivory and other luxury products in return for jewellery, wine, textiles and olive oil. Egyptian records of the 2nd millennium BC often mention Cretans (as tribute-bearers, although Crete was never a tributary to Egypt), but Minoan civilization soon developed its own highly distinctive characteristics.

A PEACEFUL SOCIETY

Notable among Minoan society's characteristics were its peaceful nature and the prominent role seemingly played by women. However, the Cretans were never total pacifists – they presumably relied mostly on their navy for defence – and ideas of a matriarchy, with women ruling the realm, are probably misplaced. Around 1700 BC all the palaces were destroyed by earthquakes.

Above: This figurine, bare-breasted in the courtly Minoan fashion, represents the earth-goddess or possibly one of her priestesses.

But, with Minoan culture vigorously resurgent, they were soon rebuilt as larger, even more elegant complexes.

Below: The Minoan Palace at Cnossus with its characteristic dark-red columns.

Cretans of the period *c.*1700–1500BC (called Late Minoan by Evans, Later Palatial by others) produced pottery, jewellery and other vivid artworks that can still delight. The palaces, especially the largest at Cnossus, had bathrooms with running water and light-wells. Their buildings had dark red columns tapering downwards and rose four or even five storeys high. On their walls, frescoes depicted a graceful, idyllic life, with bare-breasted women in long flounced dresses and with elaborate hairstyles, and clean-shaven men in kilts, surrounded by flowers, birds and dolphins. A cult of the bull was clearly central to Cretan religion, demonstrated by the many images of bulls and by the bull-dance, a game or rite common on murals. (This may underlie the Minotaur legend.) The axe was also a common motif. Paved roads connected palaces with outlying villages and villas around the island. The population of Cnossus town reached *c.*30,000 by 1500BC, making it the largest Mediterranean city of the age.

Minoan culture was not confined to Crete, however, but also flourished in the Cyclades and some other Aegean islands. Most prosperous was Thira (Santorini), which was probably independent. Murals excavated there show a fleet attacking a walled town and reveal Thira's links with north Africa. But the predominant note, as in Crete, was peaceful. After 1600BC, Minoan culture spread to the mainland, with such strong Minoan influence

Above: The cult of the bull was central to Minoan religious life. Bull-leaping, as this fresco reveals, was dangerous. It probably had religious connotations but was conducted insouciantly.

Below: This languid youth is called the Lily Prince. Like most Cretan young men, he wore only a kilt, but he also has an elaborate head-dress.

LINEAR B
Linear B was the third form of writing developed in Crete. The first was a pictographic script, the second was a syllabic system, Linear A, while Linear B was similar but in Greek. Linear B tablets record flocks of sheep, oil jars, horses, spears and chariots, not royal edicts, history or poetry, yet they shed invaluable light on this highly civilized society now ruled by Greek incomers.

---

evident in mainland sites that Evans talked of a Minoan Empire. But certainly here the rulers, Mycenaean Greeks, were independent, merely employing Minoan craftsmen to make them precious objects.

THE THIRA ERUPTION
This cheerful civilization was in its prime when an immense volcanic explosion tore Thira apart. It also devastated other islands, including Crete, while preserving part of Thira town under lava like a Bronze Age Pompeii. The date of the eruption remains debated, ranging from 1600BC, the geologists' preferred date, to 1460BC, which suits archaeologists better. Indisputably, when Cnossus Palace was rebuilt *c.*1450BC, it was occupied by Greek-speakers, for clay Linear B tablets written in early Greek have been found there. Although a warlike note emerges in the tablets, Cnossus, 'house of the double axe' (*labyrinth*), remained unwalled.

Then one day *c.*1380BC a fire, caused by accident, earthquake or human attack, gutted the palace once more and it was never rebuilt. Minoan culture now entered its less glorious 'post-palatial' phase. This may not have been totally impoverished – Homer ranked Crete as second only to Mycenae in his 'Catalogue of Ships'. But *c.*1100BC Dorian Greek invaders, who had attacked the Peloponnese already, invaded Crete, finally wrecking the Minoan world.

MYCENAEAN SOCIETY
THE FIRST GREEKS, 1600–1200BC

Above: The Lion Gate, erected c.1300BC, still guards the entrance to Mycenae. Its massive walls so impressed later Greeks that they assumed they had been built by Cyclops, one-eyed giants.

Below: The Mycenaeans were impressively skilled architects, building corbelled vaults. The 'Treasury of Atreus', which dates from c.1300BC, is not really a treasury but a tomb.

In the 3rd millennium BC a prosperous Bronze Age culture developed in the Greek peninsula in small unwalled towns such as Lerna in the Peloponnese. Around 2000BC these were destroyed by invaders, probably from the north, and for a time urban life totally disappeared. Then c.1600 BC a new civilization emerged that built amazing tombs at Mycenae in the north-west Peloponnese filled with so many gold artefacts that Homer's epithet "Mycenae, rich in gold" is justified. This was probably due to the rise of powerful new kings, not fresh invasions, but nothing is known about them.

In Greek legend, including Homer's great poems *The Iliad* and *The Odyssey* written in the 8th century BC, Mycenae was the paramount kingdom, so this first Greek civilization has been called Mycenaean. But Achaean, the term used by both Homer and by contemporary Hittite kings in Anatolia who had diplomatic dealings with them, is more apposite. Nothing suggests that Bronze Age Greece was politically united under Mycenae, although it probably controlled the Argolid plain beneath its citadel, from which paved roads radiated.

MYCENAEAN CULTURE EMERGES

In the 16th century BC the cultural influence of Minoan Crete, then far more sophisticated, was overwhelming on the Mycenaeans. Superb gold cups found near Sparta, showing bulls being tethered by long-haired youths, reveal this. But already distinctive themes favoured by Mycenaeans, such as hunting and warfare, were emerging. The early (c.1600BC) gold mask from Grave A at Mycenae shows a fierce bearded warrior most unlike clean-shaven Minoans. After 1460BC Greek warriors, possibly from Mycenae, occupied the palace at Cnossus after the Thira eruption.

Among cultural imports from Crete was writing in Linear B, subsequently adopted in Mycenaean centres. These include nearby Tiryns, Thebes and Orchomenus in central Greece, Athens (although it remained of secondary importance) and Pylos in the south-west Peloponnese. Excavations at Pylos have unearthed a complete Mycenaean palace, with elegant tapering columns and murals depicting hippogriffs and other mythical figures. Uniquely, Pylos was unwalled. Linear B tablets found there reveal a bureaucracy trying to control almost every aspect of daily life. Mycenaean culture spread north into Thessaly, across the Aegean to Miletus on the Asian shore and east to Cyprus. There was probably a palace near Sparta in the fertile Eurotas valley, but none has been found.

PROTECTING MYCENAE

Mycenaean wealth grew throughout the 14th century BC. It probably came from both trading and raiding – piracy, Thucydides noted, was socially perfectly acceptable in legendary times. Greek artefacts, mainly pottery, have been found as far west as Sicily and as far east as Syria. Mycenaean outposts replaced Minoans

SCHLIEMANN AND TROY

In the 19th century, scholars dismissed Homer's epic and other Greek myths as just that: pure invention, with no historical foundation. Ancient Greeks, who had treated such legends as fact, were considered to have been childishly naïve. Considered equally naïve, therefore, was German businessman Heinrich Schliemann (1822–90), who had been fascinated by Greek legends since childhood. Convinced that Homer had not invented the Trojan wars, Schliemann began excavating the site in north-west Turkey reputed to be Troy. He discovered nine superimposed cities, and in 1873 found a treasure that he declared was the Trojan king Priam's. In fact, it was much older. In 1876 he turned to Mycenae and soon struck gold there too – literally, for he found himself looking at the gold death mask of a Bronze Age king. "I have looked on the face of Agamemnon," he declared. Again, he was out by centuries, but he had convinced the world that legendary Greece had existed in some form.

after 1400BC, but the Mycenaeans soon ventured further north than Cretans had ever sailed.

To protect Mycenae against threats from abroad or at home, massive new walls were built around its citadel, incorporating the earlier grave circles. The towering Lion Gate, with two lions flanking a pillar, dominated the new

nobles, craftsmen and traders made up a little city. Most palaces had a pillared *megaron* (throne-room), but few reveal much evidence of planning.

THE ENIGMA OF TROY

Up by the Hellespont (Dardanelles) the Mycenaeans found the ancient trading city of Troy, controlling trade routes from the Black Sea, a good enough cause for war. The Trojan War, that epic ten-year siege variously estimated to have occurred between 1250 and 1190BC, remains one of archaeology's enigmas. There *was* a city destroyed around there and then, Troy VIIA, and recent excavations have shown that this was larger than once thought, with impressive palaces. Perhaps a Greek army led by Agamemnon, king of Mycenae, did besiege Troy to retrieve Helen, the Spartan king's beautiful wife abducted by the Trojan prince Paris, as recounted in Homer's *Iliad*. Or perhaps it did not. But the 13th century certainly ended in general wars. Mycenaean civilization, top-heavy, collapsed soon after.

Above: The gold mask that Heinrich Schliemann, the enthusiastic rediscoverer of Troy and Mycenae, called the Mask of Agamemnon is in fact far earlier than the Trojan Wars, dating from the 16th century BC. Its still half-barbaric splendour indicates that the lords of Mycenae had already grown rich, whether this wealth was from trading, raiding or fighting as mercenaries.

Left: The Minoans' cultural influence on the mainland Greeks was for a time overwhelming. This gold cup, found at Vaphio, near Sparta, and dating from the 16th century BC, is Minoan both in its naturalistic style and its bull-taming theme.

THE DARK AGES
1200–800BC

Above: Ephesus was among the greatest of the Ionian cities of Asia Minor, traditionally founded by the Ionians fleeing from the Doric invasions around 1000BC.

Around the year 1200BC almost all the major Mycenaean centres were destroyed by unknown attackers despite their strengthened fortifications. Some sites, such as Pylos, were abandoned, fires baking their Linear B tablets enduringly hard for posterity. Only Athens, secure on its rock thanks to a covered passage running down to a secret well, survived intact. (In legend this survival was due to the self-sacrifice of Codrus, its last king, who thus fulfilled a prophecy that the city would be saved if its king were killed. After such a heroic death, Athens became an aristocratic republic.) But its status as the one unsacked city reinforced subsequent Athenian beliefs that they alone were autochthonic (sprung from the land), unlike the Dorian newcomers.

Later Greeks remembered the upheavals at the end of the Mycenaean age as the 'return of the Heraclids' (sons of Heracles or Hercules – a Mycenaean prince) to reclaim their inheritance two generations after the Trojan War.

They failed to realize that there had been a complete break between the elaborate, literate world of the Bronze Age palaces and much simpler later societies. The break was not sudden – Mycenae itself was briefly reoccupied and there were actually some new settlements along the east coast of Attica. But what followed was a true Dark Age: illiterate, isolated and impoverished.

MYCENEAN REFUGEES

According to legend, Dorian tribes entered Greece from the north-west, a region untouched by Mycenaean culture, and moved down into the Peloponnese, destroying as they went. Refugees from the kingdom of Pylos sailed first to impregnable Athens and then across the seas to found new cities on the Asian coast such as Smyrna and Ephesus, soon called Ionian. Some Mycenaeans sailed further east to untouched Cyprus, where they maintained a fading Bronze Age lifestyle, complete with chariots and a

Right: By 800BC Greece had become linguistically divided into groups, whose culture as well as dialect were distinctive. Foremost among these groups were Ionian, spoken in Athens, the islands and Ionia, and Doric, spoken in Sparta, Crete and Rhodes.

simplified syllabic alphabet, down to the time of the Persian Wars (499–478BC). Others retreated into Arcadia, the mountainous heart of the Peloponnese, where they preserved their own independence, dialect and customs, albeit in rustic form.

Another explanation for the collapse of Mycenaean civilization is climate change – perhaps a prolonged drought – but there certainly were large migrations and invasions in the years after 1200.

THE DIALECTS OF GREECE

The main reason for accepting in outline the tradition of northern invaders is linguistic. While all Mycenaean Greeks seem to have spoken much the same language (judging by Linear B tablets), in historic Greece there were several major linguistic divisions. In much of the Peloponnese, including the Corinthian Isthmus (though not inland Arcadia), in Crete, Rhodes and some southern Cyclades, the Dorian dialect became the norm. Dorian Greek was in some ways old-fashioned – the *a* in *phrater* (brother) remained long, for example, while it became short in other dialects. In Attica, including Athens, the island of Euboea, many of the Cyclades, most east Aegean islands and many cities on the east coast, people spoke Ionic Greek. This became the language (in grandly poetic form) of Homer, greatest of all Greek poets, in the 8th century BC.

In central Greece – Boeotia and Thessaly, on the large island of Lesbos and the adjacent northern Asian coastline, all colonized from central Greece, Aeolic, a form of Greek closer to Ionian than Dorian emerged, which influenced the language of Hesiod, the second great Greek poet. Another dialect, now called North-western Greek, was spoken appropriately in north-western Greece, while modern scholars, recognizing the similarity between the Greek of the Arcadians and that of the Cypriots, have identified a separate Arcado-Cypriot dialect. This was not, however, recognized at the time. Although there were definite cultural as well as linguistic differences

between the dialect-speakers, all could understand each other well enough – when they wanted to.

FRESH BEGINNINGS

Clues to new developments in this illiterate society come mostly from pottery, whose shards have survived. At first potters continued to produce feebler versions of the ornate Mycenaean 'palace' styles, but *c.*1050BC a wholly new style emerged in Athens called Protogeometric. Radically simple, content with almost abstract designs and not concerned to fill every empty space, it ranks as perhaps the first truly Hellenic art. (The Greeks now began calling themselves Hellenes: 'Greek' is a Roman word.) Over the next three centuries Geometric art, using stark zigzags and triangles, developed more elaborate shapes but remained stylized. Clothing, too, changed around 1100BC, with simple Dorian cloaks replacing elaborate Mycenaean-style clothes. Almost all buildings were now made of timber, with small thatched cottages serving as temples, befitting a society hardly above subsistence level, ruled by local lords hardly richer than the peasants working for them. Few men looked far afield.

Above: In tradition and probably in reality, Athens was the 'unsacked city' that alone survived the Dorian invasions, perpetuating Mycenaean traditions.

Below: The far simpler way of life that emerged in Greece's Dark Ages (1200–800BC) resulted in a new, simpler style of pottery: Protogeometric and then, as in this 8th-century BC amphora, Geometric.

THE GREEK RENAISSANCE
800–700 BC

Above: The stories of Odysseus' wanderings, so magnificently related by Homer, gave generations of potters themes. Here Odysseus is bound to the mast of his ship to resist the fatally alluring song of the Sirens.

Below: Mt Olympus, the highest mountain in Greece and often cloud-capped, was the mythical home of the 12 Olympian gods. It also marked the northern frontier of Greece proper.

Around 800 BC this enclosed, static society began to change. The spur was increasing population, growing (if still modest) prosperity at home and renewed contacts with traders from the Levant. The traders were Phoenicians, a Semitic people from the coasts of modern Lebanon who founded Carthage near modern Tunis in 814 BC. The use of iron also spread, giving Greek farmers metal axes, ploughs and other useful implements. But Greek society remained essentially aristocratic, meaning ruled by *aristoi* (the best), as hereditary nobles modestly called themselves.

THE GREEK ALPHABET

Eastern influences first appear in art, depicting human beings and animals, often mythical such as sphinxes, in freer if not yet realistic ways. But the greatest single change was the revival of literacy. Around 770 BC some Greeks, probably poets, adopted the Phoenician alphabet,

Above: Olympia in the Peloponnese emerged as one of the holiest sites in the Greek world in the 8th century BC, famed for its quadrennial games, the greatest in the Greek calendar, and later for its temple housing an enormous statue of Zeus.

adding the vowels needed for Greek to make 24 letters and adjusting the symbols. Semitic *aleph* became Greek *alpha*, the first letter. More flexible and easier to learn than the 300-character Mycenaean system, the new alphabet spread around the Greek world. Our own Roman alphabet derives directly from it. One of the first uses of literacy was to record the works of Homer, the greatest Greek poet.

HOMER'S *ILIAD* AND *ODYSSEY*

There are no reliable details about Homer's life. He probably lived *c.*750 BC on the island of Chios or the adjacent Ionian mainland, and traditionally was blind. Whether the two great Homeric poems, *The Iliad* and *The Odyssey*, were written by the same man remains debated. Homer's theme in *The Iliad* is the wrath of Prince Achilles and its disastrous effects on the last stages of the ten-year Trojan War, of which, however, he gives only fleeting glimpses. In this grand tragedy he lauded heroic values such as *philotimon* (love of honour), *areté*

(meaning variously courage, excellence, perfection), endurance and a fiercely competitive individualism.

By contrast, in *The Odyssey*, his adventure-story-cum-comedy, Odysseus triumphs chiefly by craftiness. Homer's description of an aristocratic society led by kings, with the voices of common people such as Thersites firmly ignored, inadvertently mingles current Iron Age customs with those of the Bronze Age. His heroes ride into battle in Mycenaean chariots and carry Bronze Age giant shields, but they are cremated, not buried as Mycenaeans were. Although they live in palaces, these are simply large houses, lacking the bureaucracies and splendours of real Mycenae or Pylos. Queen Penelope, wife of wandering Odysseus, spins her own wool.

Homer's influence on later Greeks has been compared to that of the Bible and Shakespeare combined – or to Hollywood *plus* television today. All Greeks with any education could quote Homer, and he

inspired men as diverse as the philosopher Socrates and Alexander the Great. In portraying the Twelve Olympians (the chief gods on Mt Olympus) light-heartedly as super-sized humans, Homer's writing had a beneficial side effect. If even Zeus, king of the gods, could be portrayed as hen-pecked by his wife Hera, there was small danger of Greeks being totally overawed by their gods' majesty. (The Greeks never had a special priestly caste or clergy. This helped philosophy– that quest for truth by non-religious means – to spring up in Ionia two centuries later.)

Above: This vase, dating from 490BC, illustrates a typically combative scene between Achilles and Memnon from Homer's first great poem, The Iliad, about the Trojan Wars. Homer's poetry swiftly became the basis of Greek culture.

Below: Almost nothing definite is known about the life of Homer, the supreme Greek poet, but he was reputedly blind. He certainly was an Ionian Greek, for he wrote in the Ionian dialect.

THE WORK OF HESIOD

Balancing the exuberant aristocratic splendour of Homer's world are the *Theognis* and *Works and Days* of Hesiod, a poet who lived slightly later (*c.*700BC) in rural Boeotia, an area noted for its dullness. An independent small farmer, Hesiod grumbles at the rich and at the weather, but provides useful advice to his feckless brother on when to sow and plough. He has a strong distrust of seafaring and a peasant's attitude to accruing more land. In his *Theognis* he gave a systematic genealogy for the gods and an account of divine myths, darker in tone than Homer's, that also proved very influential on later generations.

THE FIRST OLYMPICS

Another vital aspect of Greek life emerged in the 8th century BC: the Olympic Games, traditionally first held at Olympia in west Greece in 776BC. At first just a Peloponnesian event, it soon attracted Greek athletes from all over Greece and overseas to its contest held once every four years. For this, the greatest athletic event in the Hellenic world, the forever-warring Greeks observed a rare truce. After an athlete's loincloth fell off, it was decided that all contestants should compete naked, like the gods. The (usually aristocratic) victor at a major contest such as chariot races was hailed as a semi-divine hero by his native city, often having a statue erected to him. At Olympia were created some of the finest temples and statues. Pheidias, the supreme Classical Athenian sculptor, made a huge statue of Zeus there.

THE EXPANSION OF GREECE 750–580BC

Above: The island of Ortygia off the south-eastern Sicilian coast became the kernel of Syracuse, founded in 734BC, ultimately the greatest and most powerful Greek city in the west. These columns of the Temple to Apollo date from the 6th century BC.

Below: Electrum and silver coins of Phocaea (Ionia) in western Anatolia.

Right: Although only founded in c.580BC, Acragas became the richest of Greek Sicilian cities, thanks to its wool trade. Its 'Vale of Temples' included this Temple to Concordia from the prosperous 5th century BC.

By 550BC Greek 'colonies' – that is, autonomous city-states, not dependent territories – stretched across the Mediterranean from southern Gaul (France) to Egypt, with many more around the Black Sea. Southern Italy was so densely settled with Greek cities that the Romans called it *magna Graecia* (greater Greece). This remarkable expansion, which occurred within about two centuries, was driven chiefly by growing land hunger.

The mountainous nature of much of the country meant that the amount of fertile land in Greece was limited. To make matters worse, the Greek custom of dividing inherited land equally between all surviving sons meant that farms often became too small to be viable. (Most Greeks were, of course, farmers.) When the population began to expand in the 8th century BC, the pressure on the available land grew.

THE FIRST COLONIES

In almost every Greek city there were persistent – if unfounded – traditions that in the legendary past a wise ruler had divided up all land equally, and so subsequent inequalities, which benefited aristocrats, were unjust. A redistribution of land was a recurring wish of ordinary people and the nightmare of ruling aristocracies. To avert revolution, cities

Above: Settlers from Megara founded Acragas (modern Agrigento), once Sicily's wealthiest city.

turned to founding colonies overseas, some of which proved very popular, some less so. The colonists sent out c.630BC from the small island of Thira to Cyrene in Libya, for example, were forbidden to return within six years no matter what happened. (In fact, Cyrene boomed thanks to its wheat and wool.)

The first colonies were founded c.750BC at Cyme and the island of Ischia, near Naples, to obtain the metals Greece lacked. These colonists were soon followed by those from other cities seeking good farmland: Chalcis, from Euboea, founded Naxos, Catania and Rhegium, now Reggio, on the Straits of Messina. Corinth founded Corcyra (Corfu) and in 734BC Syracuse, which was to become the greatest city in the west. The Achaeans of the north Peloponnese founded Sybaris, Croton and Metapontion in southern Italy, all built on prime agricultural land, where they soon grew rich. Even Sparta joined in, founding Taras (Taranto) c.700BC.

After a pause, colonies also began founding their own colonies. Megara, a Sicilian offshoot of the Megara that lies north-east of Corinth, sent settlers west to Selinus (Selinunte) c.630BC and to Acragas

Left: Starting with colonies at Ischia and Cumae around the Bay of Naples in c.750BC, the Greeks founded cities all around the Mediterranean, from the south of France (Marseilles, Antibes and Nice) to Naucratis in the Egyptian Delta, to solve problems of over-population at home.

(Agrigento), later Sicily's wealthiest city, c.580BC. Phocaea, in north Ionia, colonized southern Gaul: Nice, Antibes, Monaco and Marseilles are all in origin Greek. The westernmost Greek city was Emporiae in north-east Spain, a trade centre (as its name suggests). Trade across the whole Greek world grew thanks to such colonization.

New settlers went out not as individuals but in planned groups, at times led by an aristocrat's younger son, sometimes not taking wives with them. Instead, they married local women, often those of native inhabitants they had dispossessed. Not all colonists necessarily came from one city. Small places such as Phocaea recruited landless men from other cities. While most colonies retained strong sentimental ties with their mother city, all were fully independent from the start.

THE CITIZEN-STATE

The *polis*, city-state, was the main political unit of Greece. It is more accurate to call it 'citizen-state', for it meant the body of all politically active men, not its buildings. (Women, slaves and resident foreigners were inhabitants but not active citizens of a *polis*.) A *polis* often had citizens living outside its walls, in villages or small towns, who walked into the city centre. When Athens itself was occupied by invading Persians in 480BC, Themistocles, its leader, was taunted with being cityless. He pointed out that Athens consisted of its assembled citizens, who would sail off and found a new city if the united Greek fleet did not stand and fight. A *polis* was always fiercely autonomous.

THE BLACK SEA

There was wealth to be had in the north-east around the Black Sea, although attempts to establish colonies in the crowded Levant failed, except briefly at Al-Mina in Syria. The three-pronged peninsula of the Chalcidice was first colonized by Chalcis, and then by Megara. Megara went on to found Byzantium (Istanbul) at a superb site on the Bosphorus in 629BC. Beyond lay the wide cold waters of the Black Sea.

Here another host of colonies sprang up: Sinope, Amisus and Trebizond on the south coast, and Olbia, Panticapaeum and Tyrus around the north, in a chilly region the Greeks found strange. The Scythian hinterland (Ukraine) exported grain, gold, timber and slaves in return for Greek oil, wine and artefacts. Wheat from the Black Sea became increasingly important to Greece, especially to Athens. Ultimately, whoever controlled the Black Sea grain supply could throttle Athens.

Below: The Greeks also settled thickly around the shores of the Black Sea, to them a very alien area. They traded olive oil, artefacts and wine for wheat with the Scythian princes who ruled the steppes and were buried with resplendent grave goods such as this gold comb. Greek influence is evident in the naturalistic style.

HOPLITES AND TYRANTS
700–550 BC

Above: This 'Corinthian style' helmet, inscribed with the name Dendas, dates from c.500BC. Such helmets gave hoplites good head protection but limited visibility.

Below: The Chigi Vase (made in Corinth but found in central Italy, hence its name) is the earliest extant depiction of hoplites marching and fighting in formation. It dates from the 7th century BC, by which time the 'Hoplite Revolution' was well under way.

Although Homer's heroes had ridden to war in chariots for individual combat, in the post-Mycenaean age chariots were replaced by horsemen. As only rich aristocrats could afford armour and a horse, warfare remained dominated by noblemen. But by the 7th century BC growing wealth, in part fuelled by trade that provided cheaper metal, meant that more farmers of the middle class (in Athens called *zeugitae*, meaning owning a yoke of oxen) could afford armour and arms. This led to radical changes in warfare and in society.

THE HOPLITES' IMPORTANCE

The main battles of ancient Greece were decided by hoplites, heavy-armoured infantrymen. Fighting in close formation, with their long spears bristling in front of them, hoplites could outface cavalry and dominated warfare in Greece for almost 500 years. The oldest set of hoplite armour, unearthed near Argos, dates from *c.*700BC. It has a plumed helmet covering the head (helmets often covered the whole face, leaving mere slits for the eyes) and heavy metal cuirass covering the body. Most hoplites had metal grieves on their legs and all carried a large (about 1m/3ft in diameter) round shield on their left arm. In their right hand they carried a 3.5m/12ft spear. Each depended for protection on his open right flank on his neighbour. If one hoplite broke ranks and fled, the entire formation could be imperilled. To keep formation required practice. It also implied a novel equality between fighters: now the most blue-blooded aristocrat was equal on the battlefield to an unwashed farmer.

Hoplites made up about one-third of a typical Greek city's adult citizen population. (The majority of Greeks, who could not afford such full armour, were enrolled as light-armed troops, long considered of minor importance on or off the battle-field.) Marching and fighting together bred a new sense of camaraderie among

the hoplite class that had political and social effects. If fighting was no longer the preserve of nobles, soon politics too was seen as concerning far more of the people. As economic growth changed the Greek world, it increased the gap between the rich – who could import new luxuries – and the rest. Discontent with aristocratic rule, long taken for granted, increased. The nobles still monopolized power, but often treated politics as a frivolous if risky game between rival families, like the athletic contests they still dominated. But most Greeks wanted *eunomia* (good government), not *stasis* (the chaotic strife of aristocratic factions).

TYRANTS: NEW-STYLE RULERS

Quick to exploit such feelings were the men, often themselves rogue aristocrats, who seized power in many states as tyrants. (Greek *tyrannos* meant at first just unconstitutional ruler, boss or chief, not dictator in a modern sense.) From the mid-7th century BC tyrants appeared in cities that were often among the most

dynamic in the Greek world. Typical was Corinth, a major trading city due to its position on the Isthmus, which produced fine pottery. Cypselus threw out its ruling Bacchiad family in 657BC, expelling other aristocrats and confiscating their property. Once in power, his regime proved so efficient and popular that he dispensed with a bodyguard. After he died in power in 625BC he was succeeded by his son Periander, who ruled for another 40 years.

Cypselus' example was soon followed in other cities such as nearby Megara and Sicyon. In neighbouring Argos the hereditary king Pheidon seems to have *become* a tyrant c.675BC, rebasing his rule on popular support. He reorganized the army as a hoplite force to crush the invading Spartans at Hysaia in 668BC. Tyrants, besides providing stable government, enriched their cities with temples and monuments. In Corinth, a dry shipway was built by Periander across the Isthmus, forerunner of the Corinth Canal.

Tyrannies spread across the Greek world from Sicily to Ionia. An unusually benevolent tyrant was Pittacus in Mytilene on the island of Lesbos, who resigned after ten years to general astonishment, the one tyrant to do so.

Among the Greek cities in Asia, tyranny was taken as the Greek norm by their Lydian and later by their Persian conquerors, who granted subject cities internal autonomy.

The last independent tyrant in eastern Greece was Polycrates of Samos, who built a famous temple, befriended the poet Anacreon and was *thalassocrat* (ruler of the seas) for a few years before being betrayed to the Persians and killed in 523BC. Few tyrant dynasties lasted long. At Corinth the tyranny collapsed soon after Periander's death. Tyranny was therefore a transitional phase for most Greek cities, with only unstable Sicilian cities reverting to tyranny in later years. And one supremely important state never experienced tyranny: Sparta.

Left: This vase from c.500BC shows hoplite warriors engaged in single combat. More typically, however, they fought in the disciplined ranks of the phalanx.

Below: A bronze helmet and cuirass, found near Argos and dating from the 7th century BC, are among the finest armour retrieved from the ancient world. Most hoplites would not have been able to afford anything so elaborate, instead having cuirasses of toughened linen.

THE IONIAN ENLIGHTENMENT 650–520BC

Above: Sappho and Alcaeus were two of the greatest poets of the Lyric Age, writing intensely personal poems of love, death and loss as well as more convivial drinking songs and marriage epithalamia.

For the eastern Greeks of Ionia and the adjacent islands the period *c.*650–500BC saw an accelerating widening of horizons that was mental as much as physical. Their merchants, colonists and mercenaries explored the world from the Black Sea down to Egypt and as far east as Babylon (in modern Iraq), bringing home new ideas. This generated a novel vivacity that was different from the heroic age before and from the serious high Classical Age (480–322BC) that followed. Women at this time in Ionia probably enjoyed greater freedom than they did later (although this does not mean that they were remotely equal), while poets wrote more personally about life, love and death. The graceful Ionic column was developed at about this time, while Western philosophy and science were born together in the cities of Ionia. This time is called the Ionian Enlightenment.

THE EGYPTIAN CONNECTION

Ionian nobles, including Alcaeus, were not above contacts with foreigners. Richest and strangest of foreign kingdoms was Egypt, from which came gold, ivory, wheat and papyrus, the best writing material. To preserve Egypt's independence, only recently regained from Assyrian domination, the pharaohs began hiring Greek hoplite mercenaries, already regarded as the world's best infantry. Some of these, employed in the far south of the country, left graffiti on the colossal statues of Abu Simbel that are still legible today. Greek merchants who penetrated into this enclosed, priest-ruled land were ultimately confined to one city, Naucratis in the Delta. Here Greeks from many different cities settled and traded, maintaining Hellenic customs in a very different world.

LYDIAN RULERS

Contacts with sophisticated peoples to the east first stimulated the Greeks, but the influence did not long remain one-way. In Anatolia the rich kingdom of Phrygia (its legendary king Midas' 'golden touch' suggests his immense wealth) was taken over by its neighbour Lydia in western Asia Minor *c.*680BC. Lydia's capital at Sardis was within easy reach of the coast, and Greek cities soon felt Lydian power encroaching on their cherished autonomy. Only Miletus, the largest and

Left: The expansion of Greek trade – north to the Black Sea, west to Italy and south to Egypt – opened Greek minds to richer, more sophisticated civilizations. It also made some Greeks much richer. Commerce was not at the time regarded as socially demeaning.

richest of the Ionian cities, remained fully independent behind formidable walls. But the Lydian kings were generally benevolent rulers. They themselves came under such strong Greek influence that they were seen as philhellenes (supporters or admirers of things Greek).

Croesus, Lydia's most famous king, who came to the throne in 560BC, enriched the sanctuary at Delphi, holiest in the Greek world, employed Greek artists and helped to finance the vast Greek temple to Artemis (Diana) at Ephesus, later one of the Seven Wonders of the Ancient World. The Lydians traditionally introduced the first minted coins to the Greeks, who soon produced coins of great beauty themselves. Coinage further encouraged commerce, if widening the gap between rich and poor. Debts could now be computed, and wealth amassed, more easily. Above all, Lydia helped to shield the eastern Greeks from invading barbarians such as the Cimmerians, who ravaged Asia Minor in the 7th century, and from empires further east such as the Medes in Iran. At a battle in 585BC between Lydia and Media, an eclipse of the sun stopped the fighting. Both sides, awe-struck, agreed on the River Halys as a boundary.

PHILOSOPHY AND POETRY

This eclipse had been predicted by Thales, the first known Greek philosopher, who was also a scientist and was therefore interested in, rather than awed by, natural phenomena. A native of Miletus, he thought long and hard about what constituted ultimate reality. A school of philosophers followed him, called Milesian after Miletus. Thales' pupil Anaximander, who lived to

c.536BC, made the first map of the known world, realizing that the Earth hung unsupported in space though not that it was spherical. To show that philosophers could be practical, Thales reputedly diverted the River Halys to let King Croesus cross it.

The period also saw a flowering of a new form of poetry, not long heroic epics but short, personal lyric verses, often of poignant beauty. (*Lyric* meant 'sung to music of the lyre', a harp-like instrument.) The most famous are the love poems of Sappho of Lesbos c.600BC, many addressed to the girls whom she reputedly taught. Only fragments of her bitter-sweet poetry survive. More typical were the witty lyrics of Alcaeus, Anacreon and Archilochus. Archilochus wrote: "Some lucky Thracian has my fine shield, I had to run and dropped it in a wood. But I got clean away, thank God! So damn the shield, I'll get another just as good."

Society, while still aristocratic, grew more pleasure-loving and relaxed. *Symposia* (leisurely drinking parties with conversation and entertainment) became popular among the wealthier.

SPARTA: THE UNIQUE STATE
700–500BC

Sheltered by mountains, the valley of the Eurotas in Laconia in the south-east corner of the Peloponnese was half-isolated. The Spartans, a Dorian people who settled there *c.*1000BC, retained ancient Dorian customs, such as common messes for all male citizens and tribal education for children, that had been abandoned by other Dorians except those in even more remote Crete. But Spartans felt themselves to be different from other Greeks. They were surrounded by non-Dorian (but Greek) peoples in Laconia, whom they reduced to serfdom as *helots* (land-bound slaves who provided food and produce for the ruling class, the *homoioi*), thereby gaining a valuable, if potentially dangerous, servile workforce.

By 700BC warfare had added the large, fertile territory of Messenia to the west, which was "good for ploughing and growing fruit", as the Spartan poet Tyrtaeus wrote. Despite being fellow Dorians, the Messenians were also enslaved. This conquest made Sparta the most powerful state in Greece. But it caused problems at home.

A RADICAL NEW CONSTITUTION

The wars had been won by Spartan hoplites, not aristocratic cavalry, but wealth was increasingly concentrated in the hands of a few. This caused discontent among ordinary Spartans struggling to maintain hoplites status. The resulting military weakness was shown up in defeat by Argos in 668BC and a revolt in Messenia. With Spartan power in danger of collapse, radical changes were needed.

Later, all reforms came to be attributed to Lycurgus, a semi-mythical figure, but the Delphic oracle, which Sparta regularly consulted, possibly played a part. Probably several individuals, including King Polydorus, were involved in helping to refound the state in the mid-7th century BC, giving it a constitution that was unusually complete and rigid. It was admired but not copied by other Greeks.

In theory, land was divided into equal lots owned inalienably by 9,000 *homoioi* (the elite ruling class, also known as 'Equals' and 'Spartiates'). They never worked their land but lived off the labour of the *helots*. The broad expanses of Messenia, reconquered by 640BC, gave Spartiates the landed wealth and leisure to train full time as soldiers. The Spartan army, distinguished by its scarlet cloaks and unbreakable discipline, became the one professional force in Greece, acknowledged as the best. In return for economic security, Spartiates surrendered their whole lives to the state.

DORIC DISCIPLINE

Every Spartan baby was examined by officials for deformities. Those considered unfit were exposed on a mountain. (Infanticide was practised in other Greek cities too but never so systematically.) At the age of seven, a boy was taken from his mother to begin his *agoge* (special training in a barracks). Wearing only a

Above: Spartans attributed their unique constitution to Lycurgus, a semi-mythical figure who reputedly created Sparta's almost totalitarian regime after the city had suffered a major defeat.

Below: The Eurotas valley, in which Sparta sits, is the most fertile in southern Greece, giving Sparta rural wealth.

thin tunic and no shoes, even in winter, he never had enough to eat, being expected to steal more food yet whipped if caught. At every stage he faced ferocious competition and punishments. Such training produced tough, obedient soldiers, taciturn – hence our word laconic (from Laconian) – dour and unimaginative.

At the age of 20 Spartiates had to win election to a mess, where they ate repellent meals, mostly 'Spartan black broth' (reputedly made from pork, blood, salt and vinegar). Homosexual affairs between prefects and younger boys were common. But Spartiates also had to produce children for the state after a strange marriage ceremony in which the bride's hair was shorn and she was dressed as a boy. Girls, too, exercised nearly naked, which shocked other Greeks.

Spartans were intensely pious, revering the gods, but they built few temples – Thucydides said that no one later seeing Sparta's sparse monuments would ever guess its power. Commerce was banned and there was no coinage, iron bars remaining the currency. Only *perioeci* ('dwellers around' in small towns who had to serve in the Lacadaemonian army), led normal lives. Every year 'war' was declared on the *helots*, during which potential rebels were secretly killed. Spartan art and poetry soon atrophied.

THE SPARTAN SYSTEM

Sparta strangely mixed oligarchy, democracy and monarchy. There were two kings – from the dynasties of the Agiads and

Eurypontids, both claiming descent from Hercules – who alone escaped the *agoge*. Only when leading the army abroad did they have real power. At home, they were constantly checked by the *ephors* (five officials chosen annually who wielded huge, if shadowy, power in this police state). The assembly consisted of all Spartans over the age of 30, who voted only on proposals put to it, one side shouting down the other – a method that Aristotle called childish but Jean-Jacques Rousseau later admired. There was also the *gerousia*, or senate, of men aged over 60 chosen for life.

Such a system was meant to defy all change, but some Spartiates finally became more equal than others. In the 4th century BC, as wealth from its newly gained empire flooded in, Lycurgus' system broke down. But by 550BC Sparta had forged a league of most states in the Peloponnese who were happy to follow it now that it seemed invincible.

Above: From the age of seven Spartan boys faced endless military-style drill. Toughness, conformity and unflinching obedience to orders were the aim, all individualism being stamped out.

Below: In its earlier days, before the grim regimentation of the Lycurgic system, Spartan potters produced fine vases such as this of Prometheus.

THE SPARTANS AND ALCOHOL

Despite worshipping many gods, the Spartans never worshipped Dionysus, the wine god. Also no *symposia*, the convivial drinking parties, were held in Spartan high society. Instead, young Spartiates were warned of the evils of alcohol by having *helots* paraded grotesquely drunk before them, to demonstrate the effects of inebriation.

ATHENS: REFORMERS AND TYRANTS 620–514BC

Above: Olives, which grow extremely well in Greece, became the main crop of Attic agriculture after Solon prohibited the export of wheat.

Right: The olive, mythical gift of the goddess Athena to her favoured city, is normally harvested in late winter or early spring. The trees take about 15–20 years to start bearing fruit and far longer to reach their prime. The olive-harvest was a popular topic for vase-painters, many of whose products were used to export the oil around the Mediterranean.

In the 7th century BC Athens was a relative backwater. Although Attica, long united under Athens, was one of Greece's largest states, it lacked wide fertile valleys. It was still ruled by a clique of Eupatrid (hereditary noble) families, who controlled the Areopagus, the supreme council. But Athens was not immune to wider economic and social changes. In 632BC Cylon, an aristocratic Olympic victor, seized the Acropolis, aiming to establish a tyranny. The *archon* (head official) Megacles – of the Eupatrid Alcmaeonid family, the most famous in Athenian history – tricked Cylon and his supporters out of the Acropolis and killed them. But the problems remained. Many poorer Athenians were *hektemoroi* (small farmers who owed their noble overlords a sixth of their produce). Falling into debt, they and their families might be enslaved and sold abroad, a fate that was even worse than a *helot*'s in Sparta.

In 625BC the Eupatrid law giver Dracon drew up and published the first written law code. Athenians could now see how severe their laws actually were: someone could supposedly be executed for stealing a cabbage. (These laws were later called 'draconian'.) Popular discontent with the status quo grew until in 594BC Solon, a Eupatrid but a noted critic of the rich, was given special powers. With them he launched a reform programme designed to avert tyranny by satisfying both the people and the rich.

"FREEING THE BLACK EARTH"
Solon's first measure was a "shaking off of debts". He stopped debt being secured on a person's liberty and freed all those already enslaved – he even tried to buy back Athenians sold abroad. He ordered the pulling up of stones that marked off land for aristocrats' tithes, "freeing the black earth" as he put it, and forbade the export of wheat from Attica to keep it cheap. Instead, he encouraged the export of olive oil; olive trees grow well in Attica, although they need 30 years to mature. Solon reformed the constitution, allowing all free citizens to attend the Assembly, which elected the *archons*. He established the Council of the 400, drawn by lot from the Assembly, as a preliminary debating body to balance the Areopagus. But he made both Eupatrids and rich commoners eligible for *archonship*

and so of the Areopagus, the highest court. Social divisions were now based on wealth not birth, allowing greater mobility. But Solon did not redistribute land as hoped or feared, for he was no revolutionary. "I gave the people such recognition as they deserved," he wrote, and went off on his travels. He left Athens more equitable if not democratic, for it was still led by aristocrats.

This was the problem. There were fierce divisions between the 'Coast' party of commercial interests and the 'Plain' of landowners, both led by irresponsible aristocrats. Years of bloody 'anarchy' – without elected *archons* – in the 580s saw war against next-door Megara go badly. Pisistratus, a noble who as *polemarch* (commander) had conquered Salamis island in 565BC, finally made himself tyrant in 561BC with other nobles' backing. They soon fell out and he was exiled. But he then made not one but two comebacks, the second time permanently in 546BC, supported by a third party of poorer farmers.

THE PISISTRATID REGIME

Once finally in power, Pisistratus surprised everyone by his moderation. Although some political rivals went into exile, there were no purges. Pisistratus helped Solon's reforms to take root by shielding the state from aristocratic faction. Unbullied by the rich, ordinary Athenians learnt to play a part in running their city. Pisistratus raised a tax of 10 per cent on farm produce, made loans on easy terms to help small farmers and started an economic boom, underpinned by stable government. To export Attic olive oil, superb new-style red figure vases

(on a black ground) were produced from 520BC onwards. Now Athenians became the finest of all Greek ceramicists. Athens also gained outposts on the Hellespont (Dardanelles), essential for protecting grain imports from the Black Sea. It still had only a tiny navy, however.

Athenian culture was not neglected either. Reputedly the first definitive edition of Homer's poems was compiled under Pisistratus' patronage, with ceremonial readings. He also inaugurated the Festival of the Great Dionysia, whose contests between different choruses later gave birth to Athenian tragedy and comedy. Among other monuments, he started a gigantic temple to Zeus, king of the gods, not completed for six centuries. When Pisistratus' sons Hippias and Hipparchus jointly succeeded their father at his death in 527BC and continued his policies, theirs must have seemed the most stable regime after Sparta's in Greece. But it was not to last.

ATHENS: THE DEMOCRATIC REVOLUTION 514–490BC

Above: The bright colours and alert energy of this koure *(young girl) embody Athens' optimistic energy at the dawn of democracy in the late 6th century* BC.

The trigger for the downfall of the Pisistratids was a quarrel over a boy's affections (very Greek, some might think), but the real causes went much deeper. By 514BC many states considered tyrants to be oppressive and turned for help to Sparta, which generally disliked tyrants as upstarts. When Harmodius and Aristogeiton, who were later hailed as liberators and had fine statues erected in their memory, killed Hipparchus for personal reasons in 514BC, the surviving Pisistratid, Hippias, became paranoid. Retreating to the Acropolis, he started a truly tyrannical reign of terror that brought about his own downfall.

THE ROLE OF CLEISTHENES

Cleisthenes, head of the Athenian Alcmaeonid family but currently in exile, had been wooing the Delphic oracle with gifts. Whenever Sparta consulted the oracle, it now replied: "Free Athens!" Finally, in 510BC a Spartan army under King Cleomones marched into Attica, and Hippias, with his supporters, fled to the Persians. Cleomones, like most Athenian nobles, then expected a return to the good old days of noble-led factions with the people merely acting as supporters. But Athens had changed. It was the genius of Cleisthenes to realize and take advantage of this.

After losing the election for the *archonship* in 508BC to Isagoras, who had Cleomones' backing, in 507BC Cleisthenes proposed a radical reordering of the whole state. In future the Assembly of all citizens voting together would be completely sovereign, and a new Council of 500, chosen by lot from all citizens irrespective of wealth, would act as a *probouletic* (preliminary debating) body. Cleisthenes abolished the old Ionic 'tribes' and replaced them with ten new ones, all artificial despite being named after ancient heroes. Each new tribe had three electoral wards called *trittyes*

Right: The 'Treasury of the Athenians' at Delphi, lavishly built all in marble, was probably erected in the 490s BC *by the young democracy in gratitude to Apollo's oracle, whose utterances "Free Athens!" had so aided the city.*

(thirds). These 30 *trittyes* contained *demes* from all three old rival factions: the city and its environs, the coastal lands and the uplands. Every citizen had to register afresh in a *deme*, which was administered by a locally elected *demarch*, and was then allocated to a *trittyes* and so to a tribe. This vastly reduced the influence of local squires on elections.

Unsurprisingly, Isagoras and his aristocratic supporters were horrified. They called on Cleomones, who had no time at all for such dangerous democratic ideas and marched straight back to Athens. There he exiled 700 Alcmaeonid supporters – Cleisthenes had already wisely withdrawn – re-established the old regime and then retired with Isagoras to the Acropolis for a celebratory dinner. But while they were celebrating, the noise of a popular rising in the streets below first amazed and then alarmed them. Although lacking any proper nobles to lead them, the people of Athens had risen in revolt on their own – an unprecedented event. Soon, surrounded and without provisions, the Spartans and Isagoras had to surrender ignominiously and leave Athens.

Cleisthenes now returned in triumph, and his reforms were implemented. He probably at this time added some immigrants from Ionia, long resident in the city, to the citizen list. As a safety valve, to stop any individual growing too powerful, he also introduced a novel scheme: ostracism. Every year there would be a vote on whether to hold an ostracism. If 6,000 citizens wanted to, they would inscribe names of men to be ostracized on bits of pottery (*ostrakia*). The man most frequently named had to go into exile for ten years but did not lose his citizenship or property. Cleisthenes, whether or not he had intended to, had created the first full democracy in Greece, probably in the whole world.

DEMOCRACY TRIUMPHANT

The new democracy in Athens was soon to be put to the test. Cleomones, burning for revenge, summoned the armies of Sparta and her Peloponnesian allies, much the most formidable force in Greece, to invade Attica, while the Thebans and Chalcians, Athens' old enemies, attacked from the north-east. The Athenians were undaunted and marched out to Eleusis to meet the Spartan-led army – which promptly turned back. The Corinthians, a major contingent of the army, had changed their minds, while the other Spartan king, Demaratus, for reasons unknown, also went home, forcing Cleomones to follow suit. The Athenians then swung east to defeat the Thebans before crossing to Euboea, where they routed the Chalcians. They subsequently settled 4,000 colonists on Chalcian land.

Below: Harmodius and Aristogeiton are shown here as the heroic liberators who killed the tyrant Hipparchus in 514BC. Although only a Roman copy of the second pair of original Greek bronzes – which were made in the 470s BC, the first pair having been carried off by the Persians in 480BC – the original group's dynamic strength shines through. It typified Athenian democracy's abounding new self-confidence.

CHAPTER II

THE PERSIAN WARS

499–478 BC

The great war between the Greeks and Persians was the pivotal event in Greek – and ultimately European – history. If the Greek cities had been defeated and become subjects of the Persian Empire, Greek brilliance could have been dimmed. The emergence of full democracy might have been stunted, even aborted. Defeating the Persians gave the Athenians the confidence to experiment further with radical democracy at home. But the Persians, although defeated, were never despicable foes. Although the Greeks liked to mock their enemies' 'Asiatic luxury', the Persians proved both formidable fighters and wise rulers.

In a few decades Persia had created the world's first global empire, stretching from the southern Balkans to the fringes of India. It governed its subjects with enlightened imperialism, tolerating different customs and gods. Life for the conquered had advantages, which some Greeks appreciated, but it lacked the one element that Greeks prized above all: total freedom. The Persian invasion of Greece achieved what nothing else did: it united enough of the feuding Greek states to defeat the invaders. Even Spartans and Athenians fought side by side. But the final glory must go to Athens, whose citizens were prepared to see their temples and homes burnt twice by the Persians rather than surrender. Greek victory over Persia made classical Athens, and so Western civilization, possible.

Left: The grand staircase of the palace of Darius I at Persepolis, ceremonial capital of the immense Persian Empire.

THE RISE OF PERSIA
550–494BC

Far to the east of the Greek cities, in mountainous southern Iran, lived a people whom the Greeks called Persians. Unsophisticated but indomitable, fine archers and horsemen, they long paid tribute to their cousins the Medes, whose empire rivalled Lydia's and Babylon's.

CYRUS, KING OF KINGS

In 550BC Cyrus of Persia unexpectedly overthrew the king of Media, becoming ruler of a united Achaemenid empire (named after Cyrus' ancestor). He enrolled the Medes as such almost-equal partners that Greeks later called collaborators or supporters of Persia 'medizers'.

Such distant events meant little to most Greeks. They seldom ventured far into Asia, although some knew of Babylon, the greatest city on Earth. But in 546BC Croesus of Lydia, an Asian monarch the Greeks knew well, attacked this new empire. He had been encouraged by the Delphic oracle's typically ambiguous pronouncement: "If you cross the River Halys [the frontier] you will destroy a mighty empire."

The first battle proved inconclusive, so Croesus withdrew west to his capital at Sardis, considering the fighting over as winter approached. He was wrong: Cyrus followed him and outside Sardis defeated Lydia's fine cavalry by sending ahead a line of camels, whose smell panics untrained horses. Sardis, with its treasures, fell to Cyrus.

Harpagus – Cyrus' best general and a Mede – then moved along the coast with a huge army, with most cities submitting to the new superpower. Some, however, preferred flight to surrender: the whole population of Phocaea sailed off. But life

for the conquered was tolerable. Persia usually reappointed existing tyrants, democracy being still almost unknown, although the tribute paid – 400 talents of silver– was siphoned to the distant Persian capital at Susa, not spent locally. Susa was linked to Sardis by the Royal Road of 2,400km/1,500 miles, down which messengers galloped bearing royal edicts. Some Persian nobles soon acquired estates in Asia Minor.

In 539BC Cyrus took Babylon 'the mighty city', liberating the captive Jews, most of whom returned to Jerusalem to rebuild their Temple. Persian rulers, though Zoroastrians (precursors of

Above: The Persian kings acquired unprecedented wealth as well as power. But they only minted coins, such as these gold and silver pieces, for use in cities in their western provinces.

Right: Croesus' notoriously rich kingdom of Lydia (in Asia Minor) was the first to fall to the Persian advance. Defeated by Cyrus the Great of Persia in 546BC, Croesus was, according to Herodotus, put on a pyre to be burnt, a scene shown on this vase of c.480BC.

the Parsees), were religiously tolerant. Cyrus' conquest of the Fertile Crescent made him the mightiest monarch yet, aptly hailed as Great King or King of Kings. Cambyses, succeeding his father in 530BC, annexed Egypt, but when he died mysteriously in 522BC the empire was shaken by rebellions. Darius, an Achaemenid relative, had to crush numerous rivals after proclaiming himself Great King.

DEMOCRACY AND REVOLT

Once secure on the throne, Darius determined to march into Europe. In 512BC he had pontoon bridges (mostly of Ionian ships) built for his army to cross the Bosphorus and River Danube. He vanished into the Scythian steppes for so long that some Greeks suggested destroying the Danube bridge – an idea wisely rejected, for Darius finally reappeared. Despite failure in Scythia, Persian power advanced along the north Aegean coast until even Macedonia acknowledged it. In 500BC Aristagoras, tyrant of powerful Miletus, suggested to Artaphernes, the Persian satrap (governor), a joint attack on Naxos, a Cycladic island. Artaphernes agreed, but it proved a fiasco. Aristagoras, fearing for his power if not his life, now jumped from being a pro-Persian tyrant to an anti-Persian democrat.

It was a popular move. The democratic revolt he started spread fast across an Ionia tired of foreign kings and domestic tyrants and inspired by Athens' new democracy. The geographer Hecataeus suggested seizing temple treasures to finance the revolt – in vain, but a Pan-Ionian Council was called, a common currency adopted and appeals sent to Greece for help. The Spartans, typically parochial, ignored them, but Athens and Eretria sent ships. In 498BC allied forces marched inland to capture and burn Sardis. Returning, they were caught in open country by Persia's superb cavalry and routed. While the Athenians returned home chastened, other Greeks from Byzantium to Cyprus now joined the revolt.

THE FALL OF MILETUS

Fierce fighting in Cyprus in 497BC, with the Greeks victorious at sea but defeated on land, led to the island's reconquest. Soon Persia subdued the northern states too. Yet the Ionian fleet, 350 ships strong, remained intact at Lade off Miletus, itself now besieged. On the fleet hung the fate of the revolt, but Ionian sailors chafed at the discipline needed to maintain battle-readiness while Persian gold won over many doubters.

When battle finally came in 494BC, the powerful Samian and Lesbian sections deserted and the other Ionians were defeated. The fall of Miletus followed and Persian revenge was brutal: most Milesians were killed or enslaved, their boys castrated, their daughters sold into harems, their city repopulated with others. The revolt was over. But, in a statesmanlike act, Persia allowed some Ionian cities to remain as democracies.

Right: Under Darius I, king 522–486BC, Persia's empire reached its greatest extent, stretching from the Danube to the Aral Sea. In the west, Darius' armies clashed with the mainland Greeks. He ordered the building of the first palaces at Persepolis, whose impressive ruins survive.

Above: The tomb of Cyrus the Great at Pasargadae in the Iranian mountains. Cyrus (reigned c.560–530BC) was the founder of the Persian Empire, defeating the Medes, Lydians and Babylonians and annexing their kingdoms.

THE BATTLE OF MARATHON 490 BC

Above: The tumulus at Marathon, raised over the bodies of Athenian soldiers killed in 490 BC, commemorates the first real Greek victory over the Persians.

With the Ionian revolt crushed, the Great King Darius turned to the Greek cities that had helped the rebels burn down Sardis, his western capital: Eretria on the island of Euboea and Athens. Envoys went around Greece demanding earth and water as tokens of submission. Many states, overawed by Persian power, complied. But at Sparta the envoys were thrown down a well and told to find earth and water there, which enraged Darius.

Athens responded no better, trying and executing his envoys. However, Cleomenes, that unusually dynamic Spartan king, was deposed around then, having gone mad – or so it was said officially. He was probably killed in secret.

In spring 490 BC a Persian fleet commanded by Datis sailed into the Aegean carrying an army around 25,000 strong, including the cavalry victorious earlier in Ionia. Crossing the Aegean, the Persians took and sacked Naxos, enslaving its citizens. However, Datis offered incense at Apollo's shrine on nearby Delos, prudently honouring the Greek god. As the armada sailed on, island after island submitted, giving hostages. In July the Persians reached Euboea, where Datis landed his army. After five days of fighting, Eretria was betrayed by some nobles who preferred Persian rule to local democracy. The city was burnt to the ground in awful warning. Then the fleet turned south-east for Attica.

On board was Hippias, the now aged ex-tyrant of Athens, who hoped that the Persians would restore him to power and still had some contacts in the city. He told the Persians the best spot to land: the open beach at Marathon 40km/25 miles north of Athens, which made perfect cavalry country.

MILTIADES' STRATEGY

Miltiades, who was an Athenian but also the tyrant of the Chersonese with a fine military reputation, had persuaded the Athenians that the army should go out to meet the Persians when they landed, rather than skulk behind Athens' walls. So almost the whole hoplite force, *c.*10,000 men, marched north in time

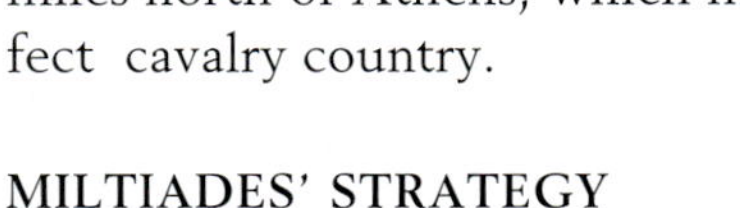

Left: The Persians relied mainly on skilled archers and fine cavalry, for their infantry was generally inferior to the Greeks'.

THE FIRST MARATHON
The race against time back to Athens by the hoplites involved marching "as fast as their legs could carry them". They covered 40km/25 miles in six hours – the very first 'marathon'.

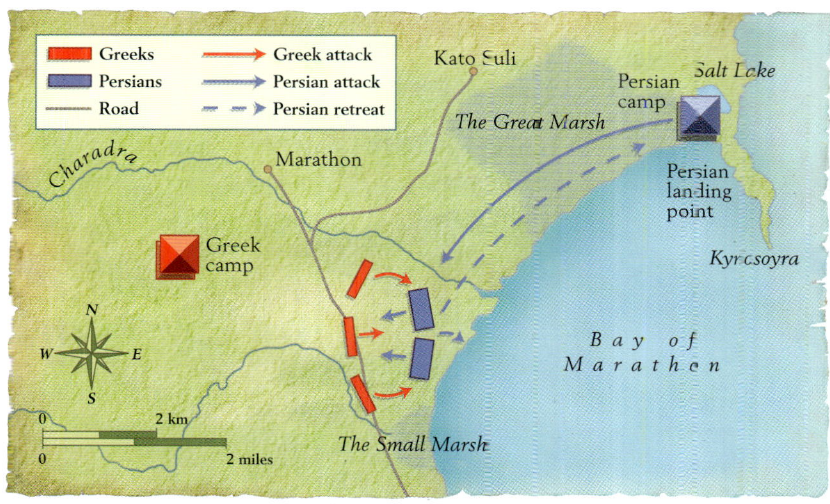

to stop the Persians break out of their beachhead. The *polemarch* (commander) was Callimarchus, but Miltiades, though only a divisional general, actually decided the tactics. A messenger, Philippides, was sent running the 225km/140 miles to beg the Spartans for help.

The Spartans, although pledged to assist, were celebrating their festival of Carnea and would not move before the full moon of 12 August. Their religious scruples were probably genuine; so was their reluctance to leave the Peloponnese. Philippides was returning disappointed when, in the blinding midday heat, he encountered the god Pan – an encouraging meeting, for Pan announced that the Athenians would triumph.

Dug in above the plain of Marathon, the Athenians watched and waited for the enemy to make the next move. So, looking up from their camp by the marshes or beached ships, did the Persians. The Athenians were encouraged by the arrival of 800 hoplites from their ally Plataea, the little city's whole army. But, still outnumbered, the Athenians were really awaiting the Spartans, due to march within a week. The Persians, who had spies everywhere, knew this too. Finally, Datis secretly re-embarked his cavalry at night to take Athens by surprise. But an Ionian, slipping across the lines, warned the Athenians: "The horsemen have left!" After persuading his co-generals, Miltiades ordered a general attack.

NOVEL BATTLE TACTICS
Even without cavalry, the Persians outnumbered the Greeks, so Miltiades decided on a novel tactic: the Greek centre would thin out to match the length of the Persian line. As the sun rose, the Athenian citizen-army began its descent to the plain, a mass of armour bristling with spears. In its centre Themistocles and Aristides, political opponents reconciled, commanded their respective tribes. As the Greeks closed to within 160m/175yd of the Persians – within bowshot – they broke into a run to crash into the lighter-armed enemy.

Soon their wings had driven back the Persians but their centre, facing the best Persian troops, was itself repelled. With remarkable discipline for amateur soldiers, the Greek wings swung to close in on the Persian centre. After fierce fighting, the battle turned, Persians fleeing in panic towards their ships or into the marshes, pursued by the Athenians. At the battle's end, reputedly only 192 Greeks had been killed for 6,400 Persians.

But Miltiades knew that the Persian fleet was still sailing for an unprotected Athens. Covered in dust and blood, the weary hoplites raced back to the south coast, arriving just in time. The Persians, seeing the armoured ranks drawn up again above the beach, abandoned their attempted landing and sailed off. Athenian hoplites had beaten the hitherto invincible Persian army in open battle. The Spartans, arriving belatedly, could only applaud, visit the battlefield and then go home quietly impressed.

Above: After the Persians had landed and made camp by the Bay of Marathon, the Athenian army occupied the hills above – an ideal spot for keeping watch, and for launching a charge downhill.

Below: The stele (funerary carving) on the tumulus at Marathon portrays an Athenian citizen-soldier, one of the hoplites who routed the Persians.

COUNTDOWN TO WAR
488–481BC

Above: View of Mount Athos in northern Greece, off whose rocky shore Mardonius was once shipwrecked in a storm.

Below: Persia's lengthy preparations for the grand invasion of 480BC included cutting a canal through the flat peninsula north of Mt Athos shown below. This was intended both to protect the invasion fleet from storms that had wrecked earlier ships and to overawe the Greeks.

For ordinary Athenians victory at Marathon appeared to lift the threat of Persian invasion for good. But they were wrong. The Great King was only marginally harmed by what was to him a peripheral battle. It needed to be avenged, of course, but the Persian Empire was still expanding and the conquest of Greece – an invitingly divided if troublesome land – remained on the agenda. This time, however, the Achaemenid Empire would act like the superpower it was. Overwhelming force on land and sea would be gathered to shock and awe Greece into submission.

PREPARATIONS FOR WAR

Mardonius, a noted young Persian general, had earlier been shipwrecked in a storm off Mt Athos in northern Greece, so it was decided to dig a canal through the Athos peninsula's neck to avoid this recurring. Thousands of conscripted workers took three years to dig a canal deep enough for galleys, while a pontoon bridge was built over the Hellespont (Dardanelles) to let the invading army cross speedily. Destroyed by storms, it was hastily rebuilt. Supplies were stockpiled in forts along the Aegean north coast and the River Strymon in Macedonia bridged. The death of Darius in late 486BC interrupted preparations only briefly. The throne passed smoothly to Xerxes his son (and Cyrus' grandson) and a revolt in Egypt was suppressed. Finally, in April 481BC, the Great King left Susa – a date marked by a solar eclipse – and slowly marched west with imposing majesty.

Forces had been levied from across his huge empire, from Africa to India. Some were not experienced fighters but many, such as the Persian cavalry, Saka mounted archers, fish-mail-armoured Medes and the 10,000 Immortals (royal guards), were. The fleet, too, was mostly professional, its 300 Phoenician galleys being thought the best afloat. Herodotus, seldom reliable on figures, totals the Persian forces at 1.7 million – an absurd figure, such a horde being impossible to feed. Modern estimates suggest *c.*250,000 soldiers, supported by a fleet of *c.*600 warships.

To deny Greece help from its Western compatriots, Carthage was urged to attack Greek Sicily. Meanwhile, Persian gold subverted north Greece. Macedonia was already a client state; Thessaly, due south, seemed ready to 'medize', its nobility favouring Persia. Beyond, central Greece looked open to pressure, while many Aegean islands were Persian-controlled. The stage was set.

THE GROWTH OF DEMOCRACY

Back in Athens, at first it was politics as usual. Celebrated for his Marathon victory, Miltiades led an attack on Paros

in 489BC. He failed abysmally, was prosecuted, heavily fined and died in debt. After 487BC election to the *archonship* was replaced by sortition (lottery). This reduced the powers of the *polemarch* (military *archon*), for a randomly chosen leader might lack military experience. Instead, the ten *strategoi* (generals) elected from each tribe became the real commanders. This marked a further advance in democracy: anyone, not just the rich, could be elected *strategos*. Ostracism was now used, probably for the first time. Among those ostracized were Pisistratids and Alcmaeonid aristocrats. The young democracy was flexing its muscles.

THE BIRTH OF ATHENS' NAVY

Athens remained mainly a land power, proud of its hoplites' victory at Marathon. One man, however, saw danger in this. Themistocles, a stocky, energetic man, brilliant at courting the people, was a radical democrat though aristocratic himself, at least on his father's side. Elected *archon* in 493BC, he began in the 480s to press for a huge expansion of the navy. This was unpopular with the rich, who would have to pay for it, while small farmers saw no need for it. When in 483BC a huge new lode of silver was found at the mines of Laurium, a great debate began in Athens about how to spend it. Aristides, now the conservative leader (nicknamed 'the Just' because he famously took no bribes), suggested distributing the windfall among the citizens, an idea with obvious appeal.

But Themistocles urged that it all be spent on a massive new navy of triremes the triple-tiered galleys that now dominated sea war. As a Persian threat still seemed remote, Themistocles pointed instead to the island of Aegina visible across the water, whose ships raided Attic coasts with impunity. His eloquence finally carried the day.

A crash programme was begun that within three years gave Athens the largest navy in Greece: 200 triremes, each requiring a crew of 170 citizen rowers and 30 sailors and marines. This vast new force – which committed ordinary Athenians to learn to fight at sea, sweating at their oars rather than parading proudly on land – was built and trained within three years. Just in time, it turned out. Meanwhile Aristides was ostracized, Athenians sickening of hearing him forever called 'the Just'.

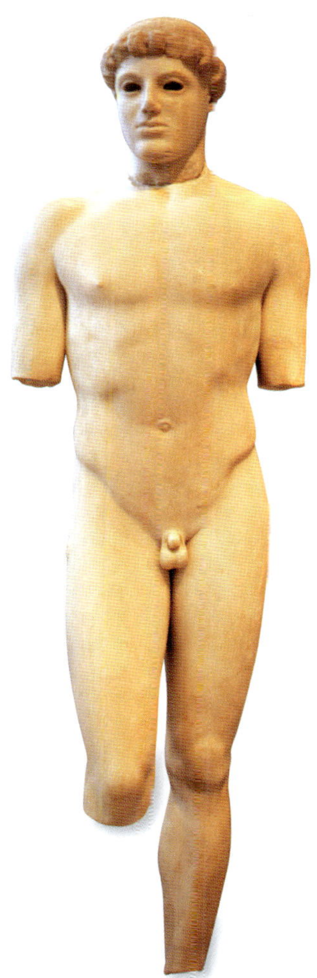

THERMOPYLAE: LEONIDAS AND 'THE 300' 480BC

In late 481BC news of the huge army being assembled at Sardis reached the Greeks, its numbers amplified by the Persians, masters of psychological warfare. Persian envoys again went around Greece demanding submission. Most cities prevaricated. Athens and Sparta called a Panhellenic League at the Corinth Isthmus, to which 40 states sent envoys. Many did not. Swallowing its pride, Athens accepted Spartan leadership on land *and* sea, despite providing most of the fleet. The League warned possible 'medizers' that their lands would be 'tithed to Apollo', i.e. taken from them, if they collaborated.

THE ORACLE'S PROPHECY

But Greeks were hardly encouraged to resist by the Delphic oracle's doom-laden prophecies. "Fly far, far away; Leave home, town and castle and do not stay," it told the Athenians. When they asked again, they were told to trust in the "wooden walls" but warned: "Divine Salamis will destroy the children of women." This perplexed the Athenian Assembly, some taking "wooden walls" to mean those once surrounding the Acropolis, others their new fleet. Themistocles, now Athens' effective leader, favoured the latter, and his advice was accepted: if necessary, the Athenians would evacuate their entire city and take to the sea. Their decision saved Greece.

DEFENCE POSITIONS

In June, when some Thessalians suggested holding the pass at Tempe beneath Mt Olympus, 10,000 hoplites went north. They had to return hastily, for Persian forces found other undefended passes. The next possible line was in central Greece, where the mountains almost touched the sea at Thermopylae. A fleet based at Artemisium in west Euboea could support an army there. King Leonidas, with an elite of 300 older Spartiates ("all with living sons") set out, collecting allied troops en route until 7,000 hoplites manned an old wall at Thermopylae. Meanwhile, 200 triremes sailed north to Artemisium. There they awaited the Persian colossus.

This took its time, advancing in slow splendour, "drinking rivers dry" as June passed into July. From Delphi came a cryptic last message: "Pray to the winds!" Greeks knew that, as summer advanced,

Above: In Persia's grand army of c.250,000 men – the largest force yet assembled – were tens of thousands of archers. They fired so many arrows at the Spartans at Thermopylae that the Greeks joked at being able to fight in the shade.

Left: As news of Persian invasion plans reached the Greeks in late 481BC, a pan-hellenic Congress was called at Corinth, the wealthy city controlling the Isthmus. There 40 cities agreed plans that gave the Spartans overall command at land and sea and threatened retribution for any state that 'medized' (collaborated with the Persians).

sudden storms could arise in the Aegean, knowledge the Persians lacked. Leonidas was not on a suicide mission, although the Delphic oracle had prophesied Sparta's fall unless one of its kings died. He hoped to hold the Persians long enough for the main Spartan army, due to celebrate the Carnaean and Olympic festivals in August, to march north. Thermopylae was a splendid position provided it was not outflanked by sea or by land. But there were paths over the mountains, known to locals.

In August, the ground shaking beneath their feet, the Persians finally arrived. So did the winds, gales scattering both Greek and Persian fleets. Xerxes, who could not let his huge army stand still for long or it would starve, ordered a direct assault on the Greeks by heavy infantry – Medes used to mountain warfare. They were not, however, used to Spartiates in their killing prime and made no progress against the hoplite wall. Then the elite Immortals joined the battle, to be repelled also.

THE BATTLE OF THERMOPYLAE

At sea things were more equal, both fleets reassembling, only battered. But a Persian force, sent to outflank the Greeks by

Above: The narrow pass at Thermopylae, where the mountains then nearly touched the sea, provided the best defensive position in Greece. Sparta's king Leonidas, with an advance army of 7,000 men (including 300 Spartiates) marched north to hold it. But a mountain path allowed the Persians finally to circumvent the Greeks.

Below: This statue of a Spartan warrior – possibly even a portrait of King Leonidas himself – reveals the disciplined determination that made Spartans the best soldiers in Greece.

sailing down Euboea's east coast, was wrecked, removing a major threat. The running naval battle that followed in the straits off Artemisium saw the Greeks, though outnumbered, undefeated. Then catastrophic news from Thermopylae changed everything.

Persian gold had found a local traitor, Ephialtes, to guide them over a mountain path to behind the wall. The Immortals followed him on a night march up to a pass where some Phocians, stationed as guards, panicked, withdrawing to let the Persians descend.

Leonidas had just enough warning to send off most troops before being surrounded. Then, with his Spartiates some unwilling Thebans and dogged Thespians, probably about 1,500 men in all, he took his last stand. This was bloody and desperate, the Spartans fighting with bare hands after their swords and spears had splintered, until finally all lay dead around the corpse of their king. The road to Athens lay open.

DELPHI'S AMBIGUOUS ROLE

The Delphic oracle, sacred to Apollo, was the most revered in Greece. Its site, stunningly beautiful, was called *omphalos* (navel of the world). There the Greeks dedicated many of their finest shrines and offerings. Its Pythean priestess gave famously obscure replies to questions, so preserving its reputation for infallibility. But Delphi's role in the great Persian wars was so inglorious that it could be suspected almost of 'medizing'. In 480BC Persian power appeared overwhelming to the shrewd priests, and Greece's chance of victory looked poor. The Delphic priests were realists: better to maintain a diplomatic neutrality than openly support either side.

SALAMIS: VICTORY AT SEA 480BC

Above: The island of Aegina, with its fine temple to Apheia, sheltered many Athenian refugees forced to evacuate their city.

Below: Themistocles, the Athenian leader, realizing that the Greek fleet lacked the skill and numbers to win a battle in open waters, tricked the Persians into sending their fleet into the narrows between Salamis island and the mainland. There the Greeks' heavier but inexperienced galleys crowded the Persians together so that they could not move, and so won decisively.

Days after Thermopylae, Themistocles reached Piraeus with the battle-scarred Greek fleet. It had slipped away by night after hearing of Leonidas' end. He now oversaw the evacuation of Athens. Women and children trundled possessions down to embark for Aegina, Troezen (which offered to educate all Athens' children) and Salamis. This small island became the Panhellenic League's headquarters and the tented city of Athens-in-exile. Not everything could be taken – one dog swum loyally after his master's boat, dying as he reached land. Nor would every Athenian leave. Some diehards retreated to the Acropolis behind wooden barricades. After firing the barricades, the Persians captured the citadel. They killed everyone there, including priests and priestesses, burning the temples – an act of sacrilegious terror never forgotten. They looted the main city while the Great King deliberated.

A NAVAL WAITING GAME

At Salamis there was disagreement among the Greeks. Some Peloponnesians, alarmed as the Persian fleet neared Salamis, wanted to retreat to the Isthmus, where a wall had been built. This option, which abandoned all hope of regaining Athens quickly, was rejected by Themistocles. He also realized that the Greek fleet, outnumbered and less skilled than the Persian, must not fight in open waters. Battle at Artemisium had shown that only at close quarters could Greeks hope to defeat Persians. It was necessary to be bottled up at Salamis, as the other Greek leaders finally accepted. A waiting game followed, but neither side could afford to wait for too long.

Persia had problems too. It had lost so many ships – through storms and enemy action – that it no longer had overwhelming naval superiority. The Greeks had about 300 seaworthy triremes by this time, the Persians probably only around 100 more. Persia could not afford to divide its fleet again, as it had before Artemisium, to threaten the Peloponnese simultaneously. Also, it was already mid-September. The campaigning season would not last much longer before autumnal storms began.

GREEK TRICKERY

Xerxes was therefore delighted when a secret messenger arrived from Themistocles, claiming his master was really the king's friend. The Greeks, he announced, were divided – the Persians must have found this credible – with many planning to flee. All the Persians had to do was to send ships to the straits' exits to catch the Greek triremes sailing off at dawn. (Sails were stowed away while fighting.) Xerxes gave his orders: Egyptian galleys would guard the western exit while the main fleet, of Phoenicians, Ionians and Carians, would enter the eastern straits to seize the fleeing Greeks. He himself would oversee his forces from a throne on the shore.

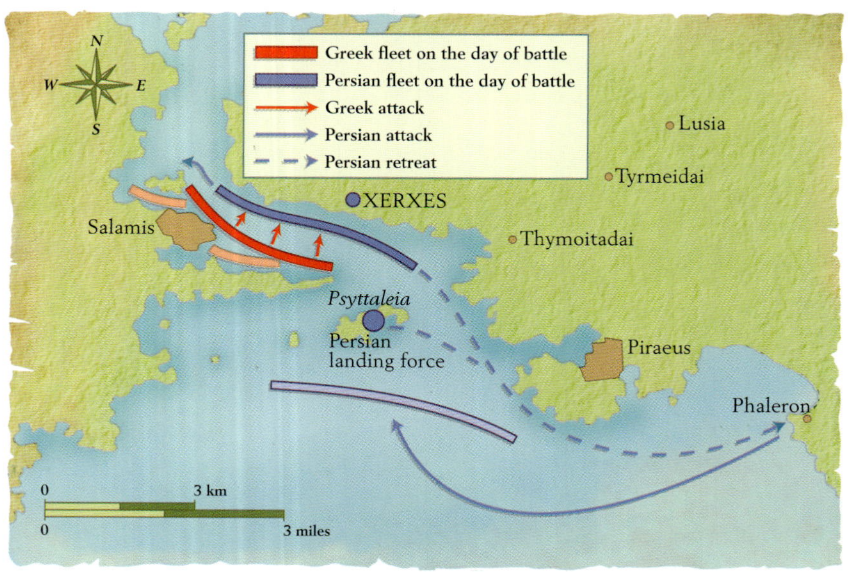

That fateful night, probably the 19/20 September, Persia's ships sailed to their appointed stations.

THE BATTLE OF SALAMIS

However, Themistocles' message was a ruse designed to lure the Persians into the Salamis straits, less than 1.6km/1 mile wide. According to Herodotus, only Themistocles knew of this. But the overall plan must have been widely discussed and agreed earlier, for the Greeks reacted swiftly to news of the Persian advance brought by Aristides. (All Athenian exiles had been recalled.)

As the sun rose over Salamis, the Persian ships nosed forward under their royal master's eye: Phoenicians on the right closest to the mainland, Carians and Cilicians in the centre, Ionians on the left. Facing them were Athenian ships on the Greek left, seemingly recoiling in fear, Peloponnesians in the centre and Aeginetans on the right. Xerxes saw far-off Corinthian ships hoisting sail and appearing to flee westward, as he had been told. All seemed to be going to plan.

Suddenly the Athenians stopped backing to surge forward, singing a *paean* (hymn). The Phoenician flagship, commanded by a brother of Xerxes, was attacked first, its royal admiral killed as he boarded an Athenian trireme. Soon the heavier Greek galleys were smashing into Persians ships too tight-packed to manoeuvre or even row properly. Some of the Persian fleet beached their crippled ships on the mainland. Several Phoenician captains were summarily executed by Xerxes for cowardice. Queen Artemisia of Halicarnassus, a Persian vassal, won his praise for apparently sinking a Greek ship, but this was actually a Persian galley blocking her escape. The Corinthians now returned to join the battle while the Egyptian squadron still waited idly.

By the afternoon the Greeks had won a crushing victory: only 40 of their own ships sunk for 200 of the Persians, the Phoenicians suffering especially. Persian power had experienced its first momentous defeat. King Xerxes returned ignominiously but safely by the same road to Asia. He left behind his unbeaten and large professional army, and much of Greece still in Persian hands.

Above: A 19th-century artist's colourful vision of the Greek fleet re-entering the Piraeus, the port of Athens, in triumph after the Battle of Salamis.

Below: King Xerxes ordered the building of a pontoon bridge over the Hellespont (Dardanelles) to allow the immense Persian army to cross from Asia to Europe early in 480BC. Jean Adrien Guignet's 19th-century painting shows Xerxes at the Hellespont.

PLATAEA: VICTORY ON LAND 479BC

The Athens to which its citizens returned that autumn was a gutted, half-ruined city. There was little comfort there. Nor were the Persians safely distant. The Persian general Mardonius, allowed to cherry pick an army for his command in Greece, had chosen the best – Persian and Median heavy cavalry, Sakae mounted archers – before settling in Thessaly for the winter. There began a war of words and nerves. Paradoxically, the Athenians, whose navy had effectively won the war at sea, now needed the Peloponnesians to fight a land battle in Boeotia to prevent Persia re-invading Attica.

AN OFFER OF PEACE…
The Spartans, with the immediate threat to the Peloponnese removed, were again loath to commit themselves, being worried about Argos, their old enemy close to home. Further arguments at League headquarters in Corinth produced no agreement, although when Themistocles visited Sparta he was fêted as a hero. This did him little good in Athens, where his opponents, the formerly ostracized Aristides and Xanthippus, were elected *strategoi* (generals) for 479BC. To Athens that spring came Alexander I of Macedonia, a wily monarch who had involuntarily entertained the Great King earlier. The offer he brought to the Athenians amid the rubble of their wrecked Agora sounded tempting: the Great King would, if Athens changed sides, not only forgive all of Athens' past acts against him but grant it special self-governing status, like Tyre or Sidon, money to restore her temples and support against her enemies. "Why be so mad as to resist the King?" asked Alexander. "You can never beat him and cannot hold out forever."

… REJECTED
The Athenians responded magnificently: "We know well enough that Persian power is many times greater than ours… But we want above all to be free, so we will never surrender… Tell Mardonius that the Athenians say: 'While the sun takes his normal path, we will make no agreement with Xerxes, but will defend ourselves, trusting in the gods and heroes who fight for us and whose temples he has burnt'." With this Alexander was dismissed. In June, Mardonius swept south again into Attica, and Athenians evacuated their city for the second time in ten months. Their envoys in Sparta warned that they might be forced to accept Mardonius' peace offers after all.

Above: A statuette from the 6th century BC showing a typically tough Spartan hoplite soldier, the victor of Plataea.

Left: After their victories, the Greeks liked to portray their opponents as unmanly cowards. This vase shows a Persian almost running away. In fact Persians were fine soldiers, despite defeats on land and sea in 480–479BC, but Greek hoplites made far better infantry than their Persian counterparts, as Plataea showed.

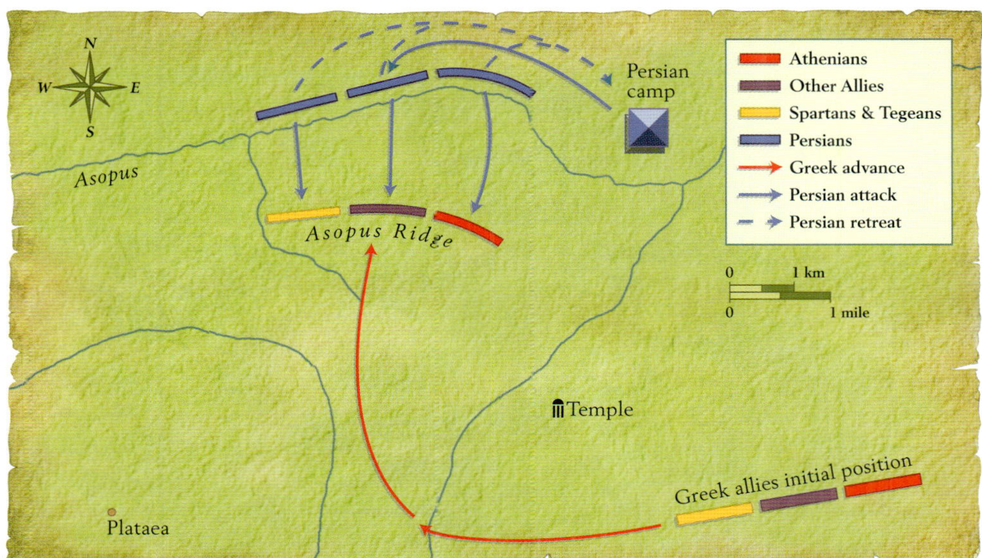

Left: The Persian army under Mardonius met the massed Greek hoplite forces in 479BC in a plain near Plataea in central Greece for the deciding land battle. When the Greeks began moving from their hilltop positions, Mardonius attacked their exposed flanks. But the disciplined Spartan phalanx routed the Persians.

Suddenly, Sparta's *ephors* announced that their army was already on the march with its allies. Mardonius, on hearing the news, torched all Athens before withdrawing to Boeotia, good country for horsemen.

PLATAEA: THE SPARTAN VICTORY

Mardonius' army numbered 60–70,000, judging by the size of the fortified camp he built near Thebes. In contrast, Aristides led 8,000 Athenian hoplites to join the League forces at Eleusis under the command of the Spartan regent Pausanias. (One Spartan king was still a boy, the other was Leotychides, commanding the fleet.) There they swore the Oath of Plataea, which began: "I will fight to the death and will not count my life more precious than freedom." It marked the high point of Greek unity. Then the hoplite army of *c.*40,000 men, the largest yet assembled, crossed into Boeotia. It took up a position on the slopes of Mt Cithaeron above the River Asopus.

There both sides waited, for Pausanias would not descend into a plain favourable to cavalry, while Mardonius would not attack uphill against hoplites. Mardonius finally sent mounted archers under Masistius, a famed nobleman, to make the Greek hoplites break rank. But Masistius was himself killed, his death grieving the Persians. The Greeks were less well supplied with food and water than the Persians, however, especially after their main spring had been destroyed, and so Pausanias decided to move the Greek army east toward Plataea, on to the low Asopus ridge, to improve water supplies. They began moving by night – a difficult manoeuvre that found many not in their new posts at dawn.

Mardonius, seeing the Greeks in such confusion, ordered an attack on the Spartans and their Tegean allies: *c.*11,000 men cut off from the main army. While Theban hoplites attacked the separated Athenians, Persian archers hailed arrows at the Spartans, killing many men. Yet Pausanias would not give the order to charge until the omens from sacrifices predicted victory. When they finally did, the Spartans rolled forwards, their disciplined phalanx carving a way through the Persians. After Mardonius, conspicuous on his white horse, was killed, his troops broke and fled toward their fort. But this proved to be no refuge, as the Athenians stormed it, killing all inside.

This battle marked the end of the Persian invasion. On the same day traditionally, the Greek fleet across the Aegean defeated the Persian fleet at Mycale (a naval battle fought on land), to which the Persians had retreated. This completed the Greeks' triumph.

Below: Spartan hoplites such as this soldier proved superior to the Persians at Plataea, the greatest land battle yet fought in Greece.

VICTORY IN THE WEST
480–474 BC

Above: Coins such as this, showing a triumphant charioteer and a winged nike (victory), were struck to celebrate victories such as Gelon's defeat of Carthage at Himera in 480 BC.

Above: The Greeks produced the world's first really beautiful coins, often to celebrate victories. This fine piece commemorates the Syracusan victory over the Carthaginians in 480 BC.

Right: Greek colonies spread throughout all Sicily except the north-west, where the Carthaginians maintained some fortified trading posts. The island's fertile interior remained inhabited mostly by native Sicels, who only slowly became Hellenized.

Before 500 BC, Carthaginian and west Greek colonies had only occasionally fought each other. The Carthaginians, who were primarily merchants, established what were really large fortified trading posts, while the Greek colonies often became rich trading and farming cities. Carthage had prevented Greek colonization of southern Spain, Corsica and Sardinia, but in Sicily it had only three small cities in the north-west around Palermo. By contrast, Greek cities lined Sicily's south, east and north-east coasts.

Tyrants were a common form of government in western Greek cities around 500 BC, gaining power easily amid the recurrent political crises. Gelon, already tyrant of Gela, seized Syracuse in 485 BC and made it capital of what became Sicily's most powerful state. He forcibly moved to Syracuse the luckless populations of Gela and other nearby cities. Linked by marriage to Theron, tyrant of Acragas (Agrigento), the second largest Greek city in Sicily, Gelon treated his poorer citizens badly, enfranchizing only the rich. His power lay in his many mercenaries, whom he made citizens of Syracuse. But his fleet of 200 triremes and army of 20,000 hoplites made him formidable, and in 481 BC the Panhellenic League begged for his assistance against Persia. In return, he demanded leadership of the League either by land or sea, both unacceptable to Sparta. As it turned out, Gelon had pressing problems close to home and gave no help to the League.

CARTHAGINIAN EXPANSION

The Persians, probably using the Phoenicians as intermediaries, had urged Carthage to distract the western Greeks.

Divisions among the Greeks now gave Carthage, already larger and richer than any other city in the Mediterranean, an opportunity to gain control of all Sicily. Theron of Acragas had just expelled Terillus, ruler of Himera, a Greek city on the north coast. Terillus appealed to the Greek tyrant of Rhegium (Reggio), who in turn asked Carthage for help. With this excuse, an armada that must have been long prepared sailed from Carthage. It was reportedly 300,000 strong, with Carthaginian infantry plus Libyan, Iberian, Sardinian and Ligurian mercenaries, transported in 3,000 ships and guarded by 200 galleys. Only the last figure seems credible, but it was certainly a huge force meant, like its Persian equivalent, to overawe the Greeks.

THE SIEGE OF HIMERA

Hamilcar, the Carthaginian *shophet* (commander), embarked this army and sailed for Palermo, where he landed only after losing most of his horse-transports in storms – a loss that was to prove crucial. He then marched on Himera and besieged it, building a fortified camp by the sea and another one inland for his vast army. Theron, reaching Himera just ahead of the Carthaginians, ordered its gates to be walled up (to prevent surrender as well as to keep out the enemy) and sent urgent messages to Syracuse. Gelon, who had been waiting, marched with 50,000 men to join forces with the Himerans, who reopened their gates.

Both sides then waited – showing that the Carthaginians lacked overwhelming superiority – until Gelon had a stroke of luck. Hamilcar had asked Selinus, a half-Greek Sicilian ally, to send him some cavalry. Their reply fell into Gelon's hands, and Syracusan cavalry entered the Carthaginian camp disguised as Selinans. Then they attacked the Carthaginians from the rear.

The battle that followed was confused and bloody. The Syracusan infiltrators killed Hamilcar as he was sacrificing to the gods (in one story he immolated himself on a pyre), but the battle was nearly lost when many Greeks stopped to plunder the Carthaginian camp. Final victory was total, however, with few Carthaginians returning home. Instead, many mercenaries ended their lives as slaves, building grand temples for the Greek cities, especially in Acragas' Vale of Temples. But the Carthaginian colonies in north-west Sicily were left alone.

A TIME OF PEACE AND PLENTY

The following decades were prosperous ones for Greek Sicily. Hieron peacefully succeeded his brother Gelon as ruler of Syracuse in 478BC. He won a great naval victory four years later when the Etruscans, still expanding their power in central Italy, were defeated at Cumae (Cyme) in 474BC. This marked the end of Etruscan expansion as definitively as Himera and the east Greek victories had checked Carthaginian and Persian power.

After 460BC democracy replaced tyranny in most major Sicilian cities. They then enjoyed a long period of unusual domestic accord and foreign peace in what came to be seen as Sicily's golden age, marked by splendid temple-building at Acragas and other cities.

Above: The Valley of the Temples at Agrigento in Sicily is one of the world's most important archaeological sites.

Below: Syracuse became the richest and most powerful Greek city in Sicily, spreading up the hills from the original island colony of Ortygia. The fortifications at Euryalus high above were to prove crucial to its defence.

CHAPTER III

ATHENS AT ITS ZENITH
478–431BC

Between its triumph in the Persian Wars in 479BC and the Peloponnesian War that began in 431BC, Athens enjoyed a golden half-century, becoming the greatest of all Greek cities. Although the era ended in political and military disasters, these do not diminish Athens' achievements, as classical culture and full democracy came of age. When Pericles, the democratic statesman, claimed that Athens was 'the educator of Greece', he was not bragging. Athens demonstrated that democracy, in its deepest (if not widest) sense, could encourage intellectual and cultural excellence: not dumbing down but climbing up. In drama, art, philosophy, architecture and politics it was a brilliant age.

Classical Athens had its downsides. Its brilliance came partly from ruling and taxing its supposed allies; full enjoyment of democracy was denied to women, foreign residents and slaves; and Athens could behave brutally toward its enemies. Athenians, intoxicated by their city's primacy, often displayed an arrogance that finally united many other Greeks against them. Despite such faults, Athens incarnated so much of what is noblest about ancient Greece that we still see it mainly through Athenian eyes. Even other Greeks acknowledged this quality. Pindar of Thebes, a city often hostile to Athens, wrote: "O shining city, violet-crowned and famous in song/Bastion of Hellas, glorious Athens, city of godlike men".

Left: The Parthenon, built between 447BC and 432BC, embodies Athenian genius at its most radiant.

SPARTAN FAILINGS, ATHENIAN INITIATIVE 479–478BC

Above: The Parthenon in Athens, set high on the Acropolis. Without its city walls, Athens was very vulnerable to attack.

Below: Under the guidance of Themistocles, one of Athens' greatest democratic statesmen, the city hastily rewalled itself after the wars, now incorporating the Diplyon and Kerameikos areas.

After the defeats of the Persian army at Plataea and the Persian fleet at Mycale in Ionia in 479BC, many Greek islands and cities in Asia happily threw off their Persian allegiance. This meant that the Panhellenic League had to decide how to defend the eastern Greeks against Persia. The Spartans suggested moving their populations to Greece proper, where they could have the territory of cities that had 'medized' (collaborated with Persia). After this impractical plan was vetoed, the League decided to continue the war against Persia, which still controlled many strongholds in the north and east Aegean.

The League's fleet sailed north to the Hellespont in the summer of 479BC, where they found that Xerxes' grandiose pontoon bridge linking Europe and Asia had been demolished. The Peloponnesians, reassured that the Persians could not easily reinvade, now sailed home. But Xanthippus, the Athenian *strategos*, led the Athenian and Ionian ships to attack Sestos on the Hellespont. This was no easy task, as it turned out, for its Persian

GREEKS AND BARBARIANS

Greeks called all non-Greek peoples *barbaros*, from which comes our word barbarian. This did not originally mean barbaric – uncouth, uncivilized – in the modern sense, for it applied equally to peoples such as the Persians, Phoenicians and Egyptians, who were generally more sophisticated than the Greeks. It just meant people who spoke another language that, to Greeks, sounded like 'bar-bar'. But after the Persian Wars, Greeks began to feel psychologically and morally superior to 'the barbarians who dwell in Asia', in other words within Persia's vast empire. Greeks alone enjoyed freedom, not having to grovel before a despot. Of course, there were plenty of tyrannies in Greece, and Athens or Sparta could behave despotically towards other cities, but this did not upset the Greek view of the world as divided into two: Greeks, free and independent; and barbarians, perhaps rich, often decadent, but always servile. Aeschylus' play *The Persians*, first produced in 472BC, exemplifies this attitude.

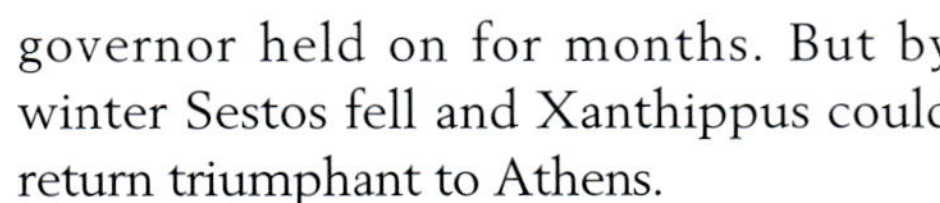

governor held on for months. But by winter Sestos fell and Xanthippus could return triumphant to Athens.

REFORTIFYING ATHENS

Back in Attica, the Athenians had returned for the second time to their city, most of which had been systematically destroyed by the Persians in their brief reoccupation that June. Only a few fine houses, where Persian grandees had stayed, were left intact. The city resounded to the noise of frantic rebuilding as Athenians tried to get roofs over their heads before the autumn rains.

Booty from Plataea may have helped some hoplite heroes, but for most citizens it was a very hard time. This did not mean that they could afford to neglect politics, however.

In its unwalled state, Athens was acutely vulnerable to any enemy. Sparta, which famously had no walls but relied on its armies, suggested that a rewalled Athens might become a base for the Persians if they returned. Instead, they suggested that Athens use the Peloponnese as its natural fortress. This argument convinced no one in Athens. But to keep the Spartans away while the Athenians were re-fortifying their city, Themistocles returned to Sparta, where he convinced the Spartans that the Athenians were only rebuilding their homes, not their walls, inviting them to send envoys to see for themselves. The envoys were promptly seized as hostages until the walls were finished.

The crash rebuilding of Athens' walls (only around the inland city itself at this stage) produced walls 8m/25ft high and 2.5m/8ft thick, enough to deter any Greek army at the time. The Spartans, grudgingly accepting this fait accompli, agreed to continue the war against Persia in 478BC.

DISGRACE OF PAUSANIAS

The League's forces were still nominally commanded by Pausanias, the Spartan victor at Plataea. His fleet of 20 triremes joined a much larger Athenian and Ionian force, and they sailed to Cyprus, conquering part of it. They then moved north to Byzantium on the Bosphorus and ejected the Persian garrison from that strategically important city. Meanwhile King Leotychidas of Sparta and Themistocles led a joint land campaign to reconquer northern Greece. Spartan proposals to punish 'medizing' states were blocked by Themistocles, however, who realized that this could boost Sparta's own power.

During 478BC it became apparent that Pausanias, like many Spartans away from domestic austerity, was highly corruptible. At Byzantium he began behaving in ways that were not only autocratic, offending the allies, but also non-Greek. He adopted Persian clothes (wearing purple trousers) and other luxuries and began corresponding secretly with Persian satraps. The Spartan *ephors* finally recalled him, but it was too late for Sparta's prestige. Pausanias' subsequent career ended with his deposition and death by starvation, again revealing Sparta's systemic shortcomings. Meanwhile, the Ionian Greeks had turned to Athens for leadership in the continuing war against Persia.

Above: The Hellespontine region (Dardanelles) became a crucial campaigning area after the Persian Wars. Wheat supplies for Athens from the Black Sea had to pass through its straits.

Right: The victorious commander at Plataea, Pausanias, later fell out with other Greeks and his own government. Recalled to Sparta and accused of treason, he was starved to death in a temple where he had sought sanctuary.

THE CONFEDERACY OF DELOS 478–460BC

Above: This serpentine column (now in Istanbul) was set up at Delphi to celebrate Greek victories. Initially it had Pausanias' name on it but this was soon removed.

Below: The new anti-Persian confederacy, formed in 477BC and led by Athens, chose the small Cycladic island of Delos, sacred to the god Apollo, for its headquarters and treasury. The alliance was known as the League or Confederacy of Delos.

In the winter of 478–477BC the Confederacy of Delos was established under Athenian leadership. Its aim was to continue the war against Persia. Despite the great victories of the previous two years, there was reason to fear a resurgence of Persian power. The Ionians, repelled by Pausanias' arrogant behaviour, had realized Sparta's inadequacy as a leader in overseas warfare, hence the setting up of the Confederacy. Sparta faced serious problems with Argos and the *helots* (Sparta's slave class) that preoccupied it over the next 20 years.

The Confederacy's headquarters was the small island of Delos in the Cyclades, sacred to Apollo. In its Council, every member from the smallest to the largest (Athens) had one vote. All swore to have "the same friends and enemies', a standard Greek oath, and to continue the alliance "until iron should float" (i.e. forever). This clause later created problems because it offered members no way out.

PAYMENT OF PHOROS

Every city – which from the start included most Ionian and Aeolian cities in Asia Minor, the adjacent islands, most

Above: The Athenian Aristides, nicknamed 'the Just' for his honesty, became the Delian League's first treasurer, impartially assessing the contributions due from each state.

of the Cyclades and all the Euboean cities except Carystus – made contributions in kind (ships) or in cash. Aristides, renowned for his honesty, became the first treasurer and assessed every *phoros* (contribution). The total revenue from 200 cities came to 460 talents, enough to maintain 100 Confederate galleys. Athens' own fleet had 200 triremes. Revenue was counted by ten Athenian officials called *hellenotamiae* (treasurers of the Greeks). Smaller members found it easier to pay in cash than to equip a trireme. Gradually, more and more members, and especially any who rebelled, moved from being independent contributors of ships to mere *phoros*-payers – a demotion that often caused resentment.

THE LEAGUE

Athens generally favoured democracy in League members, but seldom imposed it without invitation by an internal faction.

Democracy alienated some states' richer citizens, who faced higher taxes under such regimes. Athens later began interfering in tributary members' legal affairs, transferring disputes between members to Athenian law courts. This had advantages – distant Athenian juries were often impartial – but further infringed cities' cherished independence.

THE RISE AND FALL OF CIMON

The Confederacy's early years were glorious. Cimon, son of Miltiades, the victor of Marathon, was Aristides' protégé. As an Athenian *strategos* and leader of the Confederate fleet, he ousted Pausanias from Byzantium, which was enrolled in the Confederacy. In 476BC he attacked the Persian fort of Eion on the River Strymon. Its commander, Boges, held out for a long time, finally immolating himself, his wives and children on a pyre rather than surrender. Cimon failed to take Persian-held Doriscus east along the coast, but in 474BC captured the island of Scyros, a pirate stronghold. The bones of Theseus, the legendary Athenian king, were found there and brought back in triumph. Athens settled *cleruchs* (colonists) on Scyros, the first of many such colonies. Carystus to the south was compelled to join the alliance, and Naxos, when it tried to secede in 469BC, was forcibly prevented – justifiably perhaps, as the war against Persia was continuing.

A convivial, conservative nobleman, Cimon dominated Athenian politics alongside Aristides after Themistocles was ostracized in 471BC. (The great radical had made many enemies, and Athenian politics were volatile.) In 467BC Cimon sailed east and destroyed a new Persian armada of 200 galleys and the accompanying army on the River Eurymedon in southern Asia Minor, ending Persia's hopes of revenge. The cities of Pamphylia and Lycia were then enrolled in the Confederacy. Spoils from the allied victory were used partly to rebuild the Athenian's Acropolis walls – a contentious use. Possibly this provoked

Thasos, a rich island controlling gold mines in Thrace, to revolt in 465BC. It took two years' hard fighting before Cimon could suppress it. Meanwhile, Athenian squadrons, fully professional after years of campaigning, sailed across the eastern Mediterranean unchallenged.

EARTHQUAKE IN SPARTA

When an earthquake in Lacademon a in 464BC triggered a *helot* revolt that the Spartans could not repress, Cimon argued that Athens should help Sparta, still her ally. He led 4,000 Athenian hoplites, noted for their siege skills, to help take Mt Ithome, the *helots'* walled refuge. But the Spartans soon grew alarmed at having democrats inside their country, while Athenians were dismayed to discover the true nature of *helot* serfdom. Humiliatingly, Sparta told the Athenians to leave, but retained its other allies.

This rebuff led to Cimon's ostracism on his return in 461BC, for Ephialtes a new radical, had emerged to dominate Athenian politics. Soon after, a war broke out that came to divide much of Greece into two hostile camps: Athenian and Spartan. But Athenian power was still expanding. When a fleet campaigning in Cyprus was asked to assist an Egyptian revolt against Persia, it sailed south.

Above: Cimon, son of Miltiades and ally of Aristides, emerged as the chief Athenian strategos after 478BC. He led the League of Delos in a triumphant series of campaigns, driving the Persians from the Aegean. At home Cimon kept open house, entertaining lavishly like the old-style aristocrat he was.

Below: The Portara Gateway, Naxos. In 469BC Naxos tried to quit the League. This attempt was suppressed by Athens, which argued that the war against Persia required all members to keep fighting.

DEMOCRACY'S COMPLETION
462–458BC

Above: An ostrakon, *a small potsherd used in the yearly poll to decide who should be ostracized (sent into temporary exile), one of the 'safety valves' of democracy devised by Cleisthenes. Many popular leaders, including Themistocles, were ostracized.*

Below: Naval power in the ancient Mediterranean depended on galleys crewed mainly by citizens. Athens' trireme fleets, which ruled the Aegean for much of the 5th century BC, *gave seasonal employment to c.30,000 citizen-rowers, forging a link between democracy at home and an aggressive anti-Persian policy abroad that soon became imperialist.*

Little is known about Ephialtes, the radical democrat who in 462BC clipped the powers of the Areopagus, the ancient court, and so helped to finalize democracy. His person or policies evidently roused violent passions, however, for he was assassinated soon afterwards – something rare even in Athens' often turbulent democracy.

But Ephialtes' legacy lived on in the career of Pericles, the Alcmaeonid aristocrat who became the greatest democratic statesman in ancient history. Pericles was to guide Athens through nearly 30 years of unparalleled brilliance.

CURBING THE AREOPAGUS
The Council of the Areopagus (named after the Hill of Ares, or Mars, where it sat) retained some powers as well as immense prestige in 462BC. Composed of ex-*archons* after their year in office (all chosen by lot since 487BC), it heard charges against elected officials after their year in power. It also probably supervised the whole body of the law, with wide if undefined powers as the main court of appeal. All this made it a force to be reckoned with, and, with its members still recruited from only the two upper classes, it tended to be conservative.

Ephialtes had paved the way for the reduction of its old powers by accusing several Areopagites of corruption. With Cimon out of sight in Sparta and then ostracized, Ephialtes was able to persuade the Assembly to transfer almost all the Areopagus' powers to itself, either in the form of the Council of Five Hundred or, when constituted as a *heliaea*, as a jury court. (The council and juries were chosen by lot so that they accurately represented the *demos*, the people.) All that was left to the Areopagus was its jurisdiction in murder and arson cases and care of the sacred olive trees. *Archons*, however, still gave a preliminary hearing to lawsuits.

DRAMA AS PROPAGANDA
Powerful propaganda in defence of this democratizing reform came in the form of Aeschylus' last, perhaps greatest, dramatic trilogy, the *Oresteia*, first produced c.458BC. The final play in the trilogy, *Eumenides* (The Kindly Ones), focuses on the plight of Orestes. Guilty of the hideous crime of matricide, Orestes flees from Delphi to Athens pursued by the Furies. On the hill of the Areopagus, Athena, patron goddess of the city, appears, rescues him and founds the Areopagus as a court specifically to deal with such cases. The Furies are then tamed and become the Kindly Ones.

Since Aeschylus, the first of the three great Athenian dramatists, came from the hoplite class (his proudest boast on his tombstone was that he had fought at the Battle of Marathon), his views probably voiced those of many ordinary Athenians.

Ephialtes and Pericles were therefore not considered to be dangerous extremists, but were merely completing the constitutional reforms begun by Cleisthenes 50 years earlier.

*Left: Attica in c.450*BC *showing the demes (the basic unit of Athenian political and social life after Cleisthenes' democratic reforms of 508*BC, *on membership of which citizenship depended), the Long Walls and the silver mines at Laurium. Even at its greatest extent, Attica was always hemmed in by potential enemies.*

SPREADING POWER

Around the same time, Pericles introduced pay for jurors, probably at a rate of one obol a day, later raised to two. As this was less than the average daily wage, it was not an inducement to idleness (as Aristophanes, the comic playwright, suggested in his play *The Wasps* of 426BC), but it did mean that poverty would not prevent poorer citizens from acting as jurors. In 458BC the *archon*ship also became a paid office to which the *zeugitae* (the middle or hoplite class) now became eligible, so robbing this ancient office, descended from the royal council, of its last aristocratic distinction. There were by this time ten *archons* (literally rulers) of whom three went back to Athens' legendary past: the *basileos*, or king *archon*, with a priestly role; the *polemarch*, originally the military commander; and the eponymous *archon*, who gave his name to the year of his election. Athens dated events by referring to the 'year so-and-so was *archon*'. Later, dating from the first Olympiad (776BC) was adopted.

THE LOTTERY IN ATHENS

Athenians made extensive use of sortition (lotteries) to choose men for many important public posts. In Athens *archons* were chosen by lot after 487BC, as were jurors and other officials. Such a way of choosing officials seems odd today, but there were sound reasons behind it: it reduced the chances of undue influence being brought to bear on anyone; it spread the benefits and the burdens of active citizenship widely; and it was regarded as incorruptible. Elaborate mechanisms for rattling the tokens with the citizens' names on them have been recovered.

Below: The approach to the Acropolis, with the ruins of the temple of Athena Nike (Athena the Victor) standing out to remind citizens of the glories of their city, as it approached its zenith.

THE FIRST PELOPONNESIAN WAR 460–446BC

Above: An Athenian citizen-soldier bidding farewell to his family, a typical scene from the First Peloponnesian War.

Cimon's humiliation broke Athens' last ties with Sparta, though in truth they had long been fraying. While Sparta's growing jealousy of Athens was held in check by *helot* revolts, Athenian power had yet to acknowledge its limits. For a short but intoxicating time, it seemed that Athens would replace Sparta as *hegemon* (leader) of mainland Greece as well as of the Delos Confederacy, while gaining a major role in Egypt. Although such hopes were shattered, Athens emerged undefeated from this First Peloponnesian War, though strained by her experiences.

In 459BC Athens allied with Argos, Sparta's traditional enemy, and settled some Messenians (rebel *helots* whom the Spartans had allowed to leave Mt Ithome) at Naupactus on the Gulf of Corinth. Both were hostile acts. War finally broke out in 459BC when Megara, Athens' small neighbour, quarrelled with Corinth and left the Spartan Alliance, allying instead with Athens. An Athenian expedition helped Megara to build long walls down to its port at Nisaea, so cutting the Isthmus road. Aegina, Athens' old

Left: A relief of Athena mourning by a gravestone. Nobly restrained in her grief, the goddess epitomizes the heroism of Athens at the time, when many citizens died fighting for their city as far away as Egypt.

naval rival, was invaded in 458BC and, after a year-long siege, capitulated, being forced into the Confederacy. A diversionary attack by Corinth on Megara was meanwhile repelled by a scratch Athenian force of old men and youngsters.

At around the same time Athens built her own double Long Walls. These linked the city to the booming port of Piraeus, making her almost impregnable to direct land attack, something that Themistocles had long ago wanted. In late 457BC a land battle at Oenophyta saw Boeotia come under Athenian control also – an unusual situation that continued for ten years. Athens then won over Phocis and Locris, small states to the west of Boeotia, parts of Thessaly and the city of Troezen in the Peloponnese.

DIFFERING ALLIANCES: SPARTA VERSUS ATHENS

The Confederacy of Delos, which by 450BC had become an Athenian empire, had started with a formal council. Always dominated by Athens as the strongest member, this disappeared by *c.*450BC. What we term the Peloponnesian League, but which was at the time called 'Sparta and her allies', seems to have had no official organization. Sparta proudly used to point out that – unlike the Athenian Empire, where finally only the two large islands of Lesbos and Chios kept their own fleets – all Sparta's allies maintained their fleets and/or armies, central to their independence. But it would have been awkward for Sparta to levy financial contributions, for most Peloponnesian states supplied armies, not fleets. Sparta could in practice be just as bullyingly exploitative toward her allies as Athens was to hers.

She seemed set to become the hegemon, the dominant Greek power, on land as well as at sea.

DISASTER IN EGYPT

What offered Athens even greater prospects of power in the 450s BC was Egypt. Immensely rich, Egypt was never happy under Persian rule. Its Persian satrap was killed in an uprising in 459BC led by Inarus, a Libyan who appealed to the Confederate fleet then in Cyprus. The Athenians dispatched a fleet of 200 ships to expel the Persians from Egypt. In return, shiploads of wheat went north to Athens. But in 456BC a Persian army drove the Athenians from Memphis (Cairo), blockading them on an island in the Nile Delta for 18 months. The whole expedition was lost, only a few of its 40,000 men escaping to distant Cyrene.

CIMON'S RETURN TO VICTORY

Alarmed at this disaster, and fearing a Persian fleet might enter the Aegean, the Athenians transferred the Confederate Treasury to Athens. (Once there, even after the panic ended, it was run conveniently in Athens' interest, one sixtieth of the total revenue being deducted annually as administrative costs.) Cimon, back from his ten-year ostracism, helped to arrange a truce with Sparta and was elected *strategos* to lead another fleet east. This, again 200 ships strong, sailed to Cyprus in 450BC, with 60 ships going to help another revolt in Egypt. They returned for a last victory at Cypriot Salamis, won by Cimon on his deathbed. In 449BC the Peace of Callias was signed between Athens and Persia. By it Athens gave up southern Asia Minor and Cyprus but the Great King agreed to keep Persian troops 80km/50 miles away from the west and north-west coasts, acknowledged as being in the Athenian sphere.

THE 30 YEARS' PEACE

Most of Athens' gains in central Greece soon unravelled: defeats in Boeotia led to its loss, and Phocis and Locris broke

away about then, as did Megara and Euboea. (Oligarchs in these states were generally anti-Athenian).

Only Euboea was finally reconquered after hard campaigning by Pericles. Athens had to abandon all its conquests except Aegina in the 30 Years' Peace of 446BC. This was signed between the Athenian Empire, as it had now become, and the Peloponnesian League.

Above: The Lions at Delos, where the Confederate Treasury was first kept.

Below: Control of the grain routes from the Black Sea remained vital to Athens' survival. Among Greek colonies in the area was Istrus near the Danube Delta.

TOTAL DEMOCRACY
ATHENIAN DEMOCRACY IN ACTION

Democracy (*democratia*) means the rule of the *demos* (the people) – literally so in Athens and other Greek cities that followed her pattern of total democracy. The twin aims of Athenian democracy were to give power to the whole populace and to avoid any individual or group gaining undue power by holding office repeatedly. In practice, this meant elevating amateurism to unprecedented heights – with results that were far less disastrous than might have been expected. Socrates the philosopher was only one of many critics of such amateurism, however. The great exceptions to this avoidance of experts were the ten annually elected *strategoi* (generals) – posts therefore sought by ambitious men, whether or not they had military talent. Pericles was re-elected as a *strategos* repeatedly (14 times after 443BC), but his power still rested ultimately on votes in the Assembly, which could censure all officials. Thucydides, a superb historian but indifferent general, was exiled for losing a crucial battle.

THE ROLE OF THE ASSEMBLY

The Assembly of all eligible male citizens (*ecclesia*) was literally the government, not a chamber of elected representatives. It listened to foreign ambassadors and voted for war or peace. Any citizen could speak, voicing his opinions before the massed citizenry. (This put an obvious premium on public speaking, and richer citizens from the mid-5th century BC began paying to acquire the rhetorical skills needed to sway the crowds.) Decrees passed by majority vote at each meeting of the Assembly were prominently displayed on whitened boards in the Agora in central Athens so "that any who wished could read them". (Most male Athenians could read a bit, though fewer could write well.)

The Assembly met in the open, originally in the Agora, the commercial and social centre. After 500BC it moved to the Hill of the Pnyx, west of the Acropolis. Speakers stood on a plinth to address the Assembly below on a platform built *c.*400BC, whose remains can be seen today. The Assembly met every nine days on average, although heralds could summon emergency meetings often initiated by

a strategos. To convene a quorum of 6,000 citizens for it, Scythian policemen would drag a cable covered in red dye through the Agora from the north, thus herding citizens into the Assembly. Any citizen who was found outside the Assembly that day with red dye on his clothes faced possible punishment.

THE DISENFRANCHISED

This quorum number of 6,000 was, however, probably only one tenth of the total citizenry at Athens' zenith. Many citizens lived in distant parts of Attica, too far from the city to attend Assembly frequently, while *cleruchs* (settlers) on the islands would seldom have voted. No provision was made to represent these absent citizens in Athens. (Our best information about Athens' constitution comes from Aristotle, the learned, perceptive, sometimes critical 4th-century BC philosopher.)

Metics (registered foreign residents settled in Athens) had no vote, nor could they own property. *Metics* rarely won citizenship, even for outstanding services to the state. Women, slaves and children never had the vote either. So Athenian democracy had its definite limits, especially after the Citizenship Law of 451BC, which stipulated that both of a citizen's parents had to be Athenian. (This would have excluded Themistocles from citizenship.) The law, proposed by Pericles, met a public desire to limit the benefits of Athenian citizenship and also encouraged *cleruchs* resident overseas to marry Athenian women left at home.

THE COUNCIL AND HELIAEA

Although the Assembly was the sovereign government, the *Boule* (the Council, originally composed of 400 men, though

increased to 500 after Cleisthenes' reforms in 507BC) discussed beforehand what matters would (usually) be debated. The Council was in theory a microcosm of the Assembly. Its members were chosen by lot from all citizens aged 30 or more; none could serve in it more than twice and never in consecutive years, so keeping it amateur. The Council was therefore only a filter for the Assembly, although in the 4th century BC it took a more active role in guiding it. This *probouletic* business was time-consuming, and from 460BC councillors received modest pay. Most citizens served as councillors at least once. No one, as Pericles said, was excluded from politics because of poverty.

Many would have served more often in a *heliaea* (jury court). These were large courts (at least 201, sometimes 2,001 men), again chosen by lot to avoid corruption or intimidation. As professional lawyers or judges did not exist, plaintiffs and defendants made their cases in person. The jurors, after deciding the verdict by majority voting, decided the penalty too, although the defendant would suggest his own. Almost all Athenian citizens played a civic or political role at some stage, just as all fought in the army or fleet. This was total democracy in action.

ATHENS THE EDUCATOR
THE IDEAL CITY

Above: The Parthenon in Athens, erected in the 5th century BC, became the most famous temple in the world.

Right: The great statue of Zeus fished from the sea off Cape Artemisium seems to incarnate Pericles' ideal vision of Athens, revealed in his famous address to the city of 431BC. Probably cast some 30 years earlier, when Pericles was starting his political career, the statue's heroic, almost overpowering nobility typifies the Periclean Age.

"Our city is open to the world, we do not at fixed times deport people to prevent them discovering our military secrets. This is because we rely not on hidden weapons but on our courage and loyalty…. We do not let our love of beauty make us extravagant, nor our love of wisdom soften us. We see wealth as something to be used properly, not to boast about. No one needs to be ashamed of being poor: shame lies in doing nothing about it. Everyone is interested not only in their own affairs but in public affairs too, and is well informed about politics. We think that the man who minds only his own business has no business in the city at all. … Taking everything into account, I declare our city is the educator of Greece."

So spoke Pericles in his funeral oration for Athenian soldiers at the start of the Peloponnesian War in 431BC. He was expressing an ideal of his city, of course (Greeks were given to idealizations), but this vision had found concrete form in the art, drama and architecture of the previous 30 years. This embodiment continued for almost 30 years after his death in 429BC. As Athens flowered, she became the educator of Greece, then later of Rome, and so of the whole Western world. The temples, with their statues and other artworks, elevated Athens – celebrating her gods, heroes and herself – to an almost godlike level.

THE TEMPLES DEBATE

By ending the long war against Persia, the Peace of Callias in 449BC triggered a debate in Athens about the purpose of her empire, especially about the contributions still paid by subordinate member states. A Panhellenic congress proposed at about this came to nothing, as Sparta predictably refused to attend.

Athens had sworn in 479BC not to rebuild the Acropolis temples burnt down by the Persians until revenged. Their smoke-blackened ruins still dominated the city skyline. Pericles proposed that, with victory won, some of the money used for the fleet should go to rebuilding the city's temples. Athens had sacrificed these (and the rest of her city) for the Panhellenic cause in 480–479BC. Now it was payback time.

Not every Athenian – let alone every subject of the empire – agreed. Thucydides, son of Melesias (not the great historian but his cousin, also related to Cimon), led the opposition in Athens. Other Greeks would feel insulted, he said, when they saw money scheduled for war being used to "gild and adorn our city like a harlot, with costly statues and 1000-talent temples". But Thucydides lost the argument. Temple-building would give employment in Athens for citizens no longer rowing the fleets. Besides, Athenians felt they deserved their city's restoration and decoration. Thucydides was ostracized.

BUILDING THE PARTHENON

In 447BC work began on what has become the most famous temple in the world: the Parthenon, within which rose the huge *chryselephantine* (gold-and-ivory covered) statue of Athena Parthenos, the virgin warrior goddess renowned for wisdom. The Parthenon, which employed the latest techniques of *entasis* – by which columns curved in to appear regular in size when seen from afar – was mainly designed by Ictinus, and completed in 432BC. The statue of Athena, and the carvings of the frieze depicting the

Pan-Athenaic Procession running around the Parthenon (some of which are now the Elgin Marbles in the British Museum) were the work of Pheidias. Both men were supreme geniuses.

SUPERSTITION TO PHILOSOPHY

Pericles served prominently on the commission overseeing work on the Parthenon. Pheidias was one of his friends. So, more controversially, was Anaxagoras from Clazomenae, a philosopher who taught that the sun was only a vast hot stone – "bigger than the Peloponnese" – and not divine. This struck some people as impious, even tempting fate. Protagoras of Abdera, another philosopher, proclaimed: "Man is the measure of all things!" Pericles, who believed in an exalted equality, shrugged off old superstitions in the bright noon of Athenian democracy, as the city drew the greatest minds of the Greek world to her.

Above: The frieze depicting the grand Pan-Athenaic Procession running around the Parthenon was carved under the direction of Pheidias, one of the greatest sculptors of classical Greece.

THE RICH PAY FIRST

There was no regular income tax in Athens. Instead, most great festivals, and many of the triremes, were financed by the 1,200 or so richest citizens. These Athenians who chose, or were chosen, to finance a particular event did not simply pay for particular dramas (the *choregeia*) or religious festivals: they had to put them on, recruiting and rehearsing the actors and musicians needed. These demands were called *liturgia* (public burdens). For a *trierarch*, his liturgy meant building, equipping, crewing and commanding a trireme – a great effort. But love of honour and *areté* (excellence) drove wealthy Athenians to compete with each other in such active munificence in the 5th century BC.

Right: Theseus Diadoumenus *is the work of another brilliant but very different sculptor of High Classicism, Polyclitus of Argos, who often worked in Athens. His statue shows the legendary Athenian king as a perfectly proportioned young man. (Polyclitus was obsessed with mathematical proportion.) This is a Roman copy of a Greek original.*

Chapter IV

WORLD WAR IN MINIATURE
431–404BC

The Peloponnesian War marked a turning point in Greek history. Before it, Athens seemed set on an upward trajectory, which could have led to most of Greece becoming united under its leadership. Afterwards, although Greek civilization continued to develop, any such possibility vanished. The long wars between 431 and 404BC affected almost all the Greek world, half-ruining much of it. We call it the Peloponnesian War, showing our Athenian perspective; for Sparta and its allies, it was the Athenian War. It has been called a world war in miniature, as important as the Persian Wars.

The war divides into two stages. In the first, 431–421BC, the Spartans often tried to attack, although they had only one brilliant general, Brasidas, whose death led to an uneasy peace. This was broken by the Athenian attack on Syracuse in 415BC. Defeat in Sicily imperilled Athens' overseas empire and its democracy at home. Athenian democracy showed itself at its worst in rejecting peace offers until final defeat. Thucydides' superb history covers the war to 411BC. If not impartial, writing with only just controlled passion, he is searchingly intelligent. His words sum up the war: 'The Peloponnesian War not only lasted a long time, but brought with it unprecedented suffering for Hellas. Never before had so many cities been captured and then devastated; never before had so many people been exiled or lost their lives.'

Left: The funerary stele of Chareidemos and Lykeas, two Athenian hoplites among the many killed in the war.

Above: Pericles was Athens' master-strategist at the war's inception, but he did not foresee the devastating plague that killed thousands of citizens, including himself.

THE PELOPONNESIAN WAR
ORIGINS AND STRATEGIES

The 30 Years' Peace between Athens and the Spartan Alliance of 446BC had run only half its course before war broke out again. Both sides could be blamed for particular incidents, but these would not have led to such a prolonged, bitter war without deep underlying rivalry.

EARLY BEGINNINGS

Corcyra (Corfu) was originally a Corinthian colony, and so was expected to show some loyalty to its mother city. It was also a rich mercantile state with a large fleet of 120 triremes. Further, it lay on the important route west to Sicily. When it appealed to Athens in 433BC over a quarrel with Corinth about their joint colony Epidamnus (Durres in Albania), Athens sent ships to assist

Below: The immediate cause of war was a clash between Corinth and Corcyra (Corfu) in 434BC. Athenian support for Corcyra, seen below, soon led Corinth to urge Sparta to declare war on Athens.

it. These had orders not to fight unless necessary but ended up defeating the far larger Corinthian fleet. Suspecting that war was now inevitable, in 432BC the Athenians issued a decree excluding Megarans from Athens and its empire. By threatening Megara with economic ruin, the decree aimed to force the city, controlling the route from the Isthmus, to ally with Athens as before.

Corinth and Megara, enraged, both appealed to Sparta to fulfil its duty as Alliance leader. Sparta, led by its pacific old king Archidamus, prevaricated, calling a congress at Corinth in late 432BC. There, Corinth dramatically described Athens' boundless energy and aggressive intentions, which swayed many Spartans. Futile diplomatic exchanges followed before Sparta finally agreed to war. As Thucydides put it: "It was not so much what their allies said to them but their fear of Athens' power" that decided them. The Spartans often prove surprisingly cautious.

ATHENIAN STRATEGY

In the first years, Athenians broadly kept to the strategy proposed by Pericles: they treated their city as an island within the Long Walls linking it to Piraeus. Abandoning the Attic countryside and small towns to Spartan invasion avoided risking full-scale defeat in the open field. But it meant evacuating the population, who had to camp in unhealthy discomfort where they could between the walls. Raids on the Peloponnese aimed to divert and harry the Spartans, but Athens did little at first to exploit its naval superiority. Pericles aimed to 'win through' – to wear out the other side through greater economic and military strength. Athens had much the better and larger fleet (300 triremes besides those of Corcyra, Lesbos and Chios), while its empire made it far

the wealthiest Greek state, with 6,000 talents in reserves. A fund of 1,000 talents and 100 triremes were set aside for emergencies. Pericles' hopes that the empire's normal *phoros* (taxes) could finance the whole war proved over-optimistic, however.

SPARTAN STRATEGY

The Spartans, with much the best army but no strategic vision, stuck to old ideas of hoplite warfare: march into Attica, defeat Athens' army, make Athens sue for peace within a year. Unfortunately for them, the Athenians refused to give battle. As Peloponnesian forces seldom stayed in Attica for more than a month, they did not even close the vital Laurium silver mines at first. However, they did damage Attic agriculture, cutting down its olive trees. These, replanted after the Persian Wars, had only just reached full maturity. Their loss was a blow even to

Above: The tomb of Dexeios, a young cavalryman killed fighting near Corinth, typifies the idealized, heroic but understated attitude of Athenians throughout the classical period in fighting for their city.

Athenians living inside the city, for many town-dwelling families still owned their ancestral farms.

Sparta claimed to be 'liberating the Greeks' from Athenian rule. Slowly it realized that only with Persian gold could it acquire the large professional fleet needed to beat Athens at sea. This meant abandoning Greek cities in Asia to Persian rule in exchange – not what most of them wanted, and a blot on Panhellenic honour. Caught in this dilemma, Sparta's strategy for a long time was ineffectual, succeeding only when atypically brilliant men guided her policies: Brasidas, a general who had escaped the mental rigidity that Spartan education induced, and Alcibiades, an Athenian traitor.

> ### AN IDEOLOGICAL STRUGGLE?
>
> The Athenians generally favoured, though did not always enforce, democracy in their subject or allied states. The Spartans preferred narrow oligarchies among their allies and neighbours, to keep the democratic virus away. Oligarchs in Lesbos hoped for Spartan support when they revolted against Athens in 428BC. (It never arrived.) In the singularly bloody civil war that racked Corcyra in 427BC, the Athenians backed the island's democrats partly because the Peloponnesians supported its oligarchs. Brasidas, marching through Thessaly in 424BC, was welcomed by local oligarchs, although – as Thucydides noted – the *demos* (people) remained pro-Athenian. But when in 415BC the Athenians attacked Syracuse, the great Sicilian city, they were fighting other democrats. Generally, disputes about types of government were of secondary importance in the prolonged struggle for hegemony.

SIEGE, PLAGUE AND REBELLION
431–427BC

Below: At the outset of war in 431BC, Athens was at its political and cultural peak, as masterworks such as this Amazon, copy of an original by Pheidias, shows.

In late May 431BC, "when the wheat was ripe", an unstoppable 30,000-strong Spartan-led army rolled into Attica, burning everything in its path. It was trying to provoke the Athenians, crammed behind their walls, to come out and fight. They did not, and after a month Sparta's Peloponnesian allies, who made up most of the army but lacked *helots* (Sparta's slave class) to work their farms, insisted on returning home.

SIEGE OF PLATAEA

Over the border in Boeotia, Plataea, allied to Athens, was besieged by Thebes. Thebes had jumped the gun in March by trying to seize Plataea in a night attack. The coup failed, the Theban infiltrators were killed and Plataea's civilians evacuated to Athens. Then the Theban army arrived to besiege the tiny city. It took four years before it fell. All the surviving garrison was killed and the city razed to the ground – a vicious start to the war.

Meanwhile, Athenian ships raided the Peloponnese around Troezen to little effect. However, the speech that Pericles made at the funeral of the Athenian soldiers first exalted democracy to the sublime levels of the Elgin marbles, at least according to Thucydides who heard it. "Individual Athenians adapt to every different sort of action with versatility and grace. We have raised marvellous monuments to our power. Future ages will wonder at us, and we need no Homer and his poems to praise us.

Our courage has blazed paths across every sea and land. This is the city for which these men nobly fought and died."

THE PLAGUE

In 430BC disaster struck Athens in a way that Pericles had not foreseen: plague, reputedly from Egypt, which kept recurring. It was probably smallpox, judging by Thucydides' graphic description. (He had it himself, noted the symptoms with clinical detachment, and recovered.) It hardly affected the land-locked Peloponnese, but it killed nearly one-third of Athens' population of *c.*175,000 crowded together.

PERICLES' FALL AND RETURN

This was the first, and in some ways worst, disaster of the war, and the people turned on Pericles as a scapegoat. An attack he led had failed to capture Epidaurus, its target. He lost office,

---

INEFFECTUAL SIEGE CRAFT
Despite being supreme on land, like the Spartans, or at sea, like the Athenians, the Greeks before 400BC were oddly inept at siege warfare. Compared to contemporary Persians or to monarchs of the 4th century such as Alexander the Great, they often seemed powerless before any well-built, well-defended wall. If surprise attack or treachery (or the two combined) failed, blockading a city until it starved was the normal way to capture it. Greek armies had battering rams and ladders and might try mining under walls or even primitive flame-throwers, but they lacked other siege engines. Only in the 4th century did Sicilian and Macedonian rulers revolutionize siege warfare with powerful catapults and gigantic siege towers.

was accused of corruption and fined. Pheidias, his friend the great sculptor, was also fined. Another friend, the philosopher Anaxagoras, had to flee the city and return to his native Lampsacus after questioning the gods' existence, an act of dangerous impiety in troubled times. Even Pericles' mistress Aspasia, who had joined their dinner parties (very few women did), was prosecuted.

But Athenian public opinion swung back and Pericles was reinstated as *strategos*. Meanwhile, campaigns in two important areas, the Gulf of Corinth, and around the Chalcidic peninsula in the north, had mixed results. Demosthenes (the 5th-century BC *strategos*, not the 4th century orator) led not unsuccessful operations in the west, although he lost many men in them.

THE REVOLT OF LESBOS

In 428BC the oligarchical government of Mytilene on Lesbos revolted, taking most of the island with it. Lesbos, which was still a free ally of Athens, had no specific complaints, so its revolt was a huge shock. To pay for its suppression, a novel property tax, the *eisphora*, was introduced. Mytilene appealed to Sparta, which sent one adviser. By May 427BC the Lesbian leaders, under pressure from their starving people, finally had to surrender to the Athenian forces.

On the news of their surrender, a great debate arose in the Assembly about how to treat the rebels. Cleon, now the leading radical, urged that all male citizens on Lesbos should be executed. His proposal was carried, and a galley sailed with the grim news. But another politician, Diodotus, persuaded an extraordinary Assembly meeting to be merciful (on the

pragmatic grounds that the Lesbian *people* had not revolted) and another galley was sent in hot pursuit. It reached Lesbos just in time and only the ringleaders were executed. But most land on Lesbos was allocated to Athenian *cleruchs*. As absentee landlords, they let it out to locals, who had to pay rent for farms that had previously been their own.

SPARTA'S DEFEAT AND PEACE
425–421BC

Below: This superb Nike (statue of victory) is one of the first semi-nude female statues in classical art. The marble statue was made by Paionius for the Messenians of Naupactus, celebrating the part they played in the Athenian victory over Sparta in 425BC.

Athens had seldom contemplated attacking the Peloponnese on its south-west flanks. It had only one ally, Zacynthus, in the area, yet Achaea and Aetolia, south and north of the Corinthian Gulf, were friendly neutrals and at Naupactus near the Gulf's mouth lived *helots* from Messenia, who were fanatically loyal to Athens. A civil war, waged with appalling savagery, led to the Corcyran democrats regaining control of most of the island. An Athenian squadron sailed to Sicily in 428BC in response to Ionian cities' appeal for protection against their Doric neighbours.

SURRENDER AT PYLOS
The Athenian *strategos* Phormion had defeated larger Peloponnesian squadrons in 429BC in the Corinthian Gulf by dazzling seamanship. (On one occasion Athenian galleys forced the enemy fleet to form a defensive circle with bows pointed outward. Around this, Athenian triremes rowed closer and closer, forcing their enemies inward to foul their oars.) After Phormion's death, in 426BC Demosthenes, another adventurous *strategos*, saved the *helots* of Naupactus from a Spartan attack. Next year, although no longer *strategos*, he accompanied a fleet bound for Sicily. This was blown into the Bay of Pylos on Messenia's west coast by a storm. Demosthenes persuaded the sailors

Above: The capture of 140 Spartiates (full Spartans) on the Island of Sphacteria in 425BC by Athenian general Demosthenes horrified Sparta and led it to sue for peace – an offer Athens rejected. The war then resumed.

to build fortifications on the peninsula of Pylos. Then the main fleet sailed on west, leaving Demosthenes with a small force.

Horrified at the presence of Athenians on Spartan soil, Sparta sent a fleet and troops to eject them. Some 420 Spartiates (Spartan's elite ruling class) with *helot* attendants occupied Sphacteria island to the south. But the wooded Pylos peninsula proved very defensible. Demosthenes' men, although outnumbered, repelled Spartan attacks by land and sea. Brasidas was badly wounded while disembarking.

The Athenian fleet, which had turned back on hearing the news, swept into Pylos Bay to defeat the Spartan ships.

Now the Spartans on Sphacteria, themselves besieged, faced starvation. Among them there happened to be some very important Spartiates.

Sparta, panic-stricken, proposed a truce. To ensure that food was supplied to the marooned Spartiates, it handed over 60 ships and sent envoys to Athens, suggesting peace on the status quo ante as in 431BC. Urged by Cleon, the Assembly rejected this proposal, but kept the 60 ships on a technicality. War resumed, with well-bribed *helots* swimming across to Sphacteria by night with food for the besieged Spartiates. But when Cleon arrived with Athenian reinforcements, including light-armed troops, they stormed Sphacteria at night. Athenian archers and javelin throwers harassed the exhausted Spartans until they surrendered. About 140 Spartiates became prisoners.

This unprecedented Spartan surrender, amazing all of Greece, boosted Athenian self-confidence dangerously. They seized Cythera, an island off Sparta's south coast to which some *helots* had managed to escape. (But Athens never attempted to rouse the resentful *helot* serfs inside Lacedaemonia, which might really have crippled Sparta.) Cleon, aggressively imperialistic, produced a new, heavier tribute list for the empire, often doubling subject cities' *phoros* (taxes). But trouble was growing in the north.

THE LOSS OF AMPHIPOLIS

Amphipolis, founded by Athens in 436BC on the River Strymon by the Macedonian border, was a key city for the empire, controlling a bridge, trade routes and gold mines. In 424BC Brasidas marched north, with no Spartiates but only 700 *helots* armed as hoplites (presumably promised their liberty). He gathered other troops en route to a stunning series of victories. These were won by diplomatic as much as military means, Brasidas having great, if menacing, charm. Some Chalcidic cities were disgruntled with Athenian rule and Brasidas initially had the backing

of the Macedonian king Perdiccas. When Brasidas reached Amphipolis after marching through a winter night, he offered its citizens such easy terms that they surrendered. Thucydides, the *strategos* commanding a squadron at Thasos, hastily sent ships in support, but too late.

Back in Athens, Thucydides was exiled (not ostracized) for the loss. This was perhaps unfair, as he had lacked the forces needed to defend the long north coast against Brasidas, Sparta's best general. Athens, defeated in a hoplite battle by Thebes at Delium in 424BC, had ignored Brasidas' campaign until it was too late. Cleon finally was sent as *strategos* with a decent force – 1,200 Athenian hoplites plus allies – in 422BC to stop Brasidas from winning more cities. In a battle outside Amphipolis, Brasidas and Cleon were both killed, leading to peace in 421BC.

THE PEACE OF NICIAS

Intended to last for 50 years, the Peace of Nicias – named after the conservative Athenian politician – attempted a return to the status quo ante. Athens returned all gains except Nisaea, Megara's port, being promised the return of Amphipolis and other northern cities. (She never got them.) Sparta returned her conquests, but the peace marked a defeat for her. Her allies refused to accept it, accusing Sparta of neglecting the Alliance's interests.

Above: The Temple to Hephaestus, built c.440BC, overlooked the Agora, Athens' social centre.

Below: Trapped on the island of Sphacteria, the Spartans effectively became hostages in their own territory thanks to Athenian tactical brilliance.

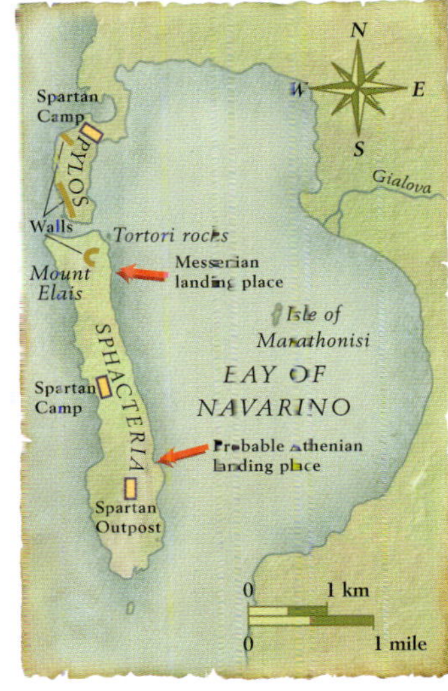

THE FALSE PEACE
421–415BC

Above: The debauched features of Alcibiades as imagined by an artist of the Roman period. The flamboyant aristocrat of the Alcmaeonid family led Athens into disastrous adventures.

When agreeing to the Peace of Nicias, Sparta was acutely conscious that the 30 Years' Peace with Argos, her old rival in the Peloponnese, was about to end. Argos had recovered from the devastating defeat in 457BC, but needed help to challenge Sparta, still the region's powerful *hegemon*. As Argos was now a democracy, Athens was the obvious choice. Many younger Athenians were bored with a peace that offered no chances of gain or glory.

ALCIBIADES' RISE TO POWER

The young Alcibiades (born *c.*450BC) now emerged as a prominent figure in Athenian public life. An Alcmaeonid and former ward of Pericles, he had been wounded while fighting bravely at the Battle of Potidaea in 430BC. His life had been saved by Socrates, a fellow hoplite who became a friend. In alliance with the radical Hyperbolus, Alcibiades, handsome, rich and flamboyant, began to dominate the Assembly. Elected *strategos* in 420BC, he persuaded Athens to ally with Argos and other states in the Peloponnese against Sparta. But the Assembly did not choose Alcibiades as *strategos* for the army sent to Argos in 419BC, which proved a mistake. After numerous manoeuvres, King Agis of Sparta defeated the joint Athenian-Argive army at Mantinea in 418BC, which, as Thucydides said, "wiped out the disgrace of Pylos". It also triggered regime change back in Argos. Under a new oligarchy, it allied with Sparta, which recovered its own self-confidence.

In Athens, Alcibiades also switched sides, forming an alliance with the conservative Nicias. Their supporters combined to ostracize Hyperbolus in 418BC, leaving Alcibiades undisputed leader of the radical democrats, who were also the keenest imperialists. This was no paradox: the prospect of employment as rowers in the fleet or of becoming *cleruchs* (colonists) made imperialism especially attractive to the *thetes* (the poorest class). Nicias, never a good general, led an unsuccessful expedition to regain Amphipolis, but Alcibiades was really looking elsewhere.

THE CAPTURE OF MELOS

Melos was a small Cycladic island that had managed to stay out of the Athenian Empire. Neither wealthy nor of much strategic value, it refused to submit to Athens, which accordingly sent a fleet and army to take it in 416BC. Athens

Left: The Erechtheum, a temple rivalling the Parthenon in ingenuity and elegance if not size, was started in 421BC after the Peace of Nicias. It was only completed in 405BC, the year of Athens' final disastrous defeat.

must have expected an easy victory, but
the Melians, though vastly outnumbered,
fought back. It took a long siege and
Athenian reinforcements before Melos
was captured. As was the custom, all
Melian men of military age were killed,
their women and children enslaved and
their land given to 500 Athenian *cleruchs*.

The most interesting aspect of
this small campaign was the debate
beforehand between the Athenian
envoys and Melian magistrates, to which
Thucydides devoted nearly a whole
chapter. In it, the Athenian envoys
propounded a nakedly self-interested
imperialism that the historian deplored
and few Greeks at the time would have
defended so openly.

THE LURE OF SICILY

Sicily has been called the Greek America.
It was a land of huge potential wealth and
power and a major grain exporter, chiefly
to the Peloponnese. But its cities were
very Greek in their constant wars. This
gave an expansionist Athens opportun-
ities. It had sent small fleets west in the
420s BC to assist Ionian cities against Doric
neighbours, and to keep the Sicilians too

Right: Nike Unlacing Her
Sandal: *this superb statue
of victory came originally
from the Temple of Athena
Nike on the Acropolis.
It was begun in* c.428BC
*and finished during the
brief peace, when Athens
indeed seemed victorious.*

preoccupied to help Sparta.
But at a congress at Gela
in 424BC, Hermocrates,
the democratic leader
of Syracuse, persuaded
the Sicilians not to
let outsiders such as
Athens meddle in
their affairs. Faced
with this 'Monroe
Doctrine' (exclud-
ing old Greece from
western Greek
affairs), the
Athenian fleet
returned home.

By the year 415BC,
however, much had
changed. Egesta, a city
in Sicily's extreme
west, appealed for
help against its neighbour Selinus, an
ally of Syracuse. Athenian envoys sent to
investigate returned with tales of Egesta's
fabulous wealth. These stirred dreams
of easy riches in the Assembly and, swayed
by Alcibiades, it voted to help Egesta.
Nicias warned against so risky a project,
but this only led to the force being
doubled and he himself was appointed a *s
trategos*, with Alcibiades and Lamachus.
Alcibiades sailed in June 415BC as one
of the three *strategoi* commanding the
armada. It had 134 triremes and transport
ships with 30,000 infantry but almost
no cavalry.

*Right: The Valley of the Temples in Sicily.
The island became a land of huge potential
wealth and power and a major exporter of
grain, chiefly to the Peloponnese.*

DISASTER IN SICILY
415–413 BC

Above: Procles Saying Farewell to his Father, *a scene typical of Athens in wartime when every citizen did his military service, and many did not return.*

Below: Crucial to besieging Syracuse were the heights of Epipolae above the city, where the fortress of Euryalus stood. Failing to hold this point, the Athenians were doomed.

Arriving in the west, the Athenians found a cool reception at Rhegium (Reggio), on the Strait of Messina. They also discovered that Egesta had deceived them about its wealth: it was not rich at all. But, as there could be no ignominious speedy return, a council of war debated the likely options. Nicias suggested arbitrating between Egesta and Selinus, then sailing around the island before returning; Alcibiades wanted a diplomatic initiative to win allies; Lamachus, the one soldier among the *strategoi*, urged an immediate knock-out blow at Syracuse, the real enemy. He failed to convince the others, however, and the Athenian fleet cruised down the coast past Syracuse, capturing one Syracusan trireme.

Athenian troops finally landed late in the season in Syracuse's Great Harbour. Here they had a success, their seasoned hoplites defeating the Syracusan army despite the latter's superior cavalry. But Nicias, senior *strategos*, then sailed the army back to Catania for the winter, which was spent trying to find allies rather than preparing for war.

ALCIBIADES' ESCAPE

While the Athenian fleet cruised past Syracuse, a state galley arrived for Alcibiades. He was being recalled to Athens, accused of profaning the Eleusinian Mysteries, the holiest and most revered of all ritual celebrations in Athens. Eluding his captors, he disappeared, to be next heard of in Sparta. His recall was the work of enemies in Athens, where he was soon tried and condemned in absentia.

WALLS AND COUNTER WALLS

In spring 414 BC the Athenians returned suddenly, caught the Syracusans off guard and seized part of Epipolae, the plateau to the west. Control of this large plateau now became central to the siege, for through it ran all roads going north. As the Athenian fleet was still superior at sea, cutting these roads would isolate Syracuse. To achieve this, the Athenians began building a double wall with a circular fort at its centre.

Syracuse was led by Hermocrates, a democratic leader comparable to Pericles, but his position remained precariously dependent on success. For the moment it looked as if the Syracusans faced defeat. Although they started to build cross walls to cut off the Athenians in turn, they were slower builders. (How much of the mainland city, as opposed to Ortygia, the citadel, was walled remains uncertain.)

Lamachus was killed during fighting around the walls, leaving the indecisive Nicias as the sole *strategos*. However, Nicias had important contacts inside Syracuse who could, he hoped, deliver the city peacefully to him. In fact, the Syracusan Assembly was about to consider peace negotiations when a Spartan force slipped past the Athenians into the city, commanded by a Spartiate of unusual energy: Gylippus.

Gylippus breathed new heart into the defence. The strong point at Labdalum was captured, an Athenian trireme taken in the Great Harbour and a new cross wall started on the northern plateau. Remarkably complacent, Nicias ignored this and instead built a fort at Plemmyrium, south of the Great Harbour, to facilitate supplies. But the Athenians were losing the cross-wall race. In a battle on Epipolae Heights, Gylippus worsted the Athenians and completed the third cross wall past the Athenians, cutting them off from land routes. Encouraged, the Syracusans began training their own large fleet, which until then had hidden from the Athenians. As fighting would be in the confined waters of the Great Harbour, the Syracusans strengthened their triremes' prows.

REINFORCEMENTS ARRIVE

Nicias wrote home despairingly in late 414BC, calling for reinforcements or recall. The Assembly, loath to abandon its Sicilian dreams, voted to send a second force of 73 ships and 15,000 men under Demosthenes. Before it arrived, the Syracusan fleet sallied forth and, in a series of engagements, damaged many Athenian ships crammed tightly together, as well as Athenian morale. This was already low because their camp on the beach near the marshes was in an unhealthy malarial spot. But the arrival of Demosthenes in 413BC with his large force alarmed the Syracusans.

RETREAT AND DISASTER

Demosthenes realized that only swift action could save the Athenians. He launched a night attack on the cross walls, which, after initial success, ended in chaotic disaster. As reinforcements could now enter Syracuse freely, Demosthenes urged immediate withdrawal by sea. Nicias hesitated until two naval defeats by the Syracusans – who closed off the Great Harbour – forced him to agree to a retreat, though by land. But an eclipse of the moon caused the superstitious Nicias to delay this plan for a month. When the

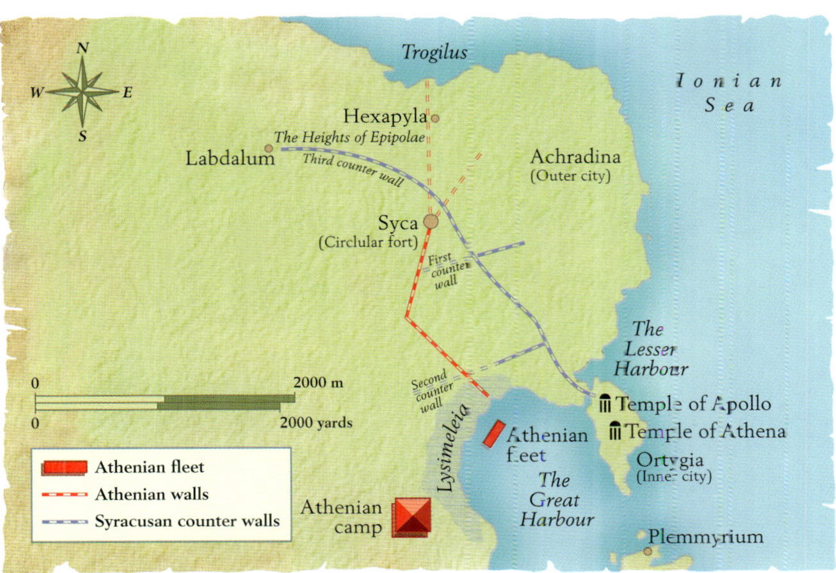

Athenians at last began their retreat, with only c.20,000 men left, the Syracusans harassed them continuously. Athenian discipline finally broke as they scrambled down to a river desperate for water, and their army surrendered.

Nicias and Demosthenes were killed, most of the rest becoming slaves in the mines. There many died in appalling conditions, although some reputedly won freedom by reciting lines from Euripides the Athenian playwright already known in Sicily. Athens' western venture had ended in disaster with the loss of 40,000 men and 200 ships.

Above: Syracuse, or its island citadel, was hard to besiege. The Athenians tried to cut the city off by building a wall, but Syracusans under the dynamic Spartan Gyllipus built a cross wall that cut them off instead.

Below: Oarsmen in a Greek trireme. Control of the seas was always vital to Athenian power. When all the Athenians' ships were destroyed at Syracuse, they were effectively doomed.

AFTER SYRACUSE
REACTION AND REVIVAL, 413–408BC

Above: From Sardis, Persia's regional capital, Tissaphernes the Persian satrap watched the Greek world, aiming to exploit Greek divisions to regain long-lost territories.

Below: The Council of 400 that replaced democracy did little except build a fort at Piraeus, whose ruins are depicted here by J.R. Herbert, a 19th-century artist. The Council soon gave way to a revived democracy.

Events in Syracuse were to have huge repercussions back in Greece. In Sparta, Alcibiades, who had evaded his Athenian captors, impressed his dour hosts, eating Sparta's revolting black broth with gusto and saying that all Athenian aristocrats considered democracy an 'acknowledged folly'. He gave the Spartans dangerously good advice: they should send an adviser to Syracuse as requested. And they should reinvade Attica, not returning home after a month's ravaging, but instead occupying the fort of Deceleia all year round.

Deceleia was 16km/10 miles north of Athens, well placed to menace the city. Its permanent Spartan garrison prevented Athenian farmers from returning to cultivate their land, as in the war's first part, and it cut the overland road to Euboea. (Some Athenian farmers had moved livestock to that island.) Worse, Deceleia attracted *c.*20,000 runaway slaves over the next decade, often skilled workers vital to the Athenian economy, as a result of which the Laurium silver mines, dependent on slaves, were abandoned.

OLIGARCHIC IDEAS

Aristocratic opposition to democracy went underground during Pericles' ascendancy. Aristocrats benefited from the empire as victorious generals or from overseas properties. But long wars led to higher taxes, such as the *eisphora* (property tax), which hit the richest. Disgusted by the rise of common (if wealthy) men such as Cleon, some aristocrats longed for the 'good old days' before democracy. One anonymous aristocrat, called The Old Oligarch although he was probably young, wrote a pamphlet, *Athenian Constitution*, *c.*424BC. Its importance lies less in sneers against democracy ("the common people dress as badly as slaves") than in showing how strong oligarchical sentiment always remained.

In Athens there was a revulsion against all regarded as responsible for the disaster, be they radical democrats or fortune-tellers – in fact almost anyone

except the ordinary Athenians who had really voted for the expedition, as Thucydides noted dryly. The emergency reserve of 1,000 talents and 100 triremes was broached, and a Board of Ten older men, including the octogenarian playwright Sophocles, supplemented the long-established Council of 500.

GROWING REVOLTS

The Athenian defeat at Syracuse roused the hopes of many people who were discontented with Athenian power. In Lesbos, oligarchs planning to revolt appealed to Sparta and this time got a response: King Agis, realizing Lesbos' proximity to the vital grain trade route, sent a fleet of 100 ships to the eastern Aegean in 412BC. Its arrival triggered revolts in Chios and Ionian cities. In addition, Sparta – again on Alcibiades' advice – had approached Tissaphernes, satrap of Sardis. Persia had adhered to the Peace of Callias, keeping away from the Aegean coast, but Athens had rashly annoyed the Great King by supporting a rebel, and Persia still wanted its former territories back. By the Treaty of Miletus in 412BC, Sparta apparently agreed to Persian claims over all Greek states ever under its rule, in return for money.

Alcibiades arrived that winter in Sardis. His seduction of King Agis' wife meant he was no longer welcome in Sparta, but he charmed Tissaphernes. Alcibiades suggested that regime change in Athens might work to Persia's advantage and he was the man to effect it. He made contacts with the Athenian fleet at Samos. This was manned chiefly by *thetes*, the poorest Athenians, but even they, disillusioned by defeat, now listened briefly to talk of oligarchy, which might win Athens vital Persian gold.

Many middle-of-the-road Athenians, led by Theramenes, felt a change in the constitution was needed, although Athens had responded vigorously to recent revolts, recruiting and dispatching a new fleet. Aristophanes' comic play *Lysistrate*, produced in spring 411BC, depicting Athens' women refusing sex to their husbands

Above: A hoplite in an elaborate helmet. All major land battles were decided by these heavy-armed infantry.

unless they made peace, voiced general war-weariness. After some democratic radicals were murdered amid growing terror, it was proposed that the franchise be restricted to the 5,000 richest citizens. Meanwhile, a Council of 400 would rule, the Assembly being abolished. "It was no small thing to destroy the Athenian people's freedom after 100 years of democracy," commented Thucydides, but the 400's rule, from June to September 411BC, revealed their shortcomings. They built a fort at Piraeus and made unsuccessful peace overtures to Sparta. Meanwhile, Alcibiades, crossing to Samos, won over the Athenian fleet. This declared itself the true Assembly and went on the offensive, led by Thrasyllus and Thrasybulus.

DEMOCRACY RESTORED

Two naval victories led to the restoration of democracy in Athens. The first at Cynossema in late 411BC restored Athenian morale; the more decisive one at Cyzicus (in the sea of Marmara) in the spring of 410BC annihilated the Peloponnesian fleet, despite Persian support. Sparta now offered peace on the basis of the status quo – an offer rejected by the Assembly, now again sovereign in Athens, as the 400 had been overthrown.

Below: The dangerous charm and charisma of Alcibiades, who long bedazzled Athens, comes over well in this Roman copy of a Greek original.

THE FALL OF THE ATHENIAN EMPIRE 408–404BC

Above: Piraeus was Athens' great port and its lifeline, for the city was crucially dependent on wheat, which was imported from the Black Sea through the Hellespont.

Alcibiades had overstated his influence with Tissaphernes to the Athenians. That devious Persian satrap, wanting to keep Sparta and Athens involved in mutually exhausting wars, dangled the prospects of Persian alliance – and gold – before both.

ALCIBIADES' RETURN TO ATHENS

Despite gaining no Persian alliance, Alcibiades retained immense appeal for the Athenians. He seemed to be the one man who might still win them the war. Gaining command of the main Athenian fleet, now in the north, he retook Chrysopolis and Chalcedon on the Bosphorus. Then in 408BC he recaptured Byzantium, usually thought impregnable on its peninsula. This ensured safe passage for ships carrying vital wheat from the Black Sea through the Bosphorus to Athens. It was time for his triumphant return to Athens.

Alcibiades reached Athens in June 407BC. His popularity surged yet higher when he gave a ceremonial military escort to the religious procession down the Sacred Way to Eleusis. (For years the Athenians had been going by ship to avoid the Spartans.) Alcibiades was now elected *strategos autocrator* (supreme commander), a role denied even Pericles. Returning to take control of the fleet, however, he made a disastrous mistake. Going off possibly to raise revenue – Athens was desperately short of cash – he left the rest of the fleet under the command of Antiochus, an experienced sailor but a drunk. Antiochus provoked and lost an unnecessary battle to the Spartans at Notion in spring 406BC. With his reputation now shattered, Alcibiades thought it wiser to retire to his castle on the Hellespont.

THE COSTLY VICTORY

The Spartan fleet was now commanded by a new *navarch* (admiral), Lysander, an unusually formidable Spartan. Lysander befriended the youthful Prince Cyrus, known as Cyrus the Younger, who had far-ranging powers. Cyrus soon committed Persia to Sparta, which in turn agreed to Persian demands. With Persian gold, Lysander could hire rowers at better rates than Athens, soon building up a large professional fleet.

But Athens was not defeated yet. Lysander was replaced as *navarch* by the less competent Callicratidas. At Arginusae, south of Lesbos, in late 406BC, the Athenians defeated the Spartan fleet again in a hard-fought battle. But in a storm at the battle's end, many Athenian sailors whose triremes had been sunk

Left: In 405–404BC, during the last winter of the war, the Spartans blockaded a starving Athens until finally the city surrendered.

were left to drown. When the news reached Athens, the Assembly, dominated by Cleophon, another demagogue, became hysterical. (Many had lost friends or relatives in this disaster.) The Assembly tried six *strategoi* for incompetence for not rescuing the drowning men. Among those found guilty and executed were Thrasyllus, Athens' best general at the time, and Pericles the Younger. Such collective trials were almost certainly illegal, but only Socrates the philosopher was brave enough to protest, nearly being lynched in the process.

Athens had now exiled or killed its last good leaders. Cleophon also persuaded the Assembly to reject further Spartan peace offers. Reportedly, he was drunk at the time.

ENDGAME AT GOAT'S RIVERS

Lysander, resuming command of the Spartan fleet in 405BC, took a 200-strong Peloponnesian fleet up to the Hellespont and captured Lampsacus on the south of the Straits, where he stationed his fleet in August. The Athenians, gathering all their ships, pulled up their fleet opposite him on the exposed beach of Aegospotami, Goat's Rivers. It was not a good spot, as supplies had to be brought every day from Sestos, 4.5km/3 miles away. Every day the Athenians rowed out to try to force a battle, but the Spartans refused. Alcibiades rode down from his castle to warn the Athenians to move, but they ignored him.

After four days, the Athenians had returned to cook lunch on the beach when Lysander's fleet sailed over. He caught the Athenians defenceless, taking their ships almost without a fight. About 4,000 Athenian sailors were summarily killed. So complete was the Spartan victory (only 20 ships under Conon escaped) that treachery was suspected. It meant the final end of Athenian seapower, and doomed the city.

The news reached Athens at nightfall, a night 'on which no man slept'. Lysander took his time, sweeping around

Right: Athena, patron goddess of Athens, thanking Hera, patron goddess of Samos. The carving of 402BC celebrates the fact that Samos was one of the few cities that remained loyal to Athens until the war's bitter end.

the Aegean and sending in all Athenian *cleruchs* to swell the numbers of starving citizens. Cleophon again vetoed the first peace proposals. Then the Assembly turned on him and had him executed for evading military service.

As the Spartan blockade tightened, a conference was called among the victors to decide on Athens' fate. Thebes and Corinth wanted the city destroyed, but the Spartans, thinking back to the Persian Wars, would not agree. Instead Athens had to pull down the Long Walls to Piraeus, give up all overseas possessions, accept the return of all exiles surrender all but 12 ships and accept a Spartan garrison.

In April 404BC Athens capitulated and Lysander entered the city in triumph The long war was over.

Below: The theatre of Dionysus, where Athens, despite the disasters – political, social, economic – of a war being lost, continued to produce some of the world's greatest tragedies.

CHAPTER V

THE GREEKS:
THE FIRST INDIVIDUALS
c.650BC–AD147

The Greeks were among the first people about whose lives, loves and luck we can still read with interest and sympathy. This is partly due to the histories and biographies that survive (many have not). But it is also because the Greek *polis* gave some men, if few women, far more scope to be individuals than earlier absolutist monarchies had. Only in Sparta, that dour militaristic state, were there few individuals. Among the Macedonian monarchs, ruling huge empires after Alexander's conquest of Asia, glamorous personalities emerged, although none competes with Alexander himself.

The Greeks tended to hero-worship great generals, athletes, artists and poets. This tendency led, from the 4th century BC, to proclaiming victorious generals or monarchs as gods. This seductive flattery continued into the Roman Empire. Average Athenians must have looked very different from the godlike beings, with perfectly proportioned bodies and features, depicted in the friezes on the Acropolis, but Greeks revered those more blessed than themselves. A Greek *polis* was composed of gods and semi-divine heroes as well as living men.

Counterbalancing such hero-worship was the strong Greek love of gossip and scandal. The Greeks loved to talk and exchange rumours in the *agora* (market place) in every *polis* and at leisurely *symposia* (dinner parties). At times they wrote down this gossip. This helps make the Greeks among the first true individuals in human history.

Left: Greek vases often depict aspects of everyday life, as in this scene of masters and pupils writing or playing the flute.

SOLON: THE GREAT REFORMER
c.640–558BC

The first Greek individuals emerge around 600BC, when myth begins to give way to more solid, if not always reliable, fact. Lycurgus, for example, legendary founder of Sparta's unique constitution, remains just that: an undatable legend. But with Greeks who lived possibly only a generation or two later, we know enough to gain a general picture. Solon of Athens was the most revered of the Seven Sages of Greece and Athens' first great statesman. He is also the first Greek about whom we can form an (almost) valid biographical picture.

Solon was a Eupatrid, from an old aristocratic family that traced its line back to the (legendary) kings of Athens. But he was not born to riches: his father, Exestecides, had spent much of his modest wealth on charitable works, according to the biographer Plutarch. Solon started life writing poetry – this was the Lyric Age of Greece, when people naturally wrote in verse – without overt political intentions, often on religious or philosophical matters. But he was soon caught up in public life amid increasing political crises. The old aristocracy still monopolized and often abused its almost feudal powers, causing tensions with other Athenians, be they the new rich or the newly pauperized.

The problem only worsened when Dracon made Athens' laws public in 622BC, for now all could see their full severity. Solon emphasized that his sympathies lay with the poor. "Often the wicked prosper while the righteous starve; Yet I would never swap my state for theirs, Or my virtue for their gold. For mine endures, while riches change everyday," he wrote. But he restored the family fortunes by trading as a merchant, then a perfectly respectable occupation, even for an aristocrat. (Early Greece was a land of inspired polymaths and amateurs.)

Above: Olive trees were important because olive oil became Attica's chief cash crop after Solon's reforms.

Below: This cup by the potter Exekias, showing the wine-god Dionysus in a vine-festooned ship, illustrates the age's general conviviality, when symposia (dinner parties) became widely established.

SOLON'S SOCIAL LEGISLATION

Solon allowed childless men to leave their property to whomever they liked, and legislated about the size of dowries to be given in marriage and about divorce. He introduced state-controlled brothels, whose transactions were taxed, while imposing restrictions on the public movement of free Athenian women. This cannot have helped their already low status.

SOLON'S GROWING REPUTATION

Solon first won fame for the conquest of Salamis, the island off the Attic coastline. He seems to have captured it from Megara more by guile than force, but the conquest was not permanent. His reputation grew further when he helped to unite Panhellenic opinion, represented by the 12 states in the Amphictyonic Council, against impious Crisa, which had tried to seize Delphi, seat of Apollo's oracle *c.*596BC. In 594BC the Athenians turned to Solon in desperation, for he alone might have the impartial wisdom to solve the problems of indebtedness and factionalism that threatened Athens with *stasis*, the civil strife that was the feared opposite of *eunomia* (social harmony).

REFORM, NOT REVOLUTION

Solon was no revolutionary, and he did not redistribute land as some expected. But nor did he make himself tyrant, as his friends had expected. Instead, his reforms aimed to avert tyranny, then common in Greece. He abolished enslavement for debt and the old aristocratic custom of taking one-sixth of every farmer's produce, and he freed Athenians sold into slavery abroad. He divided Athenian society into four classes based on wealth, not birth The top two classes alone remained eligible for public office and membership of the Areopagus, the highest council and court, but everyone could attend and vote in the Assembly or the *Heliaea*, the jury court now supplementing the Areopagus. This last privilege, Plutarch noted, was "at first worth very little but later became extremely important"

Right: After inaugurating his reforms, Solon went travelling for 10 years to allow his changes to settle down. Visiting Egypt, immemorially ancient to Greeks, he conversed with priests at Sais, learning the tale of the end of Atlantis. The philosopher Plato, Solon's descendant, later wrote down this story in Timaeus.

TRAVELS AND OLD AGE

Aware that he had annoyed almost as as many people of every sort as he had pleased, Solon ordained that his new laws, exhibited on revolving wooden blocks for all to see, should remain in force for a century. He then set off on his travels for ten years, hoping that in his absence Athens would settle down and his reforms take root.

In Egypt Solon debated with the priests of Sais and recorded their legend (or history) of Atlantis, the drowned superpower. Two centuries later Plato, one of his descendants, publicized the Atlantis story in his *Timaeus*.

According to tradition, Solon also visited King Croesus of Lydia but was totally unimpressed by the fabulous royal wealth he was shown (although this story seems implausible on grounds of chronology). Returning home, Solon warned against the tyrannical intentions of Pisistratus – in vain, for Pisistratus seized power in 561BC. The new tyrant must have retained some affection for the man who had reputedly once been his lover, however, as Solon was never arrested, dying at a ripe old age.

Above: No original portraits of Solon survive, but this fine copy of the Roman period suggests his mixture of shrewdness, farsightedness and nobility.

BENEFICIAL TYRANTS
PITTACUS, PISISTRATUS & POLYCRATES

Above: Sometimes called 'the philosopher-tyrant' because of his self-restraint and tact, Pittacus ruled Mytiline sagely for 10 years, before retiring not a penny richer.

'Tyrant' to Greeks in *c.*600BC was no more abusive than 'chief' is to us. The men who won and exercised extraordinary power could even sometimes be called 'beneficial tyrants'.

PITTACUS OF LESBOS, *c.*650–570BC
Pittacus was born in Mytilene on the island of Lesbos, then growing fast both economically and culturally. He may have been an impoverished aristocrat, although

Below: The magnificent Ionic columns of the Temple of Olympian Zeus in Athens date from the 2nd century AD, but the temple's immense ground plan was laid out in the 6th century BC by the Pisistratids, who built on a newly magnificent scale.

enemies later sneered that he was a commoner. As was so often the case, the nobility was oppressing the people economically while their dynastic feuding shook the city. Pittacus led Mytiline's forces to a great victory over the Athenians at Sigeum *c.*600BC.

Around 590BC Pittacus was chosen by the people as *aesymnetes* (supreme ruler), but only for ten years. Pittacus used his powers wisely, granting a general amnesty and ruling with moderation, if firmness, and encouraging trade. One of his most notable laws doubled the punishment for crimes committed when drunk. He was not, however, draconian. Although he had to banish both Alcaeus and Sappho, the greatest poets of the age, for their intrigues, they later returned. ("Forgiveness is better than repentance," was a saying attributed to him.) Like Solon, he wrote in rather good verse. After ten years, he resigned all power and retired to private life no richer, to general surprise. He was considered one of the Seven Sages of Greece.

PISISTRATUS OF ATHENS,
***c.*600–527BC**
Pisistratus, descended from the legendary kings of Pylos, won fame as *polemarch* (war leader) for capturing the island of Salamis from Megara *c.*565BC. He was so popular that he started a new party called the Hill (there were already Plain and Coast parties). After appearing one day in the Assembly, apparently wounded by enemies, he was voted a bodyguard of 50 men – armed with clubs, not spears. With it, he seized the Acropolis as tyrant in 560BC. According to Herodotus, he "governed the country in an excellent way", but opponents drove him out in 556BC.

By allying with Megacles, head of the Alcmaeonids, he staged a comeback by an absurd ruse: he persuaded a

PATRON OF THE ARTS

Like most tyrants, Pisistratus built to impress. He constructed a fine new *agora* with marble fountains and began an immense temple to Zeus. Under Pisistratid rule, tragedy was born on the Athenian stage, differentiating itself from the ancient 'goat dance'. Pisistratus also had Homer's works edited and given public readings.

tall young woman to dress up as Pallas Athene, the city's goddess, complete with spear and helmet, and enter Athens in a chariot acclaiming Pisistratus. ("The silliest trick in history", wrote Herodotus witheringly.) Pisistratus quarrelled with the Alcmaeonids too, going back into exile. He then established a base on the Strymon in north Greece and allied with Thebes and Argos. With their backing, he finally returned to power in 546BC.

Once established, Pisistratus, far from acting despotically, mellowed. He even dispensed with his bodyguard, walking around like any citizen. He could do this because life in Athens during the next 30 years improved all around. Solon's agrarian reforms, especially those encouraging the export of olive oil, were bearing fruit. Both farmers, and the craftsmen who made the increasingly splendid vases for the oil, benefited. Pisistratus also encouraged Solon's constitution, tactfully manipulating it to ease supporters into major posts while the Assembly slowly gained confidence.

If Pisistratus had the common touch, chatting easily with farmers, he also rebuilt links with some nobles, including Miltiades, head of the Philiads, who went off to establish a fort on the distant Hellespont (Dardanelles). This helped to safeguard grain imports. His sons Hippias and Hipparchus succeeded after his peaceful death in 527BC. They continued his policies until 514BC, when the assassination of Hipparchus induced paranoia in Hippias.

POLYCRATES OF SAMOS, REIGNED 540–523BC

Polycrates seized control of the island of Samos *c.*540BC, aided by his two brothers. Unfraternally, he killed or exiled them to rule alone, but he proved a popular ruler thanks to his massive public works. He built a huge aqueduct, tunnelled through rock for over 1.6km/1 mile; a grand temple to the goddess Hera, stupendous fortifications and a vast harbour mole. Aristotle, exaggerating for once, compared his projects to the pyramids.

Polycrates patronized poets such as Anacreon, while ruling the seas with a fleet of 100 galleys. He paid for this splendour partly by piracy – his ships raided friends and enemies alike, but he reputedly returned goods to the former – but chiefly through an alliance with Amasis, Pharaoh of Egypt. Amasis hired the Samian fleet to defend Egypt against Persia. But when Persia was about to attack Egypt, Polycrates changed sides sending ships to help the Persians.

This did him little good, for he was tricked out of his island fortress by an invitation to Sardis. There the Persian satrap had him crucified – a form of execution the Persians had pioneered but that was later adopted by other peoples.

Above: A 16th-century painting by Giovanni Fedini shows Polycrates of Samos. Growing alarmed that his good luck tempted fate, Polycrates threw his best ring into the sea to appease the gods. But a fisherman catching an unusually fine fish, offered it to the tyrant. Inside it was the ring. Polycrates realized he was doomed.

Left: Noblest of the many grand projects that Polycrates, the tyrant of Samos undertook was the Heraion, the vast new temple to Hera, Samos' patron goddess.

DEMOCRACY'S CHAMPIONS
CLEISTHENES AND THEMISTOCLES

Cleisthenes was, according to Herodotus, the founder of "democracy for the Athenians", while Themistocles saved Athens from Persian conquest. These two statesmen played a major part in the development of Athens.

CLEISTHENES: FATHER OF DEMOCRACY, *c*.570–500BC

Cleisthenes' career started in typically aristocratic factionalism, for he was an Alcmaeonid. But it ended most atypically in democratic revolution. He became chief *archon* in 525BC, when the tyrants Hippias and Hipparchus were letting other aristocrats share office. Later, he quarrelled with the Pisistratids and went into exile. He used his wealth for an abortive invasion of Attica and, more fruitfully, to win over the Delphic Oracle. This persuaded Sparta, once friendly to the Pisistratids, to expel Hippias in 510BC.

King Cleomenes and his Athenian ally Isagoras expected a return to old-style aristocratic politics, and Isagoras was elected *archon* in 508BC. At this stage Cleisthenes, refusing to play the old game, "added the *demos* (people) to his faction", in Aristotle's words. This means that he got them on his side, but his real intentions remain debated. His radical constitutional reforms created ten entirely new *phylae* (tribes), each with a *trittyes* (third) representing all three parties: Coast, Hill and City [Plain earlier]. This almost eliminated local aristocrats' hold on politics. Other reforms to the Council, and the introduction of ostracism (to banish uppity politicians), confirmed the new democracy. Athenians clearly liked it. When Isagoras returned to Athens with Spartan support and banished 700 opponents, they felt confident enough to rise against him, restoring their democracy, of which Cleisthenes was the perhaps unwitting father.

THEMISTOCLES: SAVIOUR OF ATHENS, *c*.525–459BC

Themistocles was an outsider in Athenian politics, having a non-Athenian mother and a father "of no particular mark" (in Plutarch's words). But, since he saved Athens from Persian conquest, he ranks with the very greatest Greek statesmen and generals.

Themistocles was intensely ambitious, although his father tried to warn him off politics. (Pointing to some rotting hulks on the beach, he said that *that* was how Athens treated discarded politicians.) Themistocles had no support from aristocratic networks (he did not even belong to a smart gymnasium), but in 494BC he was elected *archon*. He had chosen to live in the rundown Ceramicus (potters') area, where he got the ears and votes of poorer citizens. As news of Persia's suppression of the Ionian revolt crossed the Aegean, filling Athens

Above: The determination and ambition of Themistocles, the radical democrat and saviour of Athens in the Persian Wars, is evident in this bust.

Below: Politics in Athens was still largely a competitive sport between rival aristocrats, who might meet at friendly symposia, *until Cleisthenes changed the rules of the game.*

with dread – it had helped Ionia burn down Sardis in 498BC – Themistocles urged the fortification of Piraeus. Until then the beach of Phalerum had been Athens' port, but it was perilously exposed. By contrast, Piraeus was on a rocky peninsula with three natural harbours. Work started on its walls and docks but, while Themistocles saw Athens' future as a seapower, few Athenians yet agreed.

In 490BC Themistocles was one of the *strategoi* (generals) at the Battle of Marathon, but Miltiades masterminded that hoplite victory. After Miltiades' death, Themistocles emerged as leader of the anti-Persian radicals. The Assembly began flexing its democratic muscles, using Cleisthenes' novelty: ostracism. Meanwhile, war with Aegina, the powerful nearby island, revealed Athenian weakness at sea. As his noble opponents were ostracized, Themistocles faced one last formidable rival: Aristides the Just. Aristides was a conservative famed for exceptional honesty, but it was Themistocles who saw the growing danger from Persia. When in 483BC a new vein of silver was found at the Laurium mines, Themistocles proposed to use the wealth from this to build a fleet of 200 triremes, not distribute it as a windfall. Aristides, losing this argument, was ostracized in 482BC. But by 481BC all Greece was alarmed at the Persian threat.

SPINNING THE PROPHECIES

At the last debate in the Assembly before war in 480BC, Themistocles spun Delphi's gloomy prophecies to show that Athenians should use the fleet to win at Salamis, not flee to the west or retreat to the Acropolis. In the long negotiations with the Peloponnesians, Themistocles often swallowed his pride. He even accepted Spartan command of the allied fleet, although Athens contributed most ships. Finally, with the Greek fleet stationed at Salamis, Persian armies thronging the shore and Persian ships approaching the island, Themistocles' trick of a secret

Above: A griffin from Susa, the administrative capital of the Persian Empire, whose growing power overshadowed Greek life for decades until countered at Salamis and Plataea.

message to the Great King persuaded Xerxes to dispatch his fleet. Once lured inside the Salamis straits, it was crushed.

Themistocles showed the same cunning when, after Persia's defeat, Sparta protested at Athens rebuilding its city walls. Going to Sparta, he prevaricated masterfully, persuading Sparta to send envoys to Athens, who then became hostages until the walls were completed. Despite this, Themistocles was ostracized in 470BC, partly because of his boastfulness. Fleeing Greece, where he had many enemies, he became governor of Magnesia, a Greek city *inside* the Persian Empire. But he reputedly killed himself rather than obey Persian orders to fight Athens, for he was no traitor. Thucydides, no admirer of radicals, summed up Themistocles: "He showed indisputable, quite exceptional natural genius... he was unrivalled at doing exactly the right thing at exactly the right moment."

PERICLES: THE SUPREME DEMOCRAT c.495–429BC

Above: Aspasia, Pericles' mistress for many years, attended symposia *and talked to philosophers, which was most unusual for any woman in Athens. Highly intelligent, she bore Pericles a son who after Pericles' death was legitimized as an Athenian citizen, also a rare honour.*

Few elected politicians give their name to an age. None can match Pericles, whose name is synonymous with Athens' most brilliant decades. An Alcmaeonid by birth (he was Cleisthenes' great-nephew), Pericles was a visionary democrat by conviction, dominating Athenian politics for 30 years (460–429BC). Abraham Lincoln modelled his Gettysburg address on Pericles' funeral oration of 431BC.

Pericles entered public life by sponsoring Aeschylus' *The Persians* in 472BC. Unusually, this play dealt with recent events, praising Athenian democracy and its role in the Persian Wars. Both were policies of Themistocles, who was ostracized soon after. Pericles then joined forces with Ephialtes, a noted radical. In 462BC they persuaded the Assembly to transfer most remaining powers from the Areopagus to the Assembly or *Heliaea* (jury court). This marked Athens' transition to full democracy. After

ANAXAGORAS

Pericles spoke only rarely in public so that people did not grow tired of his voice. But when he did, he was the greatest orator of his day, with a "nobility in his speech utterly free of vulgar mob-oratory", according to Plutarch. Pericles had learned his skills partly from Anaxagoras of Clazomenae. This philosopher so impressed contemporaries that he was nicknamed 'Brain Personified'. Anaxagoras questioned conventional wisdoms. Solar eclipses, for example, were natural phenomena, not signs from the gods. This was intellectually too daring even for Athens, however, and Anaxagoras had to flee the city in 428BC, accused of impiety.

Ephialtes' murder by enraged opponents, Pericles became the radicals' leader, though still only in his thirties.

Pericles introduced modest payment for jurors and Council members – just enough to ensure that poverty stopped no Athenian from attending them. During subsequent wars he proved a competent *strategos*. After the Peace of Callias in 449BC ended the Persian Wars, Pericles urged that surplus public money be used to rebuild the temples on the Acropolis, whose blackened ruins had been left as sharp reminders of the Persian occupation. But this was a controversial use of tribute from allied states, and some Athenians, led by Thucydides son of Melesias, attacked him for it. Pericles, who exalted Athens' role as the educator of Greece, won the

Left: The Temple of Athena Nike (Victorious), although finished in 425BC after Pericles' death, derived from his vision of Athens as the 'school of Hellas', a beacon of democracy.

argument, and Thucydides was later ostracized. Athens under Pericles was, Thucydides the historian wrote, "ostensibly a democracy but actually ruled by one man". But, despite his "high note of aristocratic, even regal leadership" in Plutarch's words, Pericles remained accountable to the Assembly.

ARCHITECTURAL PATRON

Pericles' role in the building of the Parthenon, one of the world's most sublime buildings, alone would ensure his fame. He arranged for Ictinus and Callicrates to design the temple and Pheidias its superb frieze. He saw the project through to swift completion, gaining a name for financial probity, then rare. He also oversaw the building of the Odeon, a covered theatre where plays were previewed, and encouraged Herodotus to read aloud from his *Histories*, the world's very first.

WAR AIMS

Pericles was an imperialist, crushing the revolt of Samos in 440–439BC, for example. He was elected *strategos* every year from 443 to 430BC, heading the Board of Generals with increasingly

professional ease. Pericles probably felt that the Athenian Empire benefited all Greece, sheltering it from Persian power, encouraging democracy and, by keeping down piracy, boosting trade. Certainly Athens thrived, with Piraeus becoming the eastern Mediterranean's greatest port. But such open imperialism created many enemies. In 433BC Pericles claimed that he could see war "bearing down from the Peloponnese". How far his policies provoked that war remains debatable.

In the Peloponnesian War that started in 431BC, Pericles' overall strategy – retreat behind the Long Walls, let the Spartans ravage Attica, but maintain naval superiority – was approved by the Assembly. When plague hit the city in 430BC, however, the people turned on him and falsely fined him for speculation. He was reinstated only just before his death from after-effects of the plague in autumn 429BC.

Pericles married and divorced when young. Divorce was common in Athens, but it was customary to marry again. Pericles did not, preferring the company of his mistress Aspasia; he was exclusively heterosexual, unlike many Greeks. He cultivated a lofty dignity in both public and private life, restricting his social life as he grew busier. The poet Ion of Chios found him "insolent and conceited", but perhaps caught him on a bad day. He is always portrayed with a helmet because, his enemies said, he wanted to hide his odd-shaped head. He seems to have had no other obvious vices. His deathbed statement that "no Athenian ever put on mourning because of me" (i.e. his policies produced no needless casualties) may be questionable, but his place as Athens' greatest leader is not.

Left: Pericles' vision of a democracy that raises its citizens up, rather than dumbs them down, inspired later democrats such as American President Abraham Lincoln, who modelled his Gettysburg Address on Pericles' famous Funeral Oration of 431BC.

Below: The archetypal picture of Pericles in his helmet, which he would have worn as strategos (elected general). He often wore it on other occasions too – to hide the odd shape of his head, according to his enemies.

DEMOCRATS IN DEFEAT
ALCIBIADES AND DEMOSTHENES

Below: Demosthenes, orator, politician and attempted saviour of Athenian independence from Macedonia's rising power, is shown here with a scroll.

Although no two men were less alike than Alcibiades and Demosthenes, both ended their lives in exile and failure after early successes.

ALCIBIADES, *c.*450–404BC
The last Alcmaeonid to lead Athens, and almost the last aristocrat in politics, Alcibiades had a dangerous Byronic glamour. Related to Pericles, whose ward he became on his father's death in 447BC, he was notably handsome and self-willed. As a boy, when defeated in a wrestling match, he sank his teeth into his opponent's arm. Accused of fighting like a girl, he replied that he was biting like a lion. Throughout his life, he ignored all the rules.

COMMAND AND DESERTION
Alcibiades met Socrates when they were fellow soldiers at Potidaea in 430BC. The philosopher saved Alcibiades' life – a favour Alcibiades returned six years later at Delium – and tried to guide the gifted young man towards philosophical virtue. This was in vain, for Alcibiades, while flattered by Socrates' attentions (in Plato's *Symposium* he is shown trying to seduce the older man), was interested in fame and power. He cut off the tail of his dog so that everyone would comment on its state, and trailed a long purple cloak around the Agora. But when his chariot won first prize in the Olympic Games, and two others belonging to him came second *and* fourth, he could claim a truly Homeric victory. Athens loved him for it.

Standing for election as *strategos* in 420BC, Alcibiades faced little competition from Nicias, the timid conservative leader, or Hyperbolus, a radical and noted windbag. As *strategos*, Alcibiades created an anti-Spartan alliance in the Peloponnese. Not re-elected – which helped to lead to allied defeat at Mantinea – he looked elsewhere for glory, notably to Sicily, which

Below: Alcibiades enjoyed a strange second career as an adviser in Sparta after being exiled. Sparta, although notoriously austere and lacking appeal for a sybaritic nobleman, boasted some fine temples, as these ruins reveal.

was alluringly rich. Alcibiades urged an expedition there, finally with success. But he had made enemies by his arrogance and his scandalous friends who mocked the city's gods. When the herms (good-luck statues set around Athens) were found mutilated, he was accused of sacrilege. Denied a chance to clear his name, he sailed with the Syracusan expedition in 415BC but was recalled to face fresh charges. Escaping his captors, he reached Sparta. There he charmed his hosts and gave them excellent advice: invade Attica again and occupy the fort of Deceleia permanently. This devastated Athenian farming and mining.

Alcibiades' lack of patriotism was common among Greeks. But when he seduced the wife of King Agis, he had to leave Sparta fast. He crossed to Asia and persuaded Tissaphernes, a Persian satrap, that he could make Athens Persia's ally. The Athenians now wanted him back. As Aristophanes wrote, "they long for him, hate him and cannot do without him". Elected *strategos* by the fleet at Samos, he recaptured important cities on the Bosphorus in 408BC. He returned to Athens in seeming triumph in 407BC, but, when his helmsman Antiochus lost a pointless battle, he slipped away to his Hellespontine fort in disgrace. From it he rode out to warn the Athenian fleet about its dangerous position at Aegospotami – in vain. He was killed by assassins in Persian pay, who surprised him in bed with his last mistress, Timandra. She buried her lover, the last Alcmaeonid, in a foreign field.

DEMOSTHENES, 384–322BC

The finest of all Greek orators, Demosthenes was the last great Athenian democrat, who battled to save his city from foreign domination. Although he ultimately failed, his was a noble failure.

Orphaned when young, Demosthenes saw his guardians steal his inheritance and had to fight to regain it. A bookish, unathletic boy, he cured a stammer to become a superb professional speech-writer, the equivalent today of a barrister. Athens remained the foremost

Greek naval state, with interests around the Aegean. This led her to clash with Macedonia, rising fast under its ambitious new king Philip II. Demosthenes, realizing that Athens' independence could not survive Macedonian hegemony, became leader of the anti-appeasement party in Athens. To rouse his countrymen, who preferred to employ mercenaries rather than fight themselves, he made four fiery speeches: the *Philippics* ('against Philip'). In these, he urged resistance to Philip's promises and threats.

Demosthenes' great political achievement was the alliance with Thebes, Athens' old rival, against Macedonia in 338BC. He fought as a citizen-hoplite in the Battle of Chaeronea, but this proved Philip's greatest victory. Philip, who wanted Athens' fleet for his planned war on Persia, dealt with Athens gently. Later, while Alexander was conquering Asia, Demosthenes defeated Aeschines, his main Athenian enemy in a brilliant speech, *De Corona* (On the Crown), defending his right to honours for services to his city. After Athenian democracy was finally crushed by Macedonia in 322BC, Demosthenes killed himself rather than fall into Macedonian hands.

Above: Alcibiades counted among his friends the penniless philosopher Socrates, who had saved his life in battle in 430BC. Socrates discerned signs of greatness in the self-indulgent aristocrat, but Alcibiades disappointed his teacher by preferring worldly fame to areté (virtue, excellence).

Below: Demosthenes was a superbly rousing orator, but his words proved useless against Macedonian revolt.

Above: A fanciful medieval portrayal of Artemisia, the fighting queen of Halicarnassus.

EXCEPTIONAL GREEK WOMEN
ARTEMISIA, OLYMPIAS AND ASPASIA

Women in classical Greece proper played no role in public life. But in Ionia and Macedonia, where they had more freedom than in Athens, some remarkable women emerged.

ARTEMISIA OF HALICARNASSUS, ACTIVE *c.*480BC

Artemisia, queen of Halicarnassus (today Bodrum) in Caria, was an exceptional queen. She took part in Xerxes' invasion of Greece, giving him unusually forthright advice. Her father, King Lygdamis, was Carian, but her mother was a Cretan aristocrat. Widowed young, Artemisia stepped into her unknown husband's royal shoes and ruled the city even after her son grew up.

She supplied five galleys to the Persian fleet in 480BC – Herodotus thought them the best ships after Sidon's – commanding them in battles off Euboea. When Xerxes held a council of war before the battle of Salamis, she alone advised against attacking the Greek fleet, saying that the Greeks could not stay long on the small, crowded island. Her advice was not taken and the Persians advanced to their defeat. In their rout, Artemisia's galley, chased by an Athenian trireme, rammed a Calyndian ship that was on her side but in her way. The Athenian trireme, thinking that Artemisia's ship was an ally, abandoned its pursuit. Artemisia saved her life by this quick thinking and gained kudos with Xerxes, watching from his shoreside throne. Thinking she had actually rammed a Greek ship, the King exclaimed: "My men are fighting like women and my women like men!"

ASPASIA, MISTRESS OF PERICLES

Aspasia came from Miletus, the most sophisticated city in the Greek world until its sack by the Persians in 493BC.

Like many adventurous Ionians after 480BC, she migrated west to Athens in search of her fortune. She became Pericles' mistress at about the time he became Athens' effective leader *c.*450BC, their relationship lasting until his death in 429BC.

She must have been highly intelligent as well as attractive, for Pericles' circle included the greatest minds of the age. Reputedly, she joined in discussions with the philosophers Protagoras and Anaxagoras, unlike a normal Athenian wife. (Significantly, we do not even know the name of Pericles' wife, whom he divorced.)

Aspasia's son Pericles, unlike Pericles' legitimate sons, inherited some of his father's talents as well as his name. Aspasia must also have helped to bring up Alcibiades, Pericles' brilliant but troublesome ward.

Right: This Roman-era portrait bust of Aspasia, Pericles' mistress, captures the intelligence, independence and noble beauty of the great statesman's lover.

Her relationship with Pericles fed slanderous gossip. He was depicted on stage as being in her power, although this was clearly not true. She was, however, caught up in the attacks on Pericles after plague broke out in 430BC and accused of impiety, according to Plutarch. After Pericles' death, however, her son Pericles was legitimized and granted Athenian citizenship, which was a rare honour. Aspasia then married Lysicles, a wealthy politician. She may have had the luck to die before 406BC, when her son was one of the six *strategoi* unjustly condemned to death for abandoning drowning sailors after the Battle of Arginusae.

OLYMPIAS, MOTHER OF ALEXANDER, *c.*364–316BC

Even wilder than Macedonia, Epirus, on the north-western edge of Greece, was ruled by kings who claimed descent from Achilles, greatest of Homeric heroes. Olympias, daughter of King Neoptolemus, reputedly met Philip of Macedonia at the mysteries (initiation into the religious ceremonies) on the island of Samothrace and fell in love. More probably their marriage in 357BC was arranged for political motives, as Philip covered his back (literally) with marital ties.

The birth of Alexander a year later produced the essential male heir, but the marriage was unhappy. Olympias was a tempestuous character, whose wrath could rival that of Achilles. Often ignored by Philip – who had mistresses, boyfriends and six other wives – Olympias turned to religion. She took part in the Bacchic mysteries, kept a sacred snake in her bed and sacrificed thousands of animals. She helped to choose her son's first two tutors and probably encouraged his belief that a god, not Philip, was his real father. Alexander inherited her fiery temper and supported her when she left Macedonia, enraged by Philip's last marriage to Eurydice. But he was reconciled with Philip, while she was divorced by him, returning only after Philip's murder in 336BC.

Olympias honoured the corpse of the assassin Pausanias, which suggests complicity in the murder, and had Eurydice killed. Alexander left her as queen in Macedonia when he crossed to Asia, but she quarrelled repeatedly with Antipater, his general. Hearing this, Alexander exclaimed that his mother charged him dearly for nine-months' lodging in her womb. In the long wars after Alexander's death, she retreated to Epirus, then returned to kill the half-witted King Philip III in 317BC. Besieged in Pydna herself a year later, she was starved into surrender. Macedonian soldiers sent to execute her refused "out of respect for her royal rank", so relatives of her victims finally killed her.

Left: Olympias wife of Philip II of Macedonia and mother of Alexander the Great, was a passionate, fiery woman who frequently quarrelled with her husband. After her son's death she played a major role in Macedonian dynastic politics.

Below: Although classical Athens was an intensely masculine society, the city revered a female deity, Athena Parthenos, the warrior-goddess whose cult image was carried aloft in every Panathenaic Festival. This statue is a copy of Pheidias' lost masterpiece.

EXTRAORDINARY SPARTANS
CLEOMONES AND BRASIDAS

Above: The agora of the ancient city of Sparta, home of one of the most absolute military regimes the world has ever known.

Below: David's famous depiction in 1814 of Leonidas, who with his Spartans fought the Persians to the last man at Thermopylae in 480BC, typifying Spartan bravery and military competence. But a more imaginative general such as Cleomones or Brasidas with 7,000 allied troops might have held the pass indefinitely.

Although the Spartan system discouraged individuality, some kings or generals escaped the general levelling, if at considerable personal risk.

CLEOMONES I: SPARTA'S DYNAMIC KING, REIGNED *c*.520–*c*.490BC

Even Cleomones' birth was controversial. His father, King Anaxandridas, had taken a second wife after his first failed to have children, but he kept her too. His second wife duly gave birth to Cleomones, but other boys were born soon after to the king's first wife.

Cleomones, succeeding to the throne *c*.520BC, soon showed unSpartan cunning. The tiny city of Plataea in 519BC asked for Sparta's protection against her powerful neighbour Thebes. Cleomones suggested that Athens, much closer, would be a better ally, thus embroiling Athens, still friendly to Sparta, with Thebes. The two cities' resulting enmity suited Sparta. In 510BC Cleomones led a Spartan army to expel Hippias from Athens, for the Delphic Oracle had urged the Spartans to free the city. Cleomones probably knew that Cleisthenes, the Alcmaeonid, had bribed the Oracle, but he did not expect the subsequent democratic revolution. Marching back to Athens to support his aristocratic friend Isagoras, Cleomones was besieged on the Acropolis and had to bargain with the Athenian *demos* for his freedom – a humiliation for a Spartan king, which deepened when his avenging invasion of Attica collapsed in 506BC due to divisions in the army.

Cleomones' position after such a defeat may have been shaky in Sparta – he had envious half-brothers – but he ignored Ionian appeals for help against Persia in order to concentrate on Sparta's real enemy: Argos.

In a brilliant ruse in 494BC, Cleomones attacked the Argives during lunch, when they were not expecting it. After the defeated Argives took shelter in a sacred wood, he had it burnt down, killing 6,000 and crippling Argos for a generation. Cleomones may not have helped Ionia, but he still pursued a vigorous anti-Persian policy. The Athenian defeat of Persia at Marathon was partly due to his pressure on Athens' enemy Aegina to stop it 'medizing' (collaborating with Persia).

Cleomones fell from power *c*.490BC in mysterious circumstances after his part in bribing the Delphic Oracle to help depose Demaratus, his fellow-king, had been exposed. This encouraged his radical ideas about becoming sole king of a broader-based state than Sparta. For support he appealed to the Arcadians, Spartan allies in the Peloponnese. His scheme failed, he was deposed and declared mad. Locked up, he was said to have committed suicide, but was almost certainly killed. Sparta thus lost an unusually dynamic king.

BRASIDAS: A SPARTAN
BY MISTAKE, DIED 422BC

Brasidas had so many non-Spartan qual-
ities – flexibility, diplomacy, eloquence
– that he should really be considered a
Spartan by mistake. An officer's son, he
grew up brave and tough but also able to
think for himself, lacking the characteristic
Spartan arrogance. Brasidas came to prom-
inence in 431BC when he rescued Methone
in the Peloponnese from an Athenian
sea-borne attack with only 100 hoplites. He
received an official award for this. He was
less successful in 429BC in western Greece,
for the Athenian fleet was vastly superior to
the Peloponnesian.

But he shone again at the Spartan
attack on Pylos in 425BC, when he urged
on hesitant Spartans landing on the rocky
shore. Badly wounded disembarking, he
took no further part in what became
Sparta's most shameful defeat.

His real opportunity came in 424BC,
when Perdiccas of Macedonia and some
northern cities appealed to Sparta for
aid against Athens, wanting Brasidas
as the general. Sparta would not risk
Spartiates (soldiers from the ruling class)
so far north but let Brasidas recruit 700
helots (slaves), whom he armed as hoplites.
Picking up 1,000 allied mercenaries as he
went, he saved Megara from Athenian
attack. Marching swiftly through a hos-
tile Thessaly, he reached Acanthus in the
Chalcidice. This, like many Greek cities,
was divided. Most Acanthians were content
as Athens' allies, but they let Brasidas talk
to them. He was so persuasive, promising
to respect their liberties, that they went
over to Sparta's side. Thucydides noted
they were also worried about their grape
harvest, with Brasidas' troops camped in
their vineyards. Two other cities, Stagira
and Argilus, joined on similar terms.

Then Brasidas had his greatest suc-
cess, marching through the winter night
to capture the unguarded bridge over the
River Strymon. Beyond it lay Amphipolis,
hugely important to Athens. Here again
Brasidas triumphed by his oratory,
offering such easy terms that the city

surrendered, unaware that Athenian
strategos Thucydides was belatedly hurry-
ing to their rescue. History's gain was
Athens' loss, for she never recaptured
Amphipolis. Brasidas had further successes
through combined diplomacy and gener-
alship, skilfully keeping his mixed army of
mercenaries and ex-*helots* in fighting form.
He was killed in battle at Amphipolis in
422BC leading a charge against Athenians
attacking the city. With him died
Sparta's most remarkable general.

*Above: These hoplites and
chariot from Laconia, c.500BC,
show characteristic Spartan
toughness. Such an upbringing
created brave foot soldiers, not
adventurous generals, who
remained a rarity.*

*Below: Sparta occupied the
fertile Eurotas Valley beneath
Mt Taygetus, some of the
richest farmland in Greece.*

CONTRASTING GENERALS
CIMON AND EPAMINONDAS

While Cimon of Athens was a bluff aristo-crat but an excellent soldier, Epaminondas of Thebes was a general of genius who briefly made his city hegemon of Greece.

CIMON OF ATHENS, *c.*508–449BC

Cimon was the son of Miltiades, architect of the victory at Marathon in 490BC, but his mother was descended from Thracian kings. Cimon lacked "any spark of Attic cleverness and eloquence" said Plutarch, yet gave "an impression of great nobil-ity and candour". Miltiades had died in disgrace and debt, so Cimon's youth was impoverished. He long lived with his sister Elpinice, which fed malicious rumours of incest, until the rich Callias married her for love and paid off their debts. Cimon himself often fell in love with aristocratic women, including Isodice from the rival Alcmaeonids. He was also fond of drink and kept open house on his estate, where all were welcome.

In 480BC, as Persian forces approached Athens, Cimon led the young knights up the Acropolis to dedicate their bri-dles to Athena before taking up spears and shields to serve as marines, a timely gesture. He fought bravely at Salamis and with Aristides commanded the Athenian fleet in 478BC against the Persians. The allied commander was the Spartan Pausanias, but he behaved so outrageously – raping freeborn girls, dressing in Persian robes – that the Greeks turned to Athens for leadership. Cimon became com-mander of the Delian League fleet, often re-elected *strategos* and capturing Persian strongholds. He also retrieved the supposed bones of Theseus, Athens' legendary king, from Scyros. His finest triumph came in 467BC, when he led a fleet to defeat the Persians at the River Eurymedon, destroying 200 ships and its army. By then, with Themistocles ostracized, Cimon appeared supreme in Athens, but city politics was fickle.

In 464BC an earthquake hit Sparta, triggering a *helot* rising. In despair, Sparta appealed to Athens, still an ally, for help. Cimon led 4,000 hoplites into Messenia, but they were rudely dismissed by the Spartans, who distrusted democrats. Back in Athens, radicals exploited Cimon's absence to introduce political reforms. For opposing these, he was ostracized. When war broke out between Athens and the Peloponnesians, Cimon volun-teered to fight as an ordinary hoplite at Tanagra in 457BC – in vain. After his ostracism ended in 451BC, however, he

Above: Miltiades, who masterminded Athenian victory at Marathon over Persia in 490BC, was the father of Cimon, another great general.

Left: A 19th-century recreation of the sea Battle of Salamis in 480BC, in which Cimon took part.

commanded the fleet that sailed against Cyprus. There he died, but his body was brought back in honour to Athens.

EPAMINONDAS OF THEBES
c.418–362BC

Epaminondas was Thebes' greatest general and statesman, who raised it briefly to hegemony in Greece and freed the *helots* of Messenia from Spartan oppression. The son of an impoverished noble, he studied under Lysis, the last Pythagorean philosopher. Epaminondas always led a life of almost ascetic poverty, refusing all gifts and bribes.

Thebes had been Sparta's keen ally against Athens in the Peloponnesian War but Spartan postwar arrogance drove it to ally with Athens in 395BC. In 382BC a Spartan coup installed a corrupt junta backed by a Spartan garrison in the Cadmaea, Thebes' citadel. Anti-Spartan feeling in the city grew until in 378BC a group of exiles returned. Disguised as prostitutes, they assassinated the junta's leaders and declared a democracy. Epaminondas and his friend Pelopidas now radically reorganized the Theban army. Instead of the usual eight-deep line of hoplites in the phalanx, with the best soldiers on the right, they raised the number to 50 deep on the left wing and trained it to attack at an angle. The core of this new army was the crack Sacred Band, 150 pairs of homosexual lovers bound by love and honour. Even Sparta's militarized pederasty had not produced this, but Epaminondas himself was unusual in never marrying and having only male lovers.

The new-model army proved its worth by crushing the Spartans at the Battle of Leuctra in 371BC, killing 400 Spartiates, a feat that astonished Greece. Soon the Thebans invaded the Peloponnese, liberating the Messenians from centuries of slavery. Epaminondas now revealed political as well as military genius. Messene was refounded as the *polis* of free Messenia, with imposing fortifications, and Megalopolis (big city) was founded

as a new capital for Arcadia, blocking Sparta's easiest egress north. Both cities were moderate democracies heading federations, like Thebes. Deprived of its *helot* serfs, Sparta sank forever to second-rank status. But Epamonindas could not establish a lasting peace. Athens, alarmed at Thebes' new power, allied with Sparta, and war continued. Epamonindas died in 362BC at the Battle of Mantinea fought against a mixed alliance. Thebes' brief hegemony was ended, but he deserved the inscription on his statue that, through his efforts and vision, "Greece was free" at least in part.

Above: The Apadana Staircase in Persepolis, capital of the giant empire that Cimon grew up trained to fight. Temperamentally pro-Spartan, he thought Athens should solely attack Persia.

Below: Thebes' heart was the Cadmaea, a citadel dating back to the Bronze Age, in which these remnants of Mycenaean walls survive.

THE FIRST HISTORIANS
HERODOTUS AND THUCYDIDES

Although there had long been royal and
religious chroniclers, analytical history
starts with the Greeks. The term *historia*
(inquiry or research) was actually first
used by Herodotus.

HERODOTUS, 'FATHER OF HISTORY', c.490–425BC

Herodotus came from Halicarnassus
(today Bodrum), an Ionian city with
a strong Carian element. It supplied
ships for the Persian fleet in 480BC but
later joined the League of Delos, so it
stood midway between Asia and Greece
culturally and politically. This was appos-
ite, for Herodotus, almost alone among
Greek writers, viewed non-Greeks with
an inquisitive, sympathetic eye, free of
racial prejudice. Probably due to political
troubles, he moved to Athens c.446BC,
where he recited part of his *Histories* in
public, then a common practice. He was
paid 10 talents at Pericles' instigation for
this, a large sum. He probably ended his
days in Thurii in southern Italy.

Little else is known of him, but, through
his writings, the reader gets to know a
genial, intelligent, often discursive but
never boring man. For Herodotus was not
only 'the father of history' but an insati-
ably curious polymath. His researches
led him to visit Egypt, Babylonia, the
Black Sea and other areas, collating tales,
legends and often surprisingly accurate
facts, geographical, cultural and histor-
ical, about the ancient world. These make
up the long preamble to his grand theme:
the Great War between the Greeks and
Persians of 499–478BC.

Herodotus was often, in the best sense,
non-judgemental. He records different
versions of an event, leaving the reader
to decide between them. He can be
over-credulous, as when describing the
gold-digging ants of India, but he can also
be sceptical – sometimes unduly so, for
he dismissed the reports of Phoenician
sailors circumnavigating Africa from
east to west that the sun shone on them
from *the north*. He was unfair to some
individuals such as Themistocles, who
appears only late in his account of the

Above: The probing intelligence of Thucydides, greatest of Greek historians, emerges in this Roman copy of an original Greek bust.

Persian invasion of 480BC. This probably reflects the bias of his sources, which he may not have fully realized. Herodotus knew little about military matters, never having been a soldier. He also had no idea of numbers, giving the total figure for Xerxes' invasion force as 1,700,000 men – absurdly large. But he wrote in clear Ionic Greek, being one of the first great prose writers.

THUCYDIDES, *c.*457–400BC

Little more than a generation younger than Herodotus, Thucydides was very different as a man and as a writer. While the supernatural still figures in Herodotus' picturesque accounts, Thucydides ignores dreams and omens. He may have read or heard Herodotus' work but probably regarded it as more of a muddle than a model. His great theme was the calamitous war between the Peloponnesians and Athens, which he was uniquely qualified to record. "I made it a principle not to write down the first story I heard, nor even to accept my own overall impressions. Either I myself witnessed the events described or I heard them from eye-witnesses whose reports I checked as carefully as possible," he wrote. He has been called the 'historian's historian' because his analytical approach and sparse prose can often seem dry compared to other writers, but underneath boil surprising passions.

Thucydides was an aristocrat, related to Cimon and to the Thucydides who was ostracized for opposing spending tribute money on the temples. He grew up a conservative democrat but keenly admired Pericles, whose radical days were over by then. Thucydides was elected *strategos* in 424BC, but his command proved disastrous: he failed to save Amphipolis, that jewel in the Athenian Empire's crown, arriving too late with reinforcements, although he held the small port of Eion. Court-martialled, he was exiled, but being defeated by a general such as Brasidas was no disgrace, many felt. In exile, Thucydides devoted himself to his *History*, a rigorous analysis of the causes as well as course of the Peloponnesian War. After 404BC and Athens' final defeat, he returned to his city and probably died suddenly, for his history breaks off abruptly in late 411BC.

Thucydides thought that Athenian politics degenerated after Pericles' death, as demagogues such as Cleon (as he saw the man who had pressed for his banishment) came to dominate the Assembly. But although he treated the ineffectual Nicias kindly because he was a rich conservative, he also discerned the visionary statesman in Themistocles, an earlier radical. He perhaps overstated the defensiveness of Pericles' strategy, but his military judgement is sound. He also had a marked philosophical streak, shown in his account of the debate over the attack on Melos in 416BC, where the Athenians argue that 'might is right'. He was an inspiration to later historians, but none matched his intellectual rigour or brilliance.

Above: Herodotus was born in Halicarnassus (Bodrum), a city on the Greek world's edge, which helped make him unusually broad-minded.

Below: Often known as the 'father of history', Herodotus' historiae (inquiries) are the first true histories.

LATER GREEK HISTORIANS
XENOPHON, POLYBIUS & PLUTARCH

Above: The philhellenic Roman emperor Hadrian (ruled AD117–138) favoured Plutarch and Arrian, giving both imperial office.

Below: In this medieval picture, the translated works of Xenophon are presented to the French king Louis XII (reigned 1498–1515), showing the lasting fame of the ancient historian.

Later Greek historians, writing of people long dead, often had no first-hand accounts but relied on earlier histories that have vanished. But Xenophon did experience the later Peloponnesian War and his Anabasis himself.

XENOPHON OF ATHENS, *c.*430–354BC

The historian Xenophon was evidently always attracted to strong characters: first to Socrates, then the Persian prince Cyrus, and lastly Sparta's King Agesilaus. Born in Athens, Xenophon joined Socrates' circle, but it is doubtful how much of his philosophy he understood. His *Apology* defends Socrates so successfully that it seems bizarre that the philosopher was ever prosecuted for 'impiety'.

Xenophon left Athens soon after its defeat to make his fortune as a mercenary. He joined 'The 10,000' Greek hoplites whom Cyrus recruited to overthrow his brother Artaxerxes, the Great King. Xenophon immortalized their adventures in *Anabasis*, a gripping tale of the 'march upcountry' towards Babylon and their retreat under his leadership in 399BC after Cyrus' death. Xenophon reveals his generalship but conceals his enrichment from the campaign.

Returning to Greece, he settled in Sparta on a large estate given him by Agesilaus – a revealing choice. His most important historical work is his *Hellenica*, which continues Thucydides' account, starting "The next day…" and running to 362BC. Xenophon's account, while valuable, does not compare with Thucydides', especially in its later parts. In *Cyropaedia* (Boyhood of Cyrus) he gave a fictionalized account of the youth of Cyrus the Great, and also wrote on topics from horsemanship to housekeeping.

POLYBIUS, *c.*200–120BC

In the 2nd century BC Rome slowly, often brutally, took over the Greek world. By good fortune, the process found a perceptive, sympathetic historian in Polybius of Megalopolis in Arcadia. A general in the Achaean League, he was taken as a hostage to Rome in 167BC. There he met Scipio Aemilianus, one of Rome's great generals, becoming his friend and adviser on Greek affairs. He learnt Latin and came to admire Rome, seemingly still in its Republican prime. Permitted to travel, he visited Spain, Gaul and Africa. Allowed to return home in 150BC, he kept up links with Scipio, accompanying him to Carthage in the Third Punic War (149–146BC).

Polybius wrote the definitive history of Rome's rise to world power during 220–145BC in 40 books. He attempted to explain why this happened and to reconcile Greeks to Rome's new dominance by showing the excellence of its balanced

Above: Plutarch's Parallel Lives *(this is an edition of 1657) coupled 25 famous Romans with 25 famous Greeks, to demonstrate that there had once been Greek men of action and there were also Roman thinkers.*

constitution (in fact, already under growing strain). Although his style is turgid, his approach, as a former general and politician, is practical and intelligent, free of national prejudice. Most of his work survives only in paraphrases by later writers, but he deeply influenced Livy, the great Roman historian.

PLUTARCH, *c.*AD46–*c.*125

Born at Chaeronea in Boeotia, Plutarch is famous for his *Parallel Lives*, 50 biographies of paired-off famous Greeks and Romans. In them he aimed to show Greeks that Rome had produced more than just soldiers and to remind Romans that there had once been great Greek generals and statesmen. In his *Lives*, Plutarch too often accepted unreliable secondary sources as true, but his great gifts as a story-teller later influenced Renaissance writers such as Shakespeare.

Plutarch was more than just a biographer, however, being also a philosopher and a priest at Delphi and an active public citizen in Chaeronea. He was educated chiefly in Athens, where he became an adherent of Platonism. He also visited Rome several times, becoming a friend of the consul Sosius Senecio and ultimately of the philhellenic emperor Hadrian (AD117–138). The emperor made him governor of Achaea (south Greece) and gave him consular honours, but Plutarch remained devoted to his small home *polis*. Sadly aware that Greece's great days were past, one of his chief concerns was to restore the shrine at Delphi. Among his other works are a diatribe against Herodotus and his *Moralia*, essays attacking Stoicism and Epicureanism.

ARRIAN (FLAVIUS ARRIANUS), *c.*AD87–*c.*147

Arrian was an upper-class Greek, from Nicomedia in Bithynia (north-west Turkey), who lived at the zenith of the Roman Empire. His greatest achievement was to write a history of Alexander the Great that made intelligent use of primary sources such as Ptolemy and Aristobulus – officers who had served under Alexander but whose accounts have been lost. Arrian's eight-volume *Anabasis* (March Upcountry) survives, as does his *Indica*, a book about India based on accounts by Megasthenes of *c.*300BC and on the voyage made by Alexander's admiral Nearchus from the Indus. From Arrian we derive most of our knowledge of Alexander's life and character.

Arrian had a glittering career himself. His father was a Roman citizen, and he studied under Epictetus, the Stoic philosopher. Arrian rose to the highest office in the Roman state, becoming consul in AD129 (still an unusual honour for a Greek) under the emperor Hadrian, whose friend he was. He later was governor of Cappadocia (eastern Anatolia), commanding two legions with which he repelled an invasion by the Alans, a nomadic tribe from Central Asia.

Above: Plutarch, most prolific of ancient biographers, wrote 200 books in total. He was also a philosopher and priest.

Below: Sides of a gold coin of Alexander the Great, the subject of Arrian's great history, the book that gives us our definitive view of the world-conqueror.

PHILOSOPHERS IN POLITICS
THALES, EMPEDOCLES & DEMETRIUS

Early Greek philosophers were often closely involved in politics, for philosophy, a Greek invention, grew up in the streets and agoras of the *polis*, not in academic seclusion. This interaction had fruitful, sometimes surprising, results.

THALES, *c.*624–547BC

Greek philosophy was born in the Ionian port of Miletus, the largest and richest Greek city before 500BC, with Thales of Miletus, the very first philosopher. For Thales, philosophy meant *thinking* about every aspect of life, not quietly accepting old myths and legends. Speculating about the nature of the universe, he decided that water was the basic principle of everything, which was not so absurd for Greeks living by the sea. He learned enough about astronomy, possibly in part from Babylonia, to predict the eclipse of the sun in 585BC.

One day, he was walking along so lost in thought that he fell into a well and had to be rescued. Annoyed at becoming a laughing stock, Thales decided to get his own back. Shrewdly noting early indications that there would be a bumper olive harvest, he bought all the oil presses nearby and later charged his fellow Milesians extra to rent them, showing that philosophers could be practical too. Even shrewder was his suggestion that the Ionian cities, already threatened by powerful Asian monarchies, should form a league with its centre at Teos, a small, centrally sited city. Unfortunately, his advice was not taken and Ionia was conquered by Persia in 545BC.

EMPEDOCLES, *c.*495–*c.*432BC

Empedocles of Acragas was remarkably multi-talented. Not only philosopher and poet, he was also scientist, doctor and social reformer, a man of such visionary enthusiasm that some thought him a charlatan, others an almost divine hero. Both Aristotle and the Roman philosopher-poet Lucretius admired him, and it is only through their paraphrases of him that fragments of his work have survived.

In the 5th century BC Acragas (today Agrigento) was the richest city in Greek Sicily, as its splendid Doric temples show. Though born into its aristocracy, Empedocles was a radical democrat. He played a key role in overthrowing the tyrant Thrasydaeus in 471BC and establishing democracy in Acragas, which lasted until the Carthaginian attack in 406BC. He guided the city through its greatest period

Below: A view of the Sicilian volcano Mt Etna erupting. Into its fiery heart Empedocles, the democratic statesman and mystical philosopher, threw himself, according to legend. More probably, he died peacefully in Olympia.

in a way comparable to that of Pericles in Athens. He then retired in 445BC to travel. According to legend, he began proclaiming himself a god and committed suicide by throwing himself into the volcano of Mt Etna. More probably he died of natural causes in the Peloponnese, where his poems were recited at the Olympiad of 440BC.

As a philosopher Empedocles was among the most important pre-Socratic thinkers. He wrote a long poem, *On Nature*, of some 6,000 lines (only 350 of which survive) in which he propounded his general theory of the universe. He postulated four elements – earth, air, fire, water – governed by two opposed forces: Love, which attracts and unites, and Strife, which repels and divides. He realized that air was solid and suggested that light from the Sun must take some time to reach the Earth. This remarkably anticipates modern physics. All his science was tinged with mysticism, however, following the pattern set by Pythagoras, the pioneer of mystical mathematics.

DEMETRIUS, *c*.350–283BC

Plato had dreamt of a 'philosopher-king' ruling with supreme wisdom, but in the ten-year rule of Demetrius of Phalerum, Athens experienced almost the opposite. Demetrius might be called a rogue-philosopher, for he behaved outrageously, unlike any normal 'lover of wisdom' (which is what philosopher means).

A student of Aristotle but an Athenian citizen, Demetrius was installed in 317BC as a quisling ruler by Cassander, the Macedonian general ruling Greece. Demetrius' books on philosophy

have been lost, only those on cooking, hairdressing and fashion surviving. This seems apposite for his strange regime. He reputedly organized processions led by a mechanical snail that spat saliva and accepted absurdly flattering epithets such as *lampito* (brilliant). Although it was a time of economic hardship, he gave wild parties and ordered 1,500 bronze statues raised in his honour. He was not brutal, but power lay with his Macedonian masters. When Cassander lost power and Demetrius was overthrown, the Athenians melted his statues down into chamber pots, now they were at last free to express their true feelings about him.

DEMETRIUS THE LIBRARIAN

Demetrius went on to advise Ptolemy I of Egypt about establishing the Library at Alexandria. With a reputed 500,000 volumes, it became the greatest library in the Ancient World. This suggests that Demetrius made a better librarian than philosopher-king.

Above: The growing wealth of the Greek world, revealed in grand temples such as those at Acragas in Sicily, encouraged the development of philosophy.

Below: The temple of Hercules at Acragas, one of the many fine temples erected in the 5th century BC, when Acragas was a flourishing democracy under the enlightened guidance of the polymathic Empedocles.

THE STRUGGLE FOR SUPREMACY

404–322BC

With the end of the Peloponnesian War in 404BC, an era of liberty seemed at hand. But this illusion was soon shattered. Sparta showed itself unfit to lead Greece, favouring corrupt despotisms. Spartan arrogance united former enemies, Athens allying with Corinth and Thebes and even Persia. The collapse of Spartan power after 371BC saw Thebes as hegemon of Greece, but its achievements – spreading democracy, new forms of federalism – only worsened general instability. Men left home to fight as mercenaries while wars exhausted their own cities, leaving them prey to outside powers.

The first such power was Dionysius I, who created a vast Sicilian empire. Greece looked west in alarm, but Dionysius hardly interfered in its affairs and his empire soon collapsed. Jason of Pherae and Mausolus of Halicarnassus proved even more transitory as hegemons. Then a truly formidable power emerged: Macedonia. Philip II exploited Macedonian strength and Greek weakness to create a solidly based hegemony. At Corinth in 337BC he declared a war of Panhellenic revenge against Persia. Alexander, his brilliant son, far exceeded his plans, so ending the Classical Age of Greece.

Left: The Temple of Apollo Delphi, Greece's spiritual heart. Macedonia won Delphi, confirming its status as hegemon.

SPARTAN SUPREMACY
POWER AND CORRUPTION, 404–377BC

On Athens' surrender in April 404BC, Thebes and Corinth pressed for its total destruction. The Spartan commander Lysander refused, however, remembering Athens' glorious stand against Persia and aware that Athens, broken but still the largest Greek city, was a potential ally. At Athens Lysander installed an oligarchy, 'The Thirty Tyrants', backed by a Spartan garrison under a *harmost* (governor). The city, disarmed, became an ally of Sparta, as did most cities in its former empire, although the Thirty did not last long.

For smaller states, Lysander favoured *decarchies*: cliques of ten reactionaries eagerly wielding power long denied them. As Plutarch wrote: "In appointing these officials, Lysander simply handed out power to his cronies, giving them absolute powers of life and death." Backing each *decarchy* was a *harmost* and garrison. The tribute that had flowed into Athens was diverted to Sparta, but Sparta's former allies Thebes and Corinth gained nothing. This alienated them.

Below: The walls of Athens are pulled down to the sound of flutes after its final defeat in 404BC. This left the city nakedly vulnerable but, thanks to Lysander's intercession, Athens itself was spared destruction.

THE MARCH OF THE '10,000'

After the Peloponnesian Wars many Greeks, who knew only soldiering, were idle amid economic depression. In 404BC Cyrus, governor of western Asia, in rebellion against his brother Artaxerxes II, the new Persian king, required mercenaries. He hired c.10,000 Greek *hoplites*, who were seen as the world's best infantry, to boost his army. Well aware that Greeks would not march far from the sea, he claimed, on starting out in 401BC, to be fighting only Pisidian bandits inland.

Among the Greek officers was a bright young Athenian Xenophon. Marching ever further east, the Greeks grew restive when they realized that they had been misled. But they were induced by extra pay to keep going until they entered Babylonia (Iraq), then a fertile, exotic land.

At Cunaxa, Artaxerxes' army finally met his brother's. In the ensuing battle the Greeks were victorious but Cyrus was killed, leaving the Greeks lost in an alien empire. After the Persians murdered their generals in negotiations, the '10,000' chose as a new leader the persuasive Xenophon, who promised to lead them home. He kept his promise. They marched up through the unknown mountains of Kurdistan and Armenia, finally reaching the Black Sea at Trapezus, a Greek city (modern Trabizond), before turning for home.

Xenophon, a good writer as well as general, vividly recorded his adventure in *Anabasis* (March Upcountry). The Persian Empire's apparent vulnerability led many Greeks to dream of expansion east, with momentous later consequences.

TROUBLE AT HOME AND ABROAD

Although Sparta could now afford to maintain a fleet, the overall results were disastrous. On leaving their austere homeland, Spartans often became totally corrupt, and Spartan *harmosts* also became infamous for predatory sexual behaviour in Greek cities under their rule. Back home, Sparta's unique egalitarian system was undermined by its new wealth, since rich Spartans got richer, buying up land from poorer ones, who no longer qualified for Spartiate status. Sparta soon faced a shortage of Spartiates, the basis of its power, worsened by the relative freedom of its women, who preferred to marry richer Spartans. As a result, Sparta's birth-rate began to fall.

Sparta had supported Prince Cyrus' rebellion against his brother, King Artaxerxes II of Persia, but Cyrus was killed fighting at Cunaxa in 401BC. The Greek cities of Ionia, now free once more, appealed to Sparta for help against Persia, leading it into war with its former ally. The Spartan king Agesilaus, Lysander's protégé, led an army to Asia to attack Persia, gaining early successes. But, thanks to Persian gold, Athens began to revive, rebuilding its fleet. Lysander himself was killed in 395BC fighting at Thebes, which, like Corinth, had allied with Athens in the 'Corinthian War'. Documents found after Lysander's death revealed that he wanted to make Sparta's monarchy elective but more powerful, suiting Sparta's new imperial role.

In 394BC an Athenian fleet in Persian pay defeated the Spartans off Cnidus. This victory was countered by a bloody battle at Corinth when Spartan discipline crushed the armies of Thebes, Corinth and Athens. But when Athens rashly sent fleets to support anti-Persian rebels in Cyprus and Egypt, Sparta restored its ties with Persia, the new arbiter of Greece.

THE KING'S PEACE

Under the Persian King's Peace of 386BC, Sparta abandoned Cyprus and Ionia to Persia. The Peace theoretically guaranteed

all European Greek cities' autonomy and forbade all confederations except Sparta's, considered a free alliance. This let Sparta destroy the Arcadian city of Mantinea, claiming that its existence violated the Peace, and move against a newly formed Chalcidic Confederacy.

En route north in 382BC, a Spartan army helped a pro-Spartan coup in Thebes. This action, actually breaking the Peace, marked the apogee of Spartan power. Three years later, a counter-coup led by Epaminondas ejected this regime, and Thebes allied with Athens. Spartan power was again threatened. Seven years later it would be crushed at Leuctra.

Above: Funerary stele of Democlides, an Athenian soldier killed at the Battle of Corinth in 394BC, when Spartan discipline managed to defeat the combined armies of Athens, Thebes and Corinth.

Below: The '10,000', the mercenary force led by Xenophon the Athenian, finally reach the Black Sea after a gruelling march through the Asian interior. They were jubilant at being back on the coast, so central to Greek life.

Above: The 4th century BC saw artists turn away from the severe heroism of the 5th century. Their new appreciation of female beauty is epitomized by the Aphrodite of Cnidus, *one of Praxiteles' masterpieces.*

ATHENS: CRISIS AND RECOVERY 399-357BC

Few cities appeared more defeated than Athens in 404BC, deprived of its empire, fleet, walls and democracy. But within a decade it had recovered, becoming once more a great naval power, albeit one chronically short of money. Even more swiftly, it regained its democracy, which worked smoothly until extinguished by Macedonian generals in 322BC.

'THE THIRTY TYRANTS'

Athens' defeat was welcomed by its oligarchs. Under Lysander's menacing eye, in July 404BC the Assembly voted power to 'The Thirty Tyrants'. The two leaders were the extremist Critias, one of Socrates' former students, and Theramenes, a moderate. They set up a body of the Eleven, ruthless special police. Only 3,000 citizens had civil rights; the rest were powerless. Leading democrats were killed in a reign of terror, as were moderates such as Niceratus, Nicias' son, and rich *metics* (foreign residents) such as Polemarchus. Athenians were ordered to arrest their fellow citizens, so implicating themselves in the regime's crimes. Most complied, but Socrates refused – and uniquely went unpunished.

Democrats fled to nearby states, whose feelings towards Sparta were changing fast. In December 404BC Thrasybulus, Anytus and 70 other exiles seized the fort of Phyle near the border. The junta's attempt to eject them failed due to a snowstorm, and they gained support. Alarmed, the oligarchs quarrelled, with Theramenes being killed as Critias took control. He appealed for Spartan help, and a garrison of 700 occupied the Acropolis. After another attempt in May 403BC to dislodge the democrats failed, Thrasybulus captured Piraeus. Critias, trying to retake it, was killed in the fighting.

Again the oligarchs appealed to Lysander, but King Pausanias, his bitter enemy, now intervened, changing Spartan policy. Pausanias let the oligarchs withdraw to Eleusis while Athens worked out a new settlement. By late 403BC, democracy had been restored with a general amnesty. Athens was free again.

Right: The stele of Dexileos, who was killed in 393BC, during the war against the Corinthians, and buried at the Kerameikos, a quarter built up in the age of Themistocles.

THE TRIAL OF SOCRATES

In 399BC Meletus, supported by Anytus, accused Socrates of "not believing in the city's gods, of introducing new gods … and corrupting the young". No court records exist of this most famous trial, but Socrates' followers Plato and Xenophon gave vivid descriptions, although neither was present. Meletus is unknown but Anytus was a democratic hero of Phyle. Socrates had fought bravely too, but was better known as an incessant questioner of customs: religious, social and political. Among his ex-students were Alcibiades, the traitor, and Critias, leader of 'The Thirty Tyrants' – odious connections. But, due to the amnesty, Socrates could not be attacked on political grounds.

A charge of impiety, however, was different. Religion in ancient Greece was a public affair: offend the gods and you risked your city suffering. The 501 jurors – chosen by lot to represent public opinion – must have seen Socrates arguing in the streets. Back in 423BC he had been caricatured in Aristophanes' play *The Clouds* as a Sophist smart alec, who taught the young about new, amoral gods. This made audiences laugh, but by 399BC

many felt that Socrates' constant questioning was no laughing matter. Even so, the majority finding him guilty – 280 to 221 – was small. When he mockingly proposed public dinners as his punishment, the majority for the death penalty was larger, but he was expected to escape. Instead, he chose death. Martyr of philosophy or teacher of tyrants? Both are valid viewpoints, but Socrates was the *only* man executed for his beliefs in classical Athens.

Above: The trial of Socrates in 399BC was a seminal, still contentious event in Western intellectual history. In this 17th-century painting the philosopher, accused of corrupting the young and introducing new gods, kills himself by drinking hemlock, still debating with his followers.

THE SECOND CONFEDERACY

Although free, Athens was hardly prosperous, but it could draw on the gold of Persia, then hostile to Sparta. Athens was soon fighting Sparta, in 394BC allying with Corinth and Thebes. In 390BC Iphicrates, an innovatory *strategos*, defeated a Spartan force outside Corinth using *peltasts* (light-armed troops). With Persian money Athens had rebuilt its Long Walls by 393BC.

Refortified, Athens began trying to regain its empire, rashly helping Evagoras of Cyprus in revolt against Persia. This led to Spartan–Persian rapprochement in the 'King's Peace' of 386BC, reaffirming Spartan hegemony in Greece. Sparta's actions over the next years united Greece against it, however. After an attempted Spartan raid on Piraeus in 378BC, Athens allied with Thebes and formed a totally new naval confederacy of its own.

Aimed against Sparta, this tried to avoid the mistakes of the first confederacy by having a separate council for the allies that Athens could not dominate and forbidding *cleruchies* (settlements). Initially popular, with 70 states joining, its fleet defeated the Spartans at Naxos in 376BC. The *strategos* Timotheus sailed victoriously round the Peloponnese in 376BC, defeating the Spartans at Alyzia in 375BC. With further operations hampered by lack of money, however, the war dragged on.

Right: The new, almost languid sensuality of 4th-century art is apparent in Praxiteles' Hermes and the Infant Dionysus.

THEBES: A BRIEF HEGEMONY 377–362BC

Above: The sphinx, the mythical creature connected through the Oedipus legend with Thebes. Its enigma in a way illustrates the Theban dilemma: freeing the Peloponnese from Spartan tyranny brought chaos, not democratic order.

Below: Epaminondas' novel tactic at Leuctra, when Sparta was for the first time defeated in open battle, was to increase the weight of his left wing, massing Theban hoplites 50 deep so that they punched through the weaker lines of their opponents.

In the winter of 379–378BC the Spartan-backed junta in Thebes was overthrown in a dramatic coup. The plotters, led by Pelopidas, entered the junta's symposium one evening in drag as courtesans and killed them all. They then instituted a new democracy, which joined the second Athenian Confederacy. While Thebes repaired its defences against inevitable Spartan counter-attacks, Epaminondas and Pelopidas, both *polemarchs* (commanders), reorganized the Theban army, making it far more professional. They founded, or refounded, a crack corps, the Sacred Band: 150 pairs of lovers, mature men bound to each other by the strongest ties possible. Over the next 40 years they proved to be the best hoplites in Greece.

In 371BC Athens and Sparta, tired of inconclusive warfare and worried about the rise of Thebes, signed the Peace of Callias. This reaffirmed the principle of each city's autonomy, as in the 'King's Peace' 15 years earlier, but the Spartan and Athenian leagues were specifically excluded. Theban control over Boeotia was not, however, for it was contentious.

(Thebes had recaptured Plataea to the east, whose citizens fled to Athens, while to the west Phocis, another old Athenian ally, was attacked. Thebes was therefore excluded from the peace.) Just weeks later, in July 371BC, Sparta sent a Peloponnesian army under King Cleombrotus against Thebes, which had only one ally: Jason of Pherae in Thessaly.

BATTLE OF LEUCTRA

Cleombrotus, approaching Thebes from the south, found the Theban army drawn up at Leuctra. The Spartans, like everyone in Greece, expected to win, as they had won every set battle for 200 years. They probably had *c.*11,000 men, mostly allies, against *c.*7,000 Thebans. Cleombrotus with his crack troops was on the Spartan right, as usual. But Epaminondas put his best troops on his *left*, deepening the Theban phalanx from the usual 8–12 men to 50 – a vast increase in striking power. Although Sparta's slingers defeated the Thebans, the Theban cavalry worsted the Spartans. Then the Theban hoplite wedge crashed into the Spartan right, breaking it. Cleombrotus was killed, his army collapsing like dominoes. About 400 full Spartiates were killed – a huge loss – and Sparta's legendary invincibility was over.

The immediate consequences of the victory were muffled by Jason of Pherae, who rapidly marched south to Leuctra to impose a truce. The Spartans were allowed to evacuate Boeotia. But when Jason's power ended soon after with his assassination, Thebes became Greece's new hegemon. This alarmed Athens almost as much as Sparta.

THE PELOPONNESE SET FREE

As news of Sparta's unprecedented defeat – by *inferior numbers* – spread around Greece, states long subject to its heavy

Left: When Epaminondas led the Theban army into the Peloponnese in the winter of 370BC, he founded a huge new federal capital, Megalopolis, to which smaller settlements nearby, such as Bassae with its renowned temple, sent delegates.

rule began to revolt. Its loathed *harmosts* (governors) were thrown out, replaced by democracies. Among the first to revolt were the Mantineans of Arcadia, long denied a proper *polis* by Sparta. While their old city was rebuilt and rewalled, Mantinea helped form a Pan-Arcadian Federation with a new capital, Megalopolis (great city). Defended by double walls, it had a grand theatre, *agora*, temples and a federal army of 5,000 men.

Arcadia was a federal democracy, with most of its citizens having a vote. In 370BC revolution at Tegea led it to join Arcadia, but Tegea's exiled oligarchs appealed to Sparta. Arcadia in turn appealed to Athens, but, as Athens proved non-committal, it turned to Thebes. This was Epaminondas' chance.

FOUNDING MESSENE

In late 370BC a Theban army invaded the Peloponnese. Almost unopposed, it ravaged the long-inviolate territory around Sparta itself; the unwalled city seemed set to fall. But winter floods swelled the River Eurotas and the only bridge was strongly defended, so Epaminondas withdrew. This was a huge blow to Spartan pride, but worse followed. Epaminondas led his army into Messenia, whose long-enslaved *helots* rose en masse. A new city of Messene was founded in 369BC on Mt Ithome, with splendid new walls 8km/ 5 miles long. Messenian exiles flocked back to their new *polis*, whose very existence doomed Spartan power. There were only 1,500 full Spartiates left anyway.

Athens, meanwhile, allied with Sparta. This was not so odd, for in the subsequent confused years states constantly changed sides. The Arcadians themselves did not want to swap Theban hegemony for Sparta's.

Pelopidas then led expeditions to Macedonia, trying to extend Thebes' influence north, while Dionysius (tyrant of Syracuse) and Persia intervened in Greek affairs, neither to much effect.

Epaminondas made four invasions of the Peloponnese, the last ending in his fatal victory at Mantinea in 362BC – fatal both to him, for he died of his wounds, and to Theban hopes of supremacy. With Pelopidas also dead, Thebes lacked great leaders. While it had finally ended centuries of Spartan supremacy and helped to free the Peloponnesians, Thebes' hegemony left Greece more exhausted and divided than ever.

Below: The massive walls of the Arcadian Gate in Messene, a city beneath Mount Ithome that Epaminondas founded in 369BC as a secure polis for Messenians who had been long enslaved by Sparta.

THE SYRACUSAN EMPIRE
411–337BC

When Syracuse defeated the Athenian invasion in 411BC, it was a democracy like Athens, as Thucydides noted, but a very unstable one. Hermocrates, leader of the defence, took 25 ships east to help Sparta, but radicals in the city then banished him. They also embroiled Syracuse in a war with Carthage, which had colonies in western Sicily. Called in by half-Greek Segesta against its rival Selinus, the Carthaginians landed with a huge army in 409BC. This captured Selinus and took Himera, another Greek city. Reputedly 3,000 Greek captives were sacrificed to Carthage's gods afterwards.

DIONYSIUS' GROWING POWER
Hermocrates was killed soon after while trying to re-enter Syracuse. Carthage then attacked Acragas, Sicily's richest city. Acragas' notoriously soft inhabitants – guards on night duty were limited to one mattress, two pillows and a quilt each – fled in panic in 406BC after a relief effort failed. As the Carthaginians besieged Gela, the next city along, panic grew in Syracuse too. Dionysius, one of Hermocrates' officers, accused the generals of treachery, becoming a *strategos* himself. In 405BC Syracusans voted him sole powers as *strategos autokrator* (supreme general) and a bodyguard of 1,000. With it, he made himself tyrant. His rule, a demagogic dictatorship but with rich backers, lasted until his death.

Dionysius' first campaign to save Gela in 405BC failed, and he had to abandon half of Greek Sicily to Carthage. Refugees streamed into Syracuse, swelling its population, as did deportees from Naxos (Taormina) and Catania – Greek cities that he had attacked without provocation. Dionysius made Ortygia, the old island core of the city, into his citadel, expelling from it all but 10,000 trusted mercenaries. Inside its turreted walls he survived a serious revolt in 403BC. Around Syracuse city he built the strongest walls yet seen, enclosing c.1,200ha/3,000 acres.

Meanwhile, he was planning revenge on Carthage, the excuse for his tyranny. Over the next years Syracuse rang to the sound of hammers and anvils. An army 80,000 strong, and a navy of 300 galleys – including the first quadriremes and quinqueremes, larger and more powerful than triremes – were assembled. Dionysius also developed the first siege engines to hurl heavy stones 300m/325 yards. Most Syracusans, whatever they felt about his tyranny, supported this armaments programme.

THE SIEGE OF MOTYA
With his base assured, in 397BC Dionysius renewed the war with Carthage. Marching to the far end of the island, he besieged the city of Motya. This was thought impregnable on its island, but Syracusan engineers built a causeway across the bay

Above: The ruins of Motya, the Carthaginian stronghold in western Sicily that Dionysius I besieged.

Below: The ruins of Carthage near modern Tunis. The perennial threat of this great enemy of the Greeks in Sicily was invoked by Dionysius I to justify his prolonged tyranny.

to let their siege-engines attack it. The battle was long and bitter. Even after their walls were breached, the Motyans fought back street by street, until a night attack finally crushed them. For the first time since 480BC, Greeks had captured a Carthaginian city. The resulting slaughter was terrible until Dionysius, mindful of the slave market, halted it.

COUNTER-ATTACK AND FURTHER CONQUESTS

Next year Himilco of Carthage, landing with an army, recaptured but did not rebuild ruined Motya, instead founding Lilybaeum nearby. Marching east, he founded on the site of Naxos a new city for the native Sicels, Taormina, later to be a Greek city. Himilco defeated the Syracusan fleet, but his attack on Syracuse itself met with disaster, thanks to its strong walls, an outbreak of plague in his army and a brilliant counter-attack by Dionysius. Dionysius let Himilco himself escape, but he had won the war. The peace treaty he made in 392BC was Machiavellian: the Carthaginians were restricted to the north-west of the island but not ejected, remaining a potential threat to justify his continued tyranny.

Master of Greek Sicily, Dionysius now turned to Italy. He had asked for a bride from Rhegium (Reggio) on the Straits of Messina, but was offered only their hangman's daughter, an unforgivable insult. In 391BC he duly attacked Rhegium. Initially defeated at sea, he waged war on the mainland with great success. He finally captured Rhegium after a long siege in 387BC, enslaving its citizens. He now controlled both sides of the Straits and much of southern Italy, the powerful city of Taras (Taranto) being his ally. Dionysius founded colonies up the Adriatic as far north as Hadria on the River Po, with trade flourishing in this new empire. After another war with Carthage, he had to cede Sicily west of the River Halcyus in 383BC. This still left him with most of the island, the greatest power in the Greek world, both feared and courted.

TYRANNY AND CULTURE

Dionysius liked to claim that he was not a tyrant, simply a *strategos* with exceptional powers. The Assembly continued to re-elect him – it had no option – and the city prospered as an imperial capital. But Syracusans were heavily taxed, under military rule and surrounded by barbarous newcomers. Dionysius himself, like many dictators, suffered from paranoia. He had visitors to his court strip-searched for weapons and allowed no one to shave him except his daughters

Above: The Temple of Hercules in Acragas, the famously rich Greek city in western Sicily that fell without a fight to the Carthaginian advance in 406BC.

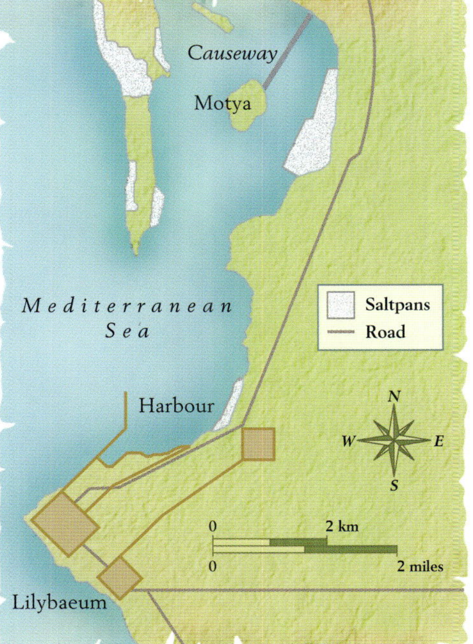

Left: Motya on its island was long considered impregnable. But Dionysius' engineers built a causeway out to it on which they brought up siege engines to take it in 397BC.

Above: The nymph Arethusa on a coin designed by Euanitos in the 4th century BC. *Syracuse produced some of the most beautiful Greek coins that have ever been minted.*

He did not even trust his wives – he had several simultaneously, for political reasons – but made them await him lying naked on a bare bed surrounded by water, so that he could see they had no daggers under the sheets. According to Plutarch, the tyrant hired prostitutes to spy on the citizens, executing 10,000 Syracusans as alleged traitors.

Dionysius harmed rather than helped Hellenism overall. He destroyed old Greek cities that opposed him, replacing their citizens with non-Greeks so that they no longer counted as true Hellenic *poleis.* But he admired Greek culture, inviting the hedonistic philosopher Aristippus, once a student of Socrates, to stay at his luxurious court. (The two men got on surprisingly well.) And Dionysius had literary ambitions himself, entering several contests.

In 367BC Athens, trying to woo him as an ally, awarded his *Ransom of Hector* first prize at the Lenaean festival. Overjoyed, Dionysius drank so much that he died. The Roman historian Cornelius Nepos acquitted Dionysius of a tyrant's three typical failings – lust, avarice and greed – but his empire rested wholly on force.

DION AND DIONYSIUS II

Dionysius I claimed to have left an "empire bound with steel", but the 29-year-old heir Dionysius II lacked his father's ruthlessness. He ended a war with Carthage, but a peaceful warlord risked seeming redundant, especially as he did not concentrate on ruling the empire. Instead, he swung between debauchery – he was excessively fond of wine and women – and philosophy, to which his relative Dion introduced him. Dion, an aristocrat who had married into the tyrant's family, meanwhile continued as chief minister.

Dion had once studied under Plato, the Athenian philosopher, and he persuaded the amenable young ruler to invite Plato to his court. Plato, already in his sixties, reluctantly agreed and made two visits to Syracuse, both of which turned out badly since Dionysius proved as idle a philosophy student as he was a ruler. Philistus, head of a rival faction, turned Dionysius against

Below: Façade of the Temple of Concord at Acragas (Agrigento), Sicily.

Below: Part of the walls around Syracuse that Dionysius extended and strengthened to cover 1,200ha/3,000 acres, making his capital much the largest Greek city.

PLATO AND THE PHILOSOPHER-KING

Although born into its ruling class, Plato turned his back on Athens' democratic politics after Socrates, his revered master, was put to death in 399BC. He spent years abroad, in Italy meeting Archytas, the Pythagorean philosopher-ruler of Taras. Visiting Sicily in 388BC, he fell foul of Dionysius I's paranoia. Arrested and deported, he was reputedly sold as a slave. Although soon rescued, it left him thinking tyranny even worse than democracy.

Plato now founded his Academy, the world's first university, outside Athens. In it a select few, including Dion, could study undisturbed. There Plato wrote *The Republic*, his blueprint for the ideal society ruled by an intellectual elite, but he did not envisage its realization on Earth. Yet when Dion begged to him to come and teach the young Dionysius II, he could hardly refuse. It was an unparalleled chance to educate this powerful ruler as a 'philosopher-king' guided by selfless wisdom of the truest kind.

Dionysius greeted Plato's arrival with delight. He dismissed his call-girls and fellow-boozers and began to study maths, which Plato thought essential for philosophy. But the young tyrant soon grew bored and, although he refused to let Plato leave, slipped back to his old ways. Finally allowed home, Plato returned to philosophy with relief. But in 357BC he was again invited, this time by Dionysius himself, who claimed to be studying philosophy seriously. Wearily, Plato again made the long voyage, only to find that the tyrant considered himself his equal as a philosopher. Plato was imprisoned and only freed because Archytas of Taras intervened. Back in Athens, he wrote *Laws*, his last and grimmest work on politics, effectively taking Sparta as his model.

Above: Plato the Athenian philosopher made three visits to Syracuse, the last two in the hope of turning its young ruler Dionysius II into a 'philosopher-king'. All ended in failure.

Dion, who went into exile in Athens. He returned to Sicily in 357BC with a tiny force, landing at Heraclea in the west. Marching on Syracuse, he was welcomed and seized the outer city, proclaiming the restoration of democracy. Dionysius, away in Italy, returned in time to save Ortygia. Murderous civil strife followed, Dionysius' Italian mercenaries slaughtering the citizens when they voted out Dion. Dion returned to save the city but ended becoming a tyrant himself, assassinated in 354BC.

TIMOLEON'S RESTORATION

A decade of civil war ravaged Sicily, numerous tyrants emerging only to be quickly murdered. Finally in 345BC Corinth, Syracuse's mother-city, sent out Timoleon, a man of exceptional integrity and ability. He drove out the tyrants and in 339BC defeated the Carthaginians, who had again attacked.

Timoleon restored a limited democracy, re-peopling Syracuse with 60,000 new immigrants from old Greece. He retired universally honoured, dying in 337BC, but his settlement unravelled within a few years.

Right: Ortygia, the island core of Syracuse that Dionysius I made into his own citadel, expelling all citizens living there.

ABORTIVE EMPIRES
JASON AND MAUSOLUS, 375–353BC

In 362BC Theban victory at the Battle of Mantinea reconfirmed that Sparta could never regain her old position but left everything else more unsettled than before. Epaminondas' death at Mantinea had ended Thebes' brief hegemony, never fully accepted by most of Greece. By then, all the traditional main powers faced varying degrees of financial and political exhaustion. The resulting vacuum in Greek affairs began to pull in more powerful rulers from the Greek fringes. Two new powers emerged that seemed poised to dominate Greek politics, but both rapidly faded after their ruler's death.

JASON OF PHERAE
REIGNED 380–370BC

Thessaly, the northernmost part of Greece proper, was unusually large and fertile – promising territory for any would-be dynast or ruler. With better-watered pastures than southern Greece, it supported superb cavalry on a scale unthinkable further south. But it had long remained divided between feuding aristocratic clans – all claiming descent from Hercules, like the kings of Sparta, the most notable being the Aleuads. Its coastal cities were intermittently independent. This disunity made Thessaly a pawn in the hands of outside powers, Persian, Athenian or Spartan. Xerxes had recruited its cavalry for his great invasion of 480BC, but their 'medizing' did not undermine Thessaly's status as fully Greek. Although by the 4th century BC life based on the *polis* was spreading throughout Thessaly, it needed a leader of genius to unite it. In Jason, Thessaly briefly found one.

The ancient title *tagus* (lord) of Thessaly had long been awarded to a ruler who could unite the land. By 380BC Jason, tyrant of Pherae in south-central Thessaly, was powerful enough to be hailed as *tagus* by other clans and cities. He began organizing a Thessalian army on a federal basis. By then he had also intervened in Euboea, if ineffectually, for Sparta opposed him. The Spartan outpost at Heraclea near Thermopylae blocked his way south into central Greece.

The news of Sparta's shocking defeat at Leuctra by Thebes in July 371BC gave Jason his opening. He raced south with his cavalry, moving so fast through hostile Phocis that it could not stop him, and reached Boeotia in time to act as armed arbitrator. The Thebans had to let the Spartans, still numerically the larger army, return to the Peloponnese. Jason did not want an over-mighty Thebes. On his way home he demolished the fortress of Heraclea, opening the way into Greece.

Jason's army, composed of Thessalian cavalry and mercenary infantry, was now the strongest in Greece, and he began building a navy. He seemed the new Greek hegemon and talked of a Panhellenic war against Persia. He aimed to dominate the next Pythian festival at Delphi, one of the greatest in

Above: The fertile plains of Thessaly, here seen from the rocks of Meteora, provided good pasture for Jason's excellent cavalry.

Below: The island, now peninsula, in Halicarnassus (modern Bodrum) on which Mausolus built his citadel with a double harbour for his growing fleet in c.360BC.

the calendar, and to become President of the Amphictyonic Council controlling Delphi. But one day, late in 370BC, as he was reviewing his cavalry he was assassinated by some Thessalian nobles for personal reasons. His son Alexander succeeded him as *tagus* but lacked his ability. Alexander was defeated by Thebes in 364BC, although that victory cost the Theban general Pelopidas his life. Alexander lived on until 358BC, his powers much reduced.

MAUSOLUS OF CARIA
RULED 377–353BC

In the 4th century BC the Persian Empire's westernmost satraps in Asia Minor gained increasing powers that made them, if not independent, in practice far more powerful, even able to pass on their titles to their heirs. Several joined the 'Revolt of the Satraps' in the 360s BC, which shook the empire. Of these dynasts, Mausolus, who succeeded his father as satrap of Caria in southwest Asia Minor in 377BC, was the most powerful. He moved his capital from inland Mylasa down to Halicarnassus, where he ruled with half-Hellenic, half-Carian splendour. He built himself a castle on the island (now peninsula) outside Halicarnassus and a double harbour for the city.

Now established, he began to expand his power: south-east toward Lycia, north toward Miletus and, most significantly, west over the islands. This meant a clash with Athens. The Second Confederacy in Athens was proving increasingly unpopular, as she reneged on initial promises not to exploit her allies, planting *cleruchs* (colonists) in Samos again, for example. In 357BC, when Chios, Cos and Rhodes revolted in the so-called 'Social War', they looked to Mausolus for support. He shrewdly let these Greek cities govern themselves, providing only small garrisons. But his mini-empire came to nothing, for he died suddenly in 353BC, after which his widow Artemisia ruled.

To commemorate her husband, she built his tomb, the Mausoleum, at Halicarnassus, which became one of the Seven Wonders of the World. Its magnificence displayed the dynasty's wealth and gave us the word 'mausoleum'.

THE 'SACRED WAR'
357–346 BC

Above: Delphi was sacred to Apollo, here seen in relaxed mode playing with a lizard, in a sculpture by Praxiteles.

Delphi was the holiest place in the Greek world, sacred to the god Apollo. For centuries cities across the Greek world, besides non-Greek kingdoms such as Lydia, had enriched the shrine with temples and statues. Set in a cleft in the mountains, Delphi itself was a tiny, powerless *polis*. Its accumulated wealth relied for protection less on awe of the god, real though that was, than on the Amphictyonic Council, backed by general Greek support. 'Sacred Wars' arose when states contested control of the shrine. The most significant resulted from Thebes' renewed attempts to dominate her immediate neighbours. It paved the way for Macedonia's domination.

PHOCIS SEIZES DELPHI

In 357 BC the Theban-dominated Amphictyonic Council fined the little state of Phocis for "cultivating sacred ground", which was an excuse for letting Thebes attack. As expected, Phocis would (or could) not pay the massive fine. Unexpectedly, their leader Philomelus seized Delphi early in 356 BC and 'borrowed' Delphic gold to hire a mercenary army of 5,000 men. He won Spartan and even Athenian tacit approval if not active support, but killed himself after having been defeated by the Thebans in 354 BC.

Thebes, thinking the war over, hired out many of its best troops to a Persian satrap, but Onomarchus (Philomelus' successor) took over the tattered Phocian army. He helped himself to more of Apollo's money, made alliances, with Pherae among other places, and recruited fresh troops. With these he won a series of startling victories, defeating Philip II of Macedonia in Thessaly in 353 BC.

At this stage Phocis ruled most of central Greece, but Philip, returning with a larger army, drove the Phocians from Thessaly in 352 BC. When Athens sent a force to hold Thermopylae, however, Philip prudently retreated to Thessaly, for the moment not wishing to offend Athens, still the greatest Greek city.

MACEDONIA'S VICTORY

The power of Phocis relied totally on Delphic money to pay its mercenaries. As this ran out, its fortunes began to decline, hastened by internal divisions. Onomarchus, killed in battle, was replaced by Phalaecus, who was soon dismissed (in theory, though in practice he still held Thermopylae, the gates of Greece). Both Athens and Sparta were

Left: The Tholos of the temple to Athena Pronoia built in Delphi, c.390 BC. Delphi was filled with sacred treasures. Melted down, these paid for a large mercenary army, so briefly making tiny Phocis hegemon of Greece.

preoccupied with events elsewhere when
Thebes appealed to Philip for help against
Phocian troops ravaging its borders. Philip
moved swiftly south, Phalaecus surren-
dering Thermopylae to him by secret
agreement. Philip now crushed Phocis,
broke it up into small villages and took its
seats on the Amphictyonic Council.

Most Greek states, Sparta excepted,
then signed the Peace of Philocrates
of 346BC, which recognized Philip's
actions. The Sacred War was finally over.
The real victor was neither Thebes nor
Phocis but Macedonia, rising fast under
its dynamic and cunning king to become
the new hegemon of Greece.

MERCENARIES

The word 'mercenary' is pejorative
today, but the Greeks called such sol-
diers either *xenoi* (foreigners) or, more
politely, *epikuroi* (helpers). The pro-
fession, if scarcely glamorous, was not
shameful. Greek mercenaries had fought
in Egypt in the 6th century BC, and in
the 5th century mercenary *peltasts* (light
troops) were used at times. But only in
the aftermath of the long Peloponnesian
War (431–404BC) did mercenaries
become important. Greeks who knew no
other trade than soldiering now sought
employment as mercenaries. Arcadia, the
impoverished heart of the Peloponnese,
was a major source of such soldiers, but
they came from other cities facing hard
times too. Probably the most famous was
Xenophon. An Athenian ill at ease with
his city's restored democracy, Xenophon,
joined the army of the rebel Prince Cyrus
in 401BC. He ended leading the '10,000',
the Greek hoplites forming the army's
core, back to the Greek world. He was
not the only such mercenary, however.

Conon, the one Athenian *strategos* to
escape the disaster at Aegospotamae in
405BC, hired himself out to the Persians
over the next years, fighting Sparta at sea.
Other 4th-century Athenian generals at
times fought as mercenaries when they
could not pay their armies, among them
Iphicrates and Timotheus, two of Athens'
best generals. More remarkably, King
Agesilaus of Sparta (ruled 399–360BC)
in his last years took service in Egypt, in
revolt against Persia, to earn money for
his now desperately strained city. This did
attract criticism, however, since Spartan
kings were descendants of the divine hero
Hercules, not just ordinary generals.

Demosthenes, Athens' great 4th-century
orator, often chided his fellow citizens for
relying on mercenaries rather than fight-
ing themselves. But in real emergencies
Athenian citizens would still fight, as they
did at Chaeronea in 338BC (if unsuccessfully).
Since many wars in the 4th century hardly
seemed worth fighting, hiring skilled
mercenaries often seemed preferable.

*Above: The tomb of
Dioscorides, a mercenary.
Mercenaries became common
in Greek armies from the
4th century BC onwards, as
ordinary citizens grew less
inclined to fight.*

MACEDONIA'S RISE TO POWER
359–336BC

North of Mt Olympus, mythical home of the gods, lay Macedonia – to other Greeks a huge, strange, half-barbaric country. It had played an often ambiguous role in Greek affairs since the reign of Alexander I, king during the Persian invasion of 480BC. Although Macedonia had had ambitious monarchs, its frequent relapses into chaos meant that it was never more than a local power. All this changed with astonishing swiftness, however, when Philip II came to the throne in 359BC.

PHILIP'S POLICY OF EXPANSION
Philip had spent his youth as a hostage in Thebes, where he admired the military skills that gave Thebes its brief hegemony. Inheriting the Macedonian throne after his brother Perdiccas III was killed fighting Illyrian tribes, he defeated the invaders decisively in 358BC, pushing inland as far as Lake Ochrid. In 357BC Philip married the Epirote princess Olympias, who bore him a son, Alexander, in 356BC. That same year he seized Amphipolis, the key city founded by Athens but lost in 424BC, which

Above: Philip II, the ruler who transformed Macedonia from chaotic backwater to Greek superpower, was only 24 when he came to the throne in 359BC.

controlled the crossing of the River Strymon. Philip conned the Athenians into thinking that he was taking it for them, but kept it himself. With Amphipolis came the gold mines of Mt Pangaeus. Philip exploited these far more energetically, its gold underpinning his growing strength. He reorganized the Macedonian army, no longer basing regiments on clans or families. He created the Macedonian phalanx, a porcupine of pikesmen, and formed the Companions, a royal guard of elite cavalry. With this

Above: The lion commemorating the Sacred Band, the elite corps of 300 Thebans killed at the Battle of Chaeronea in 338BC.

Below: At Chaeronea, Philip on the right feigned retreat, drawing on the Athenians, who exposed their flank to a cavalry charge by Alexander.

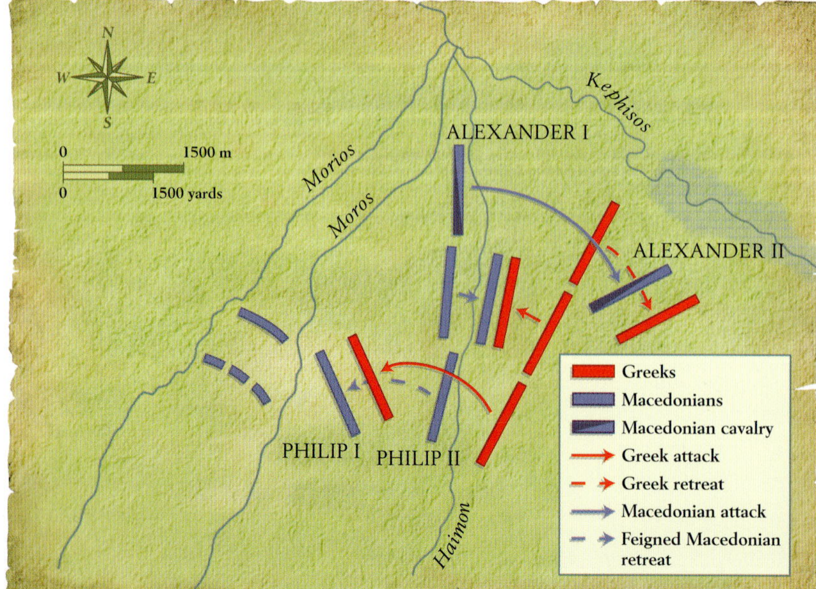

new, increasingly professional army under his excellent general Parmenion, he began his policy of relentless expansion.

In 356BC Philip took Pydna and Potidaea, cities on Macedonia's own coast, and captured Crenides inland on the Thracian border. Defeated in Thessaly by Phocis' superior numbers in 353BC, he took revenge soon after at the Battle of the Crocus Field, routing the Phocians with his cavalry. This victory made Philip *tagus* of Thessaly, controlling the port of Pagasae, and he added Thessaly's superb cavalry to his army.

A THREAT TO ATHENS

Although Philip was checked at Thermopylae in 352BC by Athens, he had transformed Macedonia's position. He could do this because the southern Greeks were distracted by other events: Thebes by the Sacred War; Athens by the revolt of its allies and Mausolus of Caria; and Sparta by problems close to home. In 349BC Philip besieged Olynthus, the main Chalcidic city. The Athenians, despite an alliance with it, did nothing to help and

Below: Demosthenes, Athens' last great democrat and a renowned orator, whose Philippics *(speeches against Philip) led his city finally to fight Macedonia.*

Philip captured it in 348BC, razing it to the ground. In 347BC Philip turned east to conquer the area around the River Hebrus in Thrace. His growing power now started to threaten Athens' vital grain supplies through the Hellespont.

When Thebes asked Philip for help against Phocis in 346BC, he marched south, finding Thermopylae unguarded. He trounced the Phocians, whom he denounced for stealing Apollo's gold, taking their seats on the Amphictyonic Council (a great coup, for it meant Macedonia's recognition as fully Greek), and presided over the Pythian games. The war-weary Greeks accepted his new status in the Peace of Philocrates. But he had grander plans: to be hegemon of Greece.

BATTLE OF CHAERONEA

In 342BC Philip marched north-east into wild Thrace. Despite falling gravely ill, he made all Thrace acknowledge his suzerainty. Alarmed at developments, which further threatened Athens' Black Sea grain, Demosthenes persuaded Perinthus and Byzantium to turn against Philip. In 339BC, when Philip besieged Byzantium, Athens declared war and sent a fleet, forcing him to withdraw. Philip, seizing 200 grain ships meant for Athens, marched into Phocis.

Demosthenes now persuaded Athens to offer an alliance to Thebes, its old rival. A tense debate in Thebes led to acceptance. The two great powers of old Greece (Sparta being now insignificant) mobilized. The two armies met at Chaeronea in August 338BC. Philip's right wing, facing the inexperienced Athenians, feigned a retreat. As the Athenians advanced, they opened a gap between themselves and the Thebans into which Prince Alexander charged with the Macedonian cavalry. Then Philip's disciplined troops turned to counter-attack. The Athenians fled. Only the Theban Sacred Band stood its ground to the last man. The Battle of Chaeronea had proved fatal to Greek liberty.

Above: A coin showing Dionysus, god of wine and ecstasy, widely worshipped in Macedonia and Thrace.

Below: This fine bronze statue of a youth, made c.340BC and found in the sea, suggests that by the mid-4th century BC Greeks lacked their earlier heroic determination.

Above: Lycurgus of Athens, in control of the city's finances for 12 years, oversaw the construction of new shipyards and the rebuilding of the Theatre of Dionysus.

Below: In 338BC Philip called a Panhellenic Congress at Corinth, chosen because it had been the seat of resistance during the Persian Wars.

THE END OF CLASSICAL GREECE
337–322BC

Philip used his crushing victory at Chaeronea wisely, as might be expected from such an astutely diplomatic leader. He treated Thebes, which had broken its treaty with Macedonia, more severely than Athens, placing a Macedonian garrison in the Cadmaea (the citadel of Thebes) but giving Athens the border town of Oropus. Philip himself returned 2,000 Athenian captives, but retained the Chersonese peninsula so that he could cut Athens' grain supplies at will. Philip then invaded the Peloponnese. Meeting no resistance, he ravaged Spartan territory, leaving it yet more isolated and powerless but still feared by its neighbours.

In 338–337BC Philip established the League of Corinth to promulgate his grand design: a Panhellenic war against Persia to avenge its attack on Greece nearly 150 years earlier. For this he ideally wanted Athenian naval help and certainly a quiescent Greece behind him. To ensure the latter, he planted Macedonian garrisons at Ambracia in the west, Chalcis on Euboea and Corinth. The Greeks voted him Captain General of the Greeks and he sent an advance force into Asia. But he himself never followed it. Returning to Macedonia, he celebrated the marriage of his daughter with her uncle Alexander of Cleopatra (he had already divorced Olympias, Alexander's mother). When in 336BC Philip was assassinated, seemingly for personal reasons, many in Greece must have expected Macedonia to relapse into its customary chaos. They were to be devastatingly disappointed.

ALEXANDER AND GREECE
Rapidly establishing himself as Philip's heir, the 20-year-old Alexander raced down into Greece – cutting steps in the sides of Mt Ossa when the Thessalians demurred at letting him pass – to squash potential revolts. Thessaly accepted him as its leader and the League at Corinth elected him General. Returning north, he swept through Thrace, crossing the Danube – the first Greek soldier to do so – before turning west to defeat the Illyrians. There he heard that the Thebans, believing a rumour of his death, had risen and massacred some Macedonian officers, besieging the rest of the garrison in the citadel. Returning south at lightning speed, he stormed Thebes in September 335BC, razing the ancient city to the ground and enslaving its inhabitants.

After this act of exemplary terror, Greece was cowed. Alexander took only 20 ships from Athens when he crossed into Asia in 334BC, mainly as hostages. He left Antipater, one of his best generals, with an army in Macedonia to control Greece. Many Greeks secretly

hoped for Alexander's defeat and death. In 331BC, Agis III of Sparta, a young monarch with more ambition than sense but financed by Persian gold, rose against the Macedonians. His army was routed at Megalopolis by Antipater and he was killed. Greece lapsed into acquiescence.

THE EXILES' DECREE

In 324BC Alexander issued an edict ordering every Greek city to take back its exiles. This caused problems for some cities, especially Athens. Its *cleruchs* had colonized Samos, so Samian exiles wanted their land back. Athens sent envoys to remonstrate with Alexander. (His other demand – that he be worshipped as a god – caused relatively few problems, for some Greeks had already been given divine honours.) Meanwhile Harpalus, a corrupt Macedonian high official, appeared at Athens with quantities of gold, some of which Demosthenes allegedly took. Then came the long-awaited news: Alexander was dead.

THE LAMIAN WAR

At first few could believe it. "If he were dead, the whole world would stink of his corpse!" declared Demades, an orator.

But as the Greeks realized that it was true, many joined Athens' revolt against the Macedonians. (Aristotle, tainted by his associations – he had been Alexander's tutor – fled from Athens to avoid it "repeating its mistake with Socrates".)

The Athenians seized Thermopylae and kept Antipater and his army beseiged in the city of Lamia through the winter of 323–322BC. But despite this initial success, the Lamian War went badly for the allied Greeks: they often did not agree on their actions, their best general, Leosthenes, was killed in the siege and Athens could not afford to man more than half of its new fleet.

After reinforcements reached Antipater from Asia, he broke out. At the Battle of Crannon in August 322BC the combined Macedonians defeated the Greeks decisively. This time peace was truly dictated. Athens' democracy became an oligarchy, with a Macedonian garrison in the Piraeus. The democratic leaders fled into exile, Demosthenes taking poison rather than face capture. With him died classical Greece.

Above: Philip in later life, arbiter of Greece, supremely experienced both as a general and a diplomat.

Below: The plains of Boeotia, called the 'dance floor of Ares (Mars)' because many battles were fought there. Alexander, by destroying Thebes in 335BC, shocked Greece into accepting his power.

PART TWO

ALEXANDER THE GREAT AND HIS HEIRS

Alexander the Great remains one of the most dramatic figures in world history. He was indisputably a military genius, overthrowing the vast Persian Empire, but views about him remain divided. The Victorian historian Thomas Carlyle called him "Macedonia's madman". More recently he has been damned as a paranoid alcoholic, murdering all in his way, perhaps before being finally murdered himself. To many at the time – be they democrats such as Demosthenes, or inhabitants of cities that he erased, such as Thebes, Tyre or Persepolis – Alexander was purely destructive. But others have seen him as transcending Greek chauvinism, trying to create a global empire that united conquered and conquerors. To romantics he has always appealed. Alexander saw himself as a reincarnation of Achilles, the Homeric hero who preferred a glorious early death to long but obscure life. In this at least he succeeded: his name has passed into legend across Asia as well as Europe.

Alexander's dazzling successes were possible only because of the achievements of his father Philip, who created the best army yet seen. If a lesser general than his son, Philip was a better politician. Alexander's successors extended Greek power and culture across Asia, creating the civilization known as Hellenistic. Although endless wars wrecked many kingdoms, Greek culture continued to spread, finally influencing art in countries as distant as India and China. The last Hellenistic monarch to resist the rise of Rome was Cleopatra VII of Egypt, a queen of legendary glamour. But Alexander's final heir proved to be Rome itself.

Left: The Alexander Mosaic *from Pompeii, showing Alexander charging towards a terrified King Darius in his chariot, probably at the Battle of Issus.*

Chapter VII

THE RISE OF MACEDONIA
359–336BC

Alexander's achievement appears so dazzling that his career has often eclipsed that of his father, Philip II. But without the patient state building of Philip, who started his reign in most unpromising circumstances and ended it as Greece's acknowledged leader, Macedonia would never have become the military powerhouse that it was by 336BC. In that year Alexander succeeded to the throne over his murdered father's body – a typical Macedonian scenario, some may have thought.

Several earlier monarchs had tried to unify this kingdom – which remained closer to the heroically chaotic world of Homer than a Greek *polis* – but their attempts had died with them, and their deaths had seldom been from natural causes. Although invaders and feuding nobles constantly threatened it, Macedonia was the largest, potentially wealthiest of Greek states. Endowed with fertile plains and wide pastures, it was occupied by a warlike people usually loyal to the throne if not always to the person occupying it. Yet Macedonia remained on the fringe of the Greek world until its meteoric rise under Philip II in the 350s BC. One of the most Machiavellian as well as energetic of rulers, Philip used bribes and promises as well as force to divide and conquer at home and abroad. After Philip's reign, Macedonia remained a major power in the Mediterranean world until the advent of Rome overwhelmed all the Greek states.

Left: Macedonia's new wealth and old martial energy are shown in this gold quiver from the royal tombs at Vergina.

EARLY OBSCURITY
*c.480–359*BC

Above: Mt Olympus, the highest peak in Greece and mythical home of the gods, lay within the Macedonian kingdom. Greek-style games were held at Dion on its slopes.

North of Mt Olympus, the greatest mountain known to the Greeks, stretched a realm that few southern Greeks knew at all and one about which even fewer normally cared, even though Macedonia had long protected Greece proper from barbarians to the north.

It was a land-bound kingdom, the coastal cities such as Olynthus being mostly colonies of southern cities. Aegae, Macedonia's ancient inland capital, overlooked the rich Emathian plain near the sea. Macedonia itself could be roughly divided into two zones: the lowland area, where royal authority ran, and the highland zone to the north and west, which was mountainous and thickly forested. Here, in walled, isolated villages, Macedonians still lived lives that Homer might have recognized but that would have struck classical Greeks as uncouth. Often herdsmen or subsistence farmers, these 'Upper Macedonians', such as the Lyncestids, acknowledged only clan chieftains. Beyond these tribal communities lived real barbarians: Illyrians and Celts to the north and north-west and Thracians to the north-east, waiting their chance to descend on the rich lowlands. Their invasions mattered much more to clans such as the Lyncestids or Orestids than to any distant king down on the coast, who could give them little protection against such attackers.

MACEDONIAN KINGS

The first notable Macedonian monarch was Alexander I (reigned 498–454BC). His reign coincided with the great Persian invasion of 480BC, in which he played an ambivalent role. He first accepted Persian overlordship in 491BC, when the Persian general Mardonius led an army west along the Aegean's north coast. In 480BC he involuntarily entertained King Xerxes himself – an expensive guest with his huge court and army – before accompanying the Persians south. But Alexander also sent ambiguously worded warnings to the Greeks. In spring 479BC he acted as the Persian's go-between, trying to persuade Athens to change sides. This was an ignominious role, but other more typical Greek states, most notably Thebes, also 'medized' (supported Persia).

Despite this, Alexander I was called the 'Philhellene' and was credited with expanding the kingdom, his reign later being remembered as a golden age. He was succeeded by his son Perdiccas II (reigned 453–413BC), who steered a delicately neutral path between the great powers of his age, Athens and Sparta. Brasidas, Sparta's greatest general in the Peloponnesian War, was invited north in 424BC by Perdiccas (among other states) to counter Athenian power. However, king and Spartan soon quarrelled as their aims diverged.

Left: Pella was Macedonia's new capital, chosen by king Archelaus in 413BC. In it the young Alexander III later was born and grew up, while Philip extended his power.

Above: Euripides, the Athenian playwright, was among the writers and artists invited to Macedonia by King Archelaus trying to Hellenize his rough kingdom.

A NEW CAPITAL

Under Archelaus (413–399BC), the capital was moved down to Pella near the coast, leaving Aegae as the ceremonial centre. A determined policy of modernization ensued. Archelaus half-tamed the warlords of upper Macedonia, uniting the two halves of his realm. The army was reorganized on professional lines and straight roads linking new forts were built, as Thucydides noted. The Athenian poet Agathon and playwright Euripides were welcomed to Pella – Euripides wrote his last tragedy *The Bacchae* there – while Zeuxis, the famous painter, lavishly decorated its new palace. Socrates, offered refuge in Macedonia when facing trial in Athens, quipped that none would go to Macedonia to see its king but all wanted to see his new palace. So Archelaus and his court were not barbarous. But after his death Macedonia entered a confused decade, with four kings in quick succession. Stability was restored by Amyntas III (393–369BC), father of Philip II and grandfather of Alexander, but Macedonia remained marginal to Greek politics.

Below: This stele of the young Xantos, from Pella, Macedonia's capital, was made c.400BC and is classically Hellenic in style.

MACEDONIANS: GREEKS OR BARBARIANS?

Alexander I was proud to be called *philhellene*, lover of things Greek, but this title was normally awarded to rulers who were definitely *not* Greek. Croesus of Lydia, the monarch overthrown by Persia in 546BC, was termed philhellene, as were some Asian rulers after Alexander. Many Greeks of the Classical Age thought that the Macedonians were barbarians, but the Boeotian poet Hesiod, writing c.700BC, considered them to be Greeks, speaking the same Aeolic dialect as the Boeotians. Two centuries later the Persians, encountering the Macedonians after conquering the Ionians, classed them as Greeks, albeit of a distinct hat-wearing type.

Certainly Macedonians were not savage like the Thracians and Illyrians further north. They may have had odd northern accents (they pronounced Philip as Bilip, for example) and no real *polis* or citizen-state, but the same was true of Aetolians in north-west Greece, and nobody disputed their right to attend the games reserved for Greeks. The Macedonian kings, the Temenids, claimed descent from the ubiquitous sons of Hercules, who, according to legend, had left Argos c.650BC to settle in Macedonia.

It helped Macedonia that Olympus, home of the gods, was within its boundaries and that the games held at Dion on its slopes were clearly Greek. Generally, southern Greeks accepted Macedonia's kings as true Hellenes but disdained their rough subjects, who still had to kill a wild boar or lion to be thought a proper man. After Alexander the Great, as Macedonia grew increasingly rich and sophisticated, such distinctions vanished.

PHILIP II: THE RISE TO POWER 359–334BC

Above: A silver tetradrachm (4-drachmae coin) of 354BC, when Philip II was starting to turn Macedonia into a great military power.

Below: The Roman-era amphitheatre at Philippi, the former city of Crenides refounded by Philip II as a military colony in 357BC to guard his eastern frontier.

In 359BC Perdiccas III of Macedonia, elder son of Amyntas III, was defeated and killed fighting Illyrian invaders on his western frontiers. At the same time savage Paeonians invaded from the north. These disasters eclipsed the modest yet real achievement of Perdiccas' reign: he had thwarted renewed Athenian attempts to regain Amphipolis (the crucial city on the River Strymon), briefly installing a Macedonian garrison there. Perdiccas left a son, Amyntas, aged two, but he also had a younger brother, Philip, aged only 24.

Philip had earlier been a hostage for three years in Thebes, where he had seen the training that at the time made Theban hoplites the best in the world. He reputedly also had an affair with Pelopidas, the much older general. On his brother's death, Philip took over the government, at first as regent for his infant nephew but soon becoming king himself. (Amyntas seems to have grown up weak-willed if not feeble-minded.)

Philip bought off the Paeonians – he knew the power of money – while defeating a pretender backed by the Athenians. Philip then released all the Athenian prisoners without ransom and openly renounced claims to Amphipolis, while secretly offering to swap it for Pydna, a free city in the Athenian Confederacy, so gaining Athens' support. He spent his first winter recruiting and training a new army. Early in 358BC, with 10,000 infantry and 600 cavalry, he routed the Illyrians, killing 7,000 of them.

AN EXPANDING EMPIRE

In 357BC Philip turned east and attacked Amphipolis, which appealed to Athens for help – in vain. Once Philip had captured Amphipolis, he kept it, outwitting the Athenians, who were preoccupied with revolts in their Confederacy and in Euboea. He also seized Crenides inland. He renamed it Philippi, making it a Macedonian military colony, the first of many. These moves secured his hold on Mt Pangaeus and its gold and silver mines.

He soon had the mines worked far more intensively than before to yield 1,000 talents a year – as much revenue as the Athenian Empire had enjoyed in its prime. "Money", Philip observed, "is the sinews of war". Soon after, he took Pydna and Potidaea, giving the latter to Olynthus. This, the most powerful Chalcidic city, had earlier sought Athenian support but now became his well-bribed ally, at least for a time. During his siege in 354BC of Methone, another Athenian ally, Philip lost one eye, marring his good looks.

AN ASTUTE MARRIAGE

Angered by Philip's actions, Athens encouraged his northern neighbours (Thracians, Illyrians and Paeonians) to attack Macedonia, but he beat or bought

Above: Olympia, site of the quadrennial Panhellenic games where Philip's horses won a prize in 356BC, usefully boosting his Hellenic credentials.

them off, using his trusted general Parmenion. Philip also employed non-military means to secure his position. In 357BC he married Olympias, niece of the king of Epirus to his west. Whether romance or *realpolitik* guided his choice (the two reputedly caught each other's eye at a midnight mystery rite on Samothrace), it was an astute move, and Olympias quickly bore him a son, Alexander, in 356BC. News of the birth reached him at the same time as that of a victory for his horses in the Olympic Games, making him a happy king. But events at another site sacred to all Greece were to give him his biggest chance yet.

THESSALY, EPIRUS, THRACE

Since seizing Delphi and its treasuries in 356BC, tiny Phocis had become the most powerful state in Greece, its mercenary forces defeating all armies sent against it. These included Thebes' and, in 353BC, Philip's, but his troops in Thessaly were outnumbered. While Phocian power was built on a dwindling supply of stolen gold, Philip's was based on the rising power of Greece's largest state. Returning to Thessaly in 352BC with a larger army, Philip routed the Phocians

at the Battle of the Crocus Field. He was then acclaimed *tagus* (ruler) of Thessaly, displacing the tyrants of Pherae. This gave him Thessaly's superb cavalry and brought him to the borders of central Greece itself.

Here, however, he was thwarted. Eubolus, then dominating the Athenian Assembly, sent a force to hold the crucial pass at Thermopylae. Not wishing to fight what was still Greece's greatest city just yet, Philip retreated and turned his attentions north. In 351BC he expelled King Arybbas, his wife's uncle, from Epirus, establishing her suitably grateful brother in his place. He brought further western tribes into Macedonia, extending his power to the Adriatic.

On his other flank lay Thrace, mountainous and wild. Profiting from divisions between its quarrelling princes, Philip pushed into its interior, founding Philippolis, his second city, on the River Hebrus. His power now touched the Chersonese – that peninsula on the Dardanelles controlling Athens' essential grain supplies.

Above: A tetradrachm showing Philip on horseback wearing a kausia, *the traditional Macedonian broad-brimmed hat.*

Below: This mural depicts the Rape of Persephone. *From a royal tomb at Vergina (ancient Aegae), it shows the latest Greek artistic styles being adopted in Macedonia.*

A NEW ARMY AND A NEW STATE:
THE RULE OF PHILIP II

Philip created the most formidable army Greece had yet seen, able to defeat Greece and (under his son) conquer Asia. It outclassed Sparta's professional armies, invincible until defeated by Thebes in 371BC. Macedonia's army was huge (by Greek standards), highly professional and increasingly filled with national pride. Philip had inherited a kingdom that was oddly archaic if vigorous, looking back to the Homeric age. He left it the indisputable Greek superpower.

The Macedonian state consisted of two parts: the king himself, who was war leader, supreme judge, high priest and government; and the Assembly of adult male Macedonians, who also constituted the army and acted as a crude court. There were no elected magistrates or Council as in a Greek *polis*. Nor was there a dangerously powerful old hereditary aristocracy, in lower Macedonia at least, although family background still mattered greatly. Philip promoted many new men, whether they came from old Macedonia or elsewhere.

Above: The young Philip II, a fine Roman copy of an original Greek statue.

Below: Macedonia, lying outside Greece proper, was far bigger than any normal Greek state but dangerously exposed to northern invaders. Once united, with its frontiers secured and its potential exploited, it became the greatest power in the Greek world.

Above: This vase from the 4th century BC depicts a rather fanciful battle scene. While Philip's victories were still won by his footsoldiers, cavalry was becoming increasingly important.

MACEDONIA'S MONARCHY

Unlike Persia, Macedonia was not an absolute monarchy. Macedonians bared their heads but did not bow before their king, whom they addressed by name, not as 'Majesty'. Any Macedonian could appeal to Philip for judgement. Kings ruled by hereditary right as members of the Temenid dynasty. If there was more than one claimant to the throne, who succeeded depended on their luck and skill in winning over generals and courtiers. Their choice was then put to

TOUGHENED PROFESSIONALS

Demosthenes complained that Macedonian armies campaigned throughout winter, unlike those of Athens or Thebes, whose citizens had to return to their farms every autumn. Only the Spartans had done this in the past. Macedonians managed this because Philip, doubling his country's size, gave newly conquered lands as estates to his followers, with slaves to work them. This enabled many Macedonians to become full-time soldiers.

The slave-worked mines of Mt Pangaeus remained the basis of Philip's revenues, allowing him to mint gold coins, the first Greek to do so. Only Persia's Great King was richer. Yet Macedonia's army was even tougher than Sparta's. While most Greek armies allowed one servant per hoplite, Philip allowed only one attendant per ten hoplites and banned carriages for his officers. He forced his men to march 48km/30 miles or more a day, summer or winter, over rough tracks. They often had to carry 30 days of supplies on their backs, to keep baggage trains as small as possible.

Philip once reproved a soldier for washing with hot water, saying that only women who had given birth should be allowed such luxury. Women were anyway banned from camp. All this was highly effective, not least because Philip himself endured every hardship and danger, being repeatedly wounded. He had emulated Spartan discipline and toughness on a larger scale.

Above: Although horsemen in the ancient world had no stirrups to hold them in place, they could still fight effectively with both lance and sword, as this coin from Taranto of c.300BC shows.

the Assembly, who voiced approval by clashing their spears, not voting. So the monarchy had a broad popular base.

Once accepted, a strong king in practice had almost boundless power, while a weak one soon lost his throne and his life. A king had, above all, to lead the army. The core of this in 359BC was the Companions (*Hetairoi*), originally 600 upper-class Macedonian horsemen. Philip increased their numbers steadily until by 338BC this royal cavalry numbered 4,000. Philip invented the title 'Foot Companions' for six battalions comprising 9,000 heavy infantry and created the Shield Bearers, 3,000 crack foot guards who, on the battlefield, linked the cavalry and infantry. He reorganized the army by forming battalions more on a territorial or tribal basis than a clan one. By 338BC he could field about 30,000 infantry and 5,000 cavalry, besides garrisoning numerous forts across his empire. This was unprecedented power.

USE OF THE PHALANX

Philip's chief military development was the much-improved phalanx, now a bristling porcupine of spear-men. They carried longer spears than ordinary hoplites: sarissas up to 5m/15ft long. No one could hold such a spears in one hand, so shields became smaller, being slung on the left arm. The phalanx was often massed up to 60 men deep (Philip remembered Theban examples) and was geared for attack, its flanks being covered by other troops. Discipline was vital to maintain formation, but with experienced Macedonian pike-men the phalanx could deliver an unstoppable punch. Philip also hired archers from Crete and engineers such as Polyeidus of Thessaly, who developed siege engines and catapults.

CREATING A UNITED KINGDOM

To unite his kingdom, Philip transplanted populations, settling highlanders in his new cities such as Philippi. He forced upland barons to send their sons to become royal pages at Pella, where they learned soldiering, royal service and some Hellenic culture – and acted as hostages. Philip cemented his power by seven marriages, including one to an Illyrian princess and two to Thessalians. But his prime marriage long remained that to Olympias of Epirus, Alexander's mother.

Below: A view of Pella, the thriving Macedonian capital under Philip II and his successors, where Alexander grew up, with a fine mosaic pavement in the foreground.

CONQUEROR OF GREECE
349–336BC

Above: Philip II was notably handsome until he lost an eye at Methone in 354BC.

Below: Philip's greatest victory came in 338BC at Chaeronea, when he gained control of Greece. This lion, erected later over the mass grave, honours the Theban dead.

In 349BC Philip attacked Olynthus, his former ally and leader of the Chalcidic League. He had an excuse – Olynthus had aided a Macedonian rebel – but Philip had long had his eye on this, the richest city in north Greece. Roused by Demosthenes' impassioned oratory, Athens finally sent the city help in the form of 2,000 men, but they were too few and too late: Philip captured Olynthus in 348BC, destroying it and enslaving its inhabitants. He then annexed the other Chalcidic cities, whose leaders he had often bribed beforehand (he was as good at bribery as he was at strategy). Stagira, which resisted, was destroyed. With his power further boosted, Philip turned south. Athens, exhausted by minor wars around the Aegean and recurrent problems in Euboea, badly needed peace, but she was to get it on unfavourable terms.

AMPHICTYONIC COUNCIL LEADER
In 347BC the Thebans, unable to repulse the Phocians still ravaging western Boeotia, appealed to Philip to become leader of the Amphictyonic Council theoretically controlling Delphi. This was the opening he had long wanted. Sending friendly letters to potential supporters in Athens, Philip invited Greek envoys to Pella, kept them waiting while he conquered more Thracian forts and then proposed peace on the terms of the status quo. During the extended diplomatic exchanges, he suddenly marched south, passing Thermopylae, where the Phocian commander Phalaecus surrendered in exchange for his liberty.

Philip then invited the Athenians to send envoys to another conference to deal with Phocian sacrilege, but Demosthenes persuaded them to ignore his offer. Philip, dominating the Amphictyonic Council, dealt with the Phocians quite lightly, fining them and breaking them up into villages, taking Phocis' seat on the Amphyctionic Council for Macedonia. That year he presided over the Pythian Games held at Delphi. He could now pass as a true Hellene; he also seemed a lesser evil to many small states than Sparta, still feared for its past misrule. At the Peace of Philocrates in 346BC Athens had to accept terms that gave her nothing important while Thebes, Philip's new ally, regained its hegemony over Boeotia. This rankled with the Athenians. For Philip, the peace brought vital recognition of his position in central Greece.

OPEN WAR
Philip did not sit on his laurels. He led his army north-west into wildest Illyria, reaching the Adriatic probably near modern Dubrovnik. Then he turned north-east to annex Thrace. During this campaign, in which he reached the Danube, he fell seriously ill and was also wounded again. Demosthenes, in his 'Philippics' speeches, mocked the "limping one-eyed monster so fond of danger that to increase his empire he has been wounded in every part of his body". But Philip's campaigns had a purpose. With troops on the Black Sea and Sea of Marmara, he could now threaten Athens' crucial grain supplies from the region. Meanwhile, his envoys poured gold into the pockets of potential friends across Greece. These included Aeschines, Demosthenes' chief opponent in Athens.

In 340BC Philip, returning from the Danube, attacked Selymbria and then Byzantium on the Bosphorus. He captured neither, mainly because he lacked ships to counter Persian and Athenian naval support, but late in 340BC he seized a grain fleet destined for

Left: Gold quiver decorated with scenes of soldiers fighting, from the probable tomb of Philip at Vergina.

Below. Byzantium (today Istanbul) shown in a 16th-century Ottoman map. Philip suffered one of his rare defeats when he besieged this city, vital to controlling grain supplies through the Bosphorus, in 340BC. But he still managed to seize a grain fleet destined for Athens.

Athens.

This meant open war. In early 339BC Philip led his army swiftly down through Thessaly, bypassing Thermopylae into Phocis, where he seized the town of Elataea. His excuse was that further problems over Delphi required his assistance. In reality, Macedonia's main army now directly threatened Athens.

Demosthenes' finest hour had arrived. He had sensibly prepared for it by naval reforms, making the trierarchy, by which the richest citizens paid for triremes, more equitable. But Philip was a threat *on land*. Demosthenes now persuaded Athens to offer Thebes, its disliked neighbour but one with a powerful army, full alliance. Athenian concessions won Thebes over and Philip finally faced the two greatest powers in old Greece united. On 2 August 338BC the armies, which were roughly equal in size if not skills, met at Chaeronea.

BATTLE OF CHAERONEA

Philip placed the 18-year-old Prince Alexander on his left commanding the crack Companion cavalry opposite experienced Theban hoplites. He himself led the Footguards facing the Athenians. Philip advanced, attacked and suddenly seemed to flee. The Athenians pursued him recklessly, opening a great gap behind them. Through it charged Alexander to surround the Thebans. Philip then counter-attacked, soon routing Athens' citizen-soldiers. But the crack Theban Sacred Band, 150 pairs of homosexual lovers, fought to the last man. Athens lost over 1,000 dead with 2,000 taken prisoner. Chaeronea, a resounding triumph for Philip, was a disaster for Greek freedom.

PHILIP: TRIUMPH AND DEATH 338–336BC

As undisputed victor, Philip could dictate the peace terms. Athens had to surrender the Chersonese (Dardanelles peninsula), becoming dependent on Macedonia's goodwill for the safety of its grain fleet. But in other ways, as even Demosthenes admitted, the peace was generous to her. She regained Oropus, long disputed with Thebes, and paid no indemnity. She also got back her prisoners without ransom, while young Prince Alexander ceremoniously returned the ashes of her dead soldiers. In return, the Athenians erected a statue to the Macedonian king. Philip's leniency to Athens was based on realism: Athens still had her long walls and large fleet, which would have made a direct attack very difficult. He also wanted her naval support for his anti-Persian crusade, while attacking Athens would harm his image as a true Hellene.

To Thebes, his former ally, Philip showed no such generosity. He broke up the Boeotian federation, long dominated by Thebes, restored the destroyed cities of Plataea and Orchomenus and put a Macedonian garrison in Thebes' citadel, the Cadmaea, to support a narrow pro-Macedonian oligarchy.

MERCY TO SPARTA
To that other once great city, Sparta, Philip behaved differently. When the Spartans refused to send envoys to the Council of Corinth, saying they were accustomed to lead other Greeks, not follow them, Philip invaded Laconia. He seemed about to take Sparta – this would not have been difficult, as Sparta's army was tiny – but desisted.

He knew that Sparta still appeared to menace her smaller neighbours, making them pro-Macedonian. He simply ravaged Sparta's lands and reduced her territory still further, giving the Dentheliatis border area to Messenia, Sparta's

Above: Philip was the first Greek ruler to mint coins in gold, such as this stater showing the god Apollo. He could do so thanks to his acquisition of the gold mines of Mt Pangaeus.

worst enemy. Then he called a Panhellenic Council at Corinth in the winter of 338–337BC.

Philip chose Corinth partly because of its central position and wealth but mainly because of its associations. It had been the seat of the Panhellenic League in the Persian Wars almost 150 years earlier, when many Greek states had for once united against Persia. Isocrates (436–338BC), the Athenian writer, had long urged the Greeks to stop fighting each other and turn against Persia, inside whose supposedly decaying empire they could easily win booty and new lands. He now hailed Philip as fulfilling his idea of a Panhellenic crusade.

A COMMON PEACE
Philip, happy to accept the aged orator's praise, had his own reasons for a Persian campaign, but first he had to settle affairs in Greece. Elected Captain-General of all

Below: Found in the royal tombs at Vergina (ancient Aegae, the old capital), this silver vase reveals the wealth and sophistication Macedonia attained under Philip II.

the Greeks, a novel post, he announced a Common Peace. Superficially, this promised an end to Greece's incessant feuds by guaranteeing existing constitutions and banning any redistribution of land. Actually, as pro-Macedonian oligarchs had taken over in many places, it meant that he controlled much of Greece. To reinforce his hold, Philip stationed garrisons in Corinth, Ambracia and Chalcis (in Euboea), besides Thebes. Philip then announced a Panhellenic war of revenge, sent Parmenion ahead with 10,000 troops into Asia and returned north. Before he conquered Asia, there were domestic affairs to deal with.

ASSASSINATION

"Wounded is the bull; the end is near; the sacrificer is at hand." So spoke the Delphic oracle, ambiguous as ever, in response to Philip's enquiries. Flushed with success, Philip assumed the bull to be the Persian Empire. In the summer of 336BC at Aegae he celebrated the marriage of his young daughter Cleopatra to her uncle Alexander of Epirus. Such close ties were acceptable among royalty, for it was certainly no love match. In 338BC Philip himself, however, had fallen in love with and married 17-year-old Eurydice, niece of his general Attalus. She produced a daughter and then, in 336BC, a son. Philip promptly divorced Olympias, whose Epirote connections were no longer needed, making Alexander deeply worried about his future.

That wedding day, however, when Philip watched the procession of the Twelve Olympian gods at Aegae, followed by his own image in Olympian size, he was flanked by both Alexanders: his son and his new son-in-law. Suddenly a young guard called Pausanias rushed up and stabbed Philip, who died instantly. Pausanias, caught as he tried to escape, was summarily executed.

The true motive behind the murder died with him – conveniently for many people, perhaps. The story went that Pausanias, once Philip's lover, had been raped by servants of Attalus, whom he had slandered. Why Pausanias should then want to murder the king is unclear. What is clear is who benefited: Alexander, Philip's son by Olympias, due to be left behind in the coming war on Persia, and Olympias, Philip's ex-wife in bitter exile. Politics in Macedonia seemed to be reverting to their chaotic norm. But Philip, although cheated of winning an Asian empire, had built his kingdom so solidly that it did not fall apart. This was his real achievement.

Left: Isocrates (436–338BC) was an Athenian orator who had long called for Greece to unite against Persia. In his last years he hailed Philip as the ideal leader for such a Panhellenic crusade.

Below: This elaborately decorated gold lamax or coffin found in the royal tombs at Vergina (ancient Aegae) contains what may be Philip II's bones. Such flamboyant riches recalled the splendours of Mycenaean Greece, the age of heroes, not of the cash-strapped democracies then the norm in Greece proper.

CHAPTER VIII

THE YOUNG ALEXANDER

356–336BC

Alexander was a phenomenon from birth – or so it was later told. The stories of his early years are so colourful that they approach the legendary. But there is no doubt that he was an infant and then adolescent prodigy. Remarkably precocious, he tamed a savage stallion at the age of 12 and was left as regent of the kingdom in his father Philip II's absence when only 16, repelling an invasion and founding his first city. When just 18 he led the decisive cavalry charge at the Battle of Chaeronea and then a delicate diplomatic mission to Athens.

Life in the Macedonian court – stimulating, exhilarating, never easy – encouraged rapid developers, but only a year later Alexander's position as heir apparent seemed endangered by Philip's last marriage. A year after that, he found himself king at the age of 20, ascending the throne over his father's corpse. What involvement if any he had in Philip's murder is unknown, but his succession was far from inevitable. Yet southern Greeks who confidently expected Macedonia to collapse into its traditional feuding after the death of Philip were soon disappointed. Within weeks Alexander was master of his own kingdom; within months he had made a lightning descent through Greece, claiming Philip's powers and titles. The Balkan campaign that followed, and the blitzkrieg return that destroyed Thebes, showed the Greeks that he was already even more dangerous than Philip.

Left: Mosaic of c.310BC from Pella showing the young Alexander hunting lions.

Left: Two scenes from The Romance of Alexander, *a fantastical history written about Alexander long after his death. At the top Alexander is seen consulting the Delphic Oracle, which is a fiction, and below with his horse Bucephalus, who existed.*

BIRTH AND CHILDHOOD
356–347BC

The baby who became Alexander III, Alexander the Great, was born in July 356BC to Olympias, daughter of the royal house of Epirus, and Philip II of Macedonia. According to Plutarch, before his birth Olympias dreamt that lightning struck her womb, while Philip dreamt that her womb showed a lion's seal. Philip received the news of the birth of his first (legitimate) son on the same day his horses won at the Olympic Games and his troops took Potidaea. Philip's Temenid family claimed descent via Hercules from Zeus, whose mountain home rose abruptly on their kingdom's southern flanks. Olympias' family traced its descent from Achilles, Homer's heroic prince. So Alexander had the most illustrious forebears imaginable. This helped shape his exalted view of his destiny.

Little is known of Alexander's childhood. The young prince could not have seen much of his father, who was away on his wars. His mother chose his first two tutors. One of them was her cousin Leonidas, who emulated his Spartan namesake, hero of Thermopylae, in toughness and austerity, confiscating anything exotic or luxurious in Alexander's belongings. When Alexander later conquered the Lebanon, he sent wagonloads of incense to Leonidas, telling him not to be mean to the gods. By contrast, kind old Lysimachus became so attached to Alexander that he followed him to Asia.

Such sober male influences countered the erratic behaviour of Alexander's mother. Olympias reputedly kept sacred snakes in her bed and worshipped Dionysus, god of wine and orgies, and the sinister Hecate, goddess of suicide and the underworld. "While others sacrifice tens and hundreds of animals, Olympias sacrifices them by the thousand and tens of thousand," wrote a student of Aristotle, the philosopher who came to know the intrigues and personalities at the court at Pella well.

NEW ARRIVALS AT PELLA
Pella was growing fast at this time, as Philip's conquests attracted diplomats, courtiers, artists, merchants and exiles. Among the last was the Persian Artabazus,

Left: The birth of Alexander in 356BC, shown in this mosaic of c.310BC from Phoenicia, was preceded by portents of greatness, according to later legends. Phoenicia was a part of Asia that became Hellenized after his conquest.

a former satrap of Phrygia. He brought with him Barsine, his beautiful young daughter, ten years older than Alexander. Alexander talked to Persians with friendly curiosity, discovering the virtues of this civilized people. Years later, when they met again in Asia, Alexander reappointed Artabazus as satrap, while Barsine became one of his mistresses. Many newcomers to Pella were southern Greeks: Nearchus the sailor from Crete, Demaratus the soldier from Corinth. All became Alexander's life-long friends, although most Macedonians looked down on Greek hirelings.

THE PEACE CONFERENCE

In 346BC Philip's combination of guile, cash and force led to his triumph. Macedonian armies entered central Greece ostensibly to 'punish' Phocian sacrilege. For the resulting peace conference at Pella emissaries came from all over Greece, including two from Athens: Demosthenes and his opponent Aeschines, probably in Philip's pay.

By then Alexander was ten, old enough to appear and "play the lyre and recite and debate with another boy", according to Aeschines. A year later a row blew up in Athens amid accusations that one or other of the politicians had flirted with the already handsome young prince and been unduly influenced by him.

Whatever the truth, it suggests that Alexander was already politically alert and also reveals the prevalence of homosexuality in Greek public life.

Above: A view of Mt Olympus, the mountain home of the Olympian gods. Alexander came to believe that he was indeed the son of Zeus, king of the gods.

BUCEPHALUS: ALEXANDER'S HORSE

When Alexander was twelve, he began one of the greatest relationships in his life – with a horse. His friend Demaratus had offered Philip a huge black stallion costing 13 talents, more than three times anything paid for a horse before. Philip ordered the horse to be led out but it bucked and reared, refusing all orders. Philip was about to reject it when Alexander offered to tame him. Taking the horse by his halter, Alexander patted and quieted him. Then he mounted and galloped around to universal applause, Philip exclaiming proudly that Macedonia would never contain such a boy. Or so the story goes. Alexander had noticed that the horse was shying at its own shadow. By turning its head to the sun, he overcame its fears.

Alexander called his horse Bucephalus, 'Ox-head', because of a white mark on his black head. Bucephalus became devoted to his royal master, following him literally to the ends of the Earth. Alexander rode him in his greatest battles and taught him to kneel fully armoured before him. When hill tribes near the Caspian Sea kidnapped the horse, Alexander's anger was so terrible that they returned him at once.

Alexander last rode Bucephalus into battle against the Indian rajah Porus in 326BC. Soon after, Bucephalus died of old age, being perhaps 30. (The Greeks did not know how to tell a horse's age by its teeth, the standard method.) Alexander commemorated his beloved stallion by founding and naming a city after him in what is now northern Pakistan.

Below: Alexander astride Bucephalus, the black stallion he rode to the ends of the Earth. Alexander was above all a cavalry commander.

EDUCATION AND YOUTH
346–340BC

Above: Aristotle, once Plato's most brilliant student in the Academy in Athens, was chosen by Philip to tutor the young prince. If unheroic in appearance, Aristotle had much the greatest mind of his generation, widening Alexander's mental horizons.

Below: A medieval painting of Aristotle's school for Alexander and his companions at Mieza, where he taught the boys subjects from zoology to drama.

Alexander had only one full sibling, his young sister Cleopatra, of whom he was very fond. But Macedonian girls played little part in public life, although they were not as secluded as those in Athens. To prepare his precocious, intelligent but emotionally volatile son for public affairs, Philip encouraged noblemen's sons to join what became the select group of Alexander's close companions. Foremost among these was Hephaistion, son of Amyntor, a Macedonian aristocrat.

HEPHAISTION

Alexander, who had learned to read when very young, was obsessed throughout his life by Homer's great poems, especially *The Iliad*, which related the exploits of his irascible supposed ancestor Achilles. In *The Iliad* Achilles is passionately devoted to his friend Patroclus, a devotion that by the 4th century BC was widely seen as erotic, although this is not how Homer shows it. Almost certainly Alexander and Hephaistion were lovers, then and for

years after, although Alexander always said that only sex and sleep reminded him that he was mortal. A homosexual relationship between the two boys would have been thought acceptable, even laudable, in Macedonia's militaristic society.

The one surviving portrait of Hephaistion does not suggest great beauty, and records indicate an utterly loyal, if rather dull, subordinate, who ended his career as Alexander's Grand Vizier. Dull but devoted loyalty was what Alexander needed at this stage, however, for his parents were constantly quarrelling. Philip was taking other, younger wives, often for reasons of state, making Olympias ragingly jealous. Alexander must have been the unhappy recipient of her hysterical rants.

Olympias reputedly also introduced a Thessalian prostitute into Alexander's bedroom to test his virility. Understandably, he rejected her – all his life Alexander hated prostitution or rape.

Other boys from the Macedonian nobility, including two sons of Antipater (one of Philip's best generals) and some from Upper Macedonian clans, notably Harpalus, joined the magic circle, soon guided by the age's greatest mind. For Philip wanted the best tutor money could buy for his adolescent son. In 342BC he chose another north Greek: Aristotle.

THE PHILOSOPHER-TUTOR

Born in 384BC in Stagira, a small city in the Chalcidice recently deleted by Macedonia, Aristotle had been the most brilliant student at Plato's Academy in Athens. He had left Athens on Plato's death in 347BC, probably disappointed at not being chosen as the next head of the Academy. Going north to the Troad (Dardanelles), he joined a community of philosophers and soon married the daughter of Hermias, a local ruler, at whose court he then lived. He later joined the polymath Theophrastus on Lesbos, where he carried out zoological investigations. So he was no reclusive academic (if unheroically "thin-legged and small-eyed") when he landed in Macedonia in 342BC to teach the 14-year-old prince. He stayed for four years based at Mieza, a small coastal town.

"[Aristotle] taught him writing, Greek, Hebrew, Babylonian and Latin. He taught him the nature of the winds and sea; he explained the stars' courses, the revolutions of the heaven…. He showed him justice and rhetoric and warned him against the looser sort of women." This comes from *The Romance of Alexander*, a most unreliable biography. Neither Aristotle nor Alexander learned Hebrew, Babylonian or even Latin (Rome was still struggling for survival in central Italy, although Aristotle, hearing of it, noted that it had the institutions of a *polis*). But Aristotle did teach Alexander a huge amount about the natural world.

Bertrand Russell, the 20th-century philosopher, thought that Alexander must have been "bored by the prosy old pedant". However, Aristotle had courtly

Right: A giant bronze head of Hephaistion, Alexander's first friend and very probably first lover. Hephaistion became Alexander's trusted second-in-command, his death causing the king huge grief.

manners – his father Nicomachus had been physician to Amyntas III, Philip's predecessor – and at only 42 was not old. Alexander's interests in botany, zoology, geography and biology were fired by Aristotle, and he later sent specimens from Asia back to his old tutor. Aristotle also deepened his knowledge of Greek literature, especially of the great Athenian playwrights, Euripides becoming one of Alexander's favourite authors. But the boy grew into a man of action, not thought, probably never much interested in Aristotle's ethics or metaphysics.

GREEKS AND BARBARIANS

On one point they differed profoundly. Aristotle had the typical Greek prejudices about 'barbarians', meaning all non-Greeks, including Persians. He considered them inherently inferior, to be treated as slaves. Quite early in his career, Alexander began thinking and behaving differently. This led him ultimately to clash with his own soldiers – and with Callisthenes, Aristotle's relative who was appointed as Alexander's official historian. Yet Aristotle must have widened and enriched Alexander's view of the world. Aristotle was enriched in the worldly sense by his stay in Macedonia. When Aristotle died in 322BC, he had 18 slaves – the sign of a rich man and unusual for a philosopher.

Below: Olympias, shown here in a cameo portrait with her son Alexander, was a stormy character, frequently clashing with her unfaithful husband and trying to turn her son against the king. Alexander inherited much of her fiery temperament and mysticism.

FIRST COMMANDS AND FAMILY QUARRELS 340-336BC

Above: Epirus on the north-western fringes of Greece was even wilder than Macedonia. It was the home of Olympias, to which she retreated after being divorced by Philip. Alexander, who had quarrelled with his father, accompanied her but soon returned.

Below: Alexander and a friend, perhaps Craterus, hunting lions, in a mosaic from Pella of c.310BC. Alexander loved hunting almost as much as war, although he did neither naked.

Far more central to Alexander's upbringing than studying botany or even Homer was his military training. All Macedonian boys learned the basics of arms drill and riding while young. By his teens Alexander was as good at riding and using arms as could be hoped. He was also a passionate hunter, pursuing the bears, lions and boars found in Macedonia's forests – a traditional royal pastime. What could not be known in advance, and could hardly be taught, was the knack of leadership, vital in the informal but absolute monarchy of Macedonia.

THE DEFEAT OF THE MAEDI

While Philip was away campaigning in Thrace and then engaged in the long, ultimately fruitless sieges of Perinthus and Byzantium in 340BC, Alexander was left as regent. This reveals Philip's confidence in a son still only aged 16, although his experienced general Antipater was on hand to give advice if needed. Alexander's military skill was tested almost at once, for an invasion by a Thracian people called the Maedi (who may have heard of the king's absence) threatened Macedonia's eastern borders. Gathering the reserves, Alexander marched forth and defeated the Maedi, pursuing them back into Thrace. To commemorate his victory, he founded a city he called Alexandropolis, the first city of many to bear his name.

BATTLE OF CHAERONEA

What Philip, who had founded only two cities himself, thought of this is not recorded, but he was clearly impressed enough by his son's military skill to give Alexander command of the cavalry on the left wing of the Macedonian army at the Battle of Chaeronea in 338BC. This was a battle mostly fought between, and decided by, hoplites. Alexander's well-timed charge cut off the amateur Athenian infantry from their Theban allies. He then turned on the more experienced Thebans, keeping them surrounded until the Macedonian infantry caught up and destroyed them. (Greek cavalry was generally ineffectual against well-trained hoplite infantry.) Alexander had shown that he could lead horsemen in a dashing charge *and* control them.

Shortly afterwards Philip appointed Alexander to lead the guard of honour that ceremoniously returned the ashes of Athens' dead soldiers to the city – something essential to Greek burial rites. As a special guest, Alexander was shown around Greece's greatest city. He reputedly turned down the offer of a young boy as company for his bed, again showing his dislike of prostitution. What he thought of Athens itself, which even Plato, who had very mixed feelings about his native city, had called "the city hall of Greek wisdom", is unknown.

QUARRELS AND EXILE

This harmony between father and son did not last. Back in Macedonia, Philip married for the seventh time, his bride

this time being the 17-year-old Eurydice, great niece of Attalus, one of his leading marshals. Unlike many of Philip's politically motivated earlier marriages, this time he was in love. Worryingly for Alexander, Eurydice's children would be full Macedonians, not half-Epirote like himself. One drunken night Attalus invited the company to pray for "a true Macedonian heir". Alexander, enraged and flushed with drink, caused a row. His father, also drunk, staggered up drawing his sword but stumbled and fell. "Here is the man who would cross to Asia but falls between the couches!" jeered Alexander. Taking his mother and close companions, he went into exile in Epirus.

Olympias remained in exile until Philip's death. But Philip still needed an adult heir, while Alexander soon grew bored in provincial Epirus and wanted to return home. Accordingly, within months things were patched up between father and son, the reconciliation helped by Eurydice's first child being a girl. Alexander remained nervous about his position, however, as was demonstrated by the Carian marriage fiasco.

THE CARIAN MARRIAGE FIASCO

Arrhidaeus was Philip's half-witted son by a dancing-girl. Such a minor royal was good enough to offer as a husband for the daughter of the Carian satrap, who wanted to ally himself with Philip, thus usefully extending Philip's influence into Asia. However, Alexander, hearing of this, secretly proposed himself as a husband instead, fearing that Philip was preparing to give his kingdom to Arrhidaeus. The delighted Carians accepted the prospect of the much more impressive Alexander until Philip vetoed the plan, exiling many of Alexander's friends who had convinced him to take this action. Taking fright, the Carians then sought a match with a Persian satrap, so wrecking Philip's carefully laid plans.

Above: In this excellent copy of a bust by Lysippus, one of the age's finest sculptors, the visionary dynamism and ruthlessness of the young conqueror are apparent.

THE DEATH OF PHILIP

In 336BC Eurydice gave birth to a boy, a rival to Alexander. That same summer Philip announced the marriage of Cleopatra, his daughter by Olympias, to Alexander of Epirus, her uncle. This marriage meant that Philip no longer needed a marriage tie with Olympias, whom he promptly divorced. Alexander again felt insecure, especially as he would be left behind as regent when Philip invaded Asia. But he accompanied his father that fateful morning in August 336BC into the theatre at Aegae to watch the procession of the Olympian gods. Within minutes Philip was dead, assassinated for unknown reasons by Pausanias, a bodyguard. Alexander was the obvious heir. But his succession was by no means automatic.

SECURING THE THRONE AND GREECE 336–335BC

Above: The wealth of Macedonia is revealed by this gold-decorated breastplate found in the royal tombs at Vergina (ancient Aegae), dating from Philip II's reign.

Below: The fine Ionic columns of the royal palace at Pella, the capital where Alexander spent his last winter in Macedonia in 335–334BC.

Although he was the obvious heir, Alexander's succession was not assured. His cousin Amyntas, shouldered aside 20 years before, had a claim to the throne, as had Eurydice's infant son. But Alexander was known to the army and nobility. When Alexander of Lyncestis stepped forward to hail Alexander as king, things moved his way. The assassin Pausanias, caught as he fled to a waiting horse, was speedily executed. So too were Amyntas, Eurydice's son and two Lyncestid brothers whose loyalty was suspect. Such precautionary ferocity became a hallmark of his reign. Alexander announced a timely tax cut and organized funeral games for his father. Philip was buried in royal splendour in tombs that have only recently been discovered.

At first Alexander could count on the firm support of only one of Philip's three marshals, Antipater. Of the two others, he had Attalus, his enemy, who was commanding the advance guard in Asia, quietly murdered. This was done with the consent of Parmenion, the last of the three. (Parmenion's sons were serving with Alexander, making them useful hostages.) With the throne secure, he led the army between the two halves of a dissected dog – an old Macedonian rite – and then turned south in October.

Greece was in ferment. News of Philip's murder had raised anti-Macedonians' hopes everywhere, and Alexander had to quash them quickly. As Thessaly refused to let him pass through the Vale of Tempe, he cut steps in the side of Mt Ossa to bypass it, being duly elected leader by the astonished Thessalians, henceforth faithful allies. Bursting into Greece at the head of his army, he deflated opposition, being elected head of the League of Corinth in succession to Philip. Then he turned north. Before he left for Asia, there were Balkan tribes to be pacified.

THE BALKANS – AND THEBES

His first target in 335BC was the Triballians, a Thracian people who had ambushed Philip three years before. Finding the crucial Shipka Pass defended by Triballians holding carts poised to roll down on his army, he ordered his men to lie flat on the ground with their shields over them. The carts rumbled past harmlessly above and Alexander resumed the attack, using slingers and archers to lure the Triballians into the open, where his infantry crushed them. The Triballians then retreated to the River Danube's far banks. Alexander lacked ships to ferry his whole army across, so the Macedonians filled their leather tents with straw and crossed the river on these impromptu rafts, protected by catapult fire. Forming up on the other side, they routed the amazed Triballians, some of whom joined this increasingly polyglot army. Alexander was the first Greek commander to cross the Danube.

Alexander next moved against the Illyrians on the north-west frontier. Getting trapped in a narrow wooded valley, he formed a phalanx 120 deep. With its massed sarissas making a terrible swishing sound and its shields clashing, he routed the Illyrians by fear rather than arms. A later night attack had completed their defeat when news from the south required his rapid return: Thebes had revolted on

DIOGENES AND ALEXANDER

Diogenes of Sinope (404–325BC) was a most original philosopher. Despising worldly goods, he slept in a tub and performed all bodily functions in public. He believed that only one thing mattered: distinguishing between virtue and vice. His scorn for convention won him the nickname *cynos* (dog), and his followers were known as Cynics. Reputedly, when visiting Corinth in 336BC, he met Alexander. Was there, the king asked, anything he could do for him? Yes, said Diogenes, get out of my sunlight. Impressed, Alexander said that if he had not been Alexander, he would have chosen to be Diogenes. Certainly both men pursued their aims with single-minded extremism.

Below: At Corinth, busy organizing the Panhellenic League against Persia, Alexander encountered the ascetic Cynic philosopher Diogenes living in a tub.

rumours that Alexander was dead, killing some Macedonian officers and restoring its democracy.

In two weeks Alexander marched his army 800km/500 miles south – an unbelievably fast pace. The Thebans at first could not believe it, but when they did, they were defiant. However, no help came from any other Greeks. In the resulting battle outside the city the Macedonians were hard-pressed at first, but the Thebans left a side gate open behind them. The Macedonians pushed through it and Thebes fell to them. All its 35,000 inhabitants were killed or enslaved and all its buildings, except the temples and the house of Pindar, its famous poet, destroyed. One of Greece's most ancient cities was no more.

After this act of calculated terror, which the League of Corinth rubber-stamped, Alexander had no further problems with the Greeks. He demanded but did not get Demosthenes from Athens, instead taking 20 Athenian triremes as hostages. Then he returned home to prepare for invasion by a winter of feasting and planning.

Above: A romantic depiction of Alexander's triumphs by the 19th-century artist Gustave Moreau. Alexander's lightning conquests soon passed into legend.

AIMS AND STRENGTHS
334 BC

Above: This elaborate krater (drinking vessel) of gilded bronze from Macedonia illustrates the kingdom's recently acquired wealth.

Below: Alexander, here shown on the Sidon Sarcophagus *carved soon after his death, always led his armies from the front in battle.*

Alexander was not the first Greek ruler to conceive of attacking the Persian Empire. Jason of Pherae and Philip, his father, had both had invasion plans, aborted only by their murders, and the idea had been suggested by Isocrates among others for decades. The 'March of the 10,000' – the Greek mercenaries who had penetrated the Persian Empire and returned almost unharmed in 401 BC – appeared to show up the vulnerability of Persia. Persia had regularly hired Greek hoplites ever since, who continued to dominate warfare around the Mediterranean and in Egypt. But the unsuccessful campaigns of the Spartan king Agesilaus in Asia Minor in the 390s BC indicated that conquering Asia needed more than just a decent general and hoplites. (Problems back in Greece had called Agesilaus home anyway.) If Persia lacked heavy infantry to match Greek hoplites, it had fine cavalry in abundance.

AN UNKNOWN EMPIRE

Few Greeks had any idea of what an eastern campaign might entail. While some had visited Susa, the administrative capital (in south-western Iran) as envoys, mercenaries, captives or craftsmen, none realized the empire's true immensity. Even Aristotle, who probably knew more geography than any other Greek, vastly underestimated the distance between the Aegean and Susa. And the Iranian heartland of the empire stretched east of Susa. To traverse the broad plateaux and mountain ranges of Asia and so conquer Persia, cavalry was needed in force, backed by an army that was professional in every department.

STRONG ARMY, WEAK NAVY

Alexander had such an army, thanks to Philip, for the first time in Greek history, based on the infantry grouped into massive phalanxes that were invincible in the right conditions. He also had the commando-style Shield Bearers, siege engines and catapults, Cretan archers and Thracian and Illyrian irregulars as slingers and javelin-throwers. Above all he had excellent cavalry, including some Thessalians, who had taught the Macedonians their highly effective wedge-attack formations.

In 334 BC, after Alexander had crossed into Asia, he fielded about 43,000 infantry and 6,000 cavalry – by Greek standards a huge army, though modest by Persian. Alexander left Antipater in Macedonia with another force of 12,000 infantry and 1,500 cavalry, plus garrison troops across Greece and the Macedonian Empire. The gold and silver from Mt Pangaeus, coupled with the lands newly conquered by Philip and worked by slave labour, helped to pay for all this. Even so, he started his campaign in May 334 BC 600 talents in debt.

The one major Macedonian weakness was its lack of a decent navy: it had only 160 triremes, including its unwilling allies. Athens had 400 triremes, although it lacked the needed sailors. This was the biggest Greek fleet, but Athens was ambivalently neutral. The 20 Athenian ships Alexander took acted chiefly as hostages in the fleet, most of which he dismissed anyway in 334 BC.

Ultimately, Alexander's greatest asset was his own genius and luck, in which he believed from the start and which his men soon came to accept. Good generals and the best army in the world were his tangible strengths.

AN EVOLVING PLAN

What exact aim Alexander had in mind when he began his attack in spring 334 BC remains debatable. The historian William Tarn wrote: "The primary reason Alexander invaded Persia was that he never thought of *not* doing so. It was his inheritance." Alexander's declared purpose in 334BC was to exact revenge on Persia for its invasion of 480–479BC. Most Greeks, however, did not see this as his real reason. Probably, like some of his officers, they expected Alexander simply to conquer Asia Minor, raid the Persian heartland and return with loot. Rivalry with his dead father – a semi-conscious desire to conquer more rapidly and completely than Philip had – may have spurred Alexander on at first, but his *pothos* (longing) and an increasingly imperial vision led him ever further east.

ALEXANDER THE ASIAN EMPEROR
On landing in Asia, Alexander threw a spear on to the shore, symbolically claiming the empire by right of conquest. He then appointed a Macedonian as satrap of coastal Phrygia, the first province conquered, seeming to continue the Persian imperial system. Soon after, however, he posed as liberator of the Greek cities in Asia Minor, and later sent back to Athens from Susa the statues of Harmodius and Aristogeiton, the Athenian tyrannicides, carried off in 480BC. But he began reappointing Persians as satraps to rule their provinces after Gaugamela in 331BC, his greatest victory. By then he was seeing himself as heir to the Achaemenids, the rulers of Asia. As such, he had to impress his Asian subjects,

and he began wearing elements of Persian royal dress and adopting Persian court customs such as *proskynesis*, bowing to the throne. Such innovations proved very unpopular with his Macedonians. Alexander, who had started the war as a Greek avenger, ended it as a Greek-speaking Asian emperor.

Above: Alexander at the charge. He loved warfare more than anything else, having a truly Homeric delight in battle.

Left: The phalanx remained a vital part of Alexander's army, although in battle he usually relied on cavalry to deliver the knockout blow.

PERSIA: AN EMPIRE IN DECLINE? 404–336BC

The Achaemenid Empire created by Cyrus the Great and Darius I (556–486BC) was the greatest power on Earth, ruling all lands between the Indus and Macedonia. The Greeks in the Persian Wars (490–478BC) had managed to defeat Persia only by an unusual display of unity. It then took the fleet of Athens and her allies in the Delian Confederacy years of hard fighting to clear the Aegean islands and coasts of Persian bases. When Athens tried to interfere in Cyprus and Egypt, she was crushingly defeated.

The Peace of Callias of 449BC had definitively confirmed that the Aegean islands and cities would be Athenian, while the rest of Asia, including half-Greek Cyprus, would be Persian, as would Egypt, the second richest satrapy in the empire. Athens in truth had stripped Persia of only a few coastal cities.

Below: The Apadana Staircase at Persepolis, built under Darius I (521–486BC), showing the 10,000 Immortals, the elite royal Foot Guards so-called because when one died, he was immediately replaced.

TREACHERY AT COURT

Persian power depended finally on the Great King, who in turn depended on his courtiers. Artaxerxes III had made Bagoas, a eunuch, his Grand Vizier (first minister) – eunuchs were often employed at court because they were thought to present no threat. Bagoas had ideas of his own, however, and in 338BC he poisoned his royal master, as he did the next king. Bagoas then chose Darius Codomanus, a nobleman with only a distant claim to the throne, as the next ruler. In 336BC he became king as Darius III, the last Achaemenid. Darius promptly had the treacherous eunuch killed; but he was to prove no match for Alexander, who became king that same year.

EGYPTIAN INDEPENDENCE

Egypt, however, proudly conscious of its ancient civilization, was never happy under Persian rule. Its *fellahin* (peasants) were swayed by its powerful priests, who were angered by obvious Persian contempt for their religion. As a result, Egypt revolted frequently. It rebelled three times in the 5th century BC and after 405BC was independent for 60 years, relying on Greek hoplites to repel the Persians. These included King Agesilaus of Sparta in his cash-strapped old age.

Egyptian independence was a humiliation as well as a financial loss for the Great King, but not a serious threat to the Persian Empire in the way that the revolt of Prince Cyrus in 401BC was. Jealous of his older brother Artaxerxes, who had just succeeded to the throne, Cyrus had had unusually wide powers in Asia Minor. He used them to recruit a rebel army, whose core consisted of 'The 10,000' Greek hoplites, and 'marched upcountry' (as recorded in soldier-turned-historian Xenophon's work *Anabasis*). Persian imperial forces did not try to check his passage until he reached Cunaxa, just north of Babylon. The rebels won the ensuing battle, thanks mainly to the Greeks. But Cyrus, who had personally tried to kill his detested brother, was killed. The subsequent return home of the Greeks under Xenophon's command made a colourful tale, and it taught the Persians the need to hire Greek hoplites. This was easy for the wealthy empire. The real problem was the satraps.

THE REVOLT OF THE SATRAPS

The Persian system of government gave satraps remarkably wide powers, both financial and military, over their often large provinces. The size and diversity of the Persian Empire and the slowness of communications perhaps made such devolution inevitable, and the 'King's Ears', as the royal agents were known, acted as a check. But in the distant reaches of the empire (western Asia Minor was a full three months' march from Susa even on the Royal Road) satraps tended to establish hereditary dynasties. These semi-ducal rulers developed local ties, ambitions and rivalries that could undermine loyalty to the crown in distant Susa.

In the 360s BC many satrapies in western Asia rose in what has been called the 'Revolt of the Satraps'. Even Cappadocia in the Anatolian interior rebelled, as did Cyprus (again) and Sidon in Phoenicia. In Caria the dynast Mausolus began to extend his power, while Egypt remained independent. Since coins, essential to pay mercenaries, were minted only in western satrapies, this loss threatened the Persians' recruitment of essential Greek mercenaries and of its fleet, mostly supplied by Phoenicia, Cyprus and Caria. The whole empire west of the Euphrates appeared lost.

In 358BC Artaxerxes II, an incompetent drunk, was succeeded by Artaxerxes III Ochus, a far better ruler. Like Philip

Above: Two of the 10,000 Immortals, Persian noblemen who formed the Foot Guards. They were, however, the only heavy infantry the Great King had at his disposal.

Below: A gold winged-lion rhyton (drinking vessel) of the 6th century BC, emblematic of the splendour and wealth of the Persian empire.

II, his contemporary, Artaxerxes played off his enemies against each other, while mustering immense forces from the Persian heartlands. With them he crushed Artabazus, satrap of coastal Phrygia (north-western Asia Minor) and began regaining the whole peninsula, helped by the death of Mausolus in 353BC. In 344BC Persia recaptured Sidon after a long siege and went on to regain Egypt, although brutal reprisals there made its rule hated. By 340BC Persian power in the west had been fully restored, it seemed, and Artaxerxes could help the cities of Perinthus and Byzantium when Philip besieged them.

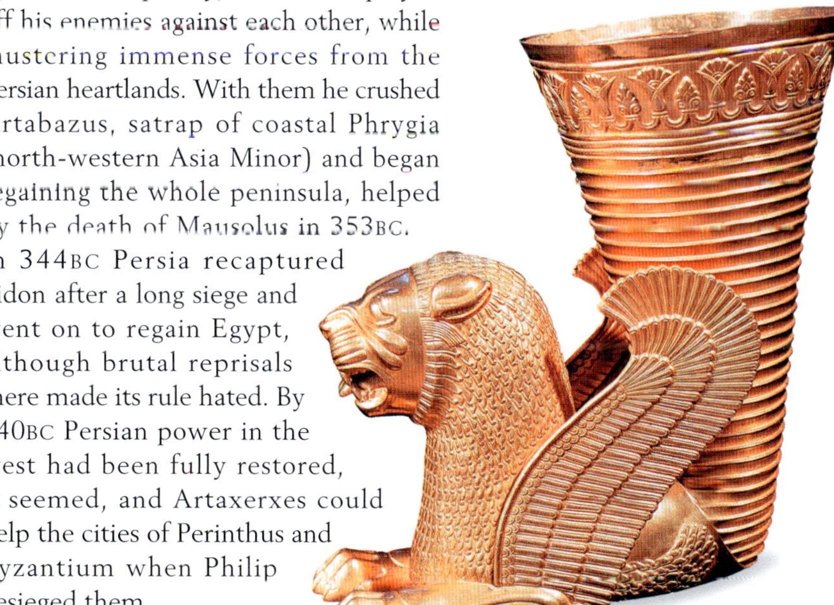

Above: By the 330s BC, the Persian empire had amassed 230,000 talents in its royal treasuries at Susa and Persepolis. Only a tiny fraction of the gold mountain was made into fine gold rhytons such as this.

PERSIA: STRENGTHS AND STRATEGY 335–333BC

The Greeks in the 4th century BC generally underestimated Persian power, their views shaped not just by the March of 'The 10,000' but by broader Greek prejudices about 'barbarians' as inherently inferior beings. But Persia remained the superpower of its age, though perhaps one with clay feet. From only a day's march inland (*c.*48km/30 miles) from the Aegean coast eastward, the Great King's power remained substantially intact. Along the Royal Road that ran nearly 2,400km/1,500 miles from Sardis to Susa messengers still galloped, bearing satraps' reports to the empire's capitals and returning with royal edicts. Babylonia, the empire's richest province and home to the earliest civilization, lay sheltered in its arc, supplying the food and other raw materials vital to the empire. The Phoenician cities, which provided professional fleets of 300 or more galleys, likewise remained under Persian suzerainty. Egypt had recently been regained and Caria and Cyprus were once more under control. None of these cities or peoples was pro-Persian but, with the exception of Egypt, none was strongly anti-Persian either. Imperial rule was usually sufficiently flexible and tolerant not to alienate its subjects.

MILITARY STRENGTHS ...

But the core of the empire remained the Iranian peoples: the Medes and the Persians, devoted to the throne and the empire, if not always to the king himself. They were united also by their common Zoroastrian religion. The Great King was not regarded as a god by the quasi-monotheistic Persians, but his role and power were seen as divinely sanctioned, under the special protection of Ahura Mazda, the Wise Lord.

Persian boys were traditionally taught to ride, shoot straight and never tell a lie. The result was very capable, tough horsemen and excellent archers. Persian archers' composite bow could fire arrows 183 metres/200 yards, and the home province of Persis (Fars) alone potentially provided 30,000 archers. On the lush pastures of Media around Ecbatana 200,000 heavy Nisaean horses grazed,

Left: A Lydian leading horses, from a relief in Darius I's palace in Persepolis. The peoples of all western Asia contributed to Persia's massive military strength.

Above: Persian power finally depended on the Great King and the last Achaemenid ruler. Darius III, pictured in his chariot, proved to be no general or even warrior when he faced Alexander.

ready to supply mounts for the 120,00 or so cavalry that the Great King could almost immediately call upon.

... AND WEAKNESSES

Only in infantry were the Persians at a disadvantage. They had 'The 10,000' foot guards, called 'Immortals' because (reputedly) the moment one died he was replaced by another. But for their other heavy infantry they had long turned to Greece. There were said, with only small exaggeration, to be 50,000 Greek mercenaries fighting on the Persian side in the two main battles, more than all the troops in Alexander's army. They proved generally loyal to their paymasters.

DISTANT THREATS

East of Media, across the high plateaux of eastern Iran and Central Asia, ran the Khorasan Highway, not a paved military road but a time-honoured route for merchants as well as armies, stretching to Bactria and Sogdiana (today Uzbekistan) in Central Asia. North of these satrapies lay arid empty steppes from which savage nomads could suddenly emerge.

These had long posed a far greater threat to the Persian heartland than Greek sailors scudding around the Aegean like pirates.

NO FIXED STRATEGY

Gold and silver tribute, from Egypt to Sogdiana, had flowed for two centuries into the royal treasuries at Susa and Persepolis, the grand ceremonial centre, creating a gigantic reserve of 230,000 talents – riches beyond the dreams of any Greek power. But all still depended vitally on the king. Darius III, although he had killed a man in single combat and was imposingly handsome, proved neither a good strategist nor a competent field commander. Twice at vital moments in battle he was to panic and flee, giving Alexander victory. Such physical cowardice or loss of nerve would have mattered less if he had remained behind the front. But Persian kings were generally expected to lead their armies

Faced with an aggressive invader, Persia could have tried to stop Alexander's army from crossing the Hellespont in 334BC, but the inferior Macedonian fleet was unopposed. They were then faced with a choice: confront Alexander head on in battle or wear him out by a scorched-earth policy, retreating east into Asia while trying to raise revolts at his rear to cut him off. Greece was full of men unhappy with Macedonian hegemony, and Persia had the money to finance revolts. This was the strategy advocated by Memnon, the mercenary from Rhodes who had risen high in Persian service, marrying a Persian wife and being granted estates in Asia Minor. His advice was at first ignored, leading to defeat at Granicus in 334BC, and then followed only half-heartedly by local satraps reluctant to devastate their provinces to deny the Macedonians food. The long defence of Halicarnassus and successful counter-attacks in the eastern Aegean were part of Memnon's policy. But Memnon died in June 333BC, and his death led to a complete change in Persian strategy that soon proved disastrous.

Above: The Cyrus Cylinder, possibly from the reign of Persia's first monarch, was found at Babylon. It proclaims Persian ideals at their highest: "I will respect the traditions, customs and religions of the nations of my empire and never let my governors and subordinates look down on or insult them as long as I shall live." Such tolerance underlay Persian imperial success.

Below: The Palace of 100 Columns, one of the palaces in Persepolis in which court conspiracies arose to hamper Persian efforts to resist the Macedonian invasion.

CROSSING TO ASIA
334BC

Above: The ruins of Troy, which Alexander visited in 334BC and refounded as Alexandria-Troas. The city thrived for centuries afterward as a polis. It was also an early tourist site.

Early in May 334BC Alexander said goodbye to his mother, who reputedly told him the 'secret' of his birth. They never met again. Then he turned his back on his homeland and marched his army through Thrace to the Hellespont. There the remnants of Parmenion's advance force were waiting – they had been driven out of Asia – and the 160 ships mostly supplied by reluctant allies.

The Straits, only 4.5km/3 miles wide, had nasty currents, but the Macedonians were more worried about the threat of a Persian fleet, superior in size and skill. None appeared, however, either because a short-lived revolt in Egypt two years before still required its presence or, more probably, due to general indecisiveness and lack of preparation on the Persian side.

SACRIFICE AND LIBATIONS
Taking the helm of the royal trireme himself, Alexander led 60 ships across the Straits, the remainder taking a different route with Parmenion. Halfway across, he sacrificed a bull to Poseidon the sea god and poured out libations to the Nereids, sea nymphs. Alexander was always meticulous about observing such rites. He then changed into full armour. As the galley grounded on the Asian shore, he hurled his spear into Persian soil, symbolically claiming it as his, and leapt ashore, the first of the Macedonians. The landing, like the crossing, went unopposed, indeed probably unremarked, by the Persians, who were slowly assembling an army inland at Dascylium.

THE TOMB OF ACHILLES
The landscape around the Macedonian army was redolent with legend and myths. It was here, according to Homer, that the Achaeans had landed almost 1,000 years before to start their ten-year siege of Troy to win back Helen, the abducted queen of Sparta. Alexander felt he was literally treading in the steps of his hero Achilles. Now was the time and place to honour him

The splendid Troy of the 'topless towers' had long decayed to a mere village when Alexander approached it. His helmsman Menoitus crowned him with a golden laurel as he entered. Then, stripping

Left: Alexander shown between Hercules (on the right), a semi-divine hero with whom he was increasingly to identify, and Poseidon, the god of the sea, whom he was always careful to propitiate with due sacrifice.

THE PROBLEM OF SOURCES

Alexander used to lament that there was no Homer then living to immortalize him like Achilles in *The Iliad*. But he took with him an official historian: Callisthenes, a relative of Aristotle. Callisthenes boasted that he would make Alexander immortal, and at first depicted him sycophantically. But Callisthenes was executed in 327BC for conspiring against the king and his history has not survived. Nor has that of three other eye-witnesses: Ptolemy, Nearchus and Onesicritus.

Ptolemy, a boyhood friend of Alexander, became one of his generals and later founded the Ptolemaic kingdom of Egypt. He wrote knowledgeably but portrayed himself favourably while depicting rivals such as Perdiccas negatively. Nearchus, another boyhood friend, who commanded a fleet, lost all influence after Alexander's death. He wrote mostly on India and his voyages. Onesicritus was a philosopher who rather magnificently tried to portray Alexander as a philosopher in arms. This required radical factual distortion. All three men's histories are lost. So too is that of Aristobulus, another eye-witness. An architect to Alexander, he compiled his history only when he was in his eighties.

Many later historians wrote about Alexander, one of the best being Quintus Curtius Rufus, a Roman of the 1st century AD. Plutarch and Arrian both wrote in the 2nd century AD under the Romans, and their biographies survive. Plutarch coupled Alexander with Julius Caesar in his *Parallel Lives*, which slanted his whole account. Arrian, who had been both a consul in Rome and a general, based his often excellent history, the best ancient account available, on Ptolemy and Nearchus. He presented a generally favourable yet not rose-tinted view of Alexander, but he was writing more than 400 years later.

Above: A coin of 323BC showing the goddess Athena from Sicyon, a Greek city in the League of Corinth, the last year of Alexander's reign.

Below: Achilles bandaging a wounded Patroclus, a scene inspired by (but not actually in) The Iliad. Alexander saw himself as a reincarnation of Achilles, the supreme hero, casting Hephaistion as a second Patroclus.

naked, he raced with his companions to the tomb of Achilles, placing a garland on it. Hephaistion ran similarly to the tomb of Patroclus. This was a very public declaration of their relationship.

At the altar of Zeus, Alexander prayed to Priam, legendary king of Troy, not to be angry with him as a descendant of Achilles, the Greek who had slain his son. Then he sacrificed at the temple of Athena, dedicating his suit of armour to the goddess. In return, he took from the temple a shield and weapons reputedly dating from the Trojan War. This set would accompany him across Asia as far as India. (Alexander, who always slept with a copy of *The Iliad* under his pillow, would face far greater challenges than his hero Achilles, however.) He granted Troy a new democratic constitution, renaming it Alexandria-Troas, under which name it flourished in subsequent centuries. Alexander then turned inland to meet and fight for the first time Persian forces.

CHAPTER IX

THE GREAT VICTORIES

334–330BC

Alexander first faced the Persian army at the Granicus in May 334BC as an unknown young general confronting a mighty empire. Three and a half years later, after he had routed the grand Persian army on open plains in the empire's heart, he was being hailed as the new lord of Asia. Alexander had won three of the most important battles in history, two of them against far larger armies. In between he had captured the island city of Tyre, long thought invincible, after one of the hardest sieges ever undertaken. All these events show his strategic vision and tactical genius. Many Persian satraps and generals now began going over to his side, accepting the new reality of power, just as he began accepting Persian noblemen as administrators of his new-gained empire.

In between these battles came one of the most mysterious episodes in the life of any world conqueror: Alexander's pilgrimage to the shrine of the Egyptian god Ammon deep in the Libyan desert. What he learned there, in the sanctuary of the deity whom the Greeks identified with Zeus, remains unknown, but it seems to have spurred him to yet further efforts. In Egypt he also founded the city of Alexandria at the mouth of the Nile, an action that in itself would have made his name immortal. If he was indeed the son of a god, as he now began proclaiming on his coins and in his speeches, then literally nothing was impossible for him.

Left: The Battle of Gaugamela, Alexander's crowning victory, depicted by the Renaissance artist Albrecht Altdorfer.

VICTORY AT GRANICUS
334BC

Above: Alexander, filled with battle lust, leading his cavalry at Granicus, a sculpture attributed to Lysippus.

Below: The Persians had taken up a defensive position above the steep slopes of the river Granicus, for some reason placing their own cavalry in front of their Greek mercenary infantry. Alexander led his right wing further out before crossing the river to attack the Persian cavalry's flank, leaving his infantry to wade across the river downstream.

The Persians had failed to prevent the Macedonian army from crossing into Asia, probably because their fleet sent to repress an Egyptian rising a year earlier had not yet returned. However, they knew of the long-planned invasion, and the slowness of their reactions came from divisions in the regional high command.

MILITARY DIVISIONS
The generals in charge of the Persian army included Memnon, a Greek general from Rhodes, and Arsites, satrap of coastal Phrygia. While Memnon was no casual *condottiere* (leader of mercenaries) – he had been 15 years in Persian service, married a Persian wife and driven the Macedonians out of Asia the year before – he was no Persian aristocrat either, unlike the other generals. Many Persians had owned large estates in the area for generations. Perhaps for this reason his advice that they should retreat, laying waste the land and luring Alexander deep into Asia Minor while threatening his communications, was rejected by the other generals. Besides, Alexander, despite recent victories in the Balkans and at Thebes, was still little known as a commander. There was no reason as yet to think him invincible.

PREPARING FOR ATTACK
So, gathering their forces, the Persian commanders decided to confront the invaders. Their troops were mostly local levies, although they included heavy armoured cavalry from Cappadocia in central Anatolia, making some 15,000 horsemen in all, plus *c.*20,000 Greek infantry. The Persians, who were not crack troops, were for once outnumbered by the Macedonians, whose forces totalled *c.*50,000. The Persians therefore needed a good position to offer a fight. They found one on the River Granicus, a small but swift river with steep banks.

Alexander's army encountered the Persians one May afternoon, unusually late in the day to start a battle. Parmenion, Philip's old general, reputedly advised waiting until dawn to attack, but Alexander replied that he would be ashamed if, after crossing the Hellespont, he let a mere stream delay him. He decided to launch a sudden attack before the Persians were fully prepared. (Or so wrote Arrian, our most reliable source.) Whatever the timing, Parmenion commanded the left wing while Alexander took the right, leading the Companion cavalry, his best troops.

ALEXANDER'S TACTIC
Plunging into the Granicus with his squadrons further upstream than expected, Alexander led his troops across obliquely, forcing the Persians to make a rapid adjustment. In Plutarch's words, Alexander "advanced through a hail of missiles towards a steep, well-defended bank, fighting the current that swept his men off their feet. His leadership seemed rash but he persisted… and reached the

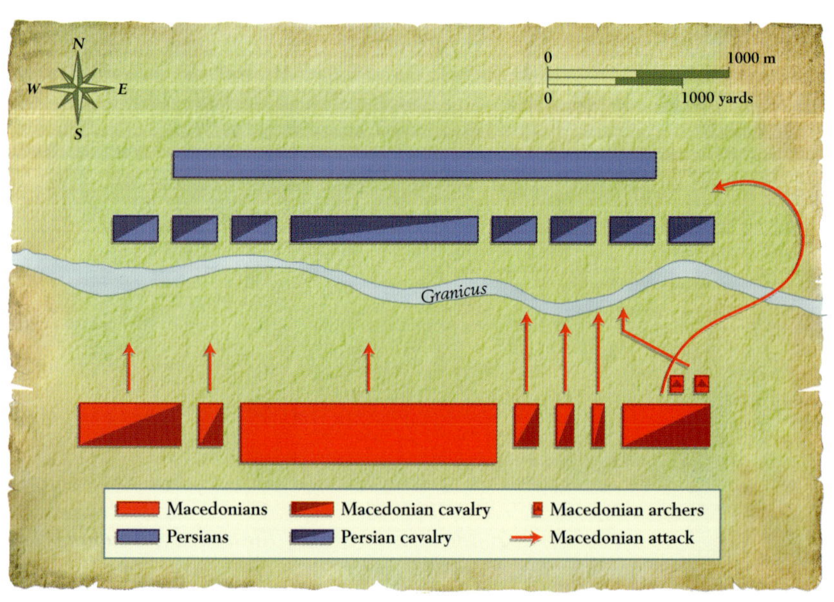

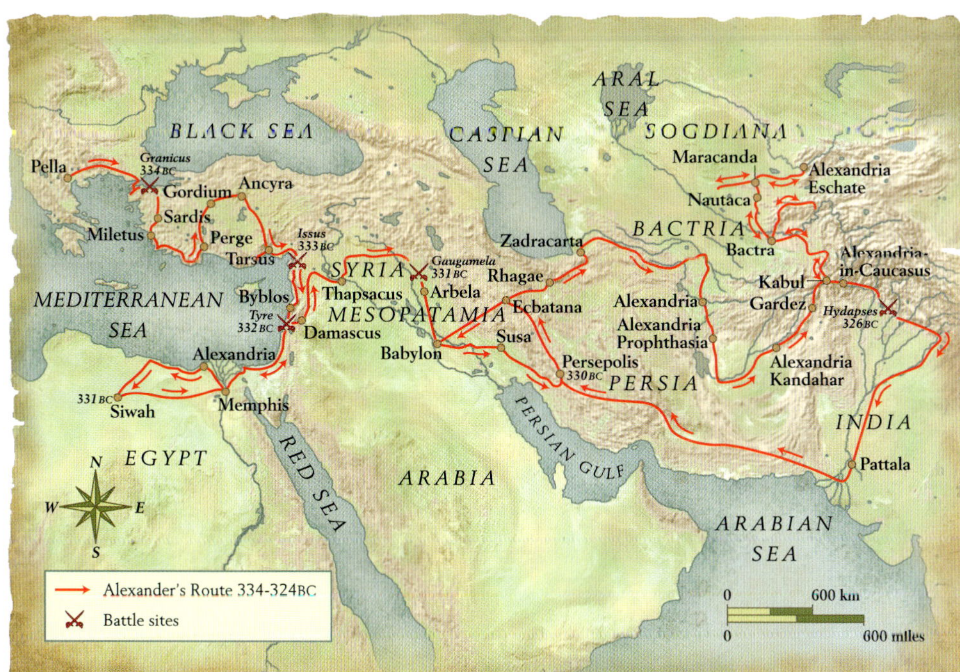

Left: Alexander's conquests took him from central Greece right across Asia to northern India, regions earlier known to Greeks only through legends and hearsay. As he marched east, he founded numerous cities to safeguard his conquests. Many have since prospered.

Below: Alexander crossing the Granicus as seen by the 17th-century artist Charles Lebrun. He was nearly killed in this, his first battle against Persia.

wet, muddy opposite bank, where he was forced to fight at once, man to man, before his supporting troops could get into formation... The Persians charged with a roar... closing in on Alexander whose shield and white-plumed helmet they recognized... the Persian generals Rhoseaces and Spithridates attacked him together." Alexander, almost killed when Spithridates split his helmet with an axe, was saved only by Cleitus 'the Black', who speared the Persian. On such timely intervention hung the whole fate of Asia.

The Macedonian cavalry, their long cornel-wood spears outreaching their opponents' lances, soon worsted the Persian cavalry. Seeing their generals killed, they turned and fled, leaving *c*.1,000 dead. The Greek mercenaries, who had been kept uselessly in the rear, now tried to make a stand. But, outnumbered and surrounded, they soon surrendered, although they managed to wound Alexander's horse.

AFTER THE BATTLE

"[Given by] Alexander, son of Philip, and the Greeks except for the Spartans, [taken] from the barbarians who live in Asia." With these words Alexander dedicated 300 suits of Persian armour taken from the defeated Persians to the goddess Athena on the Athenian Acropolis – significantly, the same number as the 300 Spartans who had fallen gallantly fighting Persia at Thermopylae in 480BC. This was brilliant propaganda. It ignored the fact that Granicus was overwhelmingly a *Macedonian* victory, stressing the tiny part played by Greeks on Alexander's side. Equally brilliant was the emphasis on the Spartans, who alone among the Greeks were not enrolled as Panhellenic allies (however reluctant) in the League of Corinth. Less clever was Alexander's treatment of the captured Greek mercenaries. He treated them all as traitors to the Panhellenic cause, killing many and sending 2,000 back to labour as slaves in Macedonia. This helped Macedonia's economy but, when other Greek mercenaries heard of it, they naturally chose to fight on rather than surrender.

Alexander visited the wounded Macedonians, recognizing many by name and praising their deeds. For the 25 Companions killed fighting, he decreed a hero's reward: Lysippus, the court sculptor, made statues of each and their families were exempted from taxation.

Above: Miletus, one of the most defensible and important Ionian cities, was surrendered to Alexander by its garrison after only a short fight.

Below: Sardis, Persia's regional capital, was handed over to Alexander by its Persian governor without a blow but with its treasury. This combination earned the adaptable Persian a post on Alexander's staff.

LIBERATING IONIA
334–333BC

After Granicus, Alexander forbade his men to plunder and marched upcountry to Sardis, the Lydian capital. The Persian commander surrendered, handing over its treasure and gaining a place on the Macedonian staff. Alexander promised to restore Lydia's old customs but made a brother of Parmenion its governor. His real concern was with the coastal Greek cities, whose liberation was among his avowed objectives.

THE RETURN OF DEMOCRACY
The cities of Ionia, some of the proudest in the Greek world, had been under Persian rule directly or indirectly since the King's Peace of 386BC. Persia had normally favoured oligarchies, finding them easier to deal with than democracies, but the result was growing political tensions between rich and poor inside these cities. This discontent erupted on the news of Alexander's victory. At Ephesus, one of the largest cities, a pro-Persian junta was expelled. Alexander restored the exiled democrats, who began taking bloody revenge until he forbade it. Other cities now welcomed him, as he "broke up oligarchies everywhere, men being given their own laws and exempted from the tribute they had paid the barbarians [Persians]". Instead Alexander, with forceful tact, asked for *syntaxeis* (contributions) to his war chest.

While Alexander favoured democracy in Ionia for essentially pragmatic reasons, elsewhere preferring other forms of government, his actions marked a true liberation, long remembered with gratitude by the cities concerned. Fifty years later, a decree from the little city of Priene, rebuilt on his orders, proclaimed: "There is no greater blessing for Greeks than freedom." Whether Greek cities on the Asian mainland were incorporated into the League of Corinth remains debatable. The non-Hellenic countryside around certainly remained unfree, with Macedonia simply replacing Persia as its feudal overlord.

SURRENDER AT MILETUS
At Miletus, strongly defended on its headland, the garrison resisted, encouraged by the large Persian fleet now nearby. Refusing to fight at sea, Alexander moved rapidly to the assault, battering his way into the city with his siege engines. The garrison, who had swum out to a tiny island, happily accepted Alexander's offer of clemency, 300 enlisting in his army.

Alexander then unexpectedly disbanded his fleet, saying he would conquer the sea by land, in other words capture the Persians' bases in Asia. In truth, he could not afford to pay 160 triremes' inactive crews of 32,000 men. He kept only 20 Athenian ships, whose crews served as hostages for their city.

SIEGE OF HALICARNASSUS

On the southern edge of Ionia lay Halicarnassus, a half-Greek, half-Carian city, its massive walls rising in a semicircle with a sea-girt citadel. Memnon, now commander of lower Asia and the whole fleet, was there with many Greek mercenaries. Alexander, marching toward it through the forests inland, was approached by Ada, widowed queen of Caria, into whose family he had tried to marry three years before. No longer a nervous adolescent but an assured king, he welcomed her surrender, reappointing her as queen with a Macedonian commander. He then became her adopted son, so winning over the Carians. (What Olympias back home thought of this is unrecorded.) But to take Halicarnassus required force.

After early skirmishing and an unsuccessful attempt to take a nearby port by surprise, the Macedonians filled in many of the city's ditches. They now turned their catapults against its walls, knocking them down in parts. But the garrison, sallying out at night, torched many of these wooden engines. Some Macedonian soldiers, getting drunk, then launched an impromptu attack on the city that ended disastrously. Memnon personally led the defenders out to repulse them. Another sortie three days later panicked the Macedonians until Alexander himself led a counter-charge, driving the defenders back. The city, shutting its gates prematurely, lost many troops and Memnon ordered a retreat to the castle. He soon sailed away, although the castle held out for a year. The two-month siege had been won more by force than great generalship.

CUTTING THE GORDIAN KNOT

It was now autumn, normally a time for rest and recuperation, but these were not concepts that Alexander recognized. Giving all recently married troops winter leave (a popular measure that also boosted the birth-rate), he entered Lycia and Pamphylia, whose steep coastlines were dotted with small Greek cities. Capturing these in his role of Panhellenic liberator – Aspendos, one of the chief cities, betrayed him and was severely fined – he turned north. Uniting with the troops back from Macedonia, he entered central Anatolia.

At Gordium he saw the Gordian Knot in the old palace of golden King Midas. Whoever untied it, said the legend, would control Asia. After fiddling fruitlessly, Alexander drew his sword and cut through it, in a way fulfilling the prophecy. But as he marched east that summer of 333BC, he left Memnon with a large fleet still threatening his rear.

Above: The lion of Didyma, one of the many Ionian cities that flourished again after Alexander's conquest, which was for them a true liberation.

Below: The Temple of Athena Polias at Priene, an Ionian city rebuilt with Alexander's help to become almost the perfect polis. *Alexander's restoration of Priene as a democracy was long remembered with gratitude.*

AN UNEXPECTED BATTLE: ISSUS 333BC

Above: Darius III, the last Achaemenid king, in growing terror as he spots Alexander hurtling towards him.

As Alexander approached the Cilician Gates, an easily defended pass through the Taurus Mountains, the Persians abandoned it. They had burnt their fields, either following Memnon's scorched-earth policy or just panicking, so the Macedonians entered Cilicia unopposed. There Alexander fell gravely ill after bathing in a river. Warned by letter that his Greek physician Philip was trying to poison him, Alexander showed Philip the letter, took his medicine – and recovered. But he had lost two months lying ill.

DARIUS' NEW STRATEGY

The general situation, meanwhile, had changed completely. Memnon, the Persians' general, died in June 333BC on campaign in the Aegean. This bad news led Darius to rethink his strategy. Deciding to go to war in person, he began massing armies, moving to Babylon. Persia had abundant cavalry even without horsemen from its further satrapies, but lacked heavy infantry apart from the 10,000 Immortals and some youthful trainees. So Darius recalled the 15,000 Greek hoplites fighting under Memnon's successors. When Persian forces left Babylon in September, they totalled *c*.150,000, far more than the Macedonians. But most had little or no military experience.

By late September, Alexander had recovered. Unaware of recent events, he decided to continue capturing Persia's naval supply ports. Parmenion went ahead to seize the Syrian Gates, the pass through the Amanus mountains, while Alexander hunted and waited on events in the Aegean. Then came startling news: the main Persian army was camped beyond the Amanus range. Alexander raced east with his army.

Leaving the sick at Issus in the corner of the gulf, he continued south down the coast, hoping to catch the Persians in the rear. But Darius, growing anxious, decided to seek out Alexander. He moved north behind the Amanus while Alexander, unaware, marched south on the coast. When the Persians crossed the Amanus, they had cut Alexander off. The Persians mutilated the Macedonian invalids they found at Issus, although some escaped to warn Alexander. He seemed trapped with no escape route.

BATTLE BY THE SHORE

Alexander, swinging around, marched his weary men back through the night until the two armies faced each other between the sea and mountains. As usual, he commanded the Companion cavalry on the right, Parmenion taking the left. Darius sent light troops up into the mountains. Alexander countered by sending archers to drive them back. He then reinforced his left wing by the coast but thinned his centre. Darius was in the Persian centre with his Greek mercenaries. Between them lay the stream of the Pinarus,

Below: At Issus, Persian numerical superiority was annulled by the narrowness of the site. Alexander thinned his centre and led his cavalry on the right in an oblique attack that crumpled up the Persian centre where Darius was.

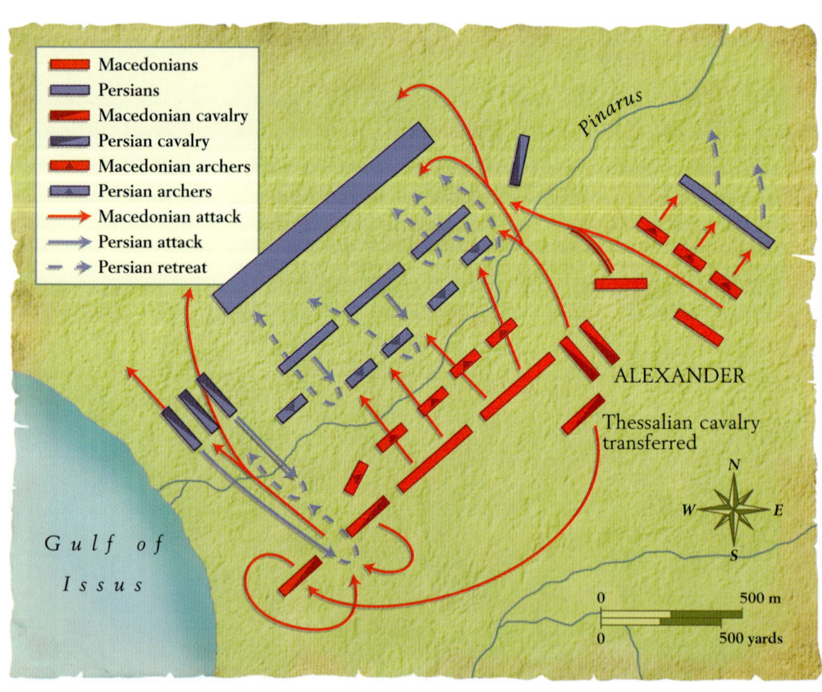

swollen by autumn floods. At noon on 1 November 333BC, Alexander gave the signal to attack.

The Macedonian cavalry surged forward, crossing the river. The infantry followed more slowly, raising their fearful war cry "ALALALAI!" The two armies met head on. Macedonian cavalry at once broke through the Persian line. Then Alexander wheeled his horsemen obliquely in toward the centre, rolling up the Persian riders on their flanks in a brilliant manoeuvre. He was heading for Darius, very visible in his chariot. Parmenion repulsed the Persian cavalry but in the centre Persia's Greek mercenaries threatened the exposed phalanx until Alexander's advance forced them to retreat. Darius, seeing Alexander cutting relentlessly toward him, turned his chariot and fled. Swapping his chariot for a horse, he abandoned even his royal cloak. Alexander pursued him until night fell, then turned back. Victory was his, won by audacious cavalry tactics.

AFTER THE BATTLE

The Persians had suffered heavily, although the traditionally cited figure of 110,000 dead is incredible. The Macedonians had 4,500 wounded and many hundreds dead. Hearing women crying nearby, Alexander discovered they were Darius' womenfolk, mourning a king they believed dead. He sent reassurances, ordering that they should be treated as royalty. Next day at their meeting, the Persian queen mother did obeisance to the taller Hephaistion. Realizing her error, she was mortified until Alexander said: "You make no mistake, madame, he too is Alexander."

He also met Barsine again, whom he had known long ago in Pella. She now became his mistress and bore him a son, forming a useful link between east and west. He then visited the wounded Macedonians, congratulating each man he had seen in battle and arranging a magnificent funeral for the dead. On the battle site he founded a city: Alexandria, today Alexandretta.

Above: After the battle, Alexander went with Hephaistion to comfort Darius' women, mourning a king they assumed dead. The queen mother saluted Hephaistion as Alexander, a mistake Alexander courteously shrugged off. This is the Renaissance artist Veronese's grand depiction of the scene.

Below: The Battle of Issus was primarily a cavalry action, which is how Jan Brueghel painted it in this dramatic canvas of 1602.

THE SPOILS OF VICTORY

Alexander was astonished by the luxury of the Persian royal tents. Even on campaign, Darius travelled in style: "When he [Alexander] saw the gold bowls, pitchers and tubs, exquisitely worked and set in a chamber fragrant with incense and spices, when he entered a tent of remarkable size and height, set with sofas and tables for his dinner, he looked at his Companions and remarked: 'This, it seems, is what it is to be king.'"

Above: The siege of Tyre in 332BC, as imagined in a medieval miniature.

Below: A battle scene between Macedonians and Persians from the Sidon Sarcophagus. *Sidon, Tyre's neighbour and rival, supported Alexander.*

THE SIEGE OF TYRE
332BC

Dispatching Parmenion to Damascus to seize the Persian war chest of 2,600 talents there, Alexander turned south to the cities of Phoenicia. Along with Cyprus, Phoenicia supplied the Persians with their fleet, although it had little love for the empire. Sidon, where a revolt had been brutally suppressed 12 years before, welcomed Alexander, as did Byblos. Tyre, the most powerful Phoenician city, invited Alexander to sacrifice to Hercules, its patron god, in old Tyre on the mainland, but refused to let Macedonians – or Persians – into the island city. Alexander, who could not leave a great naval power neutral behind him (as he told his army) decided to capture it. This was easier proclaimed than done.

CONSTRUCTING A MOLE
Tyre, a walled island 4.8km/3 miles in circumference, was reputedly impregnable, having once withstood a siege by a Babylonian king for 13 years. It lay 800m/880 yards from the mainland in sea that was 180m/600ft deep. On the land side its thick walls rose to 45m/150ft and it had a powerful fleet. Alexander had no fleet at all and his torsion catapults had a maximum range of only 270m/300 yards. Yet in January 332BC he began to build a mole, or causeway, out of the ruins of old Tyre to let his catapults and siege towers reach the city. So started the seven-month siege, the greatest in antiquity.

At first all went well, but as the mole entered deeper waters, it came within range of Tyrian catapults, while triremes sallied from the city's twin harbours to rain arrows on the workers. Alexander ordered in siege towers as protection. In response, the Tyrians secretly built a vast fireship. When a favourable wind blew, they sailed this floating bomb across the waters. On impact, it was ignited, torching the mole's towers and catapults. Undaunted, Alexander ordered the mole rebuilt, only this time wider to take more

engines and towers. Then he went north to friendly Sidon, where he had some very good news.

The fleets of Sidon and neighbouring cities had returned, deserting the Persian side. With their 100 ships, plus 120 from Cyprus and 9 from Rhodes, Alexander now had supremacy at sea. Attempts to lure Tyre's fleet out failed, however, for the Tyrians blocked their harbours. Instead, the other Phoenicians built floating battering rams protected by roofs of fireproof hide. They rowed these around to attack Tyre's weaker seaward walls.

Meanwhile, Alexander's engineers rebuilt the mole wider and at an angle to the prevailing wind. On to it rolled the tallest siege towers yet seen: 20 storeys high with battering rams and catapults on their upper decks. Soon Tyre was besieged on all sides.

LONG RESISTANCE OF TYRE

But the Tyrians were not defeated yet. Hanging leather skins stuffed with seaweed from their battlements, they cushioned their walls against missiles. They dropped rocks on to siege ships and sent underwater divers to cut the Macedonian vessels' moorage cables, after which the Macedonians switched to solid chain. Then the Tyrians used sharp poles to slice through the ropes holding the battering rams and poured red-hot sand on to the besiegers below, penetrating the attackers' armour and causing them agony. By late June, as stalemate threatened, some advised a truce; Tyre was no longer so vital since other fleets had become allies. A letter came from Darius, offering all land west of the Euphrates, his daughter in marriage and 10,000 talents. Reputedly, Parmenion urged acceptance but Alexander refused. He wanted the whole empire – and he would not leave Tyre untaken.

THE FINAL ONSLAUGHT

One hot July noon, when the besiegers were lunching or snoozing, Tyre's best ships slipped out and attacked the Cypriot fleet, sinking five galleys. Alexander broke off his lunch to lead the counter-attack, sinking all the Tyrians. By now Tyre was both starving and without allies. Alexander began an all-out attack on every side of the city at once. The fleets attacked the seaward walls and both harbours while siege towers and catapults assaulted the land side from the mole. Overwhelmed, the Tyrian defences collapsed and the Macedonians poured into the city early in August. Alexander had conquered the sea from the land.

Alexander killed 8,000 Tyrians, crucifying 2,000. The rest were sold into slavery, the customary fate. Tyre gained a Greek name and constitution but never regained its primacy. As Alexander advanced toward Egypt, only Gaza, the old Philistine stronghold, refused to surrender. Its siege took two long months, during which Alexander was twice wounded. When Gaza finally fell, Alexander dragged its commander, Batis, around the walls behind his chariot – an excruciating death. (Achilles in *The Iliad* had similarly dragged Hector around Troy, but the Trojan prince was already dead. This was gratuitous cruelty.)

Above: Totally destroying the old Phoenician city of Tyre, Alexander refounded it as a Greek city. It thrived under the Roman empire, as Roman-era ruins attest.

Below: Tyre, on its massively fortified island, was thought impregnable. Alexander proved otherwise, but it took him seven months and tested his military genius to its limits.

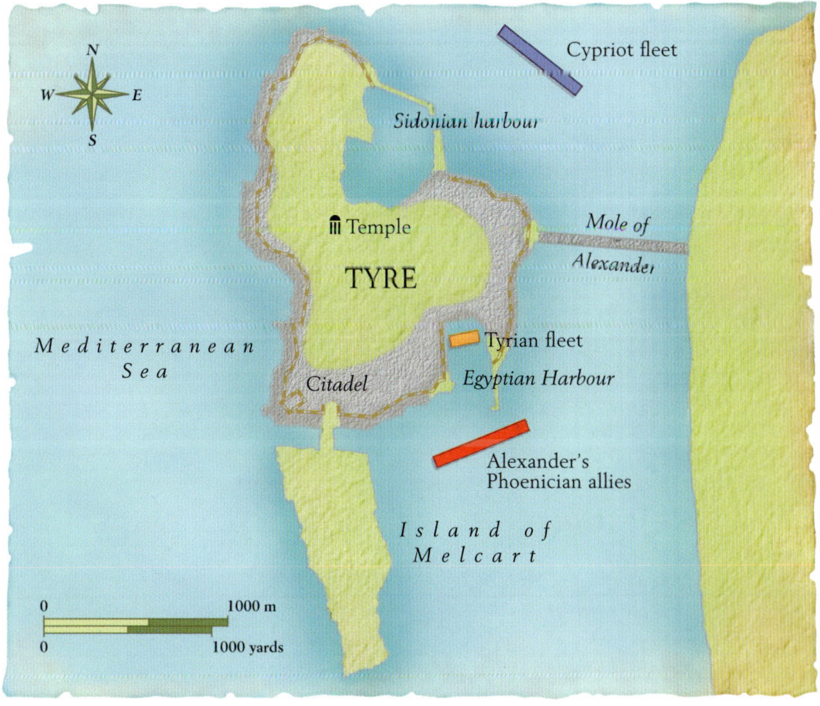

Above: Alexander, crowned as Pharaoh by Egyptian priests, was depicted on temple carvings as being greeted by Ammon Ra, greatest god of the Egyptian pantheon.

Below: A 19th-century view of the great harbour at Alexandria showing all that remains of the lighthouse, once one of the Seven Wonders of the World. For 1,000 years after its foundation in 331BC, Alexandria was one of the greatest cities on earth.

EGYPT: THE FOUNDING OF ALEXANDRIA 332–331BC

Egypt had known, and generally accepted, the Greeks as traders, mercenaries and colonists (at Naucratis in the Delta) for more than two centuries. It had even seen some of the very first tourists, such as Herodotus, that insatiably curious historian-traveller of the 5th century BC.

PERSIAN MISRULE

The country had been in intermittent revolt against Persia for 130 years, remaining independent for more than 60 years of the 4th century BC. (Its customs and religion, which intrigued the Greeks, seem to have annoyed the Zoroastrian Persians, who were essentially monotheistic.) Persian behaviour had been at best insensitive, at worst deliberately offensive, to Egypt's powerful priesthood and the devout *fellahin* (peasantry), who made up the bulk of this strongly hierarchical society.

"In Egypt," Plato had observed, "it is impossible for a king to rule without the priests." Persian soldiers had reputedly roasted and eaten the sacred bull Apis, replacing it with a donkey – an animal the Egyptians loathed. But despite this, many Egyptian temples had retained their estates under Persian rule.

When Alexander crossed the desert in late 332BC, the Persian governor of the frontier fort of Pelusium opened its gates to him. Alexander then sailed up the Nile to Memphis unopposed. In the ancient capital he was enthusiastically welcomed by the people and priests and lodged in the old palace of the pharaohs. Alexander sacrificed to the Egyptian gods, especially Apis, and was crowned Pharaoh, as inscriptions in the temples attest. As Pharaoh, he was the son of Horus, the divine son of the sun god Ra, and also beloved son of Ammon, creator of the Universe. These titles impressed him far more deeply than his Companions probably at first suspected.

CHOOSING THE SITE

Alexander held athletic games "to which the most famous performers came from all over Greece", according to Arrian, then sailed down the west branch of the Nile and around Lake Mareotis to the sea. Here he was struck by "the excellence of the site, convinced that if a city were built upon it, it would prosper. Filled with enthusiasm, he himself oversaw the layout of the new city, indicating the site

THE ROSETTA STONE

Much of what we know about ancient Egypt under the ancient Pharaohs and the Ptolemies derives from the Rosetta Stone, a trilingual slab dating from 196BC. In that year Ptolemy V (descended from Alexander's general Ptolemy I) set up the stone bearing a declaration in three languages: Greek (the language spoken in Alexandria and in government); demotic Egyptian (spoken by native Egyptians); and hieroglyphs (the sign-writing read only by priests). Ptolemy V's government, facing problems after military defeats and peasant revolts, put up trilingual stones proclaiming that Ptolemy was the truly anointed Pharaoh. This stone was found in the city of Rosetta – hence its name – during the French occupation of Egypt under Napoleon in 1799. It was taken to London after the British expelled the French. The French had made copies of the writing and after 1815 a race developed between the two countries to decipher the stone first. The brilliant French Egyptologist François Champollion finally cracked the hieroglyphs in 1822–4.

Right: The stone found at Rosetta contained trilingual inscriptions that enabled the Egyptologist Champollion finally to decipher it in 1824.

for the market square and the temples, Greek gods being chosen along with the Egyptian Isis, and the exact limits of its walls." (Arrian again.) Alexander sprinkled barley meal taken from his soldiers' rucksacks to mark out the city's walls – an omen that Aristander, his prophet, interpreted to mean that the city would enjoy the fruits of the earth.

GREATEST MEDITERRANEAN CITY

There had long been a trading post on the offshore island of Pharos, on which, according to a legend, the lovers Helen and Paris of Troy had hidden after fleeing Sparta. Sheltering the harbour from the prevailing north winds that kept the site cool in summer, Pharos became famous for the giant lighthouse on it. Alexandria itself soon became the greatest city in the Mediterranean, a trading metropolis renowned for luxury and culture, with the biggest library in the world.

Alexander could not have foreseen all this, but he certainly envisaged the city as Egypt's new capital, well sited to receive the grain and other products flowing down the Nile, while looking out toward Greece. Alexander settled Macedonian veterans, Greeks, prisoners and some Jews in his new city. Later, under the Ptolemies and Romans, Alexandria seems to have lacked a council and assembly, but its founder may have originally granted it these vital aspects of a Greek *polis*. Alexandria-in-Egypt, despite many vicissitudes, would flourish until the Arab conquest nearly a thousand years later.

Below: Sailing unopposed up the Nile to the ancient capital Memphis, Alexander was welcomed by the Egyptians, who had come to loathe the Persians.

EGYPT: THE PILGRIMAGE TO SIWAH 331BC

Above: The Temple of Horus at Edfu was built under the Ptolemies, the Hellenistic kings who succeeded Alexander. It is the best preserved of all Egyptian temples.

Below: The Temple of Ammon at Siwah, the mysterious oasis-shrine that Alexander visited in 331BC.

The most mysterious episode in Alexander's life may have originated in a simple desire to explore the vague western frontier of Egypt, his newest conquest. Early in 331BC he led a small group west along the coast from the Nile. Then he turned due south into the desert, heading for Siwah, an oasis 320km/200 miles inland with a famous shrine. Siwah had long been revered by the citizens of Cyrene, the greatest Greek city in Libya. Cyrenians worshipped its ram-headed god Ammon/Amun as Zeus. (The Greeks, being tolerant polytheists, often identified other people's deities with their own.) Through Cyrene's influence, Siwah's fame had spread to Greece proper, where it was seen as an African version of the Delphic oracle. But Siwah itself, although visited by a few Greeks remained mysterious.

UNKNOWN MOTIVES

Alexander's true motives for this long diversion, while Darius was slowly assembling Persia's Grand Army beyond the Euphrates, remain disputed. For some people his desert pilgrimage reveals his mystical belief in his destiny; others view it as an attempt to bolster his position with his new subjects, Egyptian and others. Callisthenes, the official historian, said that it was due to Alexander's "thirst for glory and because he heard Hercules and Perseus had gone there before him". These were heroes Alexander always strove to rival.

HOLY GUIDES

Riding camels, the party left the coast at Paraetonium, entering a sand desert. After four days' wandering, they had almost run out of water when a rainstorm enabled them to refill their water bottles.

From then on they travelled by night to avoid the heat through a landscape later described by the Victorian traveller Bayle St John: "A gorge black as Erebus lies across the path and on the right stands a huge pile of rocks, looking like the ruins of some vast fabulous city…. There were yawning gateways flanked by bastions of immense altitude; there were towers and pyramids and crescents and domes and dizzy pinnacles and majestic crenellated heights, all invested with unearthly grandeur by the magic light of the moon."

Unsurprisingly, Alexander's party became lost, until (according to Ptolemy) two holy serpents appeared to guide them through the Pass of the Crow and down to the first oasis. Beyond a glittering white salt desert lay Siwah itself, green and lush with palm and fruit trees.

THE ORACLE'S ANSWER

The chief priest emerged from the temple to invite Alexander into the sanctuary, not requiring him to change his travel-stained clothes, like most supplicants. The other Macedonians, left in an outer courtyard,

can have heard little of what went on in the temple's dim recesses, but Callisthenes recorded it as if he had: "The oracles were not given in spoken words as at Delphi or Miletus, but mostly given in nods and signs, just as in Homer '[Zeus] spoke and nodded assent with his dark brows.' The prophet answered for Zeus, telling the king directly that he was the son of Zeus." All Alexander would say afterwards was that he was pleased with the result.

Returning without trouble to Memphis, he divided Egypt's government between civilian and soldiers, a sensible arrangement he later repeated elsewhere. Then he moved north, gathering his armies at Tyre for the decisive encounter with Darius' Grand Army.

Above: At the remote but famous sanctuary of Ammon, whom the Greeks identified with Zeus, Alexander was reputedly told that he was the son of Zeus, an answer that fully "satisfied him".

ALEXANDER THE GOD

"Zeus is the father of all men but makes the best especially his own," Alexander once said. For Greeks, there was no sharp division between gods and men, the world being filled with gods. The kings of both Sparta and Macedonia claimed descent from Hercules, the mythical hero who became a god. Lysander, the Spartan general who defeated Athens in 404BC, was hailed as a god by grateful oligarchs. Philip II was later portrayed like an Olympian deity on some coins and statues. After his Siwah pilgrimage, Alexander began invoking Zeus as his father, implying his own divinity. (He regarded Ammon as a form of universal Zeus, not a local deity.) This later caused problems. When Alexander first tried to get Macedonians to offer him *proskynesis* – the homage Persians paid their Great King, although monotheistic Persians never *worshipped* their monarch – he had to back down.

Finally, in 324BC Alexander, at the height of his powers, ordained his deification, which produced varying reactions. The Ionian cities, already worshipping Alexander, complied happily; the Spartans replied laconically: "Alexander can call himself a god if he wishes"; and Demosthenes probably spoke for most Athenians when he said, "Alexander can be the son of Zeus – and of Poseidon also if he wants." After his death, Alexander was often depicted as divine with the horns of Ammon, and many of his successors claimed godlike attributes.

Right: Alexander with the horns of Ammon, the Egyptian deity he claimed as his father, on a coin issued by Ptolemy I (ruled Egypt 323–284BC).

THE GREAT VICTORY: GAUGAMELA 331BC

Alexander waited in Tyre through the early summer until he heard that Darius had mustered the Persian Grand Army in Babylonia. He did not want another inconclusive battle like Issus but needed to defeat the Persians totally and openly. However, Darius was based in Babylon 1,120km/700 miles from the Mediterranean, far beyond Alexander's knowledge. In July, Hephaistion went north to the River Euphrates with an advance guard. He found himself facing Mazaeus, a Persian satrap of Syria, with 3,000 troops, mostly Greek mercenaries. For some weeks the two forces faced each other, perhaps exchanging secret messages. Then, as Alexander approached with his 47,000-strong army, Mazaeus retreated, burning the fertile Euphrates valley. Alexander took the northern unburnt route towards the Tigris – a fast-flowing river hard to cross if defended.

Above: Indo-Greek tetradrachms and staters struck by the kings of Bactria.

Below: The Azara Herm, a copy of an original by Lysippus, showing the dynamic conqueror in distinctly tough-looking mode. Alexander was always keenly aware of the propaganda value of his image.

THE SITE OF THE BATTLE

Curiously, his army was unopposed as it crossed the river, the cavalry wading in upstream to shelter the infantry. The land around was invitingly unravaged. Darius probably was luring the Macedonians on to a spacious battlefield (the plain of Gaugamela) of his choosing. On 20 September the Moon eclipsed. Alexander sacrificed to the Sun, Moon and Earth, revealing astronomical knowledge learned from Aristotle and also Greek piety. Aristander, his prophet, interpreted the ominous event favourably. Then Alexander heard that Darius was camped nearby with a "force much larger than at Issus". Ignoring advice to try a night attack

Above: The archetypal image of Alexander at the charge, huge-eyed and with wind-swept hair.

(always risky), Alexander rested his army while he reconnoitred. What he saw was awesomely impressive.

Darius had mustered the largest army that he could feed, perhaps 250,000 men. He meant to win by sheer numbers. Crucially, he had *c.*40,000 cavalry, while Alexander had only 7,000. Only in heavy infantry were the Persians weaker, deprived of Greek recruits. To let their chariots charge smoothly, the Persians had cleared the ground while fixing stakes to protect their flanks. Having seen the Persian set-up, Alexander planned at leisure, keeping the Persians waiting for two nights. At noon on 1 October 331BC the battle for Asia began.

THE OPPOSING ARMIES

On the Persian left was the formidable Bactrian and Scythian cavalry commanded by their satrap Bessus. Mazaeus commanded the cavalry on the right. In the centre was Darius, protected by 15 Indian elephants (whose smell panics horses that are unused to them), 6,000 Greek hoplites, the Persian infantry and

200 scythed chariots. The Persians could potentially outflank the Macedonians on both sides.

Alexander slanted his whole army obliquely, with 10,000 sarissa-wielding heavy infantry in the centre. Their right flank was protected by 3,000 mobile Shield Bearers, linked to the Companion cavalry on the right led by Alexander himself. Ahead of them ran 2,000 archers, slingers and javelin throwers. On the left wing, under Parmenion, were the Thessalian horsemen and remaining Macedonian cavalry, forming the anchor of the slanting line. Alexander's wing actually found itself opposite Darius' centre. To counter flank attacks, Alexander ordered the infantry to face about if needed to form a square – a manoeuvre requiring perfect discipline. At the tip of each cavalry wing, infantry units of veteran mercenaries were concealed, a tactic first recommended by Xenophon.

Alexander began battle by leading his wing to the right while holding back Parmenion's troops. This drew the Persian left flank out, away from their elephants and defences. The Bactrians charged, trying to outflank the Macedonians, but the latter's cavalry wedges turned to face them, foot soldiers emerging to drive back the disconcerted Bactrians. In the centre the chariots were countered by archers. Any surviving chariots rattled harmlessly through infantry ranks that opened up.

THE FLIGHT OF THE KING

As Alexander intended, a gap appeared to the left of Darius' centre. Into it he led his Companion cavalry in a wedge-shaped attack heading for the Great King. The Shield Bearers followed on foot. Darius, again seeing nemesis on a black horse bearing down, again turned and fled – Alexander reputedly got close enough to kill the royal charioteer. Meanwhile, the Persian cavalry under Mazaeus on the right beat back Parmenion's wing, pushing past into the Macedonian camp. There they discovered the Persian Queen Mother, who looked at them in stony,

immobile silence until they withdrew. Other gaps appeared in the central Macedonian line but the troops turned to form oblongs. Then news of the Great King's flight demoralized the Persian army. Mazaeus, recalling his unbeaten cavalry, rode hard for Babylon.

The battle had been won by just 3,000 cavalry Companions supported by 8,000 Shield Bearers under Alexander's visionary leadership. He tried to pursue Darius, but swirling masses of retreating troops, thick clouds of dust and then nightfall hampered him. Darius swapped his chariot for a horse and rode into the mountains of Media. On the battlefield Alexander was hailed as lord of Asia.

Above: The Battle of Gaugamela, *as painted by the French 17th-century artist Lebrun. Macedonian victory at this huge battle determined the fate of half Asia.*

Below: Heavily outnumbered at Gaugamela, Alexander slanted his army obliquely so that the right wing, under his command, attacked the Persian centre. Alexander's wedge-shaped cavalry formations cut through the Persian ranks and Darius fled in contagious panic.

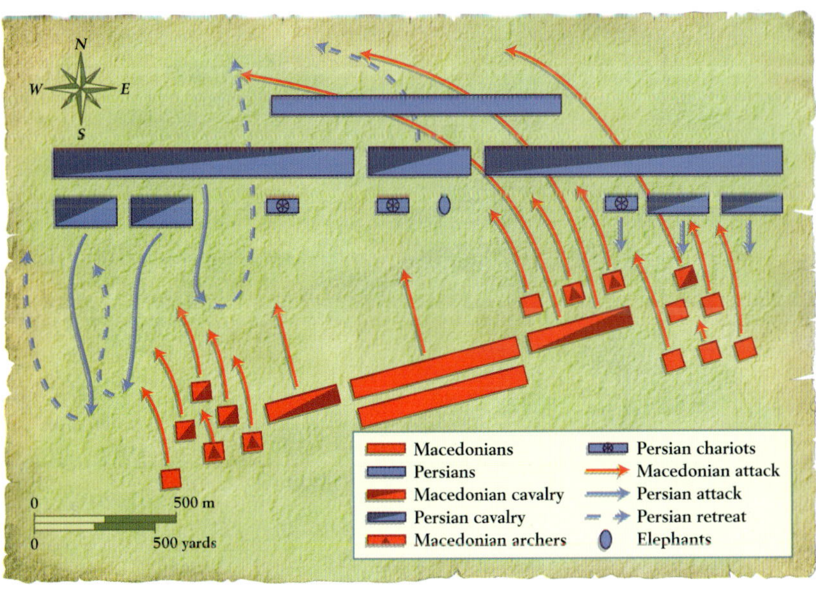

IN BABYLON
331BC

Above: The Babylonians offering Alexander the city's keys, from a French medieval painting.

Below: Alexander entered Babylon in triumph, its citizens welcoming him as a liberator from Persian misrule, depicted in an 18th-century painting by Gasparo Diziano.

When Alexander reached Arbela 112km/70 miles away, he realized that Darius had vanished into the mountains of Media (Kurdistan), at which point he abandoned the pursuit. After burying his dead (Persian casualties far outnumbered Macedonian, but even they had suffered heavily, Hephaistion being among the wounded), he quit the battlefield.

South lay Babylon, the greatest metropolis of Asia, "surpassing in splendour any city in the known world", as Herodotus had written 150 years earlier. Still a great trading city, Babylon was also an important religious centre. Its Chaldean priests were famed as mathematicians, astronomers and astrologers. Babylon's defences of brick-built walls were 60m/200ft high and wide enough for chariots to drive around two abreast, according to Herodotus. Surrounded by deep moats, they rivalled Tyre's.

ALEXANDER WELCOMED
But there was to be no siege of Babylon. As the Macedonians warily approached the city, its gates opened and the Persian general Mazaeus rode out with his sons to greet Alexander. Behind him came Babylonian priests chanting and dancing, with city magistrates bearing gifts to indicate their surrender. Alexander mounted a special chariot to ride in triumph up the dead straight avenues toward the royal palace, a complex with 600 rooms. This welcome from Babylon's elite was both politic and heart-felt.

PERSIAN MISRULE
Like Egypt, Babylon and Babylonia had often been maltreated by the Persians. While Cyrus, its first Persian ruler, had meticulously respected Babylon's customs and deities, later kings such as Xerxes had abused the city, demoting it from being the satrapy capital after a revolt and expropriating its temple treasures. Persian rule had drained Babylonia of wealth and even population (it had to send 500 eunuchs and 1,000 talents of silver to the Persian court each year). Many farms had been given to Persian nobles as estates, while the irrigation channels on which Babylonia's fertility relied were neglected.

THE HANGING GARDENS OF BABYLON

One of the Seven Wonders of the Ancient World, the Hanging Gardens of Babylon displayed Babylonian wealth and ingenuity. Traditionally they were built *c*.600BC for Amytis, the Median-born wife of King Nebuchadnezar II, homesick for her green native mountains. Babylon had many massive ziggurat temples and other buildings, often with walls 25m/80ft thick, mostly dating from that period.

Archaeologists have not yet decided which ruins in Babylon are the Gardens described by Greek writers such as Strabo and Diodorus Siculus in the 1st century BC. Diodorus wrote: "The approach to the Garden sloped like a steep hill and the several parts of the structure rose from one another tier on tier. On all this earth had been piled and was thickly planted with trees of every kind that, by their size and beauty, delighted beholders... Water machines raised abundant supplies of water from the river,

Above: The Hanging Gardens of Babylon, one of the Seven Wonders of the Ancient World, as imagined in a a 19th-century illustration.

though hidden from view." Clearly the Babylonians had pumps to raise water from the Euphrates, and they waterproofed their brickwork. Confusion comes from a translation error: the Greek word *kremastos* means *overhanging*, not hanging. Creepers and branches overhung the walls of this sky-garden.

Above: The Ishtar Gate, Babylon's grandest ceremonial gate, for which the city was renowned, was dedicated to Ishtar, the goddess of sacred prostitution.

TEMPLES AND BROTHELS

Concerned as ever to give the local gods their due, Alexander sacrificed in the temple of Bel-Marduk, Babylon's patron deity. He then clasped the hand of the golden statue to show that, like the old Babylonian kings, he received his power direct from the god. He took the old title King of the Lands and ordained the rebuilding of E-sagila, the great ziggurat-temple 60m/200ft high, damaged by Xerxes. He also ordered that Greek plants be added to the varieties growing luxuriantly in Babylon's famous Hanging Gardens, although few probably survived Babylon's heat. But he reappointed the Persian Mazaeus as satrap, with Apollodorus as Macedonian military governor – a wise balance repeated elsewhere later. Persians who knew the locality were obviously useful.

While Alexander was restoring the temples, his men were enjoying the city's equally famous brothels, helped by a

generous pay bonus. Quantities of gold, in ingots rather than coin, had been found in the city, which were minted into coins to pay the army. (Herodotus had noted the strange Babylonian custom that required *all* women to ritually prostitute themselves in temples before marriage.)

After a month's rest and recreation, the army marched out south-east towards Susa and Persepolis, the Persian capitals.

Below: Babylon's walls gleamed with enamelled bricks depicting animals such as this lion.

THE DESTRUCTION OF PERSEPOLIS 331–330BC

Susa, the eastern terminus of the Royal Road that ran west to Sardis, lay in the Elamite plain, not Persia proper. Although reputedly even hotter than Babylon, Darius I (522–486BC), the greatest Achaemenid king, had made it his growing empire's administrative capital. From Susa's huge palace, orders, threats and bribes had gone out to Greece for nearly 200 years, so its name sounded sinister to Greek or Macedonian ears. However, the army entered Susa unopposed.

WEALTH OF SUSA AND PERSEPOLIS

The Macedonians were stunned by the accumulated wealth of the Persian kings – 50,000 talents of silver, according to Arrian, plus rich carpets and furnishings. Among the treasures were the bronze statues of Harmodius and Aristogeiton, the Athenian tyrannicides carried off by Xerxes when he sacked Athens in 480BC. Alexander sent them home, reaffirming his Panhellenic credentials. Another incident pointed to new dilemmas for Alexander. Sitting on the Great King's throne, his feet did not reach the royal footstool (Persian kings were generally tall), so a table was fetched. The sight made Alexander's old friend Demaratus cry with joy but caused a Persian court eunuch to burst into tears. To reconcile such disparate groups would prove very difficult, but at present Alexander was still Panhellenic leader. Even more treasure was stored at Persepolis,

the ceremonial capital in Persia's mountainous heart. Sending Parmenion with the baggage by the slower road, in December Alexander took the direct route with the Companion cavalry and best troops. En route he crushed some tribesmen, accustomed to demanding tribute from passing Persians, by a dawn raid that shocked them into surrender and paying *him* tribute of 30,000 sheep a year. (They were shepherds, not farmers.) Finally he entered Iranian territory, unknown to the Greeks, and approached the Persian Gates.

THE PERSIAN GATES

The Gates were a formidable natural barrier of rock strengthened by walls, blocking a 2,150m/7,000ft high pass. Lining them were 40,000 troops, whose catapults showered boulders and arrows on the Macedonians. Mauled, the latter hastily retreated. For a moment Alexander seemed baffled. Then a shepherd told him of a rough path that ran high up behind the pass. Leaving some men with orders to attack when they heard trumpets, he took the others on a night march of 17km/11 miles through snow-covered forests. Dividing his forces at the summit, he sent heavier troops down to the River Araxes to cut off the retreat. Then the rest sprinted 9.6km/6 miles uphill to surprise the Persian outposts in the dark. They fled without giving the alarm. At daybreak Alexander attacked the unsuspecting Persians from the rear while Craterus launched a frontal assault. Bewildered, the Persians scattered, jumping from cliffs or being killed trying to flee. The road to Persepolis lay open.

THE WEALTH OF PERSEPOLIS

The palaces of Persia stood on an artificial mound 18m/60ft high in the valley of Mervdasht, then fertile but now arid.

Above: Two bull-headed columns among the ruins of the huge palace of Darius I in Persepolis, which Alexander burnt one drunken evening.

Left: A gold rhyton (animal-shaped drinking vessel), part of the immense treasures stored at Susa and especially at Persepolis, then considered the "richest city on earth."

Approached by magnificent staircases were the audience halls of Darius and Xerxes, with the palaces around them. Their brick walls, 20m/65ft high, were covered in gold and glazed tiles, while huge columns with bull-head capitals supported high roofs. Alexander's men had never seen anything like this "richest city under the sun". The Persian governor showed Alexander around the palaces, including the royal treasury, containing 120,000 talents of gold and silver.

Alexander ordered 10,000 pack animals to carry the treasure to Susa for safekeeping and reinstated the Persian governor. Only then did he let his men loot the palaces, producing an orgy of destruction that wrecked many great artworks. Guards and inhabitants were killed indiscriminately until the king ended it. Seeing a statue of Xerxes shattered on the ground, Alexander began to have doubts about how far he could continue a war of revenge. But the greatest destruction was still to come.

BURNING THE PALACES

Among the intrepid women who had accompanied the army from Europe was Thais, the beautiful Athenian mistress of Ptolemy, the general and historian. Ptolemy never mentioned her role in the drunken night that saw the burning of the palaces, but others did. At a banquet

where music played and women such as Thais were present, Alexander and his companions drank heavily. Thais teasingly said it was up to the women to punish the Persians finally for their attack on Greece and burn the palaces. The Companions guffawed approval and Thais, seizing a torch, led a wild procession up the great staircases. First Alexander, then Thais, threw a flaming torch on to the floor of Xerxes' Hundred-Columned Hall. Its cedar-wood columns quickly caught light, as did the other palaces. Soon all were ablaze, as archaeology has confirmed. The Panhellenic crusade had achieved its declared goal.

Above: Darius made Persepolis the ceremonial capital of his immense empire in the 6th century BC, *building palaces intended to overawe his subjects with magnificent staircases such as this.*

Below: Another view of the palace of Darius I at Persepolis, whose high walls were once covered with glazed and gold tiles.

CHAPTER X

THE LORD OF ASIA

330–323BC

The capture of Persepolis, followed by the murder of Darius, changed Alexander's policies and attitudes. He no longer saw himself as a Panhellenic leader but as king of Asia, if not exactly Great King. Increasingly, Iranian nobles accepted him as such, but this did not mean an end to his wars.

Persia had left many tribes in remoter eastern provinces unsubjugated – not something Alexander was prepared to do. To be true ruler of all the satrapies, he had to fight guerrilla wars in areas unknown to Greeks. His military genius always won through, but he faced growing problems with his own soldiers. As lord of Asia he had to appear suitably regal, adopting at least some oriental customs. These proved anathema to Macedonian veterans, still the core of his army. In an increasingly tense atmosphere, plots were discovered – or fabricated – that led to old comrades being killed. When Alexander marched yet further east into India, his men finally rebelled.

Robbed of his desire to reach the world's limits (as he understood them), Alexander returned to Babylon, where he died aged not yet 33. Unusually open-minded, he had tried to unite the Persian and Macedonian nobility to create a new ruling class. Although this new elite would speak Greek, the plan led the Macedonians to outright mutiny. Alexander, suppressing this, continued with ever more grandiose plans. These died with him, however, as did all hopes of a united empire.

Left: The pass into Kafiristan among the Hindu Kush mountains, north-western Pakistan.

FROM PERSEPOLIS TO HERAT 330–329BC

Above: Alexander's pothos *(longing) for the distant horizon shines through this bust. Wanderlust may explain part of his endless journeying.*

Below: A fertile valley in the Elburz Mountains in northern Iran, which the Macedonians crossed in 330BC. The area makes an exception to the aridity of much of the Iranian plateau.

Alexander headed north to Ecbatana (Hamadan) in May 330BC. With a 60,000-strong army, he expected to fight another battle. But although Darius had his eastern satraps' troops and a few Greek mercenaries, he again turned and ran. This time, his irresolution proved fatal. Bessus, a distant relation, seized and deposed him, determined to retreat east to Bactria, his far-off satrapy. In early June, Alexander reached Ecbatana, where he paid off his Greek allies and Thessalian cavalry, officially ending the Panhellenic crusade. Some chose to re-enlist at increased rates; the rest returned home.

DARIUS' DEATH

The moment he heard of Darius' capture, Alexander took 500 horsemen in hot pursuit east beyond Ragae (Tehran). But they caught up with the Persian rebels too late. Bessus, seeing him approach, had murdered Darius, abandoning his baggage to flee east. The body of the last Achaemenid was found in a wagon by a Macedonian, bound in gold chains. When Alexander saw his enemy's corpse, he wrapped it in his own cloak and sent it to Persepolis for royal burial – actions that reveal that he saw himself as Darius' heir.

ALEXANDER HEADS EAST

So, increasingly, did the Persian nobility. As Alexander's army passed through the wooded Elburz Mountains and down to Zadracarta on the Caspian Sea, Persian nobles, including the Grand Vizier implicated in Darius' murder, appeared. Alexander pardoned him and many others. Obviously he needed experienced Persian-speakers (few Macedonians ever spoke much Persian) to run the complex imperial administration.

Artabazus, whom Alexander had met long ago at Pella and whose daughter Barsine had been his mistress, arrived with his seven sons, some of them former satraps. Alexander welcomed them all warmly. Darius' 1,500 Greek hoplites were pardoned and enlisted in the army –

BAGOAS: THE PERSIAN BOY

Among the former Persian courtiers of Darius III who joined Alexander, one became very close to him: the young eunuch Bagoas, namesake but no relation of the evil vizier who had poisoned Artaxerxes III. Markedly good-looking, recorded as the "son of a certain Pharnuches", Bagoas joined the select list of Alexander's known lovers (which comes to four women, one man and one eunuch). How Hephaistion reacted is unknown. But Alexander thought highly enough of Bagoas' abilities beyond the bedroom to make him *trierarch* (captain) of one of the ships in the fleet he later built on the Indus. Later still, he kissed him publicly at a festival on his return from India to tumultuous cheers. Pederasty was a practice Macedonians happily accepted.

but at the old, lower rate. At Zadracarta Alexander held games by the seaside. His men, relaxing for two weeks, admired this strange sea, which hardly tasted of salt but teemed with fish. Most Greeks thought that the Caspian opened north on to the Ocean, which they regarded as encircling the world. (Unknown to them, the Caspian's north coasts had been discovered under Darius I nearly 200 years before.) The prospect of being so near the world's edge must have aroused Alexander's *pothos* (longing) again. But his route lay south-east into the heartland of old Iran.

A KING TO THE PERSIANS

As the army marched south-east from Meshed into Areia (west Afghanistan), news came that Bessus was "wearing his diadem erect", in other words had proclaimed himself Great King.

Partly in response to this, Alexander now began adopting some aspects of Persian court etiquette to impress his new subjects. These included elements of Persian dress, a sensitive subject

to Macedonians, who thought trousers laughably effeminate. Alexander avoided trousers but began wearing a diadem (cloth band) and Persian robes such as a purple-striped tunic. He also introduced Persian court ceremonial, with cup bearers, eunuchs and (reputedly) 365 concubines. Ushers and chamberlains now controlled access to the king, who sat enthroned in splendour. All this differed sharply from the informal Macedonian court, so at first Alexander ran parallel courts, one for Persians, another for Macedonians.

In Ariana, Alexander founded another city, now Herat, and reappointed its satrap Satibarzanes. There were dangers in employing Persians, however, for Satibarzanes at once revolted. Alexander, turning back, defeated his army but Satibarzanes himself escaped with some men.

PHILOTAS AND PARMENION

In September 330BC, Alexander faced trouble closer to home. A plot was uncovered to kill him in which Philotas, son of the veteran general Parmenion and a boyhood friend, was allegedly involved. Found guilty by the Macedonian army, recovering for a moment its old judicial role, Philotas was stoned to death.

Secret orders sent by special couriers then ensured the killing of Parmenion, who was commanding the reserve forces in distant Ecbatana, a position of great power on the route back home. But while Philotas had a trial and execution, if one based on doubtful charges, Parmenion's death has been considered by some historians as tantamount to murder. The elderly general, Philip's old friend, was simply stabbed to death.

Below: A winged gold ibex of the Achaemenid period. Ibex were common in the remote mountains of Central Asia, the home of independent Iranian barons, whom Alexander had to fight or charm into accepting his rule.

THE ROAD THROUGH OXIANA 329–328BC

Above: Double bull-headed pillars, typical of Persian architecture's majestic elegance, which Alexander came to appreciate.

The campaigns of the next two years reveal Alexander's unbeatable determination, as he pushed deeper into central Asia. In October 329BC he entered Drangiana (Halmand), there founding another Alexandria: Kandahar. He decided to approach Bactria, Bessus' huge central Asian satrapy, from the south-east. As always, he chose the hardest route.

As winter deepened, he led 40,000 men up the south flanks of the Hindu Kush, known to Greeks as the Indian Caucasus. Men and horses suffered terribly on the snow-covered slopes of the highest mountains they had yet seen, afflicted by altitude sickness. But, encouraged by their king, who personally helped the stragglers, they marched on. By moving in mid-winter, they surprised hibernating tribesmen who might have harried them – and wrong-footed Bessus. Alexander founded another city, Alexandria-in-the-Caucasus, near Kabul, resting his army before continuing north in early May. A gruelling climb over another spur of mountains brought the army down into Bactria and its chief city of Bactra (Balkh).

DEATH OF BESSUS

Bessus, panicking at Alexander's sudden appearance, fled north to Sogdiana, so losing almost all support. Alexander, cheered by hearing that the treacherous satrap Satibarzanes had been killed, crossed the desert to the Oxus, the huge river dissecting the land. Here he faced a problem: there were no bridges and no timber with which to build any. So he ordered his men to stuff their leather bags with hay and swim across on these. Startled, the Sogdians offered to surrender Bessus, thinking this would end the pursuit. Ptolemy, one of Alexander's chief generals, rode out to collect the captive traitor. Bessus was then displayed naked by the roadside, and repeatedly whipped as the army passed by. Later in Ecbatana he was mutilated and impaled – a punishment Arrian, the philosophically minded historian, damned as barbarous, but which the Persians expected.

DEATH IN SAMARKAND

The Sogdians had expected Alexander to turn back after Bessus' capture, but he marched on towards the River Jaxartes,

Right: Alexander's route east only occasionally followed the obvious roads, as he zigzagged north and south in pursuit of enemies, sometimes retracing his steps. He was always master of the unexpected approach.

THE CITY-FOUNDER

Alexander is credited with founding at least 18 cities, often replacing older settlements, which may make him the greatest city-founder ever. Some have thrived, others have vanished, and many survive under new names (Kabul, Khojend). Alexandria-in-Egypt, which was new, rapidly became one of the world's greatest cities. The colonists Alexander planted across Asia were mostly soldiers who, because of illness, age or wounds, could not keep up with his unending blitzkriegs. Supplementing this Greek-Macedonian core were varied locals and captives.

After Alexander's death some colonists, who had seldom been volunteers, tried to return home but were forced back to their new cities by Alexander's successors. Whatever Alexander's motives were for these foundations (some people have seen them simply as military bases to control his empire), they helped to spread Hellenism widely, if thinly, across Asia. As the ruins of Ai Khanum in Afghanistan (another Alexandria) show, such cities could boast fine theatres, gymnasia and temples – all considered essential for a Greek *polis*.

Above: Herat in western Afghanistan is one of the 18 cities Alexander founded (or refounded) as military colonies, most of which have since flourished. The cities were planned as typical Greek poleis, *each with a theatre, agora, gymnasium, temples and a council.*

Below: Often campaigning in mid-winter across high mountains to catch enemies unawares, Alexander and his army had to cross snowed-up passes. Innate Macedonian toughness and Alexander's example kept the army going.

taking Samarkand (Miracanda) en route. The river marked the outer limits of the Persian Empire. Here Alexander founded another city, Alexandria-Eschate (the Furthest), today Khojend. Then rebellion in the rear led by Spitamenes, one of Bessus' associates, forced him to send 2,000 troops to relieve Samarkand.

He himself crossed the Jaxartes to defeat Scythian horsemen on the river's far side, who were hindering his operations. These nomads, cousins of those north of the Black Sea, were duly defeated, but in the desert Alexander contracted dysentery from bad water. In other fights he was wounded twice, once severely. Then came news that Spitamenes had defeated the Samarkand relief force, and Alexander had to gallop madly with some cavalry to save Samarkand. Spitamenes simply vanished into the mountains, involving the Macedonians in protracted guerrilla warfare. When the army returned to Samarkand in late summer 328BC, tempers were frayed, despite the welcome arrival of 21,000 reinforcements (mostly Greek), led by Nearchus, an old friend.

To relax, they drank to excess even by Macedonian standards. (Water in Samarkand was notoriously brackish.) In their drunkenness they quarrelled, resentments at Alexander's orientalizing policies resurfacing. Cleitus the Black, a *hipparch* (cavalry general) who had saved Alexander's life at Granicus, taunted the king, saying that he owed his success to his soldiers and Philip. Enraged, Alexander seized a spear and ran Cleitus through. Then, overcome by remorse at murdering an old friend at dinner, he withdrew to his tent and lay for three days without eating or drinking. Finally his soldiers decided that the anger of Dionysus, the wine god whose festival they had overlooked, lay behind it. This incident reveals Alexander's touchy pride and sometimes homicidal rage.

Above: As Alexander marched ever further east, he encountered and conquered ever higher mountains, such as the Kohi-i-Baba range. On one of them he found true love in Roxane.

Below: The wedding of Alexander to Roxane, here painted by the 18th-century Italian artist Marianno Rossi, took place high up on the Sogdian Rock. There were no offspring until a son was born after Alexander's death.

MARRIAGE ON THE ROCK
327 BC

In late 328 BC, Alexander's fortunes changed for the better. In November, Spitamenes, whose guerrilla raids had been growing desperate, was killed by his own troops; his severed head was thrown into Alexander's camp. Early in 327 BC, Alexander renewed his campaign in eastern Sogdiana, ignoring snow blizzards that claimed the lives of 2,000 men, determined to quash the last opposition.

Near modern Hissar on the Koh-in-Noor mountains, rebels had found refuge in a local baron's castle on the Sogdian Rock, reputedly 3,600m/12,000ft high. When Alexander demanded their surrender, he was told to grow wings and fly. Angered, he chose 300 volunteers to climb by night up the sheer icy rock face. On the climb 30 men died. But, when morning broke, the Sogdians looked up to see what they thought was an army high above them. Overwhelmed, they at once surrendered.

LOVE AT FIRST SIGHT

Among the captives was the stunningly beautiful daughter of Oxyartes (another Sogdian baron), Roxane, whose name meant 'little star'. Alexander fell in love with her at first sight. As she was a captive, he refused to force marriage on her despite the obvious political advantages that would accrue. Luckily, politics and passion coincided, for Roxane accepted him and a sky-high wedding was celebrated on the Rock. The castle held enough supplies to feed an army for two years – a most useful dowry. The newly weds together cut a loaf of bread with Alexander's sword, each then eating one half (an old Persian custom).

Alexander took his army and bride back to Bactra, which he replanned as a splendid Greek city. He also ordered that 30,000 upper-class Persian boys be enrolled for training as soldiers. Their weapons were to be Macedonian and their language Greek. Their families were

given no choice about this conscription, the first of many measures intended to create a united Perso-Macedonian ruling class. The next measure was even more contentious.

THE PAGES' PLOT

The Persians used to offer *proskynesis* (obeisance) to superiors, especially to the Great King, in differing ways according to their own class. This did not entail Persian nobles grovelling on the ground, as later historians supposed, although common people did kowtow, as would special supplicants. Instead, Persian noblemen would bow and, extending their hands, blow a kiss. Their superior would respond by embracing them in ways varying according to rank. All Persians paid *proskynesis* to the Great King, but this implied no worship of the Persian monarch, who was no god. However, *proskynesis* was how the Greeks honoured their gods, not their monarchs, and its prevalence among Persians soon led to serious misunderstandings.

Callisthenes, the campaign's official historian and cousin of Aristotle, used to boast that Alexander would be famous only due to his history. (Ironically, it has since been lost.) In it he had often praised the king as godlike. Among other things he tutored the royal pages, Macedonian boys in their teens, in the same way that

Aristotle had taught Alexander. Most Greek rulers regarded a tame philosopher as essential to court life, and Alexander kept on good terms with Aristotle for a long time. Callisthenes, however, while indisputably learned, was also silly.

Alexander realized that it was galling to Persian nobles to see the Macedonians approach him, their king, with rough informality while they had to make formal obeisance. With some close friends such as Hephaistion he conceived a way to get Macedonians to pay him *proskynesis* too. One night at dinner in Bactra a gold cup filled with wine went around the tables. Each guest, forewarned, stood up, drank a toast and did *proskynesis*, bowing slightly, before going up to receive a kiss on the cheek from Alexander. All went smoothly until the cup reached Callisthenes. He drank, did *not* bow but still expected a kiss, which Alexander refused. "So, I go the poorer by a kiss", Callisthenes said.

Nothing more happened that night. But soon after, when the army was on the move again, a serious conspiracy was discovered among the pages. One boy, Antipater, humiliated by a flogging he had received for killing a boar before Alexander (a major breach of etiquette), formed a plot with friends to kill the king while he slept.

As it happened, Alexander stayed up drinking until dawn so the plot misfired. But, unable to keep quiet, the pages talked about it until it reached the ears of Ptolemy. Arrests, interrogations and torture followed. The pages asserted that Callisthenes had urged them on. He was arrested, tortured and hanged. Whether he was behind the plot is unknown, but he articulated the anti-Persian attitudes of Greeks and Macedonians, who, feeling they were the master-race, refused to adopt oriental customs.

Plans to introduce *proskynesis* were probably shelved for the time being, as the army turned east towards India, Alexander's next and by far his most adventurous goal.

Left: North of Sogdiana lay steppes inhabited by Scythian and other nomads, whose realm spread across Eurasia to the Black Sea. This gold vase, if Scythian in style and content, may have been made by Greek craftsmen.

Below: Aristotle, Alexander's old tutor, had suggested his relation Callisthenes as the campaign's historian and tutor to the royal pages. But Callisthenes filled the young pages' heads with lofty theories that led them to plot against Alexander and so to their arrest and execution. Callisthenes had already angered the king by refusing to salute him in the Persian manner.

Above: Alexander attacking Porus, the Indian rajah on an elephant, from a coin struck to celebrate the victory.

Below: If less famous than his battles with the Persians, Hydaspes was perhaps Alexander's finest victory. Facing a superior enemy with elephants across a monsoon-swollen river, Alexander deceived Porus by dividing his forces and taking half of them upstream to cross secretly. His other troops then crossed in direct attack. The resulting battle was hard fought but Porus became a loyal ally.

INDIA: THE WORLD'S END
327–326BC

The Persian Empire had once stretched to the River Indus, but its Indian satrapies (now north-west Pakistan) had long been lost by Alexander's time. India was known to Greeks as a wildly exotic land – filled with gold-hunting ants among other wonders, according to Herodotus – and fabulously rich. Its kingdoms were powerful but divided.

For Alexander, to have refused such a challenge was unthinkable, although he had no idea how huge and varied India really was. There was almost no contact at the time between Greece and India. One problem he did recognize was that of India's elephants, whose smell terrified horses that were unused to them.

The army Alexander led into India in late 327BC was now half-Asian. He had left many older soldiers in Bactria as settlers or garrisons, recruiting instead 30,000 Bactrian and Sogdian horsemen. While most of the Macedonian infantry had abandoned its long sarissas in favour of shorter, more manageable spears and there had been changes in the army's structure, its core and senior officers remained solidly Macedonian.

STAND-OFF AT THE HYDASPES

The direct route into India lay through the Khyber Pass. But Alexander, wanting to secure his lines of communication, turned left for an arduous campaign against mountain tribes in the wild Chitral and Swat regions.

The climax came early in 326BC with the capture of the rock-fortress of Aornus on the summit of Pirsar 1,500m/5,000ft above the Indus. Misunderstanding local legends, the Greeks identified a local god as Dionysus, the wine god who had visited India. This misread myth further fuelled Alexander's dreams of conquest.

When the army descended to the Indus, it was welcomed by the ruler of Taxila. This made an enemy in Porus, rajah of Pauravas just to the east. Porus gathered a large army, including 85 elephants, on the east bank of the River Hydaspes (now the Jhelum), which the Macedonians reached in May.

The stand-off resembled that at the River Granicus eight years earlier, but this time Alexander was heavily outnumbered and also faced the challenge of many elephants. The Hydaspes, swollen by recent rains, was also a far more formidable river than the Granicus, being at least 0.8km/½ mile wide.

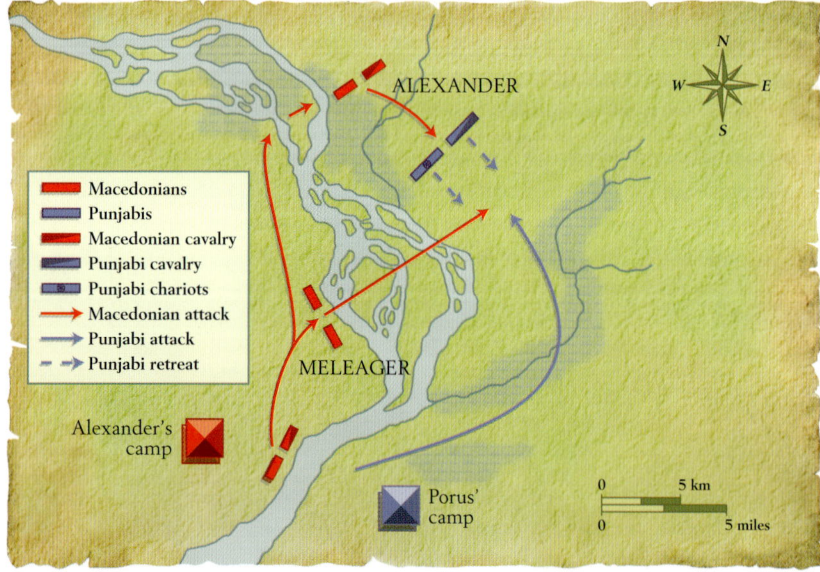

VICTORY OVER PORUS

Alexander reacted with typical ingenuity. First, he gave the impression that he was making a permanent camp. From it he launched boats on the river every night to exhaust the Indians by constant false alarms. Then, his moves disguised by rain, he secretly divided his forces, leaving Craterus with half the army. He covertly marched his force of 11,000 infantry and 5,000 cavalry 24km/15 miles upriver. Here a wooded island enabled his men to cross by night, the infantry in specially built boats, the horsemen wading through water that rose to their horses' necks. They reformed on the far bank to meet Porus' advance guard coming toward them.

Other Macedonians under Meleager now crossed halfway up the river to catch the Indians on their flank. Alexander deployed his normal oblique attack to devastating effect, his light infantry dispersing to counter the elephants while the cavalry again delivered the knockout blow. After long, hard fighting, Porus' army was routed in perhaps Alexander's most brilliant victory. Alexander, accepting Porus' surrender, restored his old kingdom and generously added to it. In return, Porus became a loyal ally. Alexander struck a series of splendid coins and medals showing elephants and Indian archers. But his appetite for Indian conquests had been sharpened, not sated.

THE ARMY'S REFUSAL

Alexander had heard that east along the Ganges lay the vast but decadent kingdom of Ksandrames, ripe for conquest. East of that must flow the all-encircling Ocean, or so Aristotle had taught. He sacrificed to the Sun and in June began planning the advance from the River Hyphasis (the Beas, east of Amritsar), where his army was encamped. As he did so, the torrential monsoon rains began.

Right: Alexander's army crossing the Indus to fight Porus, an illustration from a medieval French manuscript of Quintus Curtius, one of the ancient histories of Alexander.

The Macedonians had never experienced such deluges. Rivers rose 9m/30ft and burst their banks, and snakes emerged in terrifying numbers. In the wet heat the soldiers' clothes rotted. For men who had marched 19,000km/12,000 miles, it was too much. When Alexander tried to enthuse them about the lands waiting to be conquered, they were silent. Then Coenus, a trusted veteran, voiced the general longing to return home.

Alexander, declaring that he would force no one to follow him but advance alone if needed, retired to his tent – but to no avail. His army was adamant. Finally, reluctantly, furiously, he agreed to return, to his soldiers' tearful joy.

Above: The river Indus rises in the mountains to the north, becoming vast as it enters the plains of Porus' kingdom.

Below: Alexander crowned with victory, a contemporary coin.

THE LONG RETURN
326–325BC

Above: Although severely wounded at the siege of Multan in 325BC, Alexander still led his men from the front.

Below: The upper reaches of the Indus, the great river down which Alexander and his fleet of 800 ships sailed to the Indian Ocean.

Returning to the River Hydaspes, Alexander found the fleet that Craterus had prepared, made from timber cut in the Himalayas. Alexander had decided to return not by the same land route but by sailing down the rivers to the Ocean. He left behind two new cities – Nicaea (Victoryville) and Bucephela, commemorating his victory and his beloved horse.

In November 326BC the army, *c.*120,000 strong, embarked in a fleet of 800 ships. Alexander "stood in the bows of his ship and poured a libation [offering] into the water from a golden bowl, solemnly invoking the river… After a libation to Hercules his ancestor, Ammon and the other customary gods, he ordered the trumpets to sound and the whole fleet began to move downriver." Hephaistion marched on the left bank and Craterus on the right as flank guards.

A BRUSH WITH DEATH

At first, the voyage seemed an exploratory cruise. Local tribes, overawed by the armada, submitted peacefully, although rapids caused problems for the amateur fleet. (Several galleys, including Alexander's own, were sunk, the king having to swim for his life, but the fleet was reassembled.)

Ahead, the warlike Mallians refused to submit – to Alexander's joy. He surprised them by a cross-desert march, besieging their capital Multan in early 325BC. The outer city was soon captured but, attacking the citadel, ladders broke, leaving Alexander with three companions on top of the wall. When Alexander jumped down to fight *inside* the walls, an arrow pierced his lung. Peucestas shielded him – with the sacred shield from Troy – until other Macedonians burst in, slaughtering everyone in the city in furious grief that their king had been killed.

In fact, Alexander survived, for his doctor removed the arrowhead. But rumours of his death persisted, so a mere week later he was taken by boat to the camp. To convince his men that he was alive, he not only got up but mounted his horse, "at which the entire army applauded wildly over and over again… pressing against him, touching his hands, knees or clothes". His lung wound proved so bad that he never again walked, let alone fought, without pain. His officers berated him for recklessly endangering his life, and the army's safety, but, as Arrian said, "in truth he was fighting mad… for him the sheer joy of battle was irresistible".

THROUGH THE DESERT

Alexander reached the Indian Ocean in July. Here the fleet was almost wrecked by gigantic storms and surprised by the tides, both unknown in the Mediterranean. He founded another Alexandria as a port, with an eye to India's potential for trade. He sent his new elephant corps and 10,000 veterans back by an easy northerly route, but had different

plans for himself. He intended to march his army through the Gedrosian (Mekran) desert, something never done before even in legend. Meanwhile, Nearchus was to sail the fleet along the coast to the Euphrates, keeping in touch with the army. Like all Greek fleets, Nearchus' ships could not stay long at sea.

At first the army's journey was pleasant: myrrh trees grew so abundantly that the soldiers crushed the precious herb beneath their feet. But soon supplies, especially of water, ran short, and although they often marched by night rather than by day, the heat was overpowering. The men slaughtered and ate their animals, including horses – a lack of discipline that Alexander ignored.

At one stage in the endless desert even their guides got lost and Alexander had to lead the army back down to the coast to find the trail.

Although Alexander now always rode, he shared the hardships in other ways, especially thirst. On one occasion some soldiers, finding a trickle of water, filled a helmet and took it to Alexander, who, wrote Arrian, "with a word of thanks took the helmet and poured the water on the ground in full view of the army. The effect of this action was extraordinary, as if every man in the army had had a good drink… proof not only of his powers of endurance but of his genius in leadership." But this genius did not save thousands from dying. Only about a quarter of the original 40,000-strong army survived – the one true defeat in Alexander's career.

REUNITED WITH NEARCHUS

Reaching the desert's edge after 60 days' march, he sent camels racing ahead to order supplies. Alexander's progress then became a glorious Bacchanalian revel as the army unwound after its ordeal.

Reunited in December 326BC, Alexander wept with joy at seeing his old friend Nearchus again, thanking Zeus and Ammon, saying this reunion gave him more joy than all his conquests. Nearchus, too, had had a hard journey. His sailors had suffered from heat, thirst, hunger and alarming encounters with whales – never seen by Greeks before – and coastal tribes when the sailors had landed looking for water and food. But Nearchus had lost only one ship.

Above: A trireme, the archetypal Greek triple-tiered galley that Alexander's men built even on the shores of the Indus for his fleet that sailed down river to the Indian Ocean.

Below: Alexander decided to return to Persia marching through the centre of the vast Gedrosian (Mekran) desert shown here, an unprecedented venture. He himself survived, but three quarters of his army died in what was his worst, if scarcely known, defeat.

Above: The Heraion, the main temple on Samos. Alexander's decree that all Greek exiles should return caused huge problems for Athens, which had settled many colonists on the island.

Below: On his return to Persia Alexander found that the tomb at Pasargadae of Cyrus the Great, founder of the Persian Empire, had been desecrated. He had the tomb's priestly guardians executed, commissioning the architect Aristobolus to restore it.

THE WRATH OF THE KING
324BC

Few of the governors appointed by Alexander, Persian or Macedonian, had ever expected to see him alive again. Fourteen of the empire's 23 satrapies were showing signs of revolt. Rumours of Alexander's near-fatal wounding at Multan, then his disappearance into the desert, had encouraged even an old friend such as Harpalus, imperial treasurer, to act independently – he had begun minting his own coins in Babylon. Alexander, as the news reached him, reacted by launching a reign of terror. In it many governors, soldiers and officials were executed for corruption or disloyalty, including 600 Thracian mercenaries who had abused their power in Media. Harpalus fled to Athens, taking his two mistresses and 6,000 talents with him, money later used to finance the anti-Macedonian cause.

PASAGARDAE AND PERSEPOLIS
Dismissing Nearchus, who sailed on to the Euphrates, Alexander continued into Persia proper. Revisiting Pasargadae, the capital of Cyrus the Great, which he had seen briefly in 330BC, he was horrified to discover that Cyrus' tomb had been desecrated, allegedly by its priestly guardians. Alexander had these Magi tortured but learnt nothing. He ordered Aristobulus, an architect-biographer, to restore the tomb completely. The task took Aristobulus many years, but the restored tomb still stands. Moving on to Persepolis early in 324BC, Alexander gazed at the blackened ruins of the palaces, perhaps now regretting that drunken night's destruction six years before.

ADOPTING PERSIAN CUSTOMS
Now he was concerned to gain the loyalty of the Persian nobility, crucial to running the empire. Personal feelings were not wholly absent, however. Bagoas, his Persian lover, for some reason hated Orsines, the aristocratic Persian governor of Persis. So Alexander had Orsines stripped of his rank and crucified without trial, his place being taken by Peucestas.

Peucestas, who had saved Alexander's life at Multan, was unusual among Macedonians in having learned Persian. He followed his king in adopting modified Persian court dress. This was what Alexander wanted all his officers to do, although most grumbled about copying slavish barbarian habits. The king had his eyes on the future, however. Reunited with Nearchus and other companions in April 324BC at Susa – whose Persian governor he imprisoned for corruption – he announced perhaps his most ambitious plan, concerning love more than war.

THE MARRIAGES OF SUSA
Alexander took two new wives – the daughters respectively of Darius and of Artaxerxes III, an earlier Great King. Persia's kings traditionally were even more polygamous than Macedonia's. Hephaistion married another princess,

THE PYRE OF CALANUS

Alexander was no "stranger to the loftier flights of philosophy, though the slave of his own ambition" in Arrian's words. Intrigued in 326BC at Taxila by *gymnosophists* (ascetics who had renounced wordly ambition), he asked some to accompany his army. Only one did, Calanus, becoming Alexander's close friend. When the army reached Persepolis in 324BC, Calanus, "who had never been sick in all his life", fell seriously ill. He decided to end his life despite Alexander's entreaties. Ptolemy was told to prepare a pyre and Calanus was "escorted to it by a solemn procession of horses, men, soldiers … bearing cups of gold and silver and royal robes." As the flames mounted, Alexander, who "felt there was something indecent in witnessing such a friend's ordeal" ordered trumpets to sound and men to raise the battle-cry. Even the elephants joined in a noisy farewell to the man whose ideals were so very different from a world-conqueror's.

Left: Calanus, the Indian gymnosophist (fakir), who had followed Alexander, decided to die on a pyre when the army reached Persepolis in 324BC.

while 80 of the most senior officers also took high-ranking Persian wives. Alexander then gave official blessing to about 10,000 of his men's unions with Asian women, with sizeable gifts. Children of these mixed marriages would be Macedonian, the nucleus of a Perso-Macedonian ruling class.

RESTRUCTURING THE ARMY

He took his army north to Opis on the Euphrates in June. There he greeted as 'successors' the 30,000 Persian youths trained as Macedonian soldiers so that they could integrate seamlessly with the rest of the infantry. Noble Iranians such as Roxane's brother were enrolled in the crack Companion Cavalry, while 10,000 older Macedonians were to be honourably discharged and sent home.

These moves caused widespread mutiny. "Go and conquer with your daddy Ammon," some soldiers jeered, deeply affronting Alexander. Ordering the arrest and execution of 13 of the ringleaders, he declared they would still be "poor beggars in animal skins" without his father's and especially his leadership. Then he dismissed the whole army, appointing Persians to every post.

Stunned, the soldiers begged his forgiveness. There followed a tearful reconciliation, a banquet at which Alexander prayed for *omonia* (harmony) between Persians and Macedonians – and the veterans went off after all. Exactly as Alexander wanted.

Above: Alexander dressed as Ares, the god of war, in a fresco from Pompeii – copying a Hellenistic original – showing his marriage to Statira, daughter of Darius III, in 324BC.

Right: Back in Persia, Alexander adopted many luxurious Persian customs to impress his Asian subjects. The Persian crown had become hugely wealthy, as this gold rhyton indicates.

THE FINAL YEAR 323BC

Above: Alexander wearing a lion-head helmet, one of the many heroic images made after his death.

Below: A bull from the Gate of Ishtar in Babylon, the city to which Alexander returned in 323BC and where he died.

In August 324BC Alexander issued an edict ordering Greek cities to take back all their exiles. This decree, announced at the Olympic Games, was applauded but caused turmoil in Greece. Many exiles had been wandering the world for years, often as mercenaries. There was often no room for them at home, especially in Athens. The city faced the prospect of thousands of *cleruchs* who had settled in Samos returning if Samian exiles regained their lands. Athens sent several embassies to the king to remonstrate. But Alexander was now little concerned with the plight of distant Greek cities, as he moved among the great capitals of his empire.

More Persians than Macedonians now held prominent posts at court and in the army, both of which looked ever more oriental. Alexander sat on a golden throne wielding a golden sceptre; his royal tent was supported by golden pillars; 500 Persian Immortals matched the 500 Macedonian Companions; and bilingual ushers, staff bearers and concubines thronged the court. Balancing such oriental splendours, many Greeks –

Above: The great stone lion outside Ecbatana (Hamadan) where Hephaistion, Alexander's oldest friend, died in 324BC. It was possibly erected to honour Hephaistion.

actors, poets, secretaries, philosophers, engineers, doctors – found employment at the new king of Asia's court.

THE LOSS OF HEPHAISTION

By October the court was in Ecbatana, the old Median capital. Here Hephaistion, Alexander's vizier, or second in command, fell ill and died, probably of typhoid. No matter that their passions had cooled since boyhood love and that Bagoas was a younger, presumably more attractive rival: Alexander plunged into an orgy of grief.

He crucified Hephaistion's unfortunate physician – Hephaistion, apparently recovering, had ignored the doctor's veto on drinking wine – and threw himself on to the corpse. He shaved his head and had the manes and tails of his horses clipped. (Achilles, his hero, had similarly mourned his lover Patroclus.) He sent messengers to Siwah, sanctuary of Zeus-Ammon his father, to ask how he should honour his dead friend. The answer came back: "as a semi-divine hero". A pyre costing an unprecedented 10,000 talents was prepared in Babylon and funeral games involving 3,000 contestants were staged. A great stone lion which is still standing was probably erected outside Ecbatana in Hephaistion's memory. Then Alexander sought consolation in his favourite activity: war. A winter campaign against the Cosseans, a primitive mountain tribe, provided his last triumph.

PLANNING THE NEXT CAMPAIGN

The Persian kings had avoided Babylon's boiling summers whenever possible. But as the weather warmed up in early 323BC, the court moved down to what was still the greatest city in the empire. Alexander had decided to make Babylon his main capital, partly because it was on a navigable river leading to the sea, so well sited for trade. Unlike the Persians, he never despised commerce. He had the Euphrates and Tigris rivers cleared of dams and ordered the construction of huge docks in Babylon.

Alexander began planning his next campaign of combined discovery and conquest around the Arabian peninsula, rich in spices yet still almost unknown. He intended to lead this himself, but also planned an expedition north to explore the Caspian Sea. Envoys from across the world saluted him outside Babylon. Some were from Carthage – a city he probably intended to conquer – but there were also Scythians, Etruscans, Celts, Ethiopians, Libyans and Iberians. (Rome, however, was probably still too small to be involved.) But he ignored Chaldean priests, who warned him against entering Babylon, saying the omens were bad. He knew these priests had embezzled money that he had earmarked for rebuilding their great temples.

Alexander also proclaimed his own deification, perhaps encouraged by delegates from the Ionian Greek cities. This was unusual but not unprecedented. Lysander the Spartan general had earlier been hailed as a god, as had Dion the Sicilian politician, and Alexander had

far exceeded their achievements. At dinner parties he now began wearing divine robes, which shocked some people but again was not totally unprecedented.

DEATH IN BABYLON

But real gods are immortal. Alexander reputedly attended several hard-drinking parties in May that took him 36 hours to sleep off. At the end of the month Medius, a Thessalian noble, gave a particularly riotous party. Afterwards, Alexander took to his bed with a fever that steadily worsened. The troops, alarmed at his absence, insisted on being admitted to his bedroom. He acknowledged each with a movement of the eyes. Three days later, on 10 June 323BC, Alexander died, aged not yet 33, after a reign of 12 years and 9 months.

Rumours soon circulated that he had been poisoned, possibly with strychnine, by Macedonians alarmed at his orientalizing of the monarchy. Far more probably he died of a marsh fever hitting a constitution already much weakened by wars, numerous wounds and excessive drinking. He left behind a stunned world and no obvious heir to his huge empire. When asked who should succeed him, he had mouthed only the reply: "the strongest".

Above: Alexander's grand entry into Babylon, the destined capital of his empire, as envisaged by the French painter Charles Lebrun.

Below: Mourning for Alexander was widespread. He was long remembered as a semi-legendary figure across half Asia, as this Bukhara miniature from 1533 shows.

ALEXANDER'S LEGACY
A MILITARY GENIUS

Above: Of all the cities founded by Alexander, none surpassed Alexandria-in-Egypt, soon one of the richest and most sophisticated cities on earth. Its greatest building was the huge pharos *(lighthouse), whose light could be seen 50km/31 miles away.*

Few individuals have had as great an impact on world history as Alexander; fewer still have generated so much controversy, arousing reactions ranging from adulatory enthusiasm to stark disgust. For some 20th-century historians he was a Greek precursor of tyrants such as Hitler, Mao or Stalin. Ernst Badian, voicing such views, described Alexander at his life's end thus: "After fighting, scheming and murdering in pursuit of… power, Alexander found himself at last on a lonely pinnacle over an abyss with no use for his power and security unattainable." For other, more romantic, historians, Alexander was a chivalrous superman, spreading Hellenic civilization but free of Greek racism. With such divergent views, it can be hard to discern the man from the myth.

Above: Alexander as Helios Cosmocrator, omnipotent sun god, in this 1st century BC medallion. He became the archetypal god-king, aped but never equalled by Hellenistic and Roman successors.

THE SPREAD OF HELLENISM

One achievement is indisputable: Alexander vastly extended the Greek world. "We Greeks sit around the sea like frogs around a pond," Plato had said, noting how Greek cities, while spread around the Mediterranean, clung to the coasts. But after Alexander's conquests, Hellenism exploded eastwards, producing a cultural diffusion with profound consequences. Centuries later art in India, central Asia and even China would show the impact of Greek ideals of the human form. While Alexander did not plan this, it was the result of his conquests.

If politically Alexander wanted a Perso-Macedonian fusion – an enlightened policy abandoned by his successors – he remained Hellenic in culture. He might wear half-Persian clothes and employ Persian nobles, but he hardly (if at all) spoke Persian. In religion also he remained Greek, as in his attitude to cities and trade. The Persians had founded only a few cities as military bases, distrusting merchants. But most of Alexander's cities were founded as true *poleis*. Trade often concerned Alexander, for he founded Alexandria-in-Egypt and another city at the mouth of the Euphrates. As Greek colonists transformed the cities of western Asia, Greek became the common tongue from the Aegean to central Asia.

The immediate results of Alexander's conquest were power and riches on an intoxicating scale for his successors. He had spent 10,000 talents (ten times the annual revenue of classical Athens at its height) just on Hephaistion's funeral

Below: Alexander was portrayed in many different guises, even rather improbably here as the rustic god Pan.

– a sign of the rich new world he had opened up. The scale of his achievement long remained unrivalled.

ALEXANDER'S EXTREMISM

Personally Alexander was a man of extremes in almost everything – fighting, feasting, drinking, weeping – except sex. Probably bisexual like many Greeks, he had at least two mistresses, three wives and two male lovers, but was always keener on war than love. He was a fighter before all else.

His ambition, his *pothos* (longing) for conquest, led to the deaths of hundreds of thousands of people. If he had lived, he was not intending to rule his empire in peace. "He would not have stopped conquering even if he had added Europe to Asia and the British Isles to Europe," wrote Arrian. "On the contrary, he would have sought unknown lands beyond them, for it was always his nature… to strive for the best."

Such Homeric striving was heroic but hardly statesmanlike. His chosen role models – Achilles the pugnacious prince, the muscle-bound demigod Hercules, Dionysus the god of wine – suggest he was starting to believe in his own myths.

In the real world, Alexander can be blamed for not leaving an heir – preferably fathered in Macedonia before he even set out for Asia. He must partly be blamed for the chaos that shattered his empire after his death, which might have come far sooner. From the Battle of Granicus to the siege of Multan, Alexander recklessly endangered his life. He was lucky to have lived so long.

Above: Alexander's most notable achievement was his overthrow of the Persian Empire, epitomized in these fallen bull-head capitals. But his visionary plan to unite the Persian and Macedonian nobility in a new ruling class died with him.

THE LEGEND OF ALEXANDER

The myth of the undefeated superhero soon eclipsed any personal failings. His successors, mostly Macedonian generals, obviously looked back to him, repeating his gestures if not his brilliance. Ptolemy I of Egypt, for example, minted magnificent coins showing Alexander as Ammon. The Romans also became obsessed by Alexander's legend.

Julius Caesar wept when he saw a statue of Alexander and realized that, at the age the young conqueror was already dead, he had achieved nothing memorable. Pompey, Caesar's rival, also imitated Alexander both in his rather absurd bouffant hairstyle and in calling himself Magnus (the Great).

The less flamboyant first emperor Augustus, after defeating Cleopatra VII, last of Alexander's successors, laid a wreath on Alexander's tomb in Alexandria in 30BC in homage. In AD216 the manic emperor Caracalla, opening the same tomb, seized Alexander's armour, which he wore for his own projected attack on the east. (Caracalla was murdered soon after.) Centuries later the first Holy Roman Emperor Charlemagne (reigned 768–814) consciously copied Alexander. A thousand years after that, Napoleon I always travelled with a portrait of Alexander. He is still widely admired for his undoubted military genius, the greatest in antiquity.

Above: Posthumous coin struck for Alexander the Great after his death in 323BC.

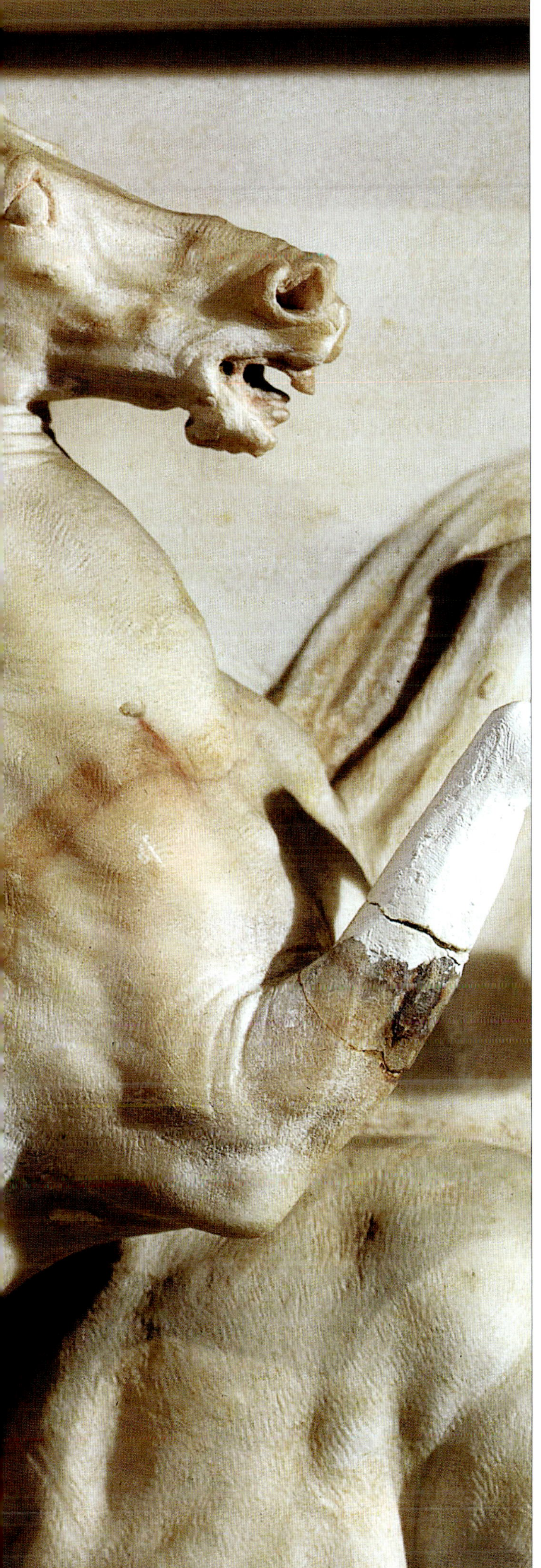

Chapter XI

ARMS AND ARMOUR
c.2000BC–AD138

Although the classical Greeks were often at war, and almost every citizen served at times as a soldier, they long remained mostly amateur fighters. War between each Greek *polis* (citizen-state) involved citizens directly, at least until mercenaries became common. So when citizens voted for a war, they were putting their own and their families' lives at risk. Even Sparta, that militaristic state, seldom started wars enthusiastically.

Greek warfare centred on set battles between hoplites, heavy infantrymen fighting in line. Hoplites were enrolled from citizen-farmers able to afford their own armour. Cavalry, recruited from the rich, was relatively unimportant. The many poorer citizens provided irregular troops and, in Athens, the crucial rowers for triremes, the triple-tiered galleys. (Slaves were seldom used for these, as they required feeding all year.) Hoplites and triremes proved a winning combination. They repelled the Persian invasions and the Carthaginians in the west. But they were later outclassed by the large professional armies of Macedonia and later monarchies, who deployed cavalry on a regal scale, elephants as tanks and more powerful catapults. They also built far bigger galleys. But ultimately even they could not counter the rising power of Rome.

Left: Alexander, shown on the Sidon Sarcophagus, *loved battles. He was primarily a cavalry commander.*

THE HOPLITE
THE ARCHETYPAL GREEK SOLDIER

Above: An Athenian ephebe (young man) pouring a libation. Military training in Athens was standardized only in the mid-4th century BC. *Earlier the Athenian army, if at times successful, was amateur compared to its fleet or to Sparta's army.*

Below: The superb discipline of Spartan hoplites repels a charge by Thessalian horsemen. Constantly drilled, Spartans for centuries made the best hoplites in Greece, distinguished by their scarlet tunics and the Lambda on their shields (for Lacedaemonia, the name of the Spartan state).

Although Homer's legendary heroes had ridden out to individual combat in chariots, chariot warfare probably never developed in Bronze Age Greece because of its mountainous terrain and small plains. Even Homer's heroes in *The Iliad* had to dismount to fight on foot. But light-armed guerilla-style troops were seldom very important either.

For the central centuries of Greek history, wars were decided by heavy infantry in set battles. These gave the victor control of the fertile farmland around each *polis*, which was absolutely vital to a city's existence.

The key Greek foot soldier through-out Greek history was the hoplite (from *hoplos*, shield or armour). These heavily armoured spear men, fighting in close formation, dominated Greek warfare from *c*.700BC until Rome's final conquest of Greece in the 2nd century BC. Hoplites, emerging as the archetypal Greek soldier by *c*.700BC, were recruited from middle-class farmers who owned 2–4ha/5–10 acres of land. Typically, these made up about 35 per cent of the *polis'* population.

ARMOUR

What hoplites wore in battle became almost a uniform. However, as each man supplied his own armour, there were considerable variations, depending on the citizen's wealth. Equipment, a major expense, was handed down from father to son if in good repair. The earliest, and among the most complete, surviving suit of hoplite armour is that of the Panoply Grave at Argos in the Peloponnese from *c*.720BC. The solid bronze cuirass (body armour) has a front and back plate, with the front plate fitting over the back plate's edges. The breastplate attempts to follow the contours of the body beneath, but a lining of some sort was presumably worn. Later, reinforced linen cuirasses became common, which were cheaper and lighter than metal cuirasses. Great men such as King Philip II had iron cuirasses decorated with gold.

Greaves (leg-guards made of bronze or iron and padded with leather) protected legs up to the knees. Helmets gradually evolved into the standard 'Corinthian type', which covered all the face, leaving mere slits for the eyes and mouth. It gave excellent protection for the head but made it hard for the wearer to hear orders. A new type of helmet, the 'Chalcidian', was therefore developed, which left openings around the ears. All helmets had crests of horsehair to make hoplites look taller and more imposing.

The hoplite carried a large shield, about 1m/3ft across and weighing at least 9kg/20lb. Made of wood but edged and faced in bronze, this was held with a single handgrip. Carried on the left arm, it defended the fighter's left side well but his right not at all. This made every hoplite dependent on his right-hand companion not breaking rank in the battle line. Training was needed to keep *en taxei* (in line – modern Greek for 'okay').

The result was a much more disciplined if less glamorous form of fighting than that of *The Iliad*.

WEAPONS

The hoplite's key weapon was his thrusting spear, called a sarissa, originally about 2.7m/9ft long. He also carried a sword about 60cm/2ft long as a cutting weapon for close fighting. Hoplites, massed eight and sometimes 16 men deep, relied on the initial shock of their charge to break the enemy line. If this failed, they turned to shoving and jostling, poking their spears either down at the neck of their opponents or up under their cuirasses.

LORDS OF THE BATTLEFIELD

Experienced hoplites were seen as lords of the battlefield, and not just in Greece. The defeat of larger Persian armies – at Marathon (490BC) and Plataea (479BC) – by Greek hoplites revealed their supremacy over Persian infantry and cavalry. Greek hoplites were employed as mercenaries by the Egyptian pharaohs in the 6th century BC and by the Persian kings after 401BC, when the rebel Prince Cyrus 'marched upcountry' with the 10,000 mercenaries, as related by Xenophon in

his *Anabasis*. The phalanx, the standard formation for hoplites, changed only slowly over the years. In later versions, which were pioneered by Epaminondas of Thebes in the 370s BC and developed by Philip II of Macedonia (359–336BC), hoplites were massed up to 50 men deep, hugely increasing their impact in full frontal charge. The length of their spears also increased – finally up to 6m/20ft – while shields shrank to about 60cm/3ft and were slung from the neck. In the Macedonian phalanx, both hands were needed to hold a sarissa, but only the first five or six ranks' spears projected. The rest remained upright, helping to shield against arrows. When sarissas were swayed en masse, they made a threatening swooshing sound. But a deep phalanx was unwieldy on uneven battlefields This led to defeats by Rome's more versatile legions at Cynoscephalae in 197BC, Magnesia in 190BC and Pydna in 168BC – battles that marked the end of hoplite warfare.

Left: A Greek hoplite and a cowering Persian archer who has lost his bow. The Greeks' recent defeat of the Persians at Plataea had boosted their military pride, but the Persians were never negligible foes.

Below: Hoplites fighting in formation, from a Corinthian vase of c.600BC. The 'hoplite revolution' had recently made these heavy infantrymen central to Greek warfare, but most armies remained glorified militias. Sparta was the outstanding exception.

CAVALRY AND IRREGULARS
HORSEMEN, SLINGERS AND ARCHERS

Aristotle thought that cavalry had once formed the Greeks' main fighting force, but he was probably mistaken. The *hippeis* (knights) retained their upper-class status in Athens, but lack of decent pasture always prevented cavalry from dominating warfare in Greece proper. Horses remained a luxurious status symbol. The Spartans, egalitarian in their militaristic way, used horses only to carry hoplites into battle. Cavalry was at times employed for scouting and to harry or pursue defeated infantry, but its total numbers remained small.

Plataea, the great land victory in 479BC over the Persians, was won despite the Greeks' lack of cavalry versus mounted Persian troops. This reinforced Greek views that cavalry hardly mattered. Even at Chaeronea in 338BC, when Philip II crushed the armies of Athens and Thebes, the main fighting was done by hoplites, despite Alexander's dashing cavalry charge. In the open spaces of Sicily, however, cavalry could prove useful: at the siege of Syracuse (415–413BC) the Athenians suffered from lack of horsemen to counter harrying Sicilian cavalry.

In Thessaly and Macedonia, lands of relatively spacious plains, cavalry always counted for far more. Significantly, these were lands where aristocracy and monarchy long survived. Alexander employed cavalry as his main aggressive arm (his 'hammer') in his Asiatic campaigns, but the infantry phalanx remained the 'anvil' needed to finish off the Persians.

LACK OF STIRRUPS
There was a technical reason why cavalry was less important in the ancient world: stirrups had not been invented. This made charging with a lance tricky, for the shock of impact threatened to unseat the rider. However, Thessalian and Macedonian horsemen used long spears, also called sarissas, with cornel-wood hearts. Such cavalry were used as his shock troops by Alexander, charging the enemy ranks directly – but Alexander's main victories were against Asian cavalry, not Greek hoplites. Highly experienced riders simply learned to grip harder with their legs to stay on. Philip II increased the number of elite Companion cavalry from only 600 at the start of his reign in 359BC to c.4,000 by the end, financed by his remarkable conquests.

THE COMPANION CAVALRY
Alexander's Companion cavalry was divided into eight squadrons (*ilai*) of 200 men commanded by an *ilarch*. The royal *ila* commanded by Alexander was larger at c.300 men. Their novel wedge-shaped formations had two advantages: they could break

through the enemy ranks and they could deploy laterally, greatly increasing their effective power.

Besides Macedonian and Thessalian horsemen, 900 light Thracian and Paeonian mounted scouts accompanied Alexander into Asia in 334BC.

CAVALRY ARMOUR

According to Xenophon, writing *c.*380BC, a cavalryman wore a cuirass with protection for the thighs, a guard for the left, unshielded, arm and a 'Boeotian' helmet giving good all-round vision. (He advocated riders slinging javelins rather than using lances.) Horses themselves were normally unarmoured. Possibly some Persian cavalry ranked as cataphracts (heavy armoured horsemen), such as the units from Bactria under Bessus at the battle of Gaugamela in 331BC, but most horsemen at this time were light cavalry.

PELTASTS, SLINGERS AND ARCHERS

At the other social extreme to the cavalry, and far more numerous, were the citizens who could not afford hoplite armour. Most served as light-armed troops called peltasts, after the pelta, the light wickerwork shield they carried. This was cut out on top to improve visibility. Their minimal armour was based on Thracian originals. In battle they fought as skirmishers, harrying the enemy with javelins before the main hoplite forces met, then falling back, sometimes sheltering behind the hoplites' shields.

Greece's mountainous terrain might have encouraged more frequent use of such mobile irregulars, but in practice they remained marginal. In 426BC the Athenian general Demosthenes saw his hoplites worsted by peltasts in remote Aetolia. He adopted their tactics to devastating effect on the island of Sphacteria in 425BC, where 292 Spartiate hoplites had to surrender to light troops. In 390BC another Athenian general, Iphicrates, defeated a regiment of 600 Spartan hoplites near Corinth by having his peltasts provoke

them to break ranks and charge out. But although this won Iphicrates great acclaim, it did not seriously affect Sparta's power.

There were also slingers and archers, considered almost cowardly fighters because they did not close up with the enemy like proper hoplites. By the late 5th century BC Cretan and Scythian archers were being employed as mercenaries, using composite bows of bone, horn, wood and sinew. These had a maximum range of *c.*140m/150 yards. Slingers were also increasingly used, those from Rhodes being considered the best. Later the Balearic Islands provided many slingers. At up to 280m/300 yards, a sling's range was greater than a bow's and it could fire stone, clay or lead shot, of which the last was particularly lethal. Philip and Alexander used numerous Thracian slingers in their armies.

Above: A mounted archer riding bareback, from an Attic red-figure plate, c.520–510BC.

Below: Lighter-armed troops such as peltasts played an increasingly important part if secondary role in battles from the 4th century BC onwards.

AMATEURS AND PROFESSIONALS
TRAINING AND FORMATIONS

Above: The 'Corinthian'-style hoplite helmet of the classical period gave excellent protection if limited vision.

Below: Hoplites fighting at close quarters. Each large round shield gave good protection to the man to the left but each man's right flank was left exposed. Hence the importance of drilling to keep en taxei, in line, something at which Spartans excelled.

Only two Greek armies before the rise of Macedonia could be called professional: the Spartans and later, for a few decades in the 4th century BC, the Thebans. All the rest were citizen-militias of varying degrees of proficiency. Two years' part-time military training on the basis of the tribe was the norm in Athens, and probably elsewhere. But adult citizen-hoplites must have practised the all-important keeping *en taxei*, 'in line', perhaps in their local gymnasium. (In Switzerland, the one modern democracy remotely comparable to ancient Greece, citizen-soldiers still train every year for a few weeks.)

EPHEBES
By the 4th century BC training for young Athenian citizens had become standard-ized. *Ephebes* (male citizens aged 18–20) were enrolled under ten *sophronistai* (drill masters), one for each tribe. A *kosmetes*

(marshal) elected by the Assembly super-vised them. *Ephebes* did a year's duty in guard-houses in Piraeus and along the borders, where they learned to fire bows and javelins and use rudimentary artillery, and to fight in line. *Ephebes* of each tribe ate together in barrack messes. At the year's end they appeared before the Assembly, where they proudly dis-played their new skills.

Athens could field about 30,000 hop-lites in the mid-5th century BC, but these were not fully professional, as defeats from Delium in 424BC by the Thebans to Crannon in 322BC, another Macedonian victory, revealed.

NAVAL STRENGTH
Athens' navy was the city's professional force, the one such navy in classical times. Although founded only in 483BC at Themistocles' urging – its half-trained crews repulsed the Persian fleet at Salamis

in 480BC however – it soon became highly professional. By the time of the victory at Eurymedon in 467BC, Athens had gained naval supremacy across the eastern Mediterranean. Each trireme's 170 rowers, recruited from the poorer citizens of Athens and other cities, became skilled, serving under professional *trierarchs* (captains) and helmsmen.

The efficiency of Athens' trireme fleets was stunningly displayed when the *strategos* Phormion defeated a larger Peloponnesian force in 429BC in the Corinthian Gulf. The Athenians drove the enemy ships into a defensive circle with bows pointed outwards. They then rowed closer and closer around until their back-paddling enemies' oars became fatally entangled with each other. Athens' thalassocracy (sea power) was unchallenged until its catastrophic defeat in Sicily in 413BC, when it lost nearly 300 ships with their skilled crews. But triremes, cramped and unseaworthy, could not stay long at sea, and needed to be beached almost every night.

SPARTAN ARMY STRUCTURE

According to Xenophon, every unit in the Spartan army, no matter how small, had its own officer. The basic unit was the *enomotia* of 36 men, four being grouped into a *lochos* (band) of 144 men, commanded by a *lochagos*. Four *lochoi* made up a division, commanded by a *polemarch* (warlord), with six such divisions in the Spartan army proper. There were few full Spartiates by the 4th century BC, however, the core army being supplemented by Peloponnesian allies. On the march, each Spartiate was accompanied by a *helot* carrying his supplies of barley, cheese, onions and salted meat. Greeks, unlike Romans, did not regularly make their camps into one-night forts.

Right: Spartan discipline had its finest hour at Thermopylae in 480BC, when Leonidas and the 300 fought to the last against the Persian hosts, a scene that inspired the French artist Jacques-Louis David 2,400 years later.

THEBES' SACRED BAND

Although little is known of early Thebes, it is clear that it too had well-trained hoplites. At Plataea in 479BC, the Thebans (fighting on Persia's side) resisted the allied Greek army stubbornly, its elite Sacred Band dying to a man.

After Thebes' liberation from Spartan occupation in 378BC, its great general Epaminondas thoroughly reorganized the Theban army, making it into a first-class fighting machine, capitalizing on an upsurge of Theban patriotism. In particular, he increased the depth of the phalanx to 50 on the traditionally weaker left wing – a stroke of genius that defeated the Spartans at Leuctra in 371BC and in several later contests.

He also resuscitated or reorganized the Sacred Band, so that it now consisted of 150 male lovers bound to each other in love and death, making them the best soldiers in Greece. Whether he changed Thebans' actual arms and armaments, perhaps anticipating the smaller Macedonian hoplite shield slung from the neck, remains unknown.

Above: Not all military service was fighting or drilling. These two hoplites are playing keritizein, *a hockey-like game, with their spears.*

Above: Quinqueremes were the largest galleys that were practical, used from the 4th century BC *onwards by Macedonian and later by Roman generals until the Battle of Actium in 31*BC.

GREEK WARSHIPS
PENTECONTERS AND TRIREMES

All Greek warships, like those of every other Mediterranean state until AD1600, were galleys. Galleys carried sails for voyages, but the nature of naval warfare meant that warships always needed huge numbers of rowers in battle. Lacking effective artillery, fights consisted of closing with the enemy and ramming or boarding. For this, concentrated bursts of powered speed, which massed rowers alone could provide, were essential. In a trireme (the archetypal Greek galley), 170 of the 200 crew were rowers, about 15 were marines (soldiers) and the remainder sailors.

PENTECONTERS AND BIREMES

Early galleys, copying Phoenician models, were penteconters rowed by 50 men. These were the ships, described by Homer, in which Odysseus sailed – and ended up shipwrecked. Built of pinewood with spruce oars, penteconters had removable masts that slotted into the keel. They were steered by oars at the stern, rudders being unknown, with bronze-clad rams. As rowers' open benches offered no comfort and galleys carried little water, ships were beached at night and during the winter. The Phoenicians added a second row of oars over the first, creating faster biremes.

TRIREMES AND LARGER GALLEYS

In the mid-6th century BC biremes were superseded by triremes with three tiers of oars, the topmost supported on an outrigger. A standard Athenian trireme was *c.*37m/121ft long and *c.*5.5m/18ft wide at outrigger level. Oars were *c.*4.5m/15ft long and almost certainly all the same length. There were 27 rowers on each side on the lowest tier, working their oars through portholes close to the water, and 27 rowers on the middle tier. The top ranks of 31 men rowed through an outrigger. Rowers were close-packed inside the hull, the lower ones having their noses almost in the bottoms of those above and in front. (The smell was noted by Aristophanes, the comedian.) Each galley had two anchors at the bow and two steering oars at the stern.

Triremes dominated Greek naval warfare until the development of quinqueremes in the mid-4th century BC. These either had two extra tiers or two men pulling each oar.

Even larger galleys are recorded, with 10, 12 and even 40 tiers of rowers. But such vessels were increasingly unwieldy. After Rome's victory over Cleopatra and Antony's mostly quinquereme fleet at Actium in 31BC, biremes again became the norm in the Mediterrenean.

Below: Another view of triremes. In battle, a galley's sails were stowed away, as triremes fought mainly by ramming enemy ships.

BATTLE TACTICS

A vital crew member was the flautist, who piped time for the rowers. Keeping time required much practice, but a well-trained trireme crew created a formidable fighting machine. This could reach maximum speeds of 24kph/15mph over short distances, accelerating very fast (for a ship). Cruising speed was only half that, but could be maintained all day. As the galley's main weapon in the classical period was its bronze ram, a favourite Athenian tactic was to shoot alongside an enemy trireme (shipping oars on the exposed side) so that the ram on the Athenian trireme's bow broke off all the enemy's oars, crippling it. Another tactic was to pass the enemy and then execute a rapid turn, ramming the enemy in his vulnerable stern. A development, perfected by the Rhodians, the finest Greek sailors after 322BC, was to dig in the front oars so that their prow dipped and their ram hit the enemy beneath the waterline, his most vulnerable point.

All such tactics required truly professional crews. Only Athens, which had about 350 triremes in its standing fleet by 431BC, could normally maintain these. Later, when Alexander's successors built huge fleets of massive vessels, galleys became primarily floating platforms for catapults and marines.

Above: A dramatic vision of a trireme at full tilt under sail and oars. The oarsmen were tiered above each other.

THE *OLYMPIAS*

In 1987 the first trireme since antiquity was launched in Greece, the *Olympias* (named after Alexander the Great's mother), commissioned into the Greek navy. The successful construction, launch and sailing of this galley answered some of the questions that have plagued historians – for example, all oars *were* the same length. But although rowed by fit young athletes from (mostly British) universities, it has not resolved all problems. It failed to reach the speeds expected, despite heroic efforts by its oarsmen. Major communication problems were solved only by electrically piping orders to the lower tiers. (This being impossible in ancient Greece, there must have been other ways of relaying commands.) Further, the 170 modern rowers could keep time only by singing together, and this is almost certainly *not* how the Greeks kept time. The ship's

trierarch could never have given orders if everyone was singing. Also, the lower tiers of *Olympias'* rowers became so thirsty that they drank all the water on board and more had to be brought in. This suggests that, splendid though the *Olympias* is, it is not the full solution to the trireme question.

Above: The Olympias, *the only recreation of an ancient galley ever built or sailed. How accurate it is remains disputed, for it proved slow and exhaustingly heavy to row for any time.*

Above: Interior of the eastern blockhouse, whose covered passages enabled the garrison to move undetected, of the massive walls of Tiryns.

Below: The Cyclopean Walls of Tiryns, built at the Mycenaean zenith, c.1300BC, long remained the most imposing walls in Greece. Awestruck Greeks later attributed such massive works to giants (Cyclopes).

FORTIFICATIONS AND SIEGES
DEVELOPMENTS IN DEFENCE

The walls of Bronze Age citadels such as Mycenae and Tiryns still rise impressively over the Greek landscape. Built before 1200BC, these Cyclopean walls (so-called because later Greeks thought that giant Cyclopes had built them) remained unsurpassed as defences for 700 years. This was due mainly to the genius of their original creators, and partly to the inadequacy of later Greek builders (and besiegers) before 500BC. When the Greeks started building in stone again after the Dark Ages that followed the collapse of Mycenaean civilization (c.1100BC), they looked back to Mycenaean models for inspiration.

EARLY CITADELS
Mycenaean builders had initially built walls with rough-cut polygonal stones, following the contours of the hills their citadels dominated. Later, cut rectangular stones were used at crucial areas, although the interior of the walls was always rubble, as it was in many later walls. The Mycenaeans, like other Bronze Age cultures, were formidable movers of megaliths. The walls of Mycenae itself are on average 5m/16ft thick, built of massive stones some weighing up to 10 tonnes/tons each. The Lion Gate at Mycenae, with its proud emblem of royal lions flanking a sacred pillar above the main gate, embodied the best defensive principles. Attackers approaching the gate would face withering fire from defenders firing through slits in the walls on both sides above.

The fortifications at nearby Tiryns are even more complex, containing galleries with vaulted roofs built into the walls. Mycenaean Athens too had walls, of which fragments remain visible. A covered path leading down the Acropolis' side to a spring helped her remain 'the unsacked city', untaken (at least in legend) by the invading Dorians.

PROJECTING TOWERS
By c.600BC Greeks were again building walls in cut stone, not rubble or mud-brick. However, the fortifications of Emporion on the island of Chios, which are among the earliest surviving, are inferior to Bronze Age predecessors. The big innovation over the next 200 years was the projecting tower. By the 5th century BC regularly spaced two-storey towers, placed not just at gateways but along walls, were common, allowing defenders to hail missiles down on attackers with their battering rams.

When, in the winter of 479–478BC, Athens hastily rebuilt its city walls after the Persians had evacuated the burnt-out city, these were c.8m/25ft high and 2.5m/8ft thick. After Athens had built the similar Long Walls connecting it to Piraeus c.4.8km/3 miles away, it became almost invulnerable to land attack – provided it always retained control of the sea. Athens' walls had towers punctuating them and recessed 'courtyard' gateways, which exposed attackers to flanking fire (useful but minor developments).

Throughout the Peloponnesian War (431–404BC) the Spartans almost never attempted to attack Athens directly.

Forts, usually on frontiers, also proved hard to capture. A typical fort such as Eleutherae on the south of the pass between Cithaeron and Mount Pastra, built *c.*400BC to guard the road from Thebes to Athens, had walls *c.*1.8m/6ft thick and *c.*4.5m/15ft high, with projecting towers with crenellations.

LINES OF FORTIFICATION

The Spartans acted differently toward the tiny town of Plataea, however, which they began besieging in 429BC. Efforts to use primitive flame-throwers or to build siege mounds for direct attacks came to nothing because the Plataeans raised their own walls. So the Spartans resorted to circumvallation, building a double wall to starve out Plataea's inhabitants. They finally succeeded, unlike the Athenians at Syracuse who spent two fruitless years (415–413BC) trying to cut the city off. Only the Persians, who took Miletus in 494BC with massive assault mounds, could normally capture walled cities.

CATAPULT POWER

Fortifications began to change when in 399BC Dionysius I, tyrant of Syracuse, started assembling forces on an unprecedented scale to attack Carthaginian bases in Sicily. Before launching his attack, he turned Syracuse itself into the best-defended large city in the Mediterranean. He extended its walls to cover the whole plateau of Epipolae, an area of nearly 1,400ha/3,500 acres. Running along the ridges, these culminated in the major fortress of Euryalus.

Its massive catapult battery, which was redesigned by Archimedes 160 years later, was raised on five solid stone pylons 11m/36ft high and carried heavy catapults able to fire down on attackers – the higher the catapult, the greater its range. These walls for long repelled even the Romans in 212BC, who finally had to capture the city by treachery.

The development of powerful catapults in the 4th century BC made platforms for defensive firepower increasingly important. Towers became taller, pierced with loopholes for artillery. The 2nd-century BC walls of Assos, in north-western Asia Minor, have slit windows for bolt-throwing catapults half-way up and broader openings on top of their fine ashlar masonry for stone-slingers. Assos' projecting towers are rounded, making them less vulnerable both to mining and to catapult attacks.

Above: The fort of Euryalus, part of the defences with which Dionysius I made Syracuse almost impregnable to land attack, the first real advance in fortifications.

Below: The development of powerful catapults in the 4th century BC soon affected Greek defences. These rounded towers at Assos could repel enemy catapult bolts.

CATAPULTS AND SIEGE TOWERS
MASSIVE MACHINERY

Above: Demetrius I, king of Macedonia 317–288BC, gained the title Poliorcetes (besieger) for his huge siege towers that rose to seven floors. But even with such giants he proved unable to capture the city of Rhodes, abandoning his towers after a long siege in 304BC.

Below: Soldiers attacking a city. (These scenes comes from the Nereid funerary monument, hence their nakedness. Greek soldiers always wore body armour while fighting.)

Until the discovery of gunpowder in the 14th century AD, catapults were the most powerful weapons any army could field. (Catapult is in origin a Greek word meaning to hurl down.) But as ways of capturing cities they were surpassed by siege towers, which in the 4th century BC, became gigantic, fundamentally altering the relationship between attacker and defender in sieges. From then on, no city could regard itself as impregnable. Although the Assyrians had used catapults in the 7th century BC, their true pioneer was Dionysius I of Syracuse, who established war laboratories for his great attack on the Carthaginians in 397BC.

NON-TORSION CATAPULTS
Two types of catapult were used: non-torsion and torsion. The first was like a far stronger hand-held bow that needed drawing back by muscle power or ratchets. The early ones were just scaled-up crossbows called *gastraphetes* (literally 'stomach bows', as the butt of the bow rested in the stomach), with trigger mechanisms. There were obvious limits as to how far even the strongest men could draw these by muscle power. But, when set on a stand and using a winch to draw the string, huge composite bows (made of horn, wood and sinew) could be very powerful. They could fire bolts – even two simultaneously – for 182m/200 yards, out-ranging normal bows.

Such catapults, mounted on siege towers, would keep defenders cowering behind their walls, as happened at Motya in Sicily. But defenders could also install catapults on walls to shoot down besiegers, as occurred during Alexander's epic siege of Tyre in 332BC. Onomarchus of Phocis had used non-torsion catapults to repulse Philip II in 354BC, an unusual defeat that taught Philip the value of these weapons. Other Macedonian rulers and later the Romans used these giant crossbows widely, the Romans deploying them even in the field. They remained expensive weapons, however. When King Archidamus II of Sparta saw one, he was stunned, exclaiming: "By Hercules, now men's courage is a thing of the past!"

TORSION CATAPULTS
Philip's great engineer Polyeidus of Thessaly developed true torsion catapults, probably only after 340BC when Philip had signally failed to capture the cities of Byzantium and Selymbria. These catapults derived their power from twisted springs (*tonoi*) made of animal sinew, hair or similarly resilient material. Their potential was much greater and they could fire heavier bolts or stones, reputedly weighing up to 82kg/180lb, at least twice as far as non-torsion catapults.

Such weapons could be used to smash down walls and buildings and lob flaming material into cities. Alexander first used them to devastating effect when attacking Halicarnassus in 334BC, and again at Tyre

and at Gaza two years later. But the fall of Tyre, which had long thought itself impregnable behind tall walls and on an offshore island, was due mainly to Alexander's siege towers.

SIEGE TOWERS

Again, Dionysius of Syracuse – whose military resources far outstripped those of any normal Greek *polis* – was the first to deploy great siege towers during his siege of Motya. These were six storeys high and could be moved on wheels. They bristled with catapults and archers on their upper floors, whose fire drove the defenders off their battlements, and with battering rams on their lower floors that smashed down the city's walls.

Alexander built even larger siege towers to capture Tyre in 332BC, reputedly *c*.37m/120ft high, so topping Tyre's walls. They were hung with sheepskins to ward off enemy missiles, with drawbridges falling from each storey to let men pour out of their many floors, battering rams swinging out at various levels and a borer on a long iron-tipped pole at their base to poke into walls. Such a tower was called, very aptly, a *helepolis* (city-destroyer).

After Alexander's death, his successors ostentatiously tried to exceed him in siege towers as in other things. Demetrius I, king of Macedon, one of Alexander's most flamboyant successors, was among the great innovatory besiegers of history, nicknamed Poliorcetes (The Besieger).

In his (finally unsuccessful) attack on the city of Rhodes in 304BC he used a *helepolis* designed by Epimachus, an Athenian. This was reputed to be *c*.45m/150ft high with a 21m/70ft square base. Its surface was covered with iron plates to protect it from missiles. Inside it had nine storeys and twin staircases, and was crammed with catapults. Its wheels were pivoted, meaning it could move sideways as well as backwards and forwards, and it was apparently propelled by 3,400 men.

Despite such massive dimensions, Demetrius' *helepolis* failed to capture Rhodes, partly because the Rhodians diverted their sewage outlets to it, causing it to sink into the mire. Demetrius abandoned both the siege and his mega-machine. By selling off its remains, the Rhodians managed to build their great statue of Helios, the sun god. Later machines were less massive and unwieldy.

Above: In siege warfare as in other matters the Romans were the direct heirs of the Hellenistic kings, using machines such as this catapult to throw large stones impressive distances.

Below: Mobile siege towers, enabling attackers to top walls, were revived in the Middle Ages in forms that often mirror those of the ancient world. This comes from De Machinis Bellicis (About War Machines) published in 1449.

GREEK WONDER WEAPONS
ELEPHANTS AND ARCHIMEDES

Above: Archimedes' most practical invention was his screw. The Archimedes' screw proved invaluable in irrigation, as this sculpture showing a North African worker using the screw to irrigate his vineyard demonstrates.

Besides conventional or obvious weapons, the Greeks and Macedonians at times resorted to more remarkable, though usually less effective, ways of making war. One was the elephant, a naturally unwarlike giant since dubbed the 'tank of antiquity', although it was much less reliable than most tanks. At the other extreme were the high-tech weapons that Archimedes, one of the greatest Greek scientists, invented for the defence of Syracuse in 212BC. Neither in the end proved very effective.

THE DEPLOYMENT OF ELEPHANTS
Elephants were unknown to the Greeks and Macedonians until Darius III, desperately mobilizing the Persian Empire's resources to repel Alexander, deployed 15 Indian elephants in the centre of the Persian line at Gaugamela in 331BC. The aim was to stampede the Macedonian cavalry (and to alarm the Macedonian infantry), because horses not used to elephants panic at their smell. Alexander, however, avoided them by

his oblique attack. The next time he encountered elephants was more serious. At the battle of the River Hydaspes in 326BC Porus, rajah of Pauravas, had 85 elephants lined up in front of his army, but Alexander's infantry opened their ranks to let them pass harmlessly by. Yet Alexander was impressed enough to form an elephant corps of his own. He sent it back by an easy route to Persia to form part of his increasingly grand and oriental court.

Within a few years of Alexander's death, Indian elephants had become a must-have weapon for any Macedonian dynast. They now had turreted howdahs carrrying archers and often carried bells and armour. In 305BC Seleucus I ceded his Indian provinces to the emperor Chandragupta Maurya in return for (reputedly) 500 elephants. With this huge elephant corps he crushed his chief rival Antigonus I at the Battle of Ipsus in 301BC, confirming both his dynasty's future and, it appeared, the importance of elephants.

The Seleucids established a stud farm for their Indian elephants outside their capital, Antioch in Syria. The Ptolemy rulers of Egypt, their great rivals, countered by capturing and training north African 'forest elephants'. These were relatives of the African elephant proper that have since become extinct. Comparatively small, they were ridden astride like a horse, and were not thought as effective as Indian elephants. (These were the elephants that Hannibal later led over the Alps.)

Left: Archimedes reportedly designed huge mirrors to focus the sun's rays on Roman ships besieging Syracuse in 212BC. Attempts to replicate this secret weapon have had no success, but he certainly designed remarkable cranes and catapults in the defence of his city.

ARCHIMEDES' INVENTIONS

Syracuse, the greatest Greek city in the west, had allied itself with Rome under Hieron II of Syracuse (ruled 265–215BC) and prospered greatly. But his successors unwisely repudiated the alliance after Hannibal's third annihilatory victory at Cannae. The Romans, alarmed, sent an army under Marcellus to capture the city.

Syracuse's formidable walls were backed up by other, more remarkable wonder weapons devised by its most illustrious resident: the mathematician and scientist Archimedes (287–212BC). When Roman quinqueremes tried to attack Syracuse on its seaward side by acting as floating siege towers, some being lashed together, a hail of missiles from catapults firing through slits forced them to retreat. Later, a Roman night attack was countered by giant cranes concealed behind the walls swinging out to drop huge stones on the Roman galleys or to grab them by their prows and so sink them. Marcellus finally had to capture the city by stealth, which was not the preferred Roman way.

Above: A Roman soldier, discovering but not recognizing Archimedes at work during the fall of Syracuse, killed him – against the orders of Marcellus, the Roman commander. Archimedes' innovatory strengthening of Syracuse's defences had made the city almost impregnable.

A DOUBLE-EDGED WEAPON

Elephants were a double-edged weapon. If terrifying charging en masse, they were prone to run amok and do as much damage to their own side as the enemy in battle. Further, ways were soon found to counter them. At the siege of Mantinea in 312BC spikes concealed under the earth by the city's defenders penetrated the beasts' soft feet to devastating effect. In reality, elephants were more prestigious than effective. In the last great battle when they were used, when the Seleucid army faced the Romans at Magnesia in 190BC, they hurt the Romans less than the Macedonians. Soon after, the Romans hamstrung the Seleucid elephants in their stud farm, ending the Seleucids' supply.

Right: After Seleucus I acquired 500 elephants in return for ceding his Indian lands, elephants became seen as super-weapons. In fact, although en masse they were formidable, they were hard to control in battle and could easily turn against friendly troops.

CHAPTER XII

FROM ALEXANDER TO HADRIAN
323BC-AD138

Alexander's sudden death without a proper heir plunged his empire into chaos. It also ended his dream of Perso-Macedonian unity. The large kingdoms that emerged from the Wars of the Diadochi (successors) are called Hellenistic, because they were essentially Hellenic (Greek) in culture and politics. But their rulers were Macedonians. Although few of these kingdoms lasted very long, Greek culture was everywhere triumphant, reaching even into India. Luxury and magnificence, epitomized by the huge statue of the Colossus of Rhodes, marked the age. Trade boomed across this wide new world, while women enjoyed greater freedom. In the end, most of it fell to the relentlessly expanding power of Rome.

One woman embodies the age and its passing: Cleopatra VII, last Hellenistic queen of Egypt. With her death in 30BC the Romans controlled the Greek world, which they had half-wrecked with their wars. The *pax Romana*, the long Roman peace that followed, allowed Greek cities to recover, while the Romans adopted and spread Greek culture. This process reached its climax under the philhellenic Hadrian (AD117–138).

Left: The Colossus of Rhodes, *a huge statue of the sun god which collapsed, painted by Louis de Caulery c.1580-1622.*

THE WARS OF THE SUCCESSORS 323–275BC

Above: Coin showing Ptolemy I, one of Alexander's generals and the first Ptolemaic king of Egypt (322–283BC). He was the ancestor of Cleopatra VII, the last and most famous Ptolemaic queen.

Below: Seleucus I, founder of the Seleucid Empire that at times stretched from the Aegean to the Hindu Kush. Seleucus had an Iranian wife, so all later Seleucid kings had some Persian genes.

Asked on his deathbed in June 323BC who should be his heir, Alexander reputedly said: "the strongest". This proved prophetic, for he had scarcely stopped breathing before his generals began fighting to control the empire. There were, however, two possible heirs of Macedonian royal blood, to whom the army turned: Alexander's idiot half-brother Arrhidaeus (who became Philip III on succeeding to the throne) and Roxane's son, Alexander IV, born in September. If Hephaistion, Grand Vizier and Alexander's oldest friend, had lived, the latter especially might have survived. As it was, the two simply became pawns of the warring generals (the Diadochi).

At first all the contestants paid lip service to the concept of an empire united under the joint kings, who reigned over rather than ruled the empire. Three men initially appeared to dominate the scene: Perdiccas, Alexander's second-in-command; Antipater, the old general left as viceroy of Macedonia, and Craterus, commanding the discharged Macedonian veterans. Perdiccas, acting as regent in Babylon, read out what he claimed was Alexander's will to the army.

This included megalomaniac plans for war against Carthage, gigantic temples and massive transfers of population between Europe and Asia. Proving as unacceptable as intended, it was unanimously rejected.

Ptolemy, one of Alexander's boyhood friends, was made governor of Egypt, to which he added Cyrene (east Libya).

Antipater, crushing the Greek states in the Lamian War in 322BC, became guardian of the young kings, establishing a Macedonian power base that he left to his son Cassander on his death. In central Asia a revolt by unhappy colonists who wanted to return home was quelled. Antigonus I, governor of Phrygia (central Asia Minor), now began extending his power south. By 316BC he had emerged as the strongest single ruler with the aid of his son Demetrius. But endless wars prevented him gaining more than western Asia, despite brilliant sieges by Demetrius, who built giant siege towers to attack Cyprus and Rhodes. Meanwhile Lysimachus carved out a kingdom in Thrace and northern Asia Minor.

THE RISE OF SELEUCUS

In 312BC Seleucus, once Alexander's infantry commander, became governor of Babylon and all lands eastward. After one of the joint kings, Alexander IV, the last of Alexander's family, was murdered by Cassander in 311BC, Antigonus took the title of king, later followed by Ptolemy and Seleucus.

In 303BC Seleucus ceded his Indian provinces to King Chandragupta Maurya in return for 500 war elephants, a huge force. Allied with Lysimachus, he used this to defeat and kill Antigonus at the Battle of Ipsus in 301BC (there were 75,000 troops on either side), gaining the title *Nicator* (victor). In all these conflicts, Macedonian soldiers remained remarkably loyal to their generals, while native populations suffered in silence the passage of the warlords with their armies. Whether their rulers were Macedonian or Persian hardly worried them.

By eliminating the only potential reunifier of Alexander's empire, the Battle of Ipsus led to the emergence of four distinct kingdoms: Ptolemy I firmly controlling

Left: Alexander's successors divided up his huge empire into constantly warring kingdoms of markedly unequal size.

Below: The winged Nike (victory) of Samothrace, one of the finest Hellenistic statues, was probably commissioned to celebrate a victory by Antigonus II, king of Macedonia, in c.250BC.

Egypt and southern Syria; Cassander shakily ruling Macedonia and parts of Greece; Lysimachus expanding his power in Thrace and Asia Minor; and Seleucus with a huge empire stretching from the Aegean to central Asia. Lysimachus' expansionism led to the next war. When in 281BC Lysimachus was defeated and killed by Seleucus, his kingdom collapsed completely. In contrast, after Seleucus' own murder in 280BC, his son Antiochus I succeeded him smoothly, ruling from the new capital of Antioch in north Syria.

Meanwhile the *poleis* (citizen-states) of Greece itself, dominated by Macedonia, periodically regained or lost their freedom. Macedonia installed garrisons at the four 'chains of Greece': Corinth, Piraeus, Chalcis in Euboea and Demetrias in Thessaly. Macedonia normally favoured oligarchies, but the distinction between democracy and oligarchy was growing blurred. All the old citizen-states were hugely outgunned by the new kingdoms.

ARRIVAL OF THE GAULS

No one was prepared for the impact of the Gauls, however. These wild Celtic invaders burst into Greece c.280BC, even threatening Delphi, the holiest shrine in Greece. Antigonus II, king of Macedonia and grandson of both Antipater and Antigonus I, triumphantly repelled them in 278BC, consolidating his position and earning some Greek gratitude.

The Gauls, crossing into Asia, were defeated by the Seleucid Antiochus I in 275BC. They were permitted to settle in central Asia Minor in the land called subsequently Galatia after them.

THE BALANCE OF POWER

These battles confirmed the new balance of power. Much the richest and most stable kingdom was that of the Ptolemies, based in Egypt but extending its power into Syria and the Aegean islands. If much the smallest kingdom, Macedonia, had excellent soldiers and the prestige of the old Macedonian crown, while the huge realm of the Seleucids at times tempted its monarchs with the (unrealizable) prospect of recreating Alexander's whole empire.

War, however, was the norm between the kingdoms and frontiers remained very fluid.

THE GREEK PHARAOHS
PTOLEMAIC EGYPT, 322–200BC

Above: Alexandria's most famous symbol was its pharos *(lighthouse), the light of which could be seen 50km/32 miles off, as depicted on a 2nd century AD Roman coin.*

Below: The tombs of Anfouchy of the 3rd century BC are among the few structures of the Ptolemaic era to survive. (Most are under water or have been destroyed.) The tombs reveal mixed Greek and Egyptian influence in the burial of the dead.

Ptolemy had noticed the potential of Egypt when Alexander annexed it in 332BC. The richest satrapy in Persia's empire after Babylonia, Egypt was unusual in being a distinct nation, defensible behind its deserts. When Ptolemy obtained its governorship in 322BC, he covertly began creating a separate state, annexing Cyrene. Kidnapping Alexander's embalmed corpse when it was en route to Macedonia, Ptolemy finally installed it at the new capital Alexandria in a grand mausoleum that also held the Ptolemies' tombs. He defeated and killed Perdiccas, who had attacked him for this act. (Perdiccas' men were reputedly devoured by Nile crocodiles after the battle.)

Ptolemy founded only one new city, Ptolemais, in the south, for the bulk of the population remained Egyptian *fellahin* (peasants). To them, the Ptolemies were pharaohs, god-kings, hailed by the priesthood, who initially welcomed them after Persian misrule. The Ptolemies

tried to impress their subjects by restoring temples damaged by the Persians and building new ones in the time-hallowed style, such as that at Edfu begun in 237BC. But no Ptolemy before Cleopatra VII (51–30BC) ever learned Egyptian. They relied instead on Greek or Macedonian administrators and *cleruchs* (Macedonian military colonists who received grants of land). When Ptolemy I died – in bed, unlike most Successors – in 283BC, he was succeeded by his son Ptolemy II (283–246BC). This cultured monarch married his sister Arsinoe, reviving the old Egyptian custom. More dynamic than her husband, Arsinoe effectively ruled Egypt until her death in 270BC.

GOLDEN AGE
The 3rd century BC was the Ptolemies' Golden Age. Ptolemy II's chief minister Apollonius perfected the fullest state bureaucracy yet seen, regulating every aspect of life. The marsh of the Fayum was reclaimed and Greek officials introduced new crops, including vines and olives. But Egypt's staple product

Above: Egypt, unlike most Persian satrapies, had had a strong national identity since the time when the first pyramids were built.

remained grain, grown by serfs tied to the land who owned nothing, not even their seed corn. All products were either heavily taxed or royal monopolies like papyrus, the precursor of paper. The resulting wealth flowed down the Nile into the royal treasury, enabling the Ptolemies to maintain huge fleets and armies. Crete, Cyprus, Samos, Cilicia and southern Syria became part of their empire, although the last was disputed by the Seleucid kings in the Syrian Wars.

ALEXANDRIA THE COSMOPOLIS

The fruits of this systematic tax-gathering were enjoyed in Alexandria, the world's first cosmopolis, welcoming Greek and other immigrants, including many Jews. With its double harbour and *pharos* – the lighthouse rising 90m/300ft, with colossal statues of a Ptolemy and his queen as pharaohs at its base – the city became the greatest in the Mediterranean. Its population probably passed 500,000 by 200BC. Its trade eclipsed that of Athens or Carthage, reaching down the Red Sea to India by 116BC. The Canopic Way, an avenue 45m/150ft wide lined with colonnades, ran west from the Gate of the Sun through the city, intersecting with similarly grand boulevards. In the west stood the Library-cum-Museum (the world's largest such building, with 500,000 scrolls, which became Hellenism's intellectual

powerhouse); a great temple to Sarapis; and an artificial hill dedicated to the god Pan. In the east royal palaces were grouped round parks, with fleets of luxurious royal pleasure barges. Beyond them stood the Hippodrome for chariot races and the Gymnasium, an important institution in Hellenistic life where men met to socialize as much as to exercise.

Alexandria was never regarded as part of Egypt proper. While it had a council of some sort, it was never a *polis* in the full Greek sense either. But Greek was the official language spoken by everyone of importance, although many races at first rubbed shoulders amiably enough.

The splendour of Alexandria depended finally on the overworked *fellahin*, for long ignored except as serfs. But in 217BC Antiochus III invaded from Syria. To repel him, Ptolemy IV conscripted Egyptian peasants en masse into the army. The resulting Egyptian victory at Raphia repelled the Seleucids for a time but revealed the growing weakness of the dynasty. Riots, strikes and rebellions began to appear in the 2nd century BC, as an increasingly corrupt bureaucracy oppressed ordinary Egyptians.

Above: Alexandria, the great port, was Ptolemy's chosen capital, replacing inland cities such as Memphis.

Below: The pylon (gateway) of the Temple of Horus at Edfu. Begun in 237BC, it exactly replicates the styles of earlier temples, for the Ptolemies posed as pharaohs to their Egyptian subjects.

SARDIS TO SAMARKAND
THE SELEUCID EMPIRE, 312–200BC

Above: A coin of Antiochus III the Great (reigned 223–187BC), who restored Seleucid power across Asia. He took southern Syria from Egypt before finally being defeated by Rome.

Below: Perge was one of the many Greek cities in Asia Minor that frequently accepted Seleucid suzerainty while keeping its internal autonomy.

More than any other successors, the Seleucids earned the right to call themselves Alexander's heirs, since they were the greatest disseminators of Greek civilization. At its peak in 280BC their empire stretched from the Aegean, where Sardis was the regional capital, to Samarcand in central Asia, an area of 3,885,000 sq km/1,500,000 square miles. It had a population of *c*.30 million – five times more than Egypt's.

To control this huge realm, Seleucus I and his son Antiochus I (281–261BC) continued Alexander's policy of founding cities at strategic points on a grand scale. (Seleucus alone reputedly founded 50). These were settled mostly with retired Macedonian troops who often married local women, their descendants later becoming full citizens. From them the Seleucids could recruit fresh generations of troops.

SELEUCIA DISPLACES BABYLON
Most new cities were called Antioch, Seleucia, Laodicea (after Seleucus I's mother) or Apamea (after Apama, his Persian wife and mother of Antiochus I). Alone of Alexander's successors, Seleucus did not repudiate his Persian wife and employed some Iranians, although the culture and politics of his new empire were Hellenistic. Seleucus I, by 312BC ruling Babylonia and eastern satrapies, founded Seleucia-on-the-Tigris on the site of Opis (now Baghdad), where a canal linked the two rivers and Alexander had once prayed for Perso-Macedonian harmony. It rapidly grew to displace Babylon as the commercial metropolis of western Asia, and Babylon fell into decay.

Seleucia's population – 600,000 in the 1st century BC, according to the geographer Strabo – remained proudly Greek, with a council, elected officials and assembly. It included people of Babylonian ancestry among its citizens.

Nearly every Seleucid foundation resembled a Greek *polis*, at least internally. Like the Romans later, the Seleucids saw their empire as a confederation of cities, to which they granted autonomy and land, sometimes with seed corn and equipment. In return, the cities were generally loyal to the dynasty, often hailing its kings as gods.

ANTIOCH THE CAPITAL
The main Seleucid capital was Antioch-on-the-Orontes in northern Syria, only 24km/15 miles from the Mediterranean, which was important to the sea-loving Greeks. Founded in 300BC, Antioch soon rivalled Alexandria in splendour and sophistication. It became the terminus for caravan routes across Asia, taxes on trade being vital to Seleucid revenues. Antioch was peopled initially with Macedonian and Athenian colonists, but many Jews

settled there later. Near Antioch the Seleucids established a stud farm to breed their renowned Indian war elephants, a pillar of their power. (Another pillar was the standing army, *c.*70,000 strong, the biggest yet seen.) The kings founded so many cities in northern Syria along the fertile Orontes valley that it resembled a second Macedonia. It was one of the few parts of Asia to be so fully Hellenized, though Phoenicia and parts of Mesopotamia were also lightly settled with Greek cities.

THE LOSS OF THE EAST

Further east, Greek cities were mere outposts of Hellenism amid an unchanged rural population. Few Iranian nobles adopted Greek customs. Generally, Graeco-Macedonian settlers' ingrained contempt for 'barbarians' doomed Seleucid attempts at mass Hellenization.

The Seleucid Empire from the start contained several types of state. Most manageable were city-states the kings themselves had founded – or, in the Aegean and Phoenicia, conquered – whose autonomy they normally respected. Far older were priestly temple-states, whose power and prestige the kings tried to curb while respecting their religious role and immunity from taxation. In Babylonia, the most civilized and richest part of their empire, the Seleucids attempted with some success to win the support of priests and merchants.

However, on the Iranian plateau, heartland of the Achaemenid Empire, they had less impact. The Seleucids had to rely on powerful satraps, or governors, who in turn depended partly on still feudal Persian nobles, who ruled their estates from their castles. The huge distances from the Seleucid heartland of Syria-Babylonia usually prevented effective royal control.

Around 255BC Diodotus, governor of Bactria, revolted against Antiochus II (ruled 261–246BC) and the huge province was lost to the empire, permanently as it turned out. But at least Diodotus was Macedonian. In 247BC, Arsaces,

Right: A bust of an unknown philosopher, one of the many Greeks attracted to the new cities that the Seleucids founded across their huge empire, which had all the trappings of a polis: agora, gymnasium, stoa, theatre.

king of the Parthians, an Iranian people, broke away from Diodotus, and his successors began expanding his kingdom in central Iran. In the 2nd century BC the Parthians over-ran ever more Seleucid territories, finally capturing Babylonia by 130BC.

To restore Seleucid power, Antiochus III made a grand military expedition through the east in 212–206BC. He forced rebel governors to recognize his suzerainty but accepted Bactria's independence, despite defeating it. He began calling himself Great King, like the Persians, but such triumphalism proved premature. Soon after he returned, the remoter satraps reasserted their independence. No later Seleucid ventured so far east. Seleucid power would soon face a new enemy: Rome.

Below: Ephesus, as the Aegean terminus of the transasiatic trade routes, flourished anew under the Seleucids, whose westernmost capital it sometimes was. These houses date from c.300BC.

Above: The fortifications of Acrocorinth high above the port, one of the 'chains of Greece' or garrisoned citadels with which the Macedonian kings controlled Greece.

Below: The dramatically sited theatre at Pergamum (Asia Minor) proclaimed the power and wealth of the Attalid dynasty, at its peak in the 2nd century BC.

MACEDONIA AND PERGAMUM
HELLENISTIC POWERS

From Macedonia, a country large only by the standards of classical Greece, came the armies that had conquered half of Asia and most of the colonists needed to establish the new Hellenistic cities. But even this combined effort did not exhaust the kingdom. It remained one of the key players in the eastern Mediterranean until Rome finally ended its existence in 168BC. Its army and navy, if relatively small, were very fine, while its much-contested control of Greece gave it both prestige and power. Macedonia itself became increasingly wealthy and fully Hellenized in the 3rd century BC. The *koine*, the common Greek dialect based on Attic (Athenian), replaced the old Macedonian dialect, as it did across the Hellenistic world.

Cassander, son of Alexander the Great's old regent Antipater, killed the boy-king Alexander IV, last of the old royal house, in 311BC. He founded the great city of Thessalonica and ruled Macedonia until his death in 297BC. Demetrius I 'the Besieger' then briefly regained control of Macedonia and Greece, but lost both when Pyrrhus of Epirus invaded from the west and Lysimachus attacked from the east. For a moment Macedonia's very existence seemed threatened until Demetrius' son – and Antipater's grandson, for the successors intermarried – Antigonus II (284–239BC) won the throne.

ANTIGONUS AND HIS SUCCESSORS
Beating the invading Gauls decisively at Lysimachia in 278BC, Antigonus reasserted Macedonia's role as a major power and ensured the future of his dynasty. He defeated the Spartans and Athenians in the Chremonidean War (267–262BC), reasserting Macedonian hegemony over Greece through the four strategic forts called the 'chains'. But he treated the Athenians tactfully. In alliance with the Seleucids, he also repulsed the Egyptian navy off Cos and Andros, checking Ptolemaic expansion in the Aegean. Antigonus encouraged Macedonia's agriculture and trade, drawing most of his revenue from his own estates without taxing his people heavily, despite extensive use of mercenaries. Private houses excavated at Pella and Thessalonica reveal Macedonian wealth and sophistication at this time.

Antigonus' successor, his son Demetrius II, was killed in 229BC fighting northern barbarians – Macedonia long acted as a breakwater against such invaders. He was succeeded in turn by his son Philip V in 221BC after a regency. Philip was handsome, energetic and ambitious. Hailed as a saviour of Greece at the Conference of Naupactus in 217BC, which attempted to find a lasting peace,

two years later he fatally allied himself with Carthage against Rome. Hannibal, its great general, appeared to be winning the Second Punic War (218–202BC). This led to the First Macedonian War. A Roman force landed in north-west Greece, did some desultory fighting and made an alliance with the Actolian League in central Greece. Peace was made on a return to the *status quo ante* in 205BC, but Rome's suspicious attention had now been turned on Macedonia.

THE RISE OF PERGAMUM

Pergamum, a previously obscure hill town in north-west Asia Minor commanding the fertile Caicus valley, became one of the great Hellenistic powers under its Attalid rulers. Its rise began in 263BC when its governor, Eumenes I, broke away. Although the Seleucids forced him to disgorge most territorial gains, he remained independent. After his successor Attalus I (241–197BC) won a dramatic victory over the Gauls, he assumed a crown and the title *Soter* (saviour) of Hellenism.

Attalus began looking west, cultivating Rome's friendship by reporting the (allegedly) dangerous ambitions of Philip V of Macedonia and the Seleucids. His son Eumenes II (197–160BC) continued this pro-Roman policy. Crucially, Eumenes supported the Romans at the Battle of Magnesia in 190BC.

Gaining a huge slice of territory in Asia Minor in return for this help, Eumenes II made Pergamum one of the architectural and artistic marvels of the Hellenistic world. Its temples and theatres rose dramatically up its hillside, while its library rivalled Alexandria's. The melodramatic splendour of the Pergamum Altar epitomizes the kingdom's flamboyant wealth. To challenge Egypt's monopoly of papyrus, parchment was reputedly invented at Pergamum, from which comes its name. Royal herds of cattle and flocks of sheep produced the hides needed for this tough, enduring writing material. In its autocratic bureaucracy,

Pergamum resembled Egypt more closely than the Seleucid realm, with many peasants working for the crown as serfs. Although the kings treated the old Ionian cities that came under their control after 189BC with diplomatic restraint, most Greeks could never forget that Attalid wealth and power stemmed from craven collaboration with Rome.

Above: The Pergamum Altar, *one of the grandest and most flamboyant in the Greek world, expresses the wealth of the Attalid kingdom of Pergamum at its 2nd-century peak. It is now in Berlin.*

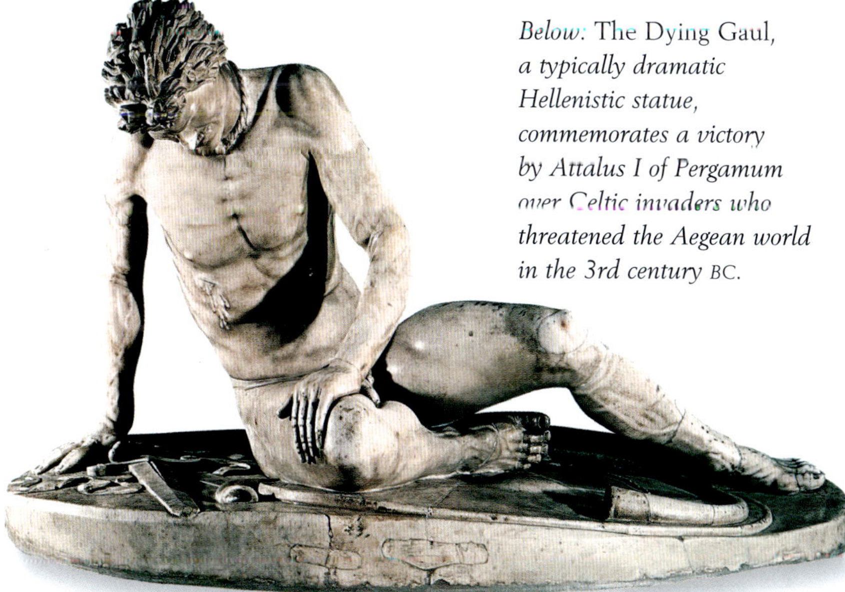

Below: The Dying Gaul, *a typically dramatic Hellenistic statue, commemorates a victory by Attalus I of Pergamum over Celtic invaders who threatened the Aegean world in the 3rd century* BC.

THE GREEKS IN THE EAST
BACTRIA AND INDIA, 350–320BC

Above: Antimachus I, Hellenistic king of Bactria (Afghanistan), conquered north-western India in c.180BC, minting fine coins such as this.

Below: Eucratides I, king of Bactria c.170–155BC, ruled a huge Indo-Greek kingdom stretching from Merv in Central Asia to Taxila in northern India.

Alexander had founded cities in Bactria (Afghanistan), Sogdiana (Tadzhikstan) and in north-western India (Pakistan) primarily for military purposes. Their often unwillingly retired soldiers would, he intended, safeguard strategically important routes and provide fresh recruits in their sons, whose mothers would be local women. Or at least that was the idea. But 20,000 settlers revolted even before Alexander's death, returning to their colonies only under compulsion from Macedonian generals.

The Seleucids continued Alexander's colonization policy but on a wider, and generally much sounder, basis. They re-established some cities and founded many further ones (often named Antioch) but made each a proper Greek *polis* with a proper council, assembly, theatre and gymnasium.

ARTISTIC FUSION

Whether or not Menander converted to Buddhism, the fusion of Greek aesthetic form with Buddhist religious ideas proved hugely influential. Chinese and even Japanese art echoes Hellenic forms, while the huge Buddhas of Bamiyan, dating from c.AD500, showed traces of Greek art. The intellectual impact proved long-lasting, too. The *Gorgi Samhita*, an astronomical work of AD230, states: "Although the Yavanas (Ionians, i.e. Greeks) are barbarians, the science of astronomy originated with them, for which they should be revered like gods."

THE RUINS OF AI KHANUM

These cities retained their Hellenic identity for a remarkably long time, as the ruins of Ai Khanum show. Originally Alexandria-on-the-Oxus in Bactria, perhaps refounded by Seleucus I, it is sited in what was then a fertile area. The ruins include a huge gymnasium, a big terraced theatre near the river, large private houses and a palace with ornate Corinthian columns around a grand courtyard. Around 300BC Clearchus, a pupil of Aristotle, brought from distant Delphi the traditional maxims for the Five Ages of Man (as the Greeks numbered them) to be inscribed in this gymnasium. A papyrus with scraps of Aristotle's philosophy has been unearthed nearby, suggesting that Greek citizens of this central Asian *polis* discussed philosophy after exercising in the gymnasium. The city reached its peak c.200BC.

THE GRAECO-BACTRIAN STATE

In 255BC Diodotus I, the Seleucid governor of Bactria, declared himself independent – perhaps despairing of

Above: Elephants, the 'tanks of ancient warfare', came mostly from India, where Alexander had encountered them in numbers at the Battle of the Hydaspes.

effective help from the Seleucids against nomadic invaders – and founded the Graeco-Bactrian state. His son Diodotus II (248–235BC) took a royal title and negotiated with the Parthians, now also independent, before being overthrown by Euthydemus I (235–200BC). An energetic king, Euthydemus retook Herat from the Parthians in the west, while in the north-east he penetrated into Sinkiang, now in China. Antiochus III defeated him but had to accept him as a subordinate ruler in 210BC. Most of what we know of these monarchs comes from their fine coinage, for there are few written records.

THE INDO-GREEKS

Seleucus I had ceded his eastern provinces to the first Maurya ruler Chandragupta in return for (reputedly) 500 elephants in 303BC. He acknowledged the rise of this new power in India by marrying one of his family to a Maurya. Although this had little political effect, it helped with Greek cultural diffusion, to which India proved very receptive. Chandragupta's grandson Asoka, possibly part-Greek by birth, became the first Buddhist

emperor (274–232BC). He had Buddhist decrees inscribed in Greek on pillars in Kandahar, showing the importance he attached both to his new pacifist religion and to his Greek-speaking subjects. He also employed Greek craftsmen.

DEMETRIUS AND HIS HEIRS

In the early 2nd century BC, with Maurya power declining, the Graeco-Bactrian monarchs began to regain lost territory and move east. Euthydemus' son Demetrius I (200–185BC) retook some Mauryan lands, founding a colony, Demetrias, near Ghazni. His kingdom was divided between three heirs. One, Antimachus I, controlled all of what is now northern Pakistan, minting resplendent coins. His successors ruled from Taxila, minting bilingual coins in Greek and Brahmi, the script of the Ganges valley. Another Indo-Greek monarch Demetrius Aniketos (Unconquered) issued coins in Greek and Prakrit. Increasingly, these coins began to bear Indian emblems, such as Lashkmi, the Hindu goddess, or a sacred tree, among Buddhism's holiest symbols. Some were even square, the preferred shape in Indian bazaars. In India the Greeks proved receptive to local culture, dropping their usual chauvinism.

Eucratides I (c.170–155BC), who took the Indian title Maharajasa (great king), ruled a united Indo-Greek kingdom from Merv in central Asia to Taxila. His successor, Menander (155–130BC), extended his power far east down the Ganges valley, his troops reaching Patna. His gold and silver coins have been found over a huge area, although he did not rule it all. He may, however, have converted to Buddhism, for he figures in Buddhist legends as Milinda the Just. Menander was succeeded by further Indo-Greek rulers, among them Queen Agathocleia, who ruled in her own right. But in the 1st century BC Scythian tribes, the Kushans, invaded Bactria and then north-western India, overthrowing the last Indo-Greek kingdom by 30BC.

Below: While all the Indo-Greek kingdoms had vanished by 30BC, Greek artistic influence persisted for centuries. It shaped depictions of the Buddha, as this Bodhisattva of Gandhara of the 4th century AD reveals.

OLD GREECE: THE AETOLIAN AND ACHAEAN LEAGUES, 320–180BC

Above: Ruins of the Tholos, Sanctuary, Athena Pronala, Delphi, control of which site signified control of old Greece.

Although Macedonian kings dominated old Greece after 322BC, garrisoning key points, Greek political – and economic and social – life was not dead. However, many smaller cities, while still cherishing their beloved autonomy, realized that they now had to band together in leagues to survive. In doing so, they showed continued Greek ingenuity and innovation until the steamroller of Roman conquest squashed all independence.

THE AETOLIAN LEAGUE
Isocrates, the 4th-century BC propagandist, had long called for Panhellenic union against 'barbarians'. Around 367BC, in the mountainous backwoods of Aetolia, many villages or cantons actually formed such a defensive league. This developed a remarkably flexible and inventive constitution, with a president and commander-in-chief annually elected who could not be re-elected on successive years. Other officials included a cavalry leader, secretary and seven financial stewards.

The League Assembly held two regular meetings a year, before and after the campaigning season, at the hilltop temple of Apollo at Thermum. Every adult male citizen had a vote and there was a federal *boule* (council) with 1,000 delegates. Each state was represented according to its population while retaining internal autonomy. Much business was later delegated to a committee of 40, for the League became less democratic as it grew. Policy remained in Aetolian hands, although some new states became honorary Aetolians.

The Aetolian League showed its strength after Alexander's death. Although its army was only 12,000 strong, it captured Naupactus on the Gulf of Corinth, repelling attacks by the Diadochi. By c.300BC it had gained control of Delphi. Although Delphi was never part of the League, the Aetolians justified their hold on the great shrine by defeating the Gauls in 279BC. In 245BC the Aetolians crushed the Boeotians, extending their power across central Greece. They generally opposed the Macedonians, posing as defenders of Greek liberty, but were also notoriously friendly to pirates.

Allied with Rome, they fought in the Second Macedonian War (200–197BC), their cavalry playing a decisive role in Roman victory at Cynoscephalae. Then, feeling that Rome had ignored them in the subsequent peace, they rashly invited the Seleucid Antiochus III into Greece in 192BC. After his defeat in 189BC, the League was reduced to Aetolia and became dependent on Rome, which later broke it up.

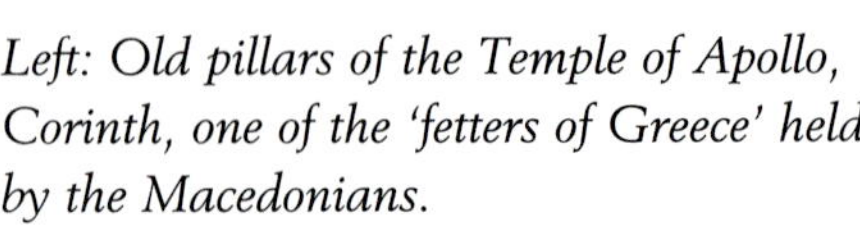

Left: Old pillars of the Temple of Apollo, Corinth, one of the 'fetters of Greece' held by the Macedonians.

THE ACHAEAN LEAGUE

Equally significant was the Achaean League, which emerged *c.*280BC on the Peloponnese's north-west coast. Originally composed of ten coastal cities, by 251BC it included Sicyon near Corinth, which was not actually in Achaea. From Sicyon came Aratus, who for a generation headed the League as president and commander-in-chief, being re-elected every other year. The League's constitution emulated the Aetolians' but with differences. Only citizens over the age of 30 could vote in the *synedos*, or Council, which made it more conservative than the Assembly. Both met four times a year at Aigion on the Gulf, once to elect League officials such as the *hipparchs* (the cavalry commanders). Votes were taken by city, not head, to stop locals swamping the vote, but only richer citizens could afford to travel to Aigion. Each city retained its internal autonomy and coinage but followed League foreign policy.

Aratus, a passionate League patriot if no great general, pursued an anti-Macedonian policy. In 243BC he surprised the Macedonian garrison of Corinth by a night attack, adding that great port to the League. Over the next years, as Macedonia faced northern invasions, the League grew to include Argos, Megalopolis and finally almost all the Peloponnese except Sparta. These new citizens were all willing League members.

But the League faced a resurgent Sparta after the reforms of Cleomones III, who captured city after city by appealing to the poor. Aratus was forced to call in the Macedonians to defeat the Spartans at Sellasia in 222BC. Under Philip V, however, the League's relations with Macedonia deteriorated and it turned to Rome. Philopoemon, its next great leader, accepted Roman help only very reluctantly. It was needed to defeat a resurgent Sparta again, which ultimately led to that proud city being enrolled in the League.

After Philopoemon's death from poisoning in Messenia in 182BC (Messenia had become another involuntary member) the Achaean League remained pro-Roman, but this did it no good at all. In 167BC, after the final defeat of Macedonia by Rome, 1,000 chiefly democratic Achaean hostages were taken to Rome, among them Polybius, the future historian. (Only 300 out of the 1,000 lived to return home.) Polybius himself came genuinely to admire Rome's unique constitution and wrote about Rome's rise to power. He also tried hard, if finally in vain, to persuade his countrymen to accept increasingly stringent Roman demands.

ATHENS AND RHODES
323–170BC

Above: Athens, liberal and usually peaceful, still attracted the greatest philosophers such as Zeno of Citium, founder of Stoicism.

Below: The entrance to Rhodes Harbour, where the Colossus once stood.

Two cities dominated the Aegean in the Hellenistic period culturally and commercially: Athens and Rhodes. The former, still the greatest Greek city in 323BC, slowly ceded economic supremacy to Rhodes, the new mercantile power, but retained its cultural primacy. Rhodes became the wealthiest independent *polis* of the age and an unusually fine example of limited democracy. Both cities ultimately fell to Rome's imperialism despite their attempts to placate it.

ATHENS: GREECE'S INTELLECTUAL AND CULTURAL CAPITAL

The Lamian War of 323–322BC, when the Greek alliance was crushed by Macedonia, ended Athenian independence and full democracy. Cassander's protégé Demetrius of Phalerum's bizarre regime in Athens mixed philosophy, authoritarianism and sybaritism. When Demetrius I the Besieger ousted Cassander's men in 307BC, democracy was partly restored, although the Assembly was no longer supreme. Exploiting Macedonian weakness during the Gaulish invasions of 280BC, Athens regained full independence but lost it to Antigonus II of Macedonia in 262BC. Antigonus, while installing a garrison in Piraeus, treated Athens tactfully – he saw it as his cultural capital, as did increasingly many Greeks.

The city's economy revived thanks to new veins of silver found at the Laurium mines, and later its control of Delos' free port. In 229BC Athens managed to buy out the Macedonian garrison, becoming effectively neutral. It cultivated Rome's friendship in the 2nd century BC, long escaping the worst wars.

Athens' importance was now overwhelmingly intellectual and cultural. It remained, most of the time, a modified democracy and it became the definitive home of philosophy. Epicurus and Zeno of Citium founded their respective schools – Epicureanism and Stoicism – c.300BC in the city, alongside the existing Platonists and Aristotelians. At the same time, Menander started the New Comedy, the origin of all subsequent 'sit coms'. Non-political in content but psychologically astute, it influenced Roman writers such as Plautus.

Hellenistic kings competed to honour Athens with fine buildings. The Seleucid Antiochus IV in the 170s BC paid for work to be restarted on the gigantic Temple of Zeus abandoned 340 years earlier, although it was not completed until the Emperor Hadrian's reign three centuries later.

On the east of the Agora in 140BC Attalus II of Pergamum built the Stoa, a huge colonnade, the last and largest of many. Beneath such colonnades the philosopher Zeno taught (so his followers were named 'Stoics'). Stoas sheltered shoppers and other citizens too. Athens' Indian summer of prosperity lasted until after 100BC, when it rashly sided with Mithradates of Pontus against Rome.

Right: Many Hellenistic monarchs endowed Athens, still the supreme Hellenic polis, with grand buildings. The Seleucid king Antiochus IV in 174BC paid for work to restart on the vast Temple of Olympian Zeus begun in the 6th century BC, but work had not gone far before Antiochus' murder cut off funds.

RHODES: A MARITIME REPUBLIC

In 406BC Rhodes' three small cities united to form a single democratic *polis*. Ruled by outside powers in the 4th century BC, after Alexander's death Rhodes declared itself free and expelled its Macedonian garrison. When Demetrius I besieged it in 305–304BC with giant siege towers, it repelled him. Rhodes enjoyed a period of great prosperity down to 166BC, displacing Athens as the hub of the Aegean. Its wealth came from its superb position at the centre of trade routes to Sicily, the Black Sea and Egypt. In 170BC its two per cent carrying tariff, primarily on wheat, yielded a million drachmas. As a result, Rhodes became the Hellenistic world's banking centre.

Rhodian democracy was limited but its aristocracy had a strong sense of *noblesse oblige*, richer citizens helping the poorer. Because of this, Rhodes enjoyed unusual social stability. All citizens served in the fleet. This albeit small fleet, comprising about 50 galleys, mostly quinqueremes,

THE COLOSSUS OF RHODES

Symbol of Rhodes' maritime wealth and one of the Seven Wonders of the World, the famous *Colossus of Rhodes* according to legend straddled the harbour entrance. In reality it did not but was impressive enough: a bronze statue of the sun god 33m/110ft high. Falling in the earthquake of 226BC, even its mighty remnants long impressed visitors. The much-copied statue of snake-strangled Laocoön and his sons marked the Rhodian school of sculpture's zenith *c.*180BC, but the whole city was adorned with artworks.

became the best in the Hellenistic world, again paid for by the rich. Rhodes suppressed piracy as Athens had once done, promulgating a maritime code later adopted by imperial Rome. When an earthquake shattered the city in 226BC, other Greek states combined to restore it, so central had Rhodes become to their political and commercial wellbeing.

Rhodes favoured a neutrality that protected its trade, but allied itself with Rome against Philip V and Antiochus III because it feared their ambitions. Its immediate reward was large: Lycia and Caria, former Seleucid territory in Asia Minor. But Rome, growing suspicious of *any* Greek state's true independence, thought it was too neutral in the Third Macedonian War (171–168BC) and made Delos a free port, so ruining Rhodes' trade. Rhodes remained culturally important, attracting poets such as Apollonius Rhodius in the 3rd century BC and philosophers such as Poseidonius (135–50BC). The future emperor Tiberius withdrew there in 6BC, actually in a sulk at being sidelined in the imperial succession but supposedly to study philosophy.

Below: In Athens, the classical tradition in art continued, creating vivid new works such as this Maenad of c.100BC.

REVOLUTION IN SPARTA
244–192BC

Above: The Vix Crater *is an unusually fine example of Laconian craftmanship.*

Since Sparta's crushing defeat by Thebes in 371BC and subsequent loss of Messenia, it had been of only minor importance, even in Greece. Its falling birth rate, coupled with the concentration of land among ever fewer rich people, meant that there were fewer full Spartiates – only 700 by 300BC – to be the hoplites that still formed the army's core. Discontent among the disenfranchised, who had lost their lands and citizenship, threatened Spartan stability. Despite this, memories of Sparta's former hegemony remained potent among both its neighbours and rulers.

AGIS THE REVOLUTIONARY
In 244BC Agis IV became king, determined on a return to the legendary excellence of the 'Lycurgan' constitution. Agis planned to divide the land into 4,500 equal lots, cancel all mortgages, allow many Spartans to regain their citizenship and enfranchise some *perioeci* (second-class citizens). This horrified conservatives: the magistrates (*ephors*), his co-monarch Leonidas II and rich citizens. Agis drove Leonidas into exile and deposed some *ephors*, but when Leonidas returned in 241BC, Agis was killed.

CLEOMENES' REFORMS
In 235BC Cleomenes III became king. Although he was the son of Leonidas, listening to Agis' widow had made him a revolutionary. He was also inspired by the teachings of the Stoic philosopher Sphaerus. Realizing that force was needed to implement reform, Cleomenes drove through revolutionary changes. He abolished debt; nationalized the land, dividing it into 4,000 lots for Spartiates and 15,000 for *perioeci*; and boosted the number of Spartiates by promoting *perioeci* or even *helots* (serfs). He also sold 1,000 *helots* their freedom, an unheard-of move.

In 229BC Cleomenes marched north and annexed some Peloponnesian cities in the Aetolian League, intending to cement domestic reform by victories abroad. Poor people in many cities flocked to him, hoping that his reforms would be emulated. This initially helped him in his war with the Achaean League. But after winning two minor victories over the Achaeans, Cleomenes returned home to pursue his revolution. He executed four conservative *ephors* and abolished their ancient office. With Sparta's army now hugely increased, he seemed

Below: This scene of martial readiness comes from the Vix Crater, *which was made in* c.500BC *when Sparta was in its austere prime – an age some reformist kings wished to revive.*

poised to conquer the whole Peloponnese. In despair, the Achaean League's leader Aratus called in the Macedonians, his bitter enemies, and their combined forces defeated Cleomenes at Sellasia in 222BC. He fled into exile in Egypt, where he committed suicide, while Sparta itself fell to invaders for the first time in its history. But the problems – principally the growing gap between rich and poor – remained, and not just in Sparta.

THE LAST SPARTAN KING

In 207BC Nabis, who was of royal blood, took the throne probably after murdering the young king Pelops. He at once re-enacted Cleomenes' reforms but in an even more drastic manner. Forming a private bodyguard of freed *helots* and mercenaries, he seized land from the rich to pay for the restoration of the common meals so important to Spartan life. Adroitly allying with Rome in the Second Macedonian War (200–197BC), he survived until a disgruntled Aetolian officer assassinated him in 192BC. Sparta was then forcibly enrolled in the Achaean League. When Rome destroyed the Achaeans in 146BC, Sparta became technically free under Rome's protection, but it was now a museum city. The emperor Augustus restored its port and it long continued its strange customs to entertain Roman tourists.

Above: Young Spartans Exercising, *painted by Edgar Degas in 1860. Among Sparta's unique features had been the way girls also exercised nude, shocking other Greeks. This way of life was in decay by the 3rd century BC, ruining Spartan strength. But some Spartan customs were long maintained to amuse Roman tourists.*

UTOPIAN REVOLUTIONS

Growing social and economic problems in many cities fused with Stoic teachings of the universal brotherhood of men to create an explosive mixture in the 2nd century BC. The founder of Stoicism, Zeno of Citium (333–262BC), had outlined in his *Republic* (now lost) revolutionary proposals for the just society, but his ideas had remained just ideas: Stoic philosophers in Athens were no firebrands. However, Zeno's ideas were elaborated by later thinkers such as Iambulus. He wrote *c.*200BC about a mythical Island of the Sun, a communistic utopia (though that word had not been invented) where all men were equal and worshipped the Sun-god. Slave risings, common at the time and always put down with great brutality, arose out of pure misery.

Especially appealing, therefore, was the utopian state that Aristonicus, the illegitimate half-brother of Eumenes II, the last king of Pergamum, attempted to inaugurate.

Eumenes had left his kingdom, which included Ionia, to the Romans in 133BC, but most Ionian cities had no wish to become subjects of Rome. When Aristonicus proclaimed his utopian City of the Sun at Pergamum, freeing slaves, many Greeks, including the Stoic philosopher Blossius of Cumae, joined him. Their army defeated a consular Roman army. It took Rome three years' hard fighting before its legions could crush the Greek utopians. Rome's revenge, typically bloody, marked the final end of Greek political experimentation.

THE WESTERN GREEKS
320–211BC

Above: Hieron II's long reign in Syracuse (269–215BC) saw unprecedented peace and prosperity. The altar he erected was the longest ever built at c.200m/650ft.

Below: Pyrrhus, king of Epirus (319–272BC), fought as a mercenary general for the Italian Greeks against Rome 280–275BC. His initial victories proved so costly they became known as 'Pyrrhic'.

In the 4th century BC Taranto (Taras) was the wealthiest city in Greek Italy, thriving on its trade, wool production and the purple dye obtained from molluscs in its lagoons. The philosopher Archytas guided its mixed democracy while also remaining on good terms with Dionysius I, tyrant of Syracuse. But after Archytas' death c.340BC, the Tarentines felt threatened by Italian hill tribes, despite having a large fleet. They summoned Alexander of Epirus, brother-in-law of the great Alexander, to help them, but he was murdered in 330BC. Then a far more formidable enemy emerged: Rome, expanding into southern Italy, founded Venosa, a military colony, only 144km/90 miles north of Taranto while extending the Via Appia, its first great military road, southeast towards Brindisi. The Tarentines grew alarmed.

TARANTO AND ROME AT WAR

In 282BC the Greek city of Thurii, Taranto's rival across the Tarantine Gulf, appealed to Rome for help against Lucanian raiders. The Romans reacted by sending a fleet into the Gulf. This broke an earlier agreement with Taranto, which in reprisal sank some Roman ships and then mocked Rome's ambassadors for speaking bad Greek.

War followed in 280BC. Taranto called in King Pyrrhus I of Epirus, the best professional general of the time, to help it. His skilled army, with its elephants and *phalanxes*, was expected to crush the amateur Romans. Pyrrhus indeed won two major victories, marching almost up to Rome's walls. But the Romans fought doggedly on, learning to counter the elephants and replacing their own losses. After one victory, Pyrrhus exclaimed that he could not afford another such – hence 'pyrrhic (unaffordable) victory'. But, seeing the Romans methodically pitching camp each night, he admitted that his enemy was "not barbarian".

After Pyrrhus withdrew from Italy in 275BC, the Romans advanced south and Taranto had to accept a Roman alliance. By 272BC all of Magna Graecia ('greater Greece', Italy's Greek cities) was in Roman hands.

THE STRUGGLE FOR SICILY

The order that Timoleon had brought to Sicily, especially Syracuse, did not long survive his death in 334BC. In 317BC Agathocles overthrew Syracuse's government with Carthaginian backing. He made himself dictator and won support from the lower classes by terrorizing the rich. Quarrelling with Carthage, he then boldly invaded Africa itself in 310BC, but had to withdraw in 307BC. But he still made himself ruler of most of Sicily, even capturing Corcrya (Corfu) and taking a royal title. After his death in 289BC his successor, Hicetas, was defeated by the Carthaginians and deposed. In the ensuing chaos, Carthage looked set to conquer the whole island until King Pyrrhus briefly intervened.

After Pyrrhus left Sicily, Hieron, one of his officers, seized power in Syracuse. He was acclaimed king as Hieron II after defeating rampaging Italian mercenaries. Hieron ruled Syracuse remarkably well for 54 years (269–215BC). He revived some of its past glories, helped by his wife Philistis' descent from Dionysius I. Shrewdly switching to support Rome in the First Punic War (264–241BC), Hieron gained most of eastern Sicily as his kingdom. He adorned Syracuse with public buildings, including the world's biggest altar (200 x 22m/650 x 74ft), and employed Archimedes, the great scientist, to fortify it. Hieron also built the largest warship yet seen, the 5,000 tonner *Alexandria*. His tax system, the Lex Hieronica modelled on the Ptolemys', took one tenth of crops grown in the kingdom – a relatively light tax, which the Romans copied. Syracusan prosperity is revealed in the fine private house recently unearthed.

The rest of Sicily was not so fortunate, being long fought over between Carthage and Rome – Acragas (Agrigento), then, Sicily's second richest city, was twice

Right: The Colosseum in Rome, the city whose fast-rising power increasingly dominated Greek politics.

sacked and once burnt. Sicily became after 241BC Rome's first *provincia* (province), ruthlessly exploited for its wheat farms. The slave gangs who worked these huge farms revolted en masse in 135–132 and 104–100BC. Meanwhile, Greek urban life decayed.

Above: Taranto (Taras) was the richest Greek city in Italy, with a fine double harbour and a thriving purple dying industry. Yet, despite hiring the finest general Pyrrhus, it fell to Rome in 272BC.

Above: Flaminius, the Roman general who defeated Macedonia at Cynoscephalae in 197BC and then promised Greeks 'freedom' at Corinth.

Below: A relief from the Temple of Neptune of c.100BC in Rome reveals Greece's growing cultural impact on the Romans.

THE SHADOW OF ROME
220–188BC

In 217BC a peace conference was held at Naupactus (Lepanto) to try to end Greece's constant wars. Agelaus of Naupactus, welcoming the delegates, pointed to the titanic struggle between Rome and Carthage then racking Italy. Now, he said, was the time when Greeks must join together like men wading through a torrent, "for if the cloud now rising in the west should spread to Greece, I fear we shall be begging the gods to give us back the chance to call even our quarrels our own."

PHILIP VERSUS ROME

His prophetic words were applauded by the delegates. However, Philip V of Macedonia, attending the conference, then made a fatal error. After Hannibal's great victory at Cannae in 216BC he, like most people, thought that Rome was doomed and so allied himself with Hannibal. He had his reasons – Roman power had been pushing down the Illyrian (Dalmatian) coast toward Macedonia. But Philip failed to realize that Rome had vast reserves of manpower and was now also the strongest *naval* power in the Mediterranean since creating its navy in the First Punic War (264–241BC). Macedonia, in contrast, had let its once fine navy decay. (On the only occasion on which a Macedonian fleet entered the Adriatic, trying to carry reinforcements to Hannibal, it fled as soon as it saw Roman ships.)

The First Macedonian War (215–205BC) was rather a non-event. Rome was too busy with events in the western Mediterranean to send large forces to Greece. The treaty of 205BC simply restored the *status quo ante*. But Rome was now increasingly interested in Greek affairs, while some Greeks thought they could call on this new power to help them in their disputes with each other.

In 200BC Rhodes and Pergamum, both of which distrusted the Seleucids and Macedonians, told the Romans that the two kings had made a secret plan agreeing to divide up the Ptolemaic Empire. They had probably not, but that year Antiochus III defeated the Egyptians at Panion, annexing southern Syria and Palestine. Meanwhile Philip seemed to be menacing Rhodes and Pergamum. Convinced, Rome declared war, sending its now battle-hardened legions east. With Aetolian cavalry to help, Philip's army was routed at Cynoscephalae in 197BC, and his power restricted to Macedonia. In particular, he gave up the 'chains of Greece', the forts that had held Greece captive. He was, however, left on the throne of Macedonia.

'FREEDOM' FOR GREECE

Greece, declared the victorious Roman general Flaminius to a congress at Corinth, would now be free. Roman troops would

be withdrawn shortly. The delighted Greeks applauded so loudly that "birds dropped from the air stunned", and they hailed Flaminius as a god, the first (not the last) Roman so honoured. But what Flaminius meant was that Greek cities could enjoy much the same limited autonomy as cities in Italy did as 'clients' of Rome, not that they would be totally free. From this misunderstanding came much later grief. But Rhodes, Pergamum and the Achaeans at the time happily accepted the peace, with only the Aetolians disgruntled at gaining nothing. In 194BC Roman troops duly left Greece.

THE FIRST SYRIAN WAR

By 200BC Antiochus III the Great seemed on top of the world. He had restored Seleucid suzerainty over the east, at least in theory; beaten the Egyptians and finally won southern Syria, long his dynasty's ambition; and pushed Pergamum back, regaining control over western Asia Minor to the Aegean. He seemed to the suspicious Romans to be about to recreate Alexander's empire. Worse, he welcomed Hannibal, Rome's arch-enemy now in exile, at his court. Antiochus even sent troops across the Hellespont, rejecting Roman protests.

In 192BC the Aetolian League invited Antiochus to intervene in Greece proper. After some hesitation and diplomatic manoeuvres, Antiochus despatched 10,000 men – enough to annoy the Romans, but not enough to impress potential Greek allies. So began what Rome called the First Syrian War. Defeated on land by the Romans at Thermopylae in 191BC (the Aetolians gave no real help), Antiochus' fleet was defeated at Myonessus after a hard battle by a combined Roman and Rhodian fleet. Finally at Magnesia in Asia Minor in 190BC, Antiochus' grand army, 70,000 strong with chariots, elephants and *cataphracts* (armoured cavalry), was routed. Scipio Africanus, who had defeated Hannibal, masterminded the Roman victory, but Pergamum's cavalry

won the day, defeating the Seleucid phalanx, which fought to the bitter end.

Pergamum had its reward two years later in the Treaty of Apamea, which gave it almost all Asia Minor west of the Taurus Mountains. Rhodes made useful gains too. The Seleucid Empire now ended at Cilicia, and it had to pay Rome 15,000 talents in reparations – a cripplingly vast sum, though payment was phased.

Above: Monument to the Battle of Cannae in which Rome suffered seemingly total defeat by Hannibal in 216BC. This led Syracuse and Macedonia to ally with Carthage.

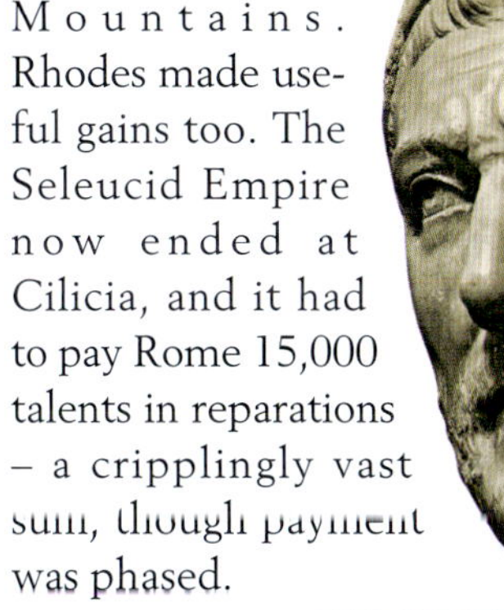

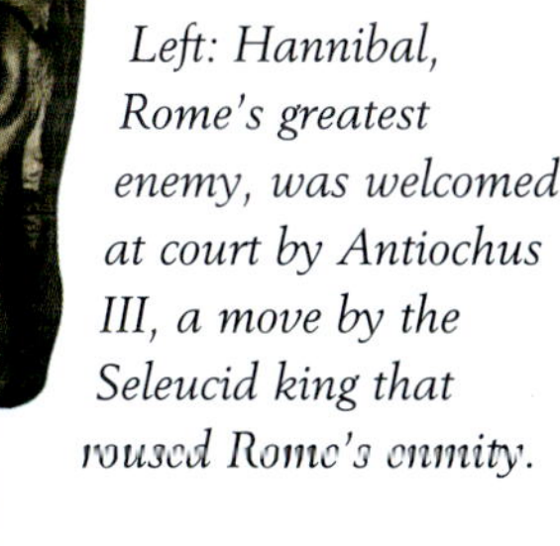

Left: Hannibal, Rome's greatest enemy, was welcomed at court by Antiochus III, a move by the Seleucid king that roused Rome's enmity.

GREECE MADE CAPTIVE
188–146 BC

Above: Perseus, last king of Macedonia, was utterly defeated by the Romans at Pydna in 168BC. His kingdom was divided into four client states before being annexed by Rome in 144BC.

Below: The circular Temple of Hercules Victor (once called the Temple of Vesta) in Rome is truly Hellenic in style. It was probably built in c.120BC by Greek craftsmen.

If Rome's policy toward the Greeks in the next decades often seemed brutal and hypocritical (promising freedom at one moment, crushing any sign of independence at the next), this reflected Rome's own ambivalence. Beyond ensuring that no power could challenge Roman hegemony, the Senate was divided. A few senators had little wish for new entanglements in Greek affairs, but other, more ambitious, Roman politicians wanted plunder and glory in the Greek east. There they could depose kings, be hailed as gods and amass unprecedented wealth. Further, Greeks themselves began coming to Rome with their quarrels, making Rome their judge.

Not all Greeks accepted this. Democrats in the cities now looked to Macedonia against Rome – many cities in Greece and Asia were still democracies, although Rome favoured oligarchies. The monarchs also had their discontents – and strengths. Although the Seleucids had lost their western lands, their empire still stretched east to Persis and south to Egypt's frontier. Macedonia likewise retained its old recruiting grounds. Ptolemaic Egypt, however, was in terminal decay although hugely rich, clinging desperately to its Roman alliance. Only Pergamum, now booming, willingly toed the Roman line.

THIRD MACEDONIAN WAR

In 179BC Perseus succeeded his father Philip V in Macedonia. Although he renewed the treaty with Rome, he forged marriage ties with both the Seleucids and the king of Bithynia (north-west Asia Minor), while overhauling the army. This alarmed Eumenes II of Pergamum, who persuaded Rome that Perseus posed a new threat. The result was the Third Macedonian War (171–168BC) in which Rome mobilized huge forces. Perseus' crushing defeat at Pydna in 168BC, despite the success of his phalanx charge, marked the end of Macedonia. Perseus was taken in chains to Rome, where he died in prison; Macedonia was split into four republics. When these quarrelled, Rome finally made Macedonia a province in 146BC.

THE SACK OF CORINTH

By 148BC the Achaeans had had enough. Filled with patriotic fervour, they declared war, hoping that the just-started Third Punic War would preoccupy Rome. It did not, and an army under Mummius was despatched. In 146BC the Romans razed Corinth, Greece's richest city, to the ground, Mummius choosing the best artworks for himself from among the ruins. Achaea became a Roman Greek province, its League dissolved.

ANTIOCHUS' AMBITIONS

In 175BC Antiochus IV seized the Seleucid throne. He was ambitious and talented, if also eccentric — he 'stood for election' like a Roman magistrate, probably mocking the new superpower's constitution if also bewildering his subjects. But he had the old Seleucid ambitions toward Egypt. In 169BC he invaded Egypt, overrunning most of it. Rome sent an envoy, Caius Popilius, who traced a circle in the sand around Antiochus and told him not to step beyond it until he agreed to withdraw. Antiochus, not wishing to challenge Rome, withdrew.

On his way home to Syria, he stopped in Jerusalem, where the Hellenizing faction of the priestly state of Judaea (one of many priestly states in his kingdom) appealed for help. Not realizing that Jews were somehow *different* from his other Semitic subjects, Antiochus despoiled the Temple, installed a Syrian garrison and erected a temple to Olympian Zeus on the site. This was probably just part of his general Hellenizing programme. But the furious revolt that broke out, led by Judas Maccabeus the High Priest, finally drove the Seleucids from Judaea for good. (Rome supported the Maccabees). However, Antiochus' attention was focused on the growing Parthian threat to the east. With his death in 164BC, the great days of the Seleucids were over, although the kingdom survived for another century.

DELOS: SLAVE ISLAND

The Romans, thinking that Rhodes had not supported them wholeheartedly against Perseus, deprived it of the tiny island of Delos in 167BC, which became a free port. It also rapidly became the greatest slave market yet seen, reputedly able to handle 10,000 slaves a day.

Right: The flow of skilled Greek captives into Rome gave rich Romans ample domestic slave labour, such as this maid. From a fresco at Herculaneum.

The slaves were mostly Greek or Hellenized people, victims of Rome's new aggressiveness. In 167BC the whole population of Epirus was enslaved, 150,000 of them glutting the slave markets. Direct taxation was abolished that year in Rome – not a coincidence.

CLASHES BETWEEN ROME AND THE ACHAEANS

In 166BC Rome took 1,000 mainly democratic hostages from the Achaean League, although the League had supported Rome. Among them was Polybius the historian. Polybius had Philopoemon, head of the Achaean League, wonder: "Should we work with our masters and not object, so that soon we get even harsher orders, or should we oppose them as far as we can, so... we can check their impulses?" The Achaeans did both after Philopoemon's death in 182BC, clashing with Rome when it demanded they relinquish not only Sparta, which was reasonable, but also Argos and Corinth both of which had been League cities for generations.

Above: If Rome conquered Greece politically, Greece conquered Rome culturally. Socrates was among the philosophers educated Romans came to revere.

CHAOS IN THE AEGEAN
150–80BC

Above: Sulla, the brutal if highly effective Roman general who sacked many Greek cities in 86BC.

Below: The Temple of Poseidon at Sunium outside Athens. It escaped the fate of the city itself, which was sacked by the Roman Sulla for supporting Mithradates, king of Pontus.

The Romans had mixed, sometimes ignoble, motives behind their actions toward the Greeks – greed and paranoid suspicion must have been the most obvious to the Greeks, if not ones that the Romans would have recognized. However, Rome can hardly have foreseen the disastrous consequences of its actions.

LARGE-SCALE PIRACY

By making Delos a 'super-port', soon very popular with Italian merchants, it wrecked the basis of Rhodes' wealth. This undermined Rhodes' fleet, which had kept down piracy. Piracy now revived on a new and massive scale. (Athens, to which Delos was theoretically restored, now lacked the strength or will to reassert its old thalassocracy, sea-power.)

By 100BC pirates were raiding right across the Mediterranean from strongholds in Crete, Lycia and Cilicia, where they lived beyond any law. Slaving was one of their main activities, and they attacked Roman as well as Greek shipping. At one stage pirates captured the young Julius Caesar. Waiting for his ransom to be paid, Caesar told his captors that he would catch them and have them crucified. They laughed; he kept his word. But although the pirates attacked far up the coasts of Italy, at one point even capturing Roman magistrates off the coast of Latium, the Romans did almost nothing to check this threat until the *Lex Gabinia* of 67BC.

TAX REFORMS AND CORRUPTION

Worse still for Greece, the concessions made by Gaius Gracchus to Roman *publicani* (tax-farmers) in 122BC to win support for his radical reforms at home led to a new venality in Roman provincial administration. The tax rates levied on provinces such as Macedonia and Achaea may have been no higher than earlier, but the *publicani* extorted vastly higher taxes from the hapless provincials for their own profits. It was a disastrous way of raising revenue.

By the 1st century BC Roman government had grown detested as the proconsuls (governors) themselves became openly corrupt, knowing they faced no real danger of prosecution back in Rome. As Cicero, the great Roman writer, said: "No words can say how deeply we are hated by foreigners because of the foul behaviour of the men we have sent out recently to govern them."

Cicero's successful prosecution of Verres, an infamously corrupt governor of Sicily who had plundered the island, was as rare as his own probity while governing Cilicia.

THE RISE OF MITHRADATES

Greek colonies had long been dotted around the Black Sea, but after Alexander's reign Greek culture began to penetrate inland also. Pontus, a fertile, well-wooded region on the south coast with abundant mineral deposits,

was ruled by kings of Iranian descent who became increasingly, if superficially, Hellenized. One of these kings, Pharnaces I (220–185BC), extended his power around much of the Black Sea.

Mithradates V Pontus (*c.*150–120BC) was the most powerful king in Asia Minor after Pergamum's end. His son Mithradates VI Eupator (120–63BC) became one of Rome's greatest adversaries and the last, rather unlikely, champion of Greek freedom. Mithradates V started by extending his power around almost the whole Black Sea, annexing the half-Hellenized kingdom of the Cimmerian (Crimean) Bosphorus in 108BC. This kingdom, which controlled Greece's vital grain supply, was threatened by Scythian tribesmen and so welcomed Mithradates' protection. He had less success with kingdoms in the Anatolian interior such as Cappadocia, but Tigranes of Armenia became his son-in-law, guarding his eastern flank. Meanwhile he built up a formidable army under a Greek general, Archelaus.

THE SACK OF ATHENS

This army was first tested in 88BC, when, reacting to an attack by his neighbour Bithynia, Mithradates swiftly overran western Asia Minor. His proclamation of liberation from the loathed *publicani* delighted the Greeks. When he crossed over to Greece itself, even long-neutral Athens rose in his support. A massacre of 80,000 Roman and Italian *publicani* and other businessmen forced Rome into a vigorous response. Sulla marched east with 100,000 men to defeat Mithradates at Chaeronaea and Orchomenus.

Athens, which he besieged through the winter of 87–86BC, surrendered too late. Sulla's army sacked the city, even removing columns from the Temple of Olympian Zeus to Rome. Other Greek cities were similarly devastated and had to pay Rome a massive indemnity. This was collected by *publicani*, who also charged interest of 50 per cent on unpaid taxes. This crippled Greece for decades.

Problems with his rivals in the Popularis party in Rome soon claimed Sulla's attention, however, and he agreed to a peace on a return to the *status quo ante* in 85BC. Mithradates surrendered all his gains and retreated to his Black Sea empire. But his strength had only been tried, not exhausted.

Above: The Agora of the Italians at Delos. After being made a free port by Rome in 166BC, Delos boomed, attracting many Italian businessmen.

Below: Mithradates VI, king of Pontus, Rome's last formidable enemy in the Hellenistic East.

THE POWER OF THE DYNASTS
84–42BC

Above: While the Romans fought each other, Parthia became Persia's successor east of the Euphrates under kings such as Mithradates I. He was still happy to be titled Philhellene, however.

From the sack of Athens in 86–85BC on, the fortunes of the Greek world were inextricably linked with those of Rome's feuding dynasts – heads of the city's noble families whose ambitions tore the Republic and its empire apart.

Sulla returned to Rome to become dictator, purge his enemies in a bloodbath, reorder the constitution on deeply reactionary lines and then suddenly, to general astonishment, retire in 80BC. While his seemingly iron-cast settlement in Rome soon started unravelling, he had also left much unfinished business in the Greek world and a thoroughly unstable situation in Asia. Egypt, although theoretically still independent, now leaned heavily on Roman support. Many Romans became tempted by the idea of annexing this, the Mediterranean's richest kingdom, but, being unable to agree on how to do so, left it shakily independent for the time being. A brief war with Mithradates in 84–83BC came to very little, but the king's strength remained undiminished.

HOSTILITIES AND MITHRADATES

In 73BC Rome faced its most serious slave revolt ever when Spartacus raised a force soon amounting to 150,000 men in Campania. The revolt took two years and a major military campaign to suppress. Meanwhile, Mithradates, alarmed at how Rome was handling its new acquisition of neighbouring Bithynia, bequeathed to it in 74BC, renewed hostilities. He invaded Bithynia, again threatening Rome's position in Asia. Lucullus, an associate of Sulla, was sent east with a large army to subdue him but faced problems with a mutinous army and Mithradates' skilful tactics. Gradually, however, he exhausted Mithradates by the usual Roman attributes – tenacity and willingness to endure high casualties. Mithradates finally had to seek refuge with his son-in-law in Armenia. (Tigranes had extended his power south to create a large kingdom.) Before Lucullus could kill either of them, he was recalled in 66BC. His command passed to a far more dashing general.

POMPEY THE GREAT

Pompey was (relatively) young and handsome and had a fine military record when the *lex Gabinia* gave him wide powers to deal with the pirates. Swiftly raising a combined land and sea force of 100,000 men, he rooted out the pirates – more by bribes and threats than military action – in only three months.

In 66BC the triumphant general was given Lucullus' command by an impatient Roman people. Over the next four years Pompey earned the title 'the Great' (which he had assumed already) by a statesmanlike mixture of diplomacy and force. Forcing Mithradates out of Pontus – to which he had returned – he drove him to his last resort: a fortress in the Crimea. There Mithradates committed

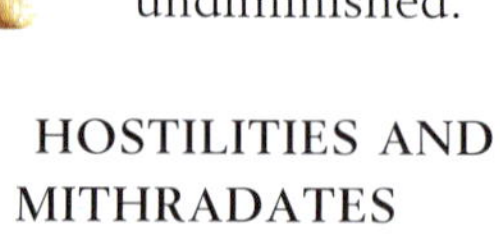

Left: Julius Caesar, charming, charismatic and unscrupulous, emerged victorious from Rome's first round of civil wars. His affair with Cleopatra VII probably produced a son and further tied Egypt to Rome.

suicide in 63BC – by the sword, after failing to poison himself. He had reputedly made himself immune to all poisons by taking a daily antidote.

With Mithradates dead, Pompey rearranged the east at leisure. Tigranes' empire was abolished but he was left in Armenia, which became a client state of Rome. So did a string of small kingdoms from the Caucasus down to the Red Sea. Pompey, capturing Jerusalem, entered the Holy of Holies in the Temple, gravely if inadvertently offending the Jews. Judaea also became a client state. The rump of the once great Seleucid Empire became the Roman province of Syria, ultimately the new centre of Roman power in the east. Pompey's settlement was brilliant: it almost doubled Rome's revenues and lasted in essence more than 100 years. Pompey returned to Rome in 62BC to celebrate another triumph.

GREECE THE BATTLEFIELD
In 49BC the first in a new round of civil wars broke out in Italy when Caesar, returning from conquering Gaul, 'crossed the Rubicon' into Italy proper without disbanding his army. (His many enemies had prevented him from standing for consul *in absentia*, and to return to Rome as an ordinary civilian would have been suicidally risky.) Greece found itself the hapless battlefield as Roman dynasts battled for supremacy. Pompey retreated to Greece where, in 48BC, he was defeated at the Battle of Pharsalus in Thessaly, being killed soon after when he landed in Egypt. Caesar, in hot pursuit, had a different encounter, with the young Cleopatra VII, co-monarch with her brother Ptolemy XIII. She had an affair with Caesar and probably a son, Caesarion. Suppressing an uprising, Caesar made her sole ruler.

Right: Pompey entering the Temple in Jerusalem. This unwittingly sacriligious act angered Jewish priests, some of whom were killed by Roman soliders when they rioted. From a medieval manuscript by Jean Fouquet.

Right: Pompey was Caesar's chief rival in the struggle for supremacy in Rome.

After Caesar's assassination in 44BC, another round of Roman civil wars racked the Greek world. The conspirators, or 'liberators' as they styled themselves, Brutus and Cassius, crossed to Greece to raise fresh armies. To pay for them, Cassius exacted yet more money from the exhausted Greek cities. Brutus and Cassius were defeated by Mark Antony at the double Battle of Philippi in November 42BC. The Roman Empire was then provisionally divided, Antony taking control of the east while Octavius Caesar, adopted son of Julius, took control of the west.

CLEOPATRA AND ANTONY
50–30BC

Above: Cleopatra, as this bust suggests, was not stunningly beautiful, but she was witty, charming and very wealthy – qualities Antony appreciated.

Below: The meeting of Antony, victorious Roman overlord of the East, and Cleopatra, last Hellenistic queen of Egypt, was one of unparalleled splendour, here envisaged by the great 18th-century painter Tiepolo.

By 50BC only Egypt, fabulously rich, remained independent of Rome. The Ptolemies relied on Roman support against external aggressors, but internally their rule was insecure. Alexandria was increasingly turbulent, while in Egypt proper the over-taxed *fellahin* and priesthood no longer supported the dynasty. Intermarriage between sister and brother, an Egyptian custom the Ptolemies adopted, may explain the feebleness of later male rulers. Their queens, in contrast, proved ruthless and dynamic. Before the dynasty's end, one great queen tried to restore Ptolemaic glory.

Cleopatra VII was born in 69BC, daughter of Ptolemy XII. From 51BC she was co-ruler with her younger brother and husband Ptolemy XIII. When Pompey, fleeing from Julius Caesar, landed in late 48BC, Ptolemy XIII's agents executed him and presented his severed head to Caesar, expecting him to be delighted.

He was not – Caesar prided himself on his clemency. He was, however, won over by the youthful charms of Cleopatra, smuggled into his chambers in a carpet. Or so legend goes.

CLEOPATRA AND CAESAR
Cleopatra was not, if contemporary portraits are honest, especially beautiful, but she was intelligent, charming and witty. She was ambitious, too, wanting to restore the Ptolemies' former empire. Her affair with Caesar, a womanizer of immense charm, probably produced a son, Caesarion. Caesar supported her against her brother, but the Alexandrian mob took against the Romans. Vicious street fighting led to part of the Library being burnt and almost to Caesar's and Cleopatra's death. They were saved by Jewish guards, which made Caesar pro-Jewish. Cleopatra duly became sole monarch. In spring 47BC, with order restored, Caesar hastened away. Cleopatra later followed him to Rome, there to witness his assassination in 44BC and the recurrence of Roman civil war. Returning to Egypt, she watched and waited on developments.

THE GREAT LOVERS
Mark Antony had been Caesar's trusted lieutenant, giving his funeral oration. Now he had to accept young Octavian, Caesar's great-nephew and adopted son, as an equal partner in the Second Triumvirate (pact) in 43BC. But he was regarded as the better soldier and, by many Romans, as a better man. He also got on well with Greeks.

After his victory at Philippi in 42BC, Antony wintered in Athens, debating and dining, before sailing east to be hailed as the god Dionysus by the Greeks of Asia. But although divine, he was still short of money. When he summoned Cleopatra

Right: Cleopatra (as painted by Cabanel, a 19th-century French artist) was not really Egyptian by culture or descent. But Roman propaganda portrayed her as a decadent oriental femme fatale, *bewitching Antony.*

to meet him at Tarsus in Cilicia, he wanted her wealth, not her body. But Cleopatra made a spectacular entry. "The barge she sat in, like a burnished throne/Burned upon the water. The poop was beaten gold:/Purple the sails and so perfumèd that/The Winds were love-sick with them" as Shakespeare, following Plutarch, later put it. Antony fell in love with her – and she probably with him – and they sailed to Alexandria for a winter of amorous luxury. They founded a club, the 'Inimitable Lives', revelling through the night, and cruised up the Nile. When he bet her that she could not eat a dinner worth a million sesterces, she dissolved a vast pearl in wine and drank it. Antony's role as Dionysus was apt, for the wine-god was the Ptolemies' patron deity. Cleopatra herself often appeared as the goddess Isis.

Meanwhile, Octavian faced major problems – revolts in central Italy, attacks on Rome's grain supplies – that forced him to ask Antony for help. This was given. In 36BC Antony's grand attack on Parthia proved a disaster, although he made Armenia a client state in 34BC. At the 'Donations of Alexandria' that year, Antony sat enthroned beside Cleopatra as she was hailed as Queen of Kings. He gave provinces from Rome's empire to their two children and hailed Caesarion, Caesar's son, as King of Kings. Cleopatra's ambitions seemed fulfilled.

DECLINE AND FALL

All this was a marvellous propaganda gift to Octavian. His poets depicted Antony as bewitched by an oriental *femme fatale*. Even so, when war was declared in 32BC, a third of the Senate went east to join Antony, whose forces were still large. But Antony alienated Roman supporters by letting Cleopatra join him in Greece,

and desertions began. The final battle at Actium in 31BC was an anticlimax, Antony and Cleopatra fled south to Alexandria for a last winter of love. Octavian followed the next year. After another defeat, Antony tried to kill himself. Dying, he was reunited with Cleopatra, who had retreated to her mausoleum. There she cheated Octavian of a triumph by poisoning herself with asps. If Antony had won, Egypt would have remained independent for longer and the Greek cities might have enjoyed more independence, but the Roman Empire would not have been radically different.

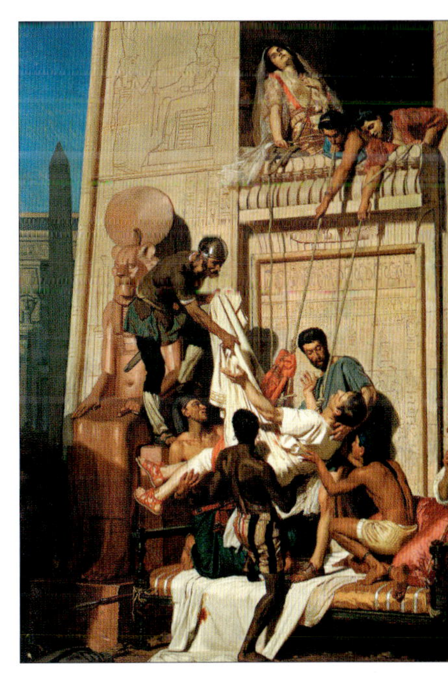

Right: Antony, dying from stabbing himself, was finally reunited in death with Cleopatra in the mausoleum to which she retreated.

AUGUSTUS AND THE PAX ROMANA 27BC–AD14

The death of Cleopatra VII marked the end of the Hellenistic age. The whole Greek world west of the Euphrates now lay under Roman control, directly or indirectly. Much of it had been ravaged by Rome's own civil wars: the grandfather of Plutarch the historian had been forced to carry sacks of grain on his back up mountains for Antony's army during the Actium campaign in 31BC; Corinth was a gutted ruin; Athens, though still a revered intellectual centre, was exhausted; so were the once brilliant cities of Ionia.

Above: The Corinth Canal, the construction of which was started under the philhellenic emperor Nero in AD66.

Below: The theatre at Taormina in Sicily, where a Roman superstructure sits on a Greek base, exemplifies how Greek and Roman cultures intermingled. Taormina was Greek in origin while Sicily itself only became fully Hellenized under Roman rule.

AUGUSTUS' RULE

Augustus, as Octavius was soon titled, had triumphed as leader of upright Romans against what he had depicted as a decadent Hellenistic world. The Greeks perhaps at first expected little from this Roman emperor (from Latin *imperator*, commander), but they were agreeably surprised. Augustus spent two years in the East, re-establishing it along lines laid out by Antony and Pompey. Herod the Great was confirmed in his Judaean kingdom, as were rulers of petty Hellenistic states fringing Rome's eastern provinces.

Roman frontier provinces such as Syria were governed by legates sent out by the emperor, often commanding legions stationed there. Egypt alone was treated differently, becoming the private fief of Augustus governed by an equestrian (knight) not a senator. This reflected imperial nervousness about giving power to a potential rival from the Senate. The first governor was Gaius Gallus, chosen because, as a poet, he might appeal to Alexandrians. (Unfortunately power went to his head and he was forced to commit suicide.) Augustus wisely refused to attack Parthia. Although he extended Rome's northern frontiers, to the Mediterranean world itself he brought peace, the long-lasting Pax Romana.

LOCAL GOVERNMENT

Most provinces in the Greek world, such as Achaea, Macedonia and Asia (western Anatolia), were governed by proconsuls appointed by the Senate if overseen by the emperor. The letters between the emperor Trajan and Pliny the Younger *c.*AD110 show just how close this supervision could be. Such governors had few troops, for much of the Roman Empire long remained lightly guarded and lightly governed. Most of the governing was done by local citizens themselves.

The empire has been called a 'confederation of cities', although the population remained mainly rural. But the local aristocracies – Rome never favoured democracies even when a republic – administered their own cities, competing to build ever grander temples, baths and theatres. (A few cities such as Tarsus had Roman rights, meaning that citizens such as St Paul were Roman citizens.) Rome's light-touch imperialism stemmed from its own lack of bureaucracy, reflecting Roman preferences. Alexandria was ruled directly, but it had long lost its council.

PROSPERITY REGAINED

Linking the cities of the newly stabilized, extended empire was a remarkable network of roads, ultimately covering 80,000km/50,000 miles. These encouraged trade, but sea routes remained far more important. Here the suppression of piracy, started by Pompey and maintained by Augustus and his successors, was crucially important.

The resulting boom saw Corinth, refounded under Augustus, become a wealthy port again. Old Ionian cities such as Miletus, Ephesus and Smyrna became unprecedentedly rich, with populations passing the 100,000 marks, as did Hellenistic cities such as Alexandria and Antioch. Athens enjoyed renewed if modest prosperity, exporting its fine Pentelic marble – and craftsmen – while educating young Roman aristocrats. Augustus built a grand new agora and *odeion* (roofed theatre) in Athens, and a small temple for the cult of Rome and Augustus on the Acropolis, stressing Roman power at the heart of Greece.

TAX REFORMS

Also of crucial importance were changes in the tax collection. Caesar had wanted to abolish the rapacious and loathed *publicani* (tax-farmers) outright, aware of their ruinous effects. The more cautious

Below: The Temple of Hera at Acragas in Sicily, whose partial recovery under the Pax Romana was typical of many Greek cities.

Augustus gradually replaced them with appointed officials whom he could trust. Taxation for most provinces was relatively small. The sales tax, for example, was only 1 per cent, and customs dues were 5 per cent. Where *publicani* survived or were introduced, their rapacity could engender revolts, as in Britain in AD61 and Judaea in AD66.

TWO CENTURIES OF PEACE

At the end of his life in AD14 the emperor Augustus was moved by a demonstration. The passengers and crew of a ship just arrived from Alexandria, greatest of Hellenistic cities, put on garlands and burnt incense to him, saying that they owed their lives and liberty to sail the seas to him.

This peaceful prosperity continued for another two centuries. The walls of most cities not actually on the frontiers, even of Rome itself, were allowed to decay in a period that was, by historical standards, phenomenally peaceful. No wonder that most Greeks were happy to honour Augustus and his successors as divine – honours that wiser emperors did not boast of in Rome itself.

Above: The Temple of Hadrian at Ephesus, one of many Ionian cities that attained its greatest prosperity in the 2nd century AD under the long Pax Romana. Hadrian was a famously philhellenic emperor.

Below: A cameo of Augustus, the first Roman emperor (27BC–AD14). Augustus admired the High Classicism of Periclean Athens.

GRAECO-ROMAN SYNTHESIS
CICERO TO HADRIAN, 80BC–AD138

Above: Bust of Cicero, the Roman orator, writer and politician who summarized and translated into Latin many works of Greek philosophy, ensuring their survival.

Below: The Maison Carrée in Nimes, a Roman colony in southern France. The temple, built in Augustus' reign, embodies Graeco-Roman synthesis, for its columns are classically Greek but its plan is wholly Roman.

In 80BC Cicero, an intensely ambitious young Roman politician, arrived in Athens to study philosophy. He was among the first in a stream of Romans who, over the next 400 years, would go to Greece to study philosophy and rhetoric. Cicero's stay had a huge impact on philosophy over the next 1,500 years in Western Europe. While his prime interest was politics, he turned to writing philosophy full-time when forced into (temporary) retirement by Caesar's ascendancy in the 40s BC. In a few years he summarized in Latin much of Greek thought, especially Stoicism, in *De Republica* and *De Finibus* ('Concerning the Highest Ends'). In his books, which survive intact unlike most ancient literature, he established Latin equivalents for basic Greek philosophical terms such as morality, quality and happiness.

Cicero's achievement in translating and synthesizing Greek thought typifies the growing Graeco-Roman fusion.

After 200BC Greek culture had flooded into Rome in the form of looted artworks and thousands of slaves, the latter often better educated than their masters. They became secretaries, librarians, doctors and tutors. While Roman nobles tended to regard Greeks politically as irresponsible, deceitful and even decadent, many admired Greek culture almost uncritically for a time.

The temple of Hercules Victor in Rome, built *c.*120BC as a perfect circle with slender marble columns, is almost wholly Greek in form. Equally Greek are the wall paintings – surviving best in Pompeii thanks to Vesuvius – probably made by Greek artists working for Roman masters. By Cicero's time, educated Romans were fluent in Greek. A century later Quintillian, the Roman grammarian who taught the sons of emperors, suggested that boys should learn Greek before they learned Latin, so essential was it to their education.

NERO'S INFLUENCE
"Greece made captive captured her conqueror and introduced the arts into rough Latium," wrote Horace, one of the emperor Augustus' chosen poets. If this Roman cultural inferiority was fading by the early 1st century AD – mainly thanks to Horace and other great Latin poets such as Virgil – Greece was still seen as the exemplar, even by some emperors.

In AD54 the 17-year-old Nero, the last of Augustus' descendants, became emperor. With genuine artistic interests if not talents, he patronized artists and architects – especially those building and adorning his vast new imperial palaces – and composed plays. Seneca, Nero's tutor and first minister, wrote philosophy and tragedies, which, if not publicly staged at the time, survived to influence later playwrights such as

Right: Admetus and Alcestis Listening
to the Oracle, *a fresco from the 'House of
the Tragic Poet' at Pompeii, reveals how
familiar Romans became with Greek myths
and how popular Greek styles were in Italy
in the 1st century* AD. *This work copies a
Hellenistic original.*

Shakespeare. But Nero performed in
public himself – something thought shame-
ful for a noble Roman – at first in Naples, a
still Greek city. He also tried to introduce
Greek athletic games without success.
When Nero entered the Olympic Games,
he won *all* the prizes, the first and last
time this happened. More positively, he
proclaimed 'freedom' for Greek cities and
ordered the digging of the Corinth Canal,
although his engineers stopped when they
hit bedrock. Nero's reign ended in civil war
in AD68, however, and there was a brief
Roman reaction under his successors.

THE PHILHELLENIC EMPEROR
In AD128 the emperor Hadrian dedicated
the Pantheon, temple to all the gods,
in Rome. This most famous and best
preserved of Roman temples exemplifies
Graeco-Roman synthesis. Behind a giant
portico essentially Greek in inspiration
rises a majestic dome wholly Roman in
conception. Hadrian, who commissioned
and possibly helped to design it, was the
most philhellenic of all great emperors,
to the point that his enemies called him
Greekling (*graeculus*).

Regarding Athens as his favourite city,
he made it head of a new Panhellenic
League and built an entire new quar-
ter, besides completing the Temple of
Olympian Zeus started 640 years before.
He was elected archon (the office still
existed), initiated into the Eleusinian
Mysteries, the holiest in Greece, and
wore a beard like a Greek philosopher.

Equally Greek was his passion for
Antinous, a youth of royal descent and so
his social equal, to be wooed not raped.
Hadrian's grief at Antinous' death struck
many as undignified, but in his reign
(AD117–138) the Greek-speaking half

of the empire finally recovered its self-
confidence and began to supply import-
ant officials. Under Hadrian too began
the systematic codification of Roman
laws, where Greek humanism and
idealism lightened Roman pragmatism.

TWO-WAY INFLUENCE
The influence did not run all one way.
The Greeks adopted some Roman archi-
tectural features, using arches and vaults
more often. Gladiatorial games, that
most Roman entertainment, were also
introduced to the Greek world. Initial
revulsion – there was a riot in Antioch
at the first – evaporated as Greeks, too,
developed a taste for these brutal thrills.
More positive and far more significant
was the emergence of a new Graeco-
Roman ruling class, exemplified by men
such as the orator Aelius Aristides or Dio
Cassius, a Greek who became a Roman
consul and historian. By AD200 a single
Graeco-Roman culture had developed
across the empire.

*Above: The emperor Hadrian,
in whose reign (*AD117–138*)
Greeks became almost equal
partners with Romans.
Hadrian himself was made
archon of Athens, a city
he loved deeply.*

Above: Desiderius Erasmus (1466–1536), the pioneering Renaissance scholar who learned Greek to translate the Bible, so unwittingly paving the way for the Reformation.

GREECE REBORN:
RENAISSANCE AND RENASCENCES

Linked to Rome, ancient Greece declined with it. The last Olympic Games was held in AD393; five years later invading Visigoths ravaged the peninsula, looting the temples. Finally, in AD529 the Academy in Athens was closed on the (east Roman) emperor Justinian's orders. Early Christianity, though Greek-speaking, was almost as hostile to Hellenism as the barbarians were. By AD600 ancient Greece was dead. All knowledge of Greek was lost in Western Europe. "*Graceum est: non legitur*" ("it is Greek, not read") medieval monks wrote besides passages in Greek. Only translations of Aristotle from the Muslim world, which retained some Greek knowledge, revived interest in that philosopher in the 13th century.

THE REVIVAL OF HELLENISM

The ghost of Greece, however, lived on. In 1438 the Council of Florence tried to reconcile differences between the Eastern and Western Churches. It had small success, but among the Eastern bishops was Bessarion, a scholar who stayed in Italy, befriending other Greek fugitives after Constantinople fell to the Turks in 1453. They brought with them manuscripts – principally by Plato, Greece's greatest

Below: The stadium at Olympia, site of the ancient Olympic Games last held in AD393.

philosopher – which Bessarion collected. In Renaissance Florence, Cosimo de' Medici founded a Platonist Academy in 1462 to study Greek and philosophy. Soon Plato's birthday was being celebrated, and Socrates was being hailed as a pagan saint, by cardinals and princes as well as scholars.

ART AND THE SCRIPTURES

The Renaissance engendered a Platonic desire (as *eros* best translates) for spiritual truth incarnate in physical beauty. Newly discovered Roman copies of Greek statues inspired Renaissance artists. Botticelli's *Birth of Venus* illustrates Greek myth in romantic Neoplatonist mode, while Michelangelo created art sublimely Platonist in aspiration. If this artistic and intellectual rebirth hardly affected politics – democracy was unthinkable in Renaissance Europe – it resurrected classical art in the Western world.

A novel use of Greek was for studying the Scriptures, originally written in Greek. Desiderius Erasmus (1466–1536) was amazed to find the Vulgate (Latin Bible) riddled with errors. His translations and biting commentaries caused a furore, paving the way for the Reformation. Erasmus spent years in England teaching Greek at Cambridge University, which became a centre of the New Learning. Queen Elizabeth I herself learned Greek. By 1600 knowing Greek was essential for any self-respecting scholar. But Hellenism's full impact had yet to be felt.

THE SHOCK OF THE OLD

In the late 18th century, travellers returned from Greece with news of a radically simple yet powerful architecture. Finds at Pompeii, that time capsule of Graeco-Roman art buried by Vesuvius, were already leading the arts toward a purer Neoclassical style. But the brutally gigantic Doric columns seen in classical Greek temples amazed Europe and America. Confronted for the first time with true classical Greek architecture, architects reinvented their own art. The British Museum in London, the Brandenburg Gate in Berlin and much of Edinburgh shows how Greek temples could be lovingly recreated in the most unlikely places.

The late 18th century was the age of revolution in America and France. While the American Founding Fathers looked mainly to Republican Rome for models, the Achaean League's federalism also inspired them. Full democracy, not thought practical in America in the 1780s, was embraced in revolutionary France ten years later. Some French Revolutionaries, following the philosopher Jean-Jacques Rousseau's primitivist ideas, looked to Sparta as an ideal state. But the full (male) franchise, fleetingly achieved by the First French Republic for the first time for 2,000 years, was inspired by democratic Athens. Not by coincidence, fashions of the time echoed those of Greece and Rome in architecture, furniture and women's clothes.

Literature reflected the new 'Hellenomania'. Poets in England such as Keats and especially Shelley, who wrote dramatic poems such as *Prometheus Unbound* modelled on Aeschylus, supported radical democratic politics. Shelley's friend Lord Byron gave his life in the cause of resurrecting Greek liberty. Politically, the Greek precedent became ever more inspiring as democracy spread in the 19th and early 20th centuries. Abraham Lincoln deliberately modelled his Gettysburg speech on Pericles' funeral oration. Later generations still feel the lure of Greece, the true birthplace of democracy.

Above: The Pantheon in Rome, Greek in its name (meaning 'for all the gods') and fusing Greek and Roman ideals, is the finest ancient classical building extant. It was an inspiration to Renaissance artists and architects, especially to Raphael (1483–1520), who is buried inside it.

Right: The Laocoön, the remarkable Hellenistic statue whose rediscovery in Rome in 1506 proved a crucial inspiration for Michelangelo, the greatest Renaissance sculptor.

INDEX

ACKNOWLEDGEMENTS

This edition is published by
Lorenz Books
an imprint of Anness Publishing Ltd
info@anness.com
www.lorenzbooks.com
www.annesspublishing.com

Anness Publishing has a picture agency
outlet for images for publishing,
promotions or advertising. Please visit
our website www.practicalpictures.com
for more information.

© Anness Publishing Ltd 2026

Publisher: Joanna Lorenz
Editor: Joy Wotton
Designer: Nigel Partridge
Illustrations and maps: Vanessa Card,
Anthony Duke, Peter Bull Art Studio
Production Controller: Ben Worley

A CIP catalogue record for this book is
available from the British Library

Page one: Fighting Hoplites
Page two: The Temple of Athena Nike,
    Athens
Page three: Statue of Pericles
Endpapers: The Temple of Poseidon,
    Sounion

PICTURE CREDITS
**The Ancient Art & Architecture
Collection Ltd**: 4.3, 8b, 9t, 14–15, 18tl &
br, 19t & b, 20t, 21t, 23t, 24tr, 25t &
b, 26tl, 27b, 28t, 29b, 30t & b, 32b, 34b,
35t, 36t, 42t, 44b, 45t & b, 47b, 50t &
b, 54–5, 58tl, 59b, 60b, 62t, 65b, 66b,
67b, 70b, 71, 73t, 76b, 78t, 86b, 88t,
91, 94l & r, 99b, 101t, 102br, 104t,
105b, 106, 107b, 112t, 114b, 124br,
135tl, 146t, 156t, 157b, 159, 164t &
b, 165t, 169t, 170t, 171t, 187t, 191t,
204b, 220t & b, 228b, 229b, 233t,
234t, 238t, 239t & b, 241t, 244t,
245br, 249t & b; /Ronald Sheridan 1,
53b, 57t, 105tr, 113b, 145b, 235t; /C.M.
Dixon 20b, 76t, 153b; /M. Williams
32t; /G.T. Garvey 33t, 46t, 102tl; /
Interfoto 4.2, 38–9, 179t; /Prisma 100b;
/D.R. Justice 116t; /C. Hellier 165b; /P.
Syder 192b; /Mike Andrews 207b; /
Prisma 248t
The Art Archive: 123b, 173b,
188t, 189m, 206t; /JFB 3, 93b; /
Archaeological Museum Naples/Dagli
Orti (A) 4br, 37, 81t, 128–9, 157tl,
162t, 166t, 174tr, 193tr, 228t; /Dagli
Orti 4.1, 6b, 24b, 36b, 41b, 43b, 46b,
48t, 56b, 63b, 75t, 83b, 89b, 101b,
112b, 116b, 118br, 122b, 124tl, 126b,
132b, 134b, 135tl, 137b, 138b, 145t,
150b, 154, 155t, 156b, 158t, 179b,
192t, 197t, 208t & b, 209t, 219b, 221b,
222t & b, 229t, 232b, 236b, 244b, 245t,
246b, 247t, 248b; /Dagli Orti (A) 41t,
78b, 80t, 108–9, 120b, 178t, 221t; /
Archeological Museum Piraeus/Dagli
Orti 4.4, 4.6, 16–17, 64, 68–9, 102; /
Archaeological Museum Salonica/ Dagli
Orti (A) 5.1, 130–1, 139t, 150t; /
Pella Museum Greece/ Dagli Orti (A)
5.2, 133b, 142–3, 148b, 190t, 196t;
/Archaeological Museum Istanbul/
Dagli Orti 5.5, 152b, 168b, 198–9, 256t;
/Musée du Louvre Paris/ Dagli Orti
5.6, 9b, 40b, 42b, 149, 183, 187b,
205b, 213b, 214–15, 234b; /Acropolis
Museum Athens/Dagli Orti 8t, 62b,
77t, 79b, 83t; /Heraklion Museum/
Dagli Orti 18tr; /National Archaeological
Museum Athens/ Dagli Orti 21b, 23b,
111t, 205t; /Museo di Villa Giulia
Rome/Dagli Orti 28b; /British Museum/
Eileen Tweedy 40t, 67t, 212t; /Musée; /
Jean Vinchon Numismatist Paris/Dagli
Orti 52t & m, 137t, 216t, 236t; /Agora
Museum Athens/Dagli Orti 60t, 200t; /
Gianni Dagli Orti 61b, 98t; /Olympia
Museum Greece/Dagli Orti 74bl; /
Neil Setchfield 93t; /Museo Tosio
Martinengo Brescia/ Dagli Orti (A) 95t;
/Musée Thomas Dobrée Nantes/Dagli
Orti 96t; /Archaeological Museum Izmir/
Dagli Orti 96b; /Archaeological Museum
Salonica/Dagli Orti 97t, 140b, 141b, 152t;
/Archaeological Museum Châtillon-sur-
Seine/Dagli Orti 99t; /Archaeological
Museum Syracuse 118t; /Museo
Capitolino Rome/Alfredo Dagli Orti
119t; /Museo Nazionale Taranto/
Dagli Orti 123t, 153t; /Chiaramonti
Museum Vatican/Dagli Orti (A) 124tr,
138t; /Museo Capitolino Rome/ Dagli
Orti (A) 124bl;/Archaeological Museum
Salonica/ Dagli Orti 127t; /Collection
Antonovich/ Dagli Orti 134t, 135tr,
225r; /University Library Istanbul/Dagli

Orti 139b; /Hellenic Institute Venice/
Dagli Orti (A) 144t; /National Museum
Beirut/Dagli Orti 144b; /British Library
146b; /Museo del Prado Madrid/
Dagli Orti (A) 147t; /Archaeological
Museum Teheran 155b; /Museo Profano
Gregoriano Vatican 158b; /Bibliothèque
des Arts Décoratifs Paris/ Gianni Dagli
Orti168t, 207t; /Bibliothèque des Arts
Décoratifs Paris/Dagli Orti 170b; /Royal
Palace Caserta Italy/Dagli Orti 186b; /
Bibliothèque Municipale Reims/ Dagli
Orti 189b; /National Archaeological
Museum Athens/Dagli Orti (A) 194tl;
/Bodleian Library Oxford 195b; /
Private Collection/Dagli Orti 196tr; /
Archaeological Museum Naples/Dagli
Orti 135b, 203b, 240b; /Museo Naval
Madrid/ Dagli Orti 206b; /Musée
Municipal Sémur en Auxois/Dagli
Orti 211t; /Bibliothèque Nationale
Paris 211b; /British Museum 218t; /
Musée du Louvre Paris/Gianni Dagli
Orti 224t; /Ephesus Archaeological
Museum Selcuk Turkey/ Dagli Orti
237t; /Archaeological Museum
Athens/Dagli Orti 237b; /Musée des
Beaux Arts Antwerp/Dagli Orti 243t;
/Musée des Beaux Arts Grenoble/
Dagli Orti 243b; /Museo della Civita
Romana Rome/Dagli Orti 246t
The Bridgeman Archive: /© Wakefield
Museums and Galleries, West
Yorkshire, UK 5mr, 163b; /Museo
Archeologico Nazionale, Naples, Italy
35b; /© Birmingham Museums and Art
Galley 65t; /© Ashmolean Museum,
University of Oxford, UK 74 tr, 203t;
/Private Collection, © The Fine Art
Society, London, UK 80b; /Galleria
degli Uffizi, Florence, Italy, Alinari
81b; /Palazzo Vecchio (Palazzo della
Signoria) Florence, Italy 89t; /Louvre,
Paris, France, Lauros/ Giraudon 92t;
100t; /Museo Capitolino, Rome, Italy,
Giraudon 95b; /Louvre, Paris, France,
Giraudon 98b; /Private Collection
105t, 127b, 192b; /Galleria degli
Uffizi, Florence, Italy, Alinari 113t; /
Stapleton Collection, UK, 114t; /British
Museum, London, UK 121b; /Private
Collection, Archives Charmet 121t,
177tm; /Vergina, Macedonia, Greece,
134b; /Bibliothèque Nationale, Paris,
France, Giraudon 140t; /Museo
Archeologico, Florence, Italy, Lauros /
Giraudon 147b; /Musee Gustave Moreau,
Paris, France, Lauros /Giraudon 151t;
Fitzwilliam Museum, University of
Cambridge, UK, 159t; /Louvre, Paris,
France 167b, 202; /Louvre, Paris,
France/Peter Willi 175t, 201b; /Musee
de la Ville de Paris, Musee du Petit-
Palais, France, Lauros / Giraudon 176t;
/Musee des Beaux-Arts, Dijon, France,
Lauros/Giraudon 176b; /Iraq Museum,
Baghdad 177tr; /National Museum of
Iran, Tehran, Iran, Lauros/ Giraudon
178b; /Private Collection, © Look and
Learn 200b; /© National Museums of
Scotland 201t; /Galleria degli Uffizi,
Florence, Italy 212b; /Hermitage, St.
Petersburg, Russia 224t; /© Bristol
City Museum and Art Gallery, UK
227t; /National Gallery, London, UK
231; /Bibliotheque Nationale, Paris,
France 240t, 241b; /Palazzo Labia,

Venice, Italy, Alinari 242b
Corbis: 182b, 185b; /© John Corbett/
Ecoscene 5.4, 180–1; /© Ric Ergenbright
186t, 190b; /© Olivier Matthys/epa
189t; /© Lloyd Cluff 191b; /© Paul
Almasy 194tr; /© Michael Nicholson
194 bl; /© Roger Wood 209b; /©
Christel Gerstenberg 213t; /© Araldo
de Luca 216b; /© James Marshall 232t
Werner Forman Agency: /Christie's,
London 90b; /Museo Ostia, Italy 90t,
**Medioimages/Photodisc/Discover
Greece/Getty Images**: 18br, 31b, 44t,
56t, 63t, 114t, 226t & b, 238b
Sylvia Kapp: 26tr, 53t, 77b, 107t,
118bl, 250, 251t, 253b, 254b
Mary Evans Picture Library: 110,
111b, 126t, 196tl, 235b
Photo12.com: /Oronoz 2, 5.3, 24tl,
33b, 49t, 57b, 70t, 72, 87t, 92b, 97b,
117t, 122t, 133t, 141t, 157tr, 160–1,
167t, 193tl, 204, 223b, 224b, 242t,
256b; /Ann Ronan Picture Library 4.5,
84–5; /Albert Arnaud 6t, 22t, 102tr,
120t, 132t, 148t; /Oasis 7b, 26br; /
ARJ 29t, 59t, 104b, 151b, 174b,
177b, 184t, 195t, 210t & b, 217b,
223t, 230t & b, 247b; /Bertelsmann
Lexikon Verlag 58tr; /JTB Photo 58bl,
82t & b, 88b; /Société Française de
Photographie 119b; /Jean Guichard
172b, 173t, 227b; /World Religions
Photo Library 182t; /Francis Latreille
185t

Frances Reynolds: 245bl, 255